Natural Groundwater Quality

Natural Groundwater Quality

250101

EDITED BY

W. Mike Edmunds

AND

Paul Shand

Blackwell
Publishing

BLACKWELL PUBLISHING
350 Main Street, Malden, MA 02148-5020, USA
9600 Garsington Road, Oxford OX4 2DQ, UK
550 Swanston Street, Carlton, Victoria 3053, Australia

First published 2008 by Blackwell Publishing Ltd

1   2008

Library of Congress Cataloging-in-Publication Data
Natural groundwater quality/edited by W. Mike Edmunds and Paul Shand.
     p. cm.
   Includes bibiographical references and index.
   ISBN 978-1-4051-5675-2 (hardcover: alk. paper) 1. Groundwater–Quality–European Union countries.
2. Water quality management–European Union countries. 3. Water quality–Standars–European Union countries. I. Edmunds, W. M. II. Shand, P. (Paul)
   TD255.N38 2008
   333.91' 04094–dc22   2007029146

ISBN: 978-14051-5675-2

A catalogue record for this title is available from the British Library.

Set in 10/13 Trump Mediaeval
by Newgen Imaging Systems Pvt Ltd, Chennai, India
Printed and bound in Singapore
by Fabulous Printers Pte Ltd

For further information on
Blackwell Publishing, visit our website at
www.blackwellpublishing.com

Contents

List of Contributors

D. Banks
Holymoor Consultancy, 8 Heaton Street
Brampton, Chesterfield, Derbyshire
S40 3AQ, UK

M. Coetsiers
Laboratory for Applied Geology and
  Hydrogeology
Ghent University
Department of Geology and Soil Science
Krijgslaan 281–S8
B-9000Gent, Belgium

M. T. Condesso de Melo
Geosciences Department
University of Aveiro
Campus Universitário de Santiago
3810-193 Aveiro, Portugal

J. Corcho-Alvarado
Climate and Environmental Physics
Physics Institute, University Bern
Sidlerstrasse
3012, Bern, Switzerland

E. Custodio
Technical University of Catalonia, DIT–UPC
Gran Capità s/n, 08034 Barcelona, Spain

D. Dimitrov
Senior Researcher
National Institute of Meteorology & Hydrology
66 Tzarigradsko Shose bul.
1784 Sofia, Bulgaria

W. M. Edmunds
Oxford Center for Water Research
Oxford University Center for the
  Environment
South Parks Road
Oxford, OX1 3QY

B. Frengstad
Norges geologiske undersøkelse
N-7491 Trondheim, Norway

I. Gaus
BRGM, France

D.C. Gooddy
British Geological Survey, Wallingford
Oxfordshire, UK, OX10 8BB

Z. Herrmann
Nerudova 939, 500 02 Hradec Kralove
Czech Republic

K. Hinsby
Geological Survey of Denmark and
  Greenland, (GEUS)
Øster Voldgade 10
DK-1350 Copenhagen K

F. Huneau
Université Bordeaux-1
GHYMAC Géosciences, Hydrosciences
Faculté des Sciences de la Terre
B18 avenue des Facultés
F-33405 Talence, France

M. Iglesias
Agència Catalana de l'Aigua
Provença 204–208
E-08036 Barcelona, Spain

J. Kania
AGH University of Science and
    Technology
Department of Hydrogeology and Water
    Protection
Al. Mickiewicza 30
30-059 Krakow, Poland

E. Kaup
Laboratory of Isotope Palaeoclimatology
Institute of Geology at Tallinn Technical
    University
10143 Tallinn
Estonia pst. 7

C. Kjøller
GEUS, Denmark

E. Kmiecik
AGH University of Science and
    Technology
Department of Hydrogeology and Water
    Protection
Al. Mickiewicza 30
30-059 Krakow, Poland

V. Kodes
Czech Hydrometeorological Institute
Na Sabatce 17, 143 00 Praha 4
Czech Republic

H. H. Loosli
Climate and Environmental Physics
Physics Institute, University of Bern
Sidlerstrasse 5, 3012 Bern
Switzerland

E. Lozano
Dept. of Geotechnics, Technical University
    of Catalonia
Gran Capità, s/n Ed. D-2
E-08034 Barcelona, Spain

M. Machkova
National Institute of Meteorology and
    Hydrology,
66 Tzarigradsko shoes bul.
1784 Sofia, Bulgaria

J. Mangion
Directorate for Water Resources Regulation
Malta Resources Authority
Millenia, Aldo Moro Rd.
Marsa MRS 9065, Malta

M. Manzano
Technical University of Cartagena
Paseo Alfonso XIII
52 E-30203 Cartagena
Spain

A. Marandi
Laboratory of Isotope Palaeoclimatology
Institute of Geology at Tallinn Technical
    University
10143 Tallinn
Estonia pst. 7

M. A. Marques da Silva
Universidade de Aveiro
Departamento de Geociências
Campus Universitário Santiago
3810-193 Aveiro, Portugal

J. Muzak
DIAMO, o. z. TUU, Machova 201
471 27 Straz pod Ralskem
Czech Republic

N. Neykov
National Institute of Meteorology &
    Hydrology
66 Tzarigradsko shose, bul.
1784 Sofia, Bulgaria

P. Neytchev
National Institute of Meteorology and
    Hydrology
66 Tzarigradsko shoes, bul.
1784 Sofia, Bulgaria

P. Nieto
Geological Survey of Spain, IGME
Ríos Rosas, 23
28003 Madrid, Spain

Marisol Manzano
Technical University of Cartagena, UPCT
Paseo Alfonso XIII, 52
30203 Cartagena, Spain

J. Novák
Depart of Risk and Reliability
Techical Univ. of Liberec
Halkova 6, 461 17 Liberec
Czech Republic

T. Paces
Czech Geological Survey, Klarov 3
118 21 Praha, Czech Republic

D. Postma
Institute for Environment and Resources
Technical University of Denmark

R. Purtschert
Climate and Environmental Physics
Physics Institute, University of Bern
Sidlerstrasse 5
3012 Bern, Switzerland

V. Raidla
Laboratory of Isotope Palaeoclimatology
Institute of Geology at Tallinn Technical
    University
10143 Tallinn, Estonia pst. 7

D. Remenarova
Czech Hydrometeorological Institute
Na Sabatce 17, 143 00 Praha 4
Czech Republic

M. Sapiano
Directorate for Water Resources Regulation
Malta Resources Authority
Millenia, Aldo Moro Rd.
Marsa MRS 9065, Malta

P. Shand
British Geological Survey
Wallingford
Oxon, OX10 8BB

E. Skovbjerg Rasmussen
Geological Survey of Denmark and
    Greenland
Øster Voldgade 10, 1350Copenhagen
Denmark

M. Søgaard Andersen
School of Civil and Environmental
    Engineering
University of New South Wales
Water Research
Laboratory, Australia

J. Szczepanska
AGH University of Science and Technology
Department of Hydrogeology and Water
    Protection
Al. Mickiewicza 30
30-059 Krakow, Poland

Y. Travi
Laboratoire Hydrogeologie
Faculté des Sciences
33 Rue Pasteur, 84000 Avignon
France

R. Vaikmäe
Laboratory of Isotope Palaeoclimatology
Institute of Geology at Tallinn Technical
  University
10143 Tallinn, Estonia pst. 7

J. Valecka
Czech Geological Survey, Klarov 3
118 21 Praha, Czech Republic

L. Vallner
Institute of Geology at Tallinn University
  of Technology
Ehitajate tee 5
19086 Tallinn, Estonia

M. Van Camp
Laboratory for Applied Geology and
  Hydrogeology
Ghent University
Department of Geology and Soil Science
Krijgslaan 281–S8
B-9000Gent, Belgium

B. Velikov
Associate Professor
St. Ivan Rilski
Sofia 1700, Bulgaria

K. Walraevens
Laboratory for Applied Geology and
  Hydrogeology
Ghent University
Department of Geology and
  Soil Science
Krijgslaan 281 – S8
B-9000Gent, Belgium

S. Witczak
AGH University of Science and
  Technology
Department of Hydrogeology and Water
  Protection
Al. Mickiewicza 30
30-059 Krakow, Poland

A. Zuber
AGH University of Science and
  Technology
Department of Environmental
  Physics
Al. Mickiewicza 30
30-059 Krakow, Poland

Foreword

The need to protect groundwater in a concerted way at European Union (EU) level has been highlighted for more than two decades. For a long time, groundwater protection regimes and associated policies have suffered from a lack of overall planning and specific instruments which could ensure concerted actions across the EU. Early groundwater legislation (Directive 80/68/EEC) rather narrowly focused on control of emissions of substances from industrial and urban sources, while provisions for control of diffuse sources came later in the legislative picture with the agriculture- and industrial emissions-related directives, and in particular with the Nitrates (96/676/EC) and IPPC (61/96/EC) Directives.

The Ministerial Seminar on Groundwater held at The Hague on 26–27 November 1991 called for an action programme to avoid long term deterioration of groundwater quantity and quality across the EU, which led to the adoption of a communication by the Commission of a Groundwater Action Programme in 1996. In this communication, clear links were established with another Commission proposal for a 'Framework Directive on Water Resources' which was the foundation of the Water Framework Directive (WFD) adopted four years later (2000/60/EC). In this context, and for the first time, groundwater has become part of an integrated water management system in which this environmental compartment naturally pertains. The WFD includes groundwater in its river basin management planning, with clear milestones concerning delineation of (ground)water bodies, economic analysis, characterisation (analysis of pressures and impacts), and monitoring and design of programmes of measures, with the aim to achieve good quantitative and chemical status for all groundwater bodies by the end of 2015.

The difficulty of defining clear criteria concerning groundwater chemical status has led the European Parliament and Council to include provisions in the WFD (Article 17) asking the Commission to come forward with a 'Daughter Directive' providing additional technical specifications. Discussions were initiated in 2001, involving many exchanges among policy makers from the Member States, stakeholder organisations and the scientific community, which resulted in the adoption of the Directive 2006/118/EC on the protection of groundwater against pollution and deterioration on 12 December 2006.

From the beginning of the debates, starting from The Hague Seminar's discussions, it has always been obvious that the concept of 'environmental quality' of groundwater was understood differently from country to country, and from various sector perspectives. In other words, the distinction between the `natural' environmental quality and the quality of groundwater for specific uses (mainly drinking water, but also irrigation, animal watering and industrial uses) was never clearly made. This was merely due to the fact that little is known about what is meant by `natural environmental quality' for groundwater. The need for undertaking research for a better understanding of environmental mechanisms was highlighted in the 1996 communication on a groundwater action programme, stressing that the fifth framework programme of research and technological development (1994–98), and in particular the 'Environment and Climate' key action, should provide advanced knowledge for a better groundwater management.

The EU BaSeLiNe project may be directly linked to this rising awareness about research needs. Funded under the Fifth Framework Programme, it led to studies of representative groundwater systems in Europe, illustrating the main processes and water quality evolution, with the aim of defining natural water quality background levels and proposing a standardised Europe-wide approach. The project timely coincided with lively debates on the development of the new Groundwater Directive, and a workshop gathering scientists from the project and policy makers from the EU Member States was called in early 2003. From this meeting and the project outputs, it has become clear that uniform groundwater quality management at a European scale would be unfeasible.

The Directive proposal published by the European Commission on 19 September 2003 highlighted this feature in its explanatory memorandum, stating 'a workshop on the BaSeLiNe project held on 27 January 2003 (and funded by the Director General for Research (DG RTD) under the Fifth Framework Programme) stressed the difficulty of setting uniform quality standards for groundwater, and emphasised the need to consider aquifer characteristics and pressures from human activities'. These recommendations are clearly reflected by the approach followed in the new Directive, namely requesting Member States to adopt `threshold values' (environmental quality standards) for groundwater at the most appropriate level (groundwater body, river basin district or national) by the end of 2008, taking into account hydrogeological characteristics, pollutant pathways and so on.

While the project yielded a solid scientific background about natural groundwater background levels, BaSeLiNe could not provide the basis for recommending a common methodology for establishing groundwater threshold values as requested in the 2003 proposal. A new project called BRIDGE (Background cRIteria for the Development of Groundwater thrEsholds) was, therefore, funded under the Sixth Framework Programme, which may be considered as a natural child of BaSeLiNe. Discussions within this project, including those on groundwater background levels, have been reflected in the way the Directive has evolved during the negotiations, resulting in an Annex II which gives clear guidelines on criteria for establishing groundwater threshold values.

However, this is not the end of the story. The knowledge about `natural quality' of groundwater is still very scarce, and on-going exchanges among scientific and policy-making communities are needed to adopt a more detailed common methodology for groundwater threshold values and put it into practice. The forthcoming years will see a considerable move in groundwater management developments, with better interactions among policy, science and other stakeholders, and a production of data and knowledge which should allow us to achieve tremendous progress in our understanding of groundwater systems.

The book `Natural Groundwater Quality' arrives in this exchange-rich period and is hence very timely. It includes 21 chapters covering the issues of inorganic and organic quality of European aquifers, timescales and tracers, baseline trends, monitoring, and various case studies from Belgium, Bulgaria, the Czech Republic, Denmark, Estonia, France, Malta, Norway, Poland, Portugal, Spain and the UK. The book concludes with a synthesis about baseline concentrations and definition of background levels.

With the publication of a parallel book on `Groundwater Science and Policy' complementing the present volume, the adoption of the new Groundwater Directive, and the publication of this book, one may say that 2006/7 has been a productive `groundwater year', providing readers from policy-making, research and industrial communities, as well as from academic circles, with a solid foundation for developing a sound groundwater system management.

Philippe Quevauviller
European Commission

Preface

The European Water Framework Directive (WFD) forms the primary legislation for the protection of the European aquatic environment. The Groundwater Directive (GD) introduced in late 2006 has been incorporated to supplement the WFD and deal with specific questions of groundwater quality and to ensure 'good status' of groundwater. There is still a poor perception and understanding of groundwater by many people involved in water management and in implementation of policy. Against this background, a consortium of European scientists conducted detailed studies of water quality in Europe, focusing on the natural 'baseline' quality of groundwater as the basis for understanding geochemical processes in aquifers and providing a basis for defining what constitutes pollution.

The aim of this book is to provide a key reference text on natural water quality of aquifers through a series of thematic chapters, together with chapters on representative groundwater systems in Europe which illustrate well the main processes and evolution of water quality.

The criteria for defining natural backgrounds of water quality are developed to provide, as far as possible, a standardised Europe-wide approach. A geochemical approach has been adopted to assess the natural variations in groundwater quality in hydrogeologically well-defined systems. This is important, since the concentrations of certain elements present naturally in groundwaters may breach guidelines for potable water quality. Baseline criteria are also needed as a reference to be able to assess quantitatively whether or not anthropogenic pollution is taking place. The timescales influencing natural processes and the rates at which natural processes are occurring are discussed as well as historical water quality trends. The extent to which pristine waters are being depleted by contaminated waters moving into the aquifer is also assessed. As well as giving the necessary scientific framework, the project results provide guidelines for new policy (specifically the Groundwater Directive), engagement with end users, including water utilities, and the general public.

A total of 25 reference aquifers from 12 European countries (UK, Denmark, Poland, Estonia, Belgium, France, Spain, Portugal, Switzerland together with Czech Republic, Bulgaria and Malta) form the basis of the book. For each aquifer, median and upper baseline (97.7%ile) values are determined for over 50 inorganic parameters. Where groundwaters have been impacted by man, criteria are suggested for determining baseline ranges. These data have then been used for a synthesis of the baseline groundwater quality in Europe. As well as major inorganic and trace constituents the occurrence and distribution of organic carbon is also considered.

Geochemical modelling carried out on three representative aquifers in Portugal, France and UK demonstrates a quantitative understanding of the major processes controlling the evolution of baseline concentrations, especially the changes occurring along a flowpath. For all three aquifers, the changes in water chemistry are well described by relatively simple reactive transport models which may be used to predict the changes in water quality in the pristine aquifer as well as when the flow rates are accelerated by orders of magnitude due to groundwater abstraction.

Isotope, noble gas and chemical tracers have been used to establish timescales in selected reference aquifers as the foundation for evaluating baseline quality evolution in old waters (^{39}Ar, ^{14}C, ^{4}He), and in particular in detecting modern inputs (^{3}H, $^{3}He/^{3}H$, ^{85}Kr, CFCs, SF_6). Combination of careful studies of chemical trends along flow lines coupled with age determination and followed by modelling presents a powerful approach for understanding the baseline conditions. The tracer studies prove to be of particular value in demonstrating mixing processes and the dimensions of flow systems in the reference aquifers.

Trends in natural baseline have been established for aquifers in most countries over a period of decades, but these data have often been lost during water industry reorganisations. Such trends usually demonstrate the interface between natural (baseline) and contaminated groundwater. Two types of trends are observed – natural trends resulting from time-dependent geochemical reactions, which are manifest mainly at the spatial scale and non-baseline trends such as marine intrusion, which are induced by pumping. A methodology is developed to relate the spatial variability of hydrogeochemical parameters with time trends.

The public perceptions of water quality were established among end users from all partner countries. Results indicate acceptance of the value of maintaining good status of pure, natural groundwater and support the tenet that the protection of supplies should be a paramount regulatory and managerial objective. There is also concern that good quality groundwater (including fossil or palaeowater) is being impacted and used too often for purposes that do not require such high quality. Support exists for greater public information, more investigation, planning and regulation, with widespread support for protecting high quality water. A strategy for baseline monitoring has been produced a) to define the natural background status of groundwater bodies and aquifer systems and b) how rapidly and in what direction the intrinsic geochemical properties may change. It is hoped that the adoption of the scientific principles developed here and more integration of the science into policy will lead to more sustainable management of our precious aquifers.

The chapters have been reviewed by at least one external reviewer and one reviewer from within the project. We would like to express our thanks to the following external reviewers:

Pierre Glyn (USA)
Dave Stonestrom (USA)
Chris Daughney (New Zealand)
Nick Walton (UK)
John Tellam (UK)
David Kinniburgh (UK)
Jürgen Sültenfuss (Germany)
Phillipe Quevauvallier (France)
Frank Wentland (Austria)

Manuel Ramon Llamas (Spain)
Andrés Sahuquillo (Spain)
Ian Cartwright (Australia)
Andrew Love (Australia)
Irina Gaus (France)
Jopep Mas Pla (Catalunya)
Willy Burgess (UK)
Pieter Stuyfzand (The Netherlands)
Eckhart Bedbur (Germany)
Laurent Cadhilac (France)
Laurence Gourcy (France)
Rasmus Jacobsen (Denmark)
Jaroslav Skorepa (Poland)
Kirsti Korrka-Niema (Finland)

The BaSeLiNe project was funded under the European Community's Fifth Framework Programme (EVK1-1999-00024) with the title 'Natural baseline quality in european aquifers: a basis for aquifer management'. We wish to thank the EU programme director Panagiotis Balabanis for his support and also Phillipe Quevauvallier who encouraged the links between the science and policy aspects of the baseline studies.

October 2007   **W. M. Edmunds**
                                                                                  P. Shand

1 Groundwater Baseline Quality

W. M. EDMUNDS AND P. SHAND

Naturally occurring groundwaters are becoming increasingly impacted by human influences. It is important to develop criteria to recognise the extent and baseline characteristics of groundwater bodies as a basis of management and specifically as a starting point to recognise pollution. The baseline concept is reviewed and a definition adopted for use in the chapters of this book. Emphasis is placed on developing a sound understanding of the spatial and temporal distribution of water quality, in relation to hydrogeological and geochemical controls. The main geochemical controls on natural baseline chemistry, which usually accounts for 98% of the mineral composition of groundwater, are outlined. The various approaches available for baseline definition are then reviewed. Statistical tools are used as a first step, but cannot explain the variance in the data for which a resort to the field situation and use of historical data is required. The median and 97.7 percentile are adopted as the main statistical parameters for initially defining baseline and comparing different data sets.

1.1 Introduction

1.1.1 Concepts and baseline definition

One of the most difficult problems in hydrogeology is to define whether groundwater quality is natural or has been affected, to a greater or a lesser extent, by human activity. The desire for pure and wholesome water is deeply rooted in our culture. This is evident from the reverence accorded to springs and underground water by our ancestors as a source of wonder to their origins and as a symbol of purity. In modern society, spring waters are still valued highly in many communities and they still embody an element of awe and mystery. The properties of pure spring water and groundwater command a high market value, and in a world where tap water is (usually wrongly) perceived as something less pure, the bottled-water image-makers go after evidence of the purity, longevity and health-giving properties of springs and groundwater.

Gross pollution of groundwater is most easily recognised by the presence of artificial organic compounds introduced only in the past half century. Yet, it remains exceedingly difficult to recognise incipient pollution in most groundwaters. From the examination of the chemical composition of a single water sample it would, as a rule, be impossible to determine whether the major ion chemistry observed was the result of natural processes or diffuse pollution related to human activity. Agricultural activity may, for example, add small amounts of major ions such as Ca, SO_4, K and NO_3 to groundwater such that, without careful monitoring, early indication of pollution could prove problematic. Nevertheless, the overwhelming origin of major ions, which constitute more than 98% of groundwater solutes, are derived from rainfall inputs and soil processes and reactions between water and bedrock. In order to be sure that pollution is taking place, knowledge is required of spatial and temporal characteristics (trends) in major ions and other elements and of a definition of the pristine hydrochemical baseline composition.

Quality changes are not only due directly to human impacts but may also be brought about indirectly by pumping – changing the aquifer system hydrodynamics (displacement of saline and redox fronts, seawater encroachment, up-coning from depth, artificial recharge variations and mixing). Dewatering and induced water table oscillation may cause oxidation of minerals, mobilisation of heavy metals and alkalinity changes. Thus the rate at which aquifers are becoming contaminated is generally difficult to estimate since no common criteria exist for such an assessment.

Although human activity may lead to water compositions which exceed drinking water standards, water quality limits may be also breached for elements such as F, As and Fe by entirely natural processes, the result of geochemical conditions existing in the aquifer or due to the specific geology of an area.

Until recently, no common approach existed for defining the natural background quality of groundwaters, which make up nearly 70% of drinking water across Europe. Definitions and guidelines are required as part of the European Commission (EC) Water Framework Directive (WFD) and, specifically, the Groundwater Directive to assess the natural variation in groundwater quality and to assess whether or not anthropogenic pollution is taking place.

The objective of this chapter is to define what is meant by baseline and to discuss the concept and its relevance to the WFD. The chapter also introduces key chemical and hydrological concepts, providing a background for the thematic and regional chapters that follow.

1.1.2 Significance of baseline studies for policy in EU and elsewhere

One of the main objectives of the current study has been to provide background scientific information pertinent to implementation of the EC WFD and its subsidiary Groundwater Directive. The WFD 2000/60/EC (EC 2000), sets out criteria for the assessment of groundwater chemical status on the basis of existing Community quality standards (for nitrate, pesticides and biocides) and on the requirement for Member States to identify pollutants (defined as substances that may occur from both natural and anthropogenic sources, and synthetic pollutants) and related threshold values that are representative of all groundwater bodies found as being at risk.

To this end, the definition of baseline chemical properties of groundwater is a necessary step in the implementation of policy at the river basin level and more specifically at the scale of individual groundwater bodies. The WFD seeks common definitions of the status of groundwater (and surface water) quality. Environmental objectives are sought to ensure that 'good groundwater status' is achieved throughout the Community. Where good water status already exists, it should be maintained and any significant and sustained upward trend in the concentration of any pollutant should be identified and reversed.

In its environmental objectives (Article 4) for groundwaters, Member States are required to implement the necessary measures to prevent or limit the input of pollutants into groundwater and prevent the deterioration of the status of all bodies of groundwater. They shall protect, enhance and restore all bodies of groundwater, with the aim of achieving good groundwater status at the latest 15 years after the date of entry into force of the Directive (currently by 2015). Finally, Member States are required to implement the necessary measures to reverse any significant and sustained upward trend in the concentration of any pollutant resulting from the impact of human activity in order to progressively reduce pollution of groundwater. This requires scientifically defendable and achievable objectives in order to recognise, stabilise and reverse upward trends and to evaluate the efficiency of remediation measures. One of the applied scientific objectives of the present book is to help generalise and transfer accumulated knowledge and establish a framework for decision making. Lack of recognition of causes of water quality problems and well intentioned but ill thought out restoration efforts result in expensive failures, that may even cause harm.

Thus, baseline criteria need to be defined and adopted as part of strategies to prevent, identify and control pollution of groundwater (Article 17), since the definition of starting points for the identification of significant and sustained upward trends are needed within 5 years of the adoption of the WFD.

As part of the implementation process (Annexe 1), all Member States are required to carry out an initial characterisation of all groundwater bodies to assess their uses and the degree to which they are at risk of failing to meet the objectives for each groundwater body (under Article 4). In specifying further characterisation of groundwater bodies or groups of bodies (Annexe 2.2), mention is made of more detailed properties of groundwater bodies, including the adsorptive properties of the soils and the stratification within aquifers with the target to establish natural background levels for these bodies of groundwater.

Although most European countries have a requirement to monitor and assess pollution, there are few which specifically address the question of defining natural quality. There are many differing existing practices which are described in Chapter 7. Without a consistent policy, remediation is difficult to establish and hence impose at a European level. The corresponding practice in the USA, for example, is for the Environmental Protection Agency (EPA) to regulate pollution-generating activities and also to set drinking water standards that water utilities must meet. At the same time, the United States Geological Survey (USGS) assesses regional aquifers and water quality though the National Water Quality Assessment (NAWQA) programme.

1.1.3 The definition of baseline

The baseline concentration of a substance in groundwater, used throughout this book, may be defined in the following manner:

> the range of concentrations of a given element, isotope or chemical compound in solution, derived entirely from natural, geological, biological or atmospheric sources, under conditions not perturbed by anthropogenic activity.

Terms such as background or threshold have sometimes been used synonymously and have often been used to identify 'anomalous' concentrations relative to typical values, for example, in mineral exploration (Hawkes and Webb 1962; Runnells et al. 1998). Additional definitions may be required for water regulation purposes, for example, when changes from the present-day status, or an arbitrarily defined timeline, of groundwater may represent the starting point of monitoring or consideration of trend reversals. The terms 'background' or 'threshold' are preferred in this context since the initial condition may include some anthropogenic component (Lee and Helsel 2005; Reimann and Garrett 2005). Precise determination of background concentrations is then needed to distinguish the anthropogenic changes in water quality caused by both pollutants *and* geogenic sources induced by intensive withdrawal of water. An additional scenario takes place when intensive withdrawal induces rapid changes in natural quality (e.g. saline intrusion or up-coning) where the baseline is changed but without addition of anthropogenic constituents.

1.1.4 *Past approaches to identifying baseline*

Published studies of groundwater chemistry specifically aimed at defining natural baseline quality are few in number, although several classic groundwater studies contain detailed information on water quality changes in entirely natural groundwater systems. In the English Chalk aquifer, for example, groundwater evolution along flow paths has been used as a means of characterising baseline groundwater quality and recognising recent inputs influenced by pollution from entirely natural trends in deeper, older waters (Edmunds et al. 1987). Studies of hydrogeochemistry in dated palaeowaters have also been carried out by Hendry et al. (1991) for the well-characterised

Milk River aquifer in Canada. Several studies have been carried out on aquifers in the USA as part of the Regional Aquifer Simulation Analysis programme, providing a sound geochemical framework for understanding natural (and polluted) water quality. One example, the central Oklahoma aquifer (Christenson and Havens 1998), has locally high levels of arsenic, selenium, chromium and uranium which are interpreted in terms of the aquifer mineralogy, and coupled hydrological and geochemical modelling has been used to understand the evolution of water quality along regional flow paths.

A first approach to defining baseline values in the UK aquifers, as a whole, is given by Edmunds et al. (2003), where a geochemical approach was favoured over a purely statistical one. Several detailed national studies of natural trace element concentrations in the UK, USA and Norway (Edmunds et al. 1987; Frengstad et al. 2000; Lee and Helsel 2005) have also been used to explore the abundance of a wide range of constituents derived largely from geological sources. More recently, the main aquifers in England and Wales were individually subject to studies to determine baseline characteristics as a basis for groundwater management and was summarised by Shand et al. (2007). Further European studies related to baseline quality have been conducted in the Netherlands (Meinardi et al. 2003), Sweden (Ledin et al. 1989) and France (Blum et al. 2002).

1.2 Hydrogeochemical controls

1.2.1 *Generation of the natural baseline*

The generation of groundwater quality can be viewed as a gigantic open system, but unevenly equilibrated chemical reactor involving rain water, soils and rock strata. The acquisition of solutes may be regarded as a series of processes and reactions taking place within the hydrological cycle which may occur over natural timescales of days to millennia. The baseline geochemical system is a response to variable recharge rates, rainfall compositions, as well as the processes that have taken place along flow lines in aquifers in response to climate and environmental change including adjustments in water levels brought about by sea-level change as water flowed toward discharge areas (Fig. 1.1).

Residence times of groundwater can typically be measured in hundreds to tens of thousands of years (Fig. 1.1). A sequence of relative timescales based largely on isotopic tracers in aquifers has been suggested (Edmunds 2001):

i. Palaeowater, that is, water recharged during or before the last glacial era,

ii. Pre-industrial Holocene water (free of any anthropogenic components),

iii. Industrial era but pre-thermonuclear era water (water free of tritium), and

iv. Modern water, that is, water younger than *ca.* 50 years, identified by presence of tritium from thermonuclear tests in atmosphere, or by presence of recent man-made contaminants such as CFCs.

Disturbances in flow and quality commenced with the onset of the Industrial Revolution and mechanised recovery of groundwater (Fig. 1.2). Before this time, natural flow regimes became established, adjusting most recently to the restoration of the present-day coastline in the early Holocene. Stratification of age and quality of groundwater would have taken place. Human impacts may, however, have been important in terms of water quality before the modern era, with forest clearances and land use changes involving land drainage, for example. The modern era, coinciding with roughly the last 200 years caused a number of adverse impacts and significant changes to the hydrochemical system, notably, deterioration in air quality, diffuse and point source pollution from agricultural and industrial sources, as well as mixing of water of different qualities as development of groundwater proceeded. The legacy of these impacts was generally slow to be recognised due to the hysteresis caused by the nature of the recharge process.

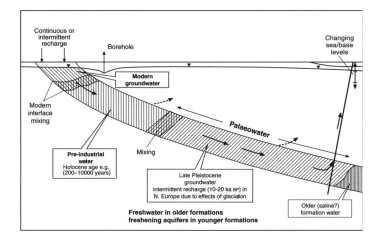

Fig. 1.1 Conceptual model of a confined groundwater system to show the definitions of modern and palaeowaters referred to in this study.

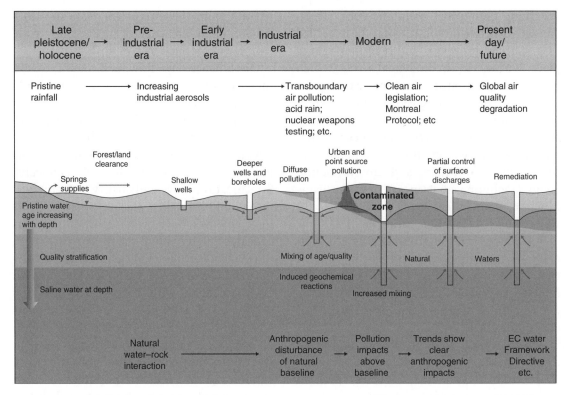

Fig. 1.2 Evolution of the natural baseline quality of groundwater and progressive impacts of human influences.

The natural stratified groundwater quality has been progressively disturbed by well and borehole drilling and pumping, especially during the twentieth century with the concomitant advance of contaminants from diffuse and point sources as well as by a changing atmospheric background from industrial sources (Fig. 1.2). Thus, it is necessary to recognise the extent and chemical character of the naturally evolved water and, as far as possible, recognise and differentiate this from the human impacts on the natural baseline. This represents the main target of this chapter and forms an important starting point for water quality management.

1.2.2 Inert and reactive tracers

In order to understand the baseline characteristics of groundwater, it is first necessary to consider which constituents of the groundwater (chemical elements/species, isotopes or gases) can be used as tracers. It is convenient to distinguish between inert and reactive tracers. Inert tracers include chemically unreactive species such as Cl and several isotopic signatures such as $\delta^{18}O$, which remain unchanged in composition over short timescales, or for which a major shift in composition indicates a new input source of groundwater. Reactive tracers include H^+ and common cations contributed by weathering and reactions taking place along flow lines.

Examples of tracer types are

Inert tracers:

Cl, Br (Br/Cl), NO_3, ^{36}Cl, $\delta^{18}O$, δ^2H, ^3H, noble gases (isotopes and ratios, e.g. ^{81}Kr).

Reactive tracers:

Major ions and ionic ratios, H^+, Si, trace elements (e.g. B, Li, F, Sr), isotope ratios ($^{87}Sr/^{86}Sr$, $\delta^{34}S$, $\delta^{13}C$, ^{14}C).

Inert tracers may be used to track the inputs to groundwater from the atmosphere, land surface and soils, which in turn can help track changes in past climates, environment and water origins, and also provide estimates of age. It is noted that nitrate may be used as an inert tracer in aerobic systems since it is stable in the presence of oxygen. Reactive tracers, on the other hand, chart the water–rock interaction and help to characterise the rock type, mineralogy, flow pathways and geochemical controls leading to the unique mineral composition of the natural water.

In order to fully understand the evolution of a groundwater body and to characterise its baseline properties it is usually necessary to adopt a multi-tracer approach. For example, a combination of inert and reactive tracers was used to deduce the evolution of groundwaters from present day to pre-Holocene in the Permo-Triassic East Midlands aquifer of England (Edmunds and Smedley 2000). This example, described later, allowed changes in baseline with time, palaeoclimate influences and the impacts of modern pollution to be assessed with some degree of confidence. A similar multi-tracer approach is adopted for the reference aquifers in this book since this

provides one of the clearest illustrations of the separation of natural trends from human influences.

1.2.3 Controls on baseline groundwater systems in the hydrochemical cycle

Rainfall modified by evapotranspiration may be a significant contributor to defining groundwater compositions. Hydration of carbon dioxide acquired during passage through the soil provides the key reactant (H_2CO_3) controlling weathering of the rocks encountered at shallow depths. The soil and unsaturated zone provide a highly reactive environment for mineral dissolution with strong changes in moisture content (water–rock ratio), recycling of solutes, pH changes neutralising acidic inputs and also high microbial populations which may catalyse geochemical reactions at shallow depths. In this way, the distinctive mineral content of many groundwaters is determined in the top 5–10 m. Lithology and mineralogy (especially the presence or absence of rapidly reacting carbonate minerals), hydrophysical properties and residence time determine the detail of the eventual groundwater quality. Saturation controls by minerals, such as calcite, may inhibit further reaction and significant modification of water quality in the deeper saturated water bodies, although redox controls are of considerable importance. The main processes whereby baseline groundwater quality is determined are shown conceptually in Fig. 1.3.

1.2.3.1 Rainfall inputs

Rainfall originates mainly in the oceans and carries with it a characteristic aerosol composition dominated by Na and Cl accompanied by high Mg/Ca ratio. As the

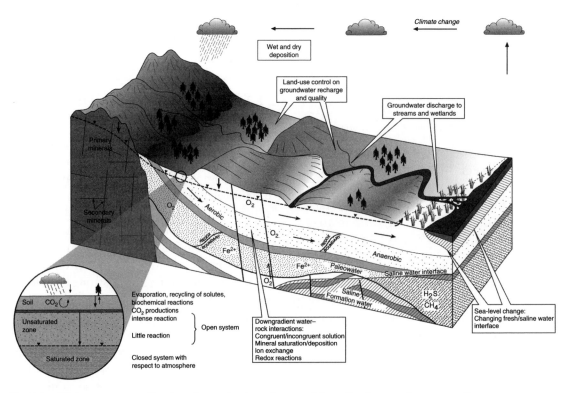

Fig. 1.3   Main processes affecting the evolution of natural (baseline) groundwater quality.

rain passes over land, rain-out of aerosols will take place so that the residual water vapour becomes depleted in Cl. Similarly, the isotopic composition of water vapour evolves as it becomes more continental (Clark and Fritz 1997). Thus, the baseline chemistry may differ significantly between coastal and inland areas in the same aquifer (as described, for example, for the Chalk of Dorset in Chapter 9). Pristine rainfall is naturally acidic with a pH of around 5.6, and, although it contains low CO_2, it can be a significant agent for water–rock interaction.

1.2.3.2   Soils

The chemistry of groundwater depends to a large extent on the output from the base of the soil zone. There is scope for considerable modification of the rainfall composition in the soil, notably by evapotranspiration causing concentration, typically by a factor of 3 in temperate regions but up to 10 or more in semi-arid parts of Europe. The diurnal and seasonal fluctuations in temperature and the variable rainfall cause wide ranges in water–rock ratio and ionic strength of the soil solutions, leading to mineral saturation, with some precipitation in dry spells and subsequent dissolution. The main impact of the soil is owing to the biological production of CO_2, which under natural baseline conditions may raise the concentration by up to 10–100 times that of the atmosphere. The resulting carbonic acid is the main control on weathering, both of carbonate and silicate minerals.

Organic soils may be naturally acidic (pH 2.9–3.3). Acidity will be neutralised to near neutral pH in alkaline soils above carbonate rocks, but in non-carbonate rocks the pH of moisture leaving the soil may remain in the range 4.0–5.0 (Dahmke et al. 1986; Edmunds et al. 1992).

1.2.3.3 Unsaturated zone

Where the water table is shallower than a few metres, capillary action will lift water to evapotranspirative sinks, concentrating the solutes; deep-rooted vegetation may also lift water from depth. Below a few metres depth, the moisture drains to the water table, typically at recharge rates of between 0.01 and 1 m yr^{-1} in porous media, but greater than this in fractured rocks where dual porosity may lead to differential rates of percolation with some fracture flow. The moisture takes with it the fingerprint of inert atmospheric inputs plus reactive tracers, solutes derived

from reactions in the unsaturated zone. Reactions taking place at the soil–bedrock interface are some of the most intense in the geochemical cycle and play a major part in determining the baseline composition.

Inert solutes, such as chloride, moving through the unsaturated zone will retain a memory of the climatic and environmental conditions at the time of recharge. Thus, past wet periods and droughts, as well as land use type and land use change will be recorded in the oscillations of Cl and other inert solutes. Superimposed on the natural baseline, the unsaturated zone then records human impacts, such as excess nitrate from fertiliser applications to arable land.

The attenuation of rainwater or soil acidity is usually completed in the unsaturated zone, inducing complementary changes in the reactive tracers, depending on the mineralogy and cation-exchange capacity. This is illustrated for a siliclastic aquifer in Fig. 1.4 using interstitial water profiles (10 m depth

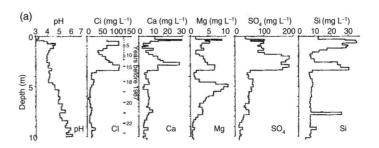

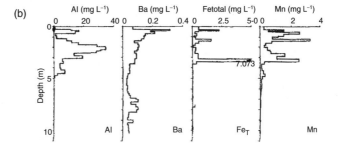

Fig. 1.4 Natural baseline conditions beneath grassland on sandstone (Triassic sandstone, English Midlands). Exchangeable cations are progressively removed by incoming low pH water; high metal concentrations occur in the upper section of the unsaturated zone.

above the water table) taken from an area of acid heathland overlying arkosic Triassic sandstone in the English Midlands (Edmunds et al. 1992). Profiles are shown for a number of elements in water obtained by immiscible liquid displacement method. This profile was drilled in a newly planted area of Cyprus pine; the high Cl concentrations above 3.5 m clearly show the period of growth (12 years) and slowing of recharge (from 95 to 22 mm yr^{-1}) as compared with the original heathland. A gradual attenuation of acidity from pH 4 to 6 takes place over the section with buffering of pH initially controlled by Al and at lower levels by base cations (Ca and Mg). A number of other metals (Fe, Mn, Cu, Zn, Ni and Be) are also mobilised in the upper 5 m of the profile. It is uncommon for acidic conditions to persist into the saturated zone and to affect overall groundwater quality, although it is evident that shallow groundwaters in non-carbonate aquifers may give rise to naturally high metal concentrations.

1.2.3.4 Hydrogeochemical processes in groundwater

Under natural conditions, groundwater quality is determined by the sum of soil-modified atmospheric inputs plus water–rock interaction taking place at the soil–bedrock interface and from longer-term reactions taking place along flow paths in the saturated zone. The initial concentrations and ratios bear a close resemblance to the parent geology for some elements. Aquifer lithology and mineral assemblage thus provide the primary groundwater signature and the groundwater will undergo geochemical reactions (dissolution–precipitation, ion-exchange or redox reactions) mixing as water moves downgradient (Fig. 1.3).

1.2.3.5 Solution–precipitation reactions

Mineral solubility and kinetics determine the release of cations to groundwater. Therefore, a consideration of the availability in the zone of weathering in the bedrock of the most soluble rock-forming minerals will be critical in establishing the initial water quality. Dissolution of carbonate and silicate minerals are largely driven by the action of CO_2 produced within the soil zone by root respiration and bacterial metabolism on solid mineral phases within the soil. A detailed treatment of the processes involved is given in several texts (Stumm and Morgan 1996; Langmuir 1997; Appelo and Postma 2005).

A number of dissolution reactions may be used to explain the release and uptake of solutes from carbonate and silicate rocks.

(1a) Congruent dissolution of calcite

$$CO_2 + \underset{\textit{Calcite}}{CaCO_3} + H_2O \rightarrow 2HCO_3^- + Ca^{2+}$$

(1b) Incongruent reaction of impure calcite

$$xCa^{2+} + \underset{\textit{Impure calcite}}{Ca_{(1-x)}Mg_xCO_3} \rightarrow \underset{\textit{Calcite}}{CaCO_3} + xMg^{2+}$$

(2a) Weathering of primary rock minerals (feldspar) to release solutes and form clays

$$2CO_2 + \underset{\textit{Albite}}{2NaAlSi_3O_8} + 11H_2O \rightarrow$$

$$\underset{\textit{Kaolinite}}{AlSi_2O_5(OH)_4} + 2Na^+ + 2HCO_3^- + 4H_4SiO_4$$

(2b) Weathering of primary rock minerals (biotite) to release solutes and form clays

$$2K(Mg \cdot 2Fe)(AlSi_3)O_{10}(OH)^{2+}10H^+ + \atop \textit{Biotite}$$

$$0.5O_2 + 7H_2O \rightarrow Al_2Si_2O_5(OH)_4$$
$$+ 2K^+ + 4Mg^{2+} + 2Fe(OH)_3 + 4H_4SiO_4$$

(3) Weathering of cation-rich clays to form cation-poor clays

$$3Na_{0.33}Al_{2.33}Si_{3.67}O_{10}(OH)_2 + H^+ + 11.5H_2O \rightarrow$$
Na-montmorillonite
$$3.5Al_2Si_2O_5(OH)_4 + Na^+ + 4H_4SiO_4$$
Kaolinite

There are common features to these reactions. All of them are hydrolysis reactions that consume CO_2 and therefore change pH. During the early evolution of the chemical composition of groundwater, the open or closed system behaviour with respect to CO_2 is important (Fig. 1.3). Up to a certain depth during recharge, the infiltrating water remains in contact with the atmospheric and soil CO_2 (open system), but below a certain depth diffusive exchange with the soil and atmosphere become negligible and the subsequent evolution takes place with a fixed quantity of atmospheric CO_2, although further *in situ* production of CO_2, for example, from breakdown of organic matter or deep crustal or mantle sources may still occur.

The reactions of silicate minerals under conditions of natural weathering, as shown above for albite and biotite, also have some common features. An unstable mineral such as albite (formed under high-temperature conditions) reacts with carbonic acid (represented above only by H^+) and forms kaolinite, a more stable mineral. In the process, cations are produced (both major and trace elements since natural minerals are generally impure) and silica is also produced along with bicarbonate. Biotite also releases some iron which precipitates under aerobic conditions as iron oxyhydroxide. It is important to note that silicate minerals are the primary source of silica in most groundwater systems and not quartz or other forms of SiO_2, which remain almost inert. The reactions of carbonate minerals are usually quite rapid whereas those involving silicates tend to be rather slow and are in general also irreversible.

Two reactions are shown for carbonate minerals. *Congruent* dissolution is the common reaction taking place involving attack by the weak acid (H_2CO_3). Calcite undergoes complete dissolution, and reaction takes place until saturation is reached, although the reaction of calcite can still occur under the dynamic (reversible) conditions of the equilibrium. During congruent solution, cations will be added in stoichiometric quantities so that the groundwater assumes cation ratios similar to those in the parent carbonate. However, as the groundwater proceeds along the flow path, *incongruent dissolution* will continue whereby the impurities in the solid phase are progressively released and a purer mineral (in this case a lower-Mg calcite) is produced. The Mg/Ca ratio in the water increases as a result of this process and may be used as an indicator of residence time and increasing maturity of groundwater – young groundwaters will generally only have low Mg concentrations.

The release of ions to solution will depend on the relative solubilites and reaction kinetics of the component minerals, the rate of supply of H^+ and temperature. Therefore, groundwater catchments with different lithologies, mineral assemblages and, especially, soil types, would be expected to give rise to different water chemistries and baseline characteristics. The baseline chemistry may reflect different land use and different recharge areas within a catchment. In carbonate areas, for example, for groundwaters at saturation with respect to calcite, the concentrations of HCO_3 are likely to reflect recharge areas which may have different $P(CO)_2$ and hence different land use and soil types (Langmuir 1971).

1.2.3.6 Redox controls

Redox reactions are very significant in determining and delineating changes in the baseline characteristics of groundwater. Most shallow groundwaters under natural conditions contain dissolved oxygen concentrations (on average around 8 mg L^{-1}) only slightly below atmospheric values. During down-gradient flow, the dissolved oxygen is progressively consumed by microbially mediated processes or inorganic reactions leading to anaerobic conditions marked by a sharp change in the redox potential (Eh) at a redox boundary (Fig. 1.3). The rate of removal of dissolved oxygen is strongly dependent on the geochemistry of the aquifer and the presence of electron donors (notably dissolved organic matter – total organic carbon [TOC] – or Fe^{2+} contained in pyrite or, to a lesser degree, silicates). In some aquifers, such as young unconsolidated sediments, there is likely to be sufficient reactive organic carbon (small molecules containing 4–8 carbon atoms) to permit microbially mediated reactions. In many aquifers such as sandstones, however, there may be a low abundance of TOC, or the carbon present may be inert; present only as humic acids or geological carbonaceous material (see Chapter 4). In such cases, the presence of interstitial pyrite may act as the dominant control on redox status.

A sequence of redox changes may be recognised along flow paths in aquifers (Champ et al. 1979; Edmunds et al. 1984) which often follow those predicted by equilibrium thermodynamics (Stumm and Morgan 1996). Thus, following the complete reaction of O_2, nitrate is the next electron acceptor, and then Mn and Fe oxides are reduced followed by SO_4 and then by methane fermentation. A similar set of oxidation half-reactions may be recognised in the sequence: organic matter>HS^->Fe^{2+}>NH_4>Mn^{2+}. For most practical purposes, it is the rate of removal of oxygen that is the most significant, since, in the reducing environment, Fe^{2+} becomes stable and may appear in significant quantities. Following the reaction of oxygen, the removal of nitrate is rapid and although under baseline conditions nitrate concentrations are low, removal of high contaminant levels are possible across the redox boundary (Parker et al. 1991). Sulphate reduction, although predicted, is inhibited by reaction kinetics in aquifers low in TOC and may persist in many reducing groundwaters (Edmunds et al. 1984).

1.2.3.7 Ion-exchange reactions

Modifications to the chemistry of groundwater by cation exchange as it moves along flow paths in aquifers are well documented (Appelo and Postma 2005). Most aquifers, and especially those formed in marine environments, are in a state of continuing evolution as freshly recharged groundwater displaces saline formation water and the parent rock undergoes adjustment including incongruent dissolution (described above) and also exchange of adsorbed cations especially sodium for calcium:

(4) Cation-exchange reactions

$$Na^+ + \tfrac{1}{2}Ca\text{-}X_2 \rightarrow Na\text{-}X + \tfrac{1}{2}Ca^{2+}$$

where X represents the exchanger.

Nearly all minerals have exchange properties and can retard the progress of charged species in groundwaters, but it is the size of the surface area of the mineral that will determine the variation in cation-exchange capacities (CECs). Clay minerals such as montmorillonite have large CECs and exceed those of, for example, kaolinite. Organic matter and various forms of iron oxy-hydroxides, however, may have the highest CECs. It is also important to note that the amount of exchangeable cations contained on mineral surfaces in many aquifers will greatly exceed the dissolved concentrations (Appelo and Postma 2005).

Examples of cation exchange taking place along flow paths are given for several of the reference aquifers featured in this book, for example, the Neogene aquifer in Belgium. It may be possible to determine the amount of Na released by cation exchange (in carbonate rocks where albite reaction is not important) from the baseline quantities entering as aerosol, by the amount the molar Na/Cl ratio exceeds the marine value (mNa/Cl = 0.86). This is illustrated for the Chalk aquifer (Chapter 9).

Cation exchange is a major control on the quality of young sedimentary aquifers, especially in coastal regions. Marine formation water is progressively removed in a freshening process which proceeds sequentially. Fresh water containing mainly Ca^{2+}, which is favoured by the natural exchangers, displaces sea water dominated by Na^+ and Mg^{2+}, first displacing the Na^+ and then the Mg^{2+} (Appelo and Postma 2005; Walraevens et al. 2007).

1.3 Timescales of groundwater movement and spatial considerations

In order to interpret the water quality variations in terms of baseline concentrations, some knowledge of water ages (timescales) within groundwater systems is required. For this purpose, the use of both inert and reactive chemical and isotope tracers is essential. Under favourable conditions,

groundwater ages can be assessed with sufficient confidence, though, very often, difficulties are encountered due to the lack of ideal tracers. To overcome difficulties, it is also sometimes convenient to combine the tracer methods with numerical models of flow and solute transport.

By investigating the evolution of water quality along flow lines, it is also possible to establish relative timescales. For instance, the identification (or absence) of marker species related to activities of the industrial era, such as elevated TOC, tritium (^3H), dissolved anthropogenic gases (chlorofluorocarbons, SF_6) and certain micro-organic pollutants may provide evidence of a recent component in groundwater. Radiocarbon (^{14}C), stable isotopes of oxygen and hydrogen in water molecules, accumulated ^4He and other noble gases are widely used for determinations of water ages. Even when absolute age determination cannot be made, it may be possible to distinguish waters of different ages in qualitative terms using reactive chemical tracers including ionic ratios such as Mg/Ca, Na/Cl and increase in trace element concentrations.

1.3.1 The East Midlands Triassic aquifer: an illustration of changing baseline

The East Midlands Triassic sandstone is highly oxidised red sandstone of arid desert and is of fluviatile origin, which at depth contains dolomite and anhydrite cements indicating a playa lake setting for deposition. It forms a major water supply source in the UK and has been the subject of detailed hydrogeological and geochemical investigations (see e.g. Bath et al. 1979; Edmunds et al. 1982; Downing et al. 1987; Edmunds and Smedley 2000; Smedley and

Edmunds 2002). It provides a good example to show how inert and reactive chemical tracers help to build a story of the changing baseline as water moves along flow paths in the aquifer (Figure 1.5). This approach is used in subsequent chapters to illustrate baseline concepts and an outline is given here using representative tracers. In Chapter 4, the East Midlands aquifer is subjected to modelling to test some of the geochemical observations, and in Chapter 5 the upper part of the aquifer is used as a case study to evaluate the use of modern tracers.

The sandstone forms a single hydraulic unit and the aquifer is confined by a thick sequence of mudrocks (Mercia Mudstone) which locally contain residual evaporite minerals. The aquifer is underlain by a sequence of Permian mudstones, marls and dolomitic limestones, and Carboniferous rocks with saline formation water. In Fig. 1.6, groundwater temperature has been used as a proxy for age or distance along the flow path, also separating waters from different depths at the same site. A selection of tracers is used to illustrate the main processes and the changes in baseline chemistry.

Groundwater age increases regularly along the flow path with maximum ages well in excess of 50 kyr and fresh waters span the period up to and beyond the range of radiocarbon dating. Groundwater recharge occurred during the Holocene, but an age gap indicates that no recharge occurred during the period c.10,000–20,000 yr BP, corresponding to permafrost in the recharge areas during the glacial maximum. The transition between groundwaters of Holocene and Pleistocene age is clearly indicated by the step in $\delta^{18}O$ of 1.7‰ and supports the evidence for recharge during a colder interstadial climate. Noble gas recharge temperatures indicate

that conditions during the interglacial period were some 6°C cooler than at the present day.

Although confined between mudstones with a potential to add salinity, the groundwater has a remarkably low total mineralisation. Chloride can be used as an inert tracer of inputs from the atmosphere and other sources (any internal sources of Cl are extremely low). Concentrations as low as 6 mg L^{-1} in older waters are likely to reflect precipitation when a more continental climate existed across northern England, when sea levels were much lower. The ratio Br/Cl has also been used to help constrain the origins of Cl, with the higher ratios reflecting the greater continentality during the Late Pleistocene. The high Cl at outcrop then defines the extent of groundwater pollution from various sources (including industrial sources of rainfall from the modern era) – inputs which, after evapotranspiration, give concentrations entering groundwater of ~27 mg L^{-1}. Sulphate (Fig. 1.6), similarly also indicates the extent of modern contaminant input, but extremely low SO_4 concentrations in water of Holocene age also reflect pristine atmospheric inputs as shown from studies of sulphur isotopes (Edmunds et al. 1995). Unlike Cl, SO_4 increases in the deeper, older waters indicating dissolution of remnant anhydrite. Cross-formational flow in the East Midlands aquifer does not occur and most of the hydrogeochemical changes are the result of internal water–rock interaction.

Sequential processes in reactive tracers reflect water–rock interaction with both carbonate and silicate minerals as well as anhydrite. The Mg/Ca ratio which is low near outcrop first indicates reaction with secondary calcite cements, but the sharp increase to values above 2 indicates the predominance of dolomite reaction; in deeper

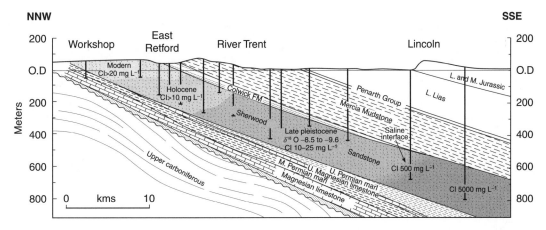

Fig. 1.5 Summary of groundwater quality in the East Midlands Triassic sandstone. Further explanation given in the text.

groundwaters the lower Mg/Ca ratios reflect the influence of Ca added by reaction of anhydrite. The molar Na/Cl ratio increases regularly along the flow path indicating reaction of albite. Several trace elements also increase regularly along the flow path and the example is shown of lithium (Fig. 1.6) which increases from 1 μg L^{-1} at outcrop to around 40 μg L^{-1} at depth. Lithium is obviously not derived from any pollution source but most probably from progressive feldspar dissolution. Like Cl, these element and ratio changes may be used as indicators of relative groundwater age as well as demonstrating the changing baseline compositions and their controls.

In this highly oxidising lithology, the groundwaters remain aerobic Figure 1.5 and oxygen persists for several thousand years (Fig. 1.6), but then its removal is marked by a change in the redox potential and the establishment of reducing conditions. This redox barrier has a strong control on the baseline concentrations of many elements, including nitrate and other redox sensitive trace elements such as Fe, Se and As. Arsenic (as arsenate) concentrations increase with residence time in the aerobic section of the aquifer highest next to the redox boundary, but then remain relatively low in the reducing section of the aquifer (Smedley and Edmunds 2002).

1.4 Baseline definition and quantification

Several approaches have been considered in the present book to help identify baseline properties and are summarised in Table 1.1. They include the use of historical data, the use of down-gradient profiles, extrapolation from adjacent areas with similar geology, geochemical modelling and the use of statistical techniques. These approaches separately or together help incipient or actual human impacts to be recognised (Shand et al. 2007).

Traditionally, the use of statistical techniques has assumed that geochemical distributions are either normal, or more generally,

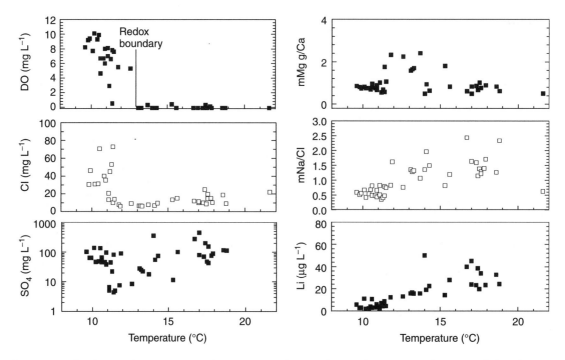

Fig. 1.6 Hydrochemical cross section for various indicators (DO, Cl, SO₄, mMg/Ca, mNa/Cl and Li along the profile shown in Figure 5 (here temperature used as proxy for distance). For further explanation see text.

log-normal (Ahrens 1954). In most natural systems, including groundwater, geochemical distributions may be polymodal and are usually heavily skewed (Shand and Frengstad 2001). Non-parametric statistical methods are advocated for geochemical distributions, which may not have a normal or log-normal distribution (Reimann and Filimoser 2000). For practical purposes, ranges in baseline concentrations are used and not simply a single average, although some normative average value may be helpful as a guideline. This range can be described in many ways, for example, by giving a mean and standard deviation, the total range in values or by describing minimum and maximum baseline concentrations after removing outliers.

The value of the commonly used statistical parameters, such as median or mode, are of use when comparing the baseline chemistry of aquifers from different areas or within the same aquifer from different regions. Use of the median value instead of the mean is preferred for comparing different datasets because it is more robust and much less affected by outlying values.

Various but rather similar approaches have been used in an attempt to define cut-off points between baseline and anthropogenic inputs (Edmunds et al. 1997; Kunkel et al. 2004; Reimann and Garrett 2005). Probability plots have been extensively used in the mining industry and such diagrams have been used to identify background and

Table 1.1 Various approaches used for the determination of baseline chemistry of groundwaters.

Method	Application to baseline studies	Advantages and limitations
Use of historical data	Important for identifying trends over lifetime of groundwater development	Highly valuable records and more effort needed to preserve these Long-term data-sets are, however, rare The range of solutes is usually small (e.g. chloride, nitrate and hardness). Analytical methods (and analysts) have changed with time, detection limits vary
Use of downgradient profiles	Profiles along flow gradients provide a useful two dimensional summary of groundwater quality evolution with time	Care needed to distinguish data from boreholes with different depths and large and small abstraction sites Limited use in unconfined groundwaters
Extrapolation from adjacent areas	Useful first approach but no substitute for examination of each water body	Lithofacies changes along strike or dip usually limit this approach. Each area has unique geological and hydrogeological properties
Geochemical modelling	Provides independent means of testing hypotheses of groundwater evolution, based on thermodynamic approach	Requires a good knowledge of the hydrodynamic situation, end member compositions and aquifer mineralogy
Statistical methods	Distinguish anomalous from typical values. Provides important summary of median and ranges of data for a water body	Cannot distinguish anomalies resulting from natural or contaminant sources Should always be used alongside hydrogechemical studies

anthropogenic populations of elements in surface water and groundwater samples adjacent to a mining and smelting area in Utah, USA (Runnells et al. 1998). Similarly, the median +2 median absolute deviation (median of the absolute deviations from the median of all data) has been used to exclude extremes (Tukey 1977; Reimann and Garrett 2005).

As a starting point, it is useful to define an upper baseline from the single data set, for example, the 95th percentile as chosen in the UK and USA baseline studies (Edmunds et al. 1997; Lee and Helsel 2005) or 97.7 percentile (mean + 2-sigma) as done by Langmuir (1997). There are no *a priori* reasons why the upper 2–10% of the data population should include

all the non-baseline samples. For example, it is obvious from examining any of the data sets that nitrate contamination is widespread and affects up to 50% of samples in some aquifers. In the present investigations, the median and the 97.7 percentile have been adopted widely to represent the average tendency and to separate outliers on a purely statistical basis.

The BaSeLiNe project (European Commission 2003), completed under the European Union Framework VI programme, forms the primary data set of information used in this book. Groundwater data sets for single aquifers (or sections of aquifers) are first treated statistically, but then examined through time series and spatial distributions (especially down-gradient profiles) where the hydrogeology and controlling hydrogeochemical processes may be taken into account. In this way, the degree of heterogeneity of a groundwater body can be interpreted in relation to groundwater flow paths, depth and residence times. Anomalies superimposed on the baseline due to pollution may normally be expected in the flow system upstream of the pristine (baseline) groundwaters.

From the foregoing discussion, it is clear that the groundwater baseline will be represented by a range of values reflecting the natural spatial variability in water quality due to geological, geochemical and environmental factors. The main purpose of the chapters in this book is to highlight the spatial variability of natural groundwater quality within groundwater systems and to provide a basis against which contamination may be recognised.

An ideal starting point to establish baseline concentrations is to locate waters where there are likely to be no traces of human impact (essentially those from the pre-industrial era), although this is not always easy for several reasons. Unconfined aquifers and confined aquifers in recharge areas usually contain, especially at shallow depths, anthropogenic components from diffuse sources (e.g. acid rain and agrochemicals). Groundwater exploitation by means of drilling may penetrate water of different ages and quality with increasing depth as a result of the stratification that invariably develops under undisturbed hydrogeological conditions. This stratification is a result of different flow paths and flow rates being established as a consequence of prevailing hydraulic gradients and the natural variation in the aquifer's physical and geochemical properties. As a result of porosity differences, groundwater is often a mixture of water flowing rapidly through fractures and older, slower-moving water held in the matrix.

The European reference aquifer studies show that many waters, recognisable as being modern and with short turnover times, may still not show other clear signs of pollution. Their composition may still overwhelmingly reflect mineralisation derived from water–rock interaction and will often be fully acceptable in terms of potability, providing that biological constituents do not pose a health risk. The most common indicator of incipient pollution in most of the European groundwaters is nitrate, which is mobile under aerobic conditions and is present above naturally low threshold values in most groundwaters. High nitrate occurrence also can be entirely natural, for example, in groundwaters from semi-arid areas where leguminous vegetation is present (Edmunds and Gaye 1997).

At the practical level, the baseline needs to relate to a specific water body and values related to the controlling geochemical processes and heterogeneity involved. Each aquifer system will be unique. A subdivision of the water body (e.g. into confined or unconfined, oxidising or reducing) may be necessary in order to properly express and define the baseline.

Once the hydrogeological and geochemical assessment has been made, a useful (single) operational definition of the baseline concentration is the median which enables intercomparison between data sets (see Chapter 2). The baseline range can be defined from 2.3% to 97.7% of sample population (95.4% of population being within that range). Another graphical presentation of data for each element or species used is also given in the form of box plots. The boxes should be in the range of 16–84% of sample population (68% of population being within that range) whereas 'whiskers' should indicate the range of 2.3–97.7% as well as outliers. The upper whisker can be regarded as the upper limit of the baseline concentration.

Care is needed in applying statistical treatment of data. For instance, when the number of observations is lower than 37, the minimal and maximal values measured will define the baseline range. In other words, there are no outliers. However, if the extremely high value(s) do not fit to any regular distribution and can be regarded as belonging to a local anomaly, then they will require special considerations on the supposed origin related either to anthropogenic pollution or geogenic influence.

Cumulative frequency plots are favoured in baseline studies as a coarse means of discriminating the natural chemistry from that due to pollution; outliers may indicate the extent of pollution but may also occur due to anomalous natural concentrations (e.g. mixing with saline groundwaters). However, there are also several types of geochemical reactions which will alter distributions by removing or limiting concentrations in solution including redox processes, adsorption onto solid mineral phases and saturation with respect to minerals that will limit the solubility of one or more elements. This is illustrated in Chapter 2, Fig. 2.2 for the main situations encountered in groundwater environments where the inferences that may be drawn from cumulative frequency plots are shown.

References

Ahrens, L.H. (1954) The lognormal distribution of the elements (a fundamental law of geochemistry and its subsidiary). *Geochimica et Cosmochimica Acta* 5, 49–74.

Appelo, C.A.J. and Postma, D. (2005) *Groundwater, Geochemistry and Pollution*, 2nd edn. Balkema, Leiden.

Bath, A.H., Edmunds, W.M. and Andrews, J.N. (1979) Palaeoclimatic trends deduced from the hydrochemistry of a Triassic sandstone aquifer. In: Proceedings International Symposium on Isotope Hydrology. Vol.II. IAEA-SM-228/27. IAEA, Vienna, pp. 454–68.

Blum, L., Chery, L., Barbier, J., Baudry, D. and Petelet-Giraud, E. (2002) Contribution à la charactérisation des états de référence géochimique des eaux souterraines. Outils et méthodologie. Rapport Final (5 vols.). BRGM Report RP/ 51549-FR.

Champ, D.R., Gulens, J. and Jackson, R.E. (1979) Oxidation–reduction sequences in groundwater flow systems. *Canadian Journal of Earth Sciences* 16, 12–23.

Christenson, S.C. and Havens, J.S. (eds) (1998) Ground-water-quality assessment of the Central Oklahoma Aquifer, Oklahoma: Results of Investigations, U.S. Geological Survey Water-Supply Paper 2357-A, 179 pp.

Clark, I. and Fritz, P. (1997) *Environmental Isotopes in Hydrogeology*. Lewis, Boca Raton, FL, 328 pp.

Dahmke, A., Matthess, G., Pekdeger, Schenk, D. and Schulz, H.D. (1986) Near-surface geochemical processes in quaternary sediments. *Journal of the Geological Society London* 143, 667–72.

Downing, R.A., Edmunds, W.M. and Gale, I.N. (1987) Regional groundwater flow in sedimentary basins in the UK. In: Goff, J.C. and Williams, B.P.J. (eds). *Fluid Flow in Sedimentary Basins and Aquifers*. Geological Society, London, Special Publication 34, pp. 105–25.

EC (European Commission). (2000) Water Framework Directive. 2000/60/EC.

EC (European Commission). (2003) Natural baseline quality in European aquifers: a basis for aquifer management. Final contract report EVK1-CT1999–0006.

Edmunds, W.M. (1994) Indicators of rapid environmental change in the groundwater environment. In: Berger, A.R. and Iams, W.J. (eds) *Geoindicators: Assessing Rapid Environmental*

Change in Earth Systems. Balkema, Rotterdam, pp. 135–50.

Edmunds, W.M. (2001) Palaeowaters in European coastal aquifers – the goals and main conclusions of the PALAEAUX project. In: Edmunds, W.M. and Milne, C. (eds) *Palaeowaters of Coastal Europe; Evolution of Groundwater since the Late Pleistocene.* Geological Society, London, Special Publication 189, pp. 1–16.

Edmunds, W.M. and Gaye, C.B. (1997) High nitrate baseline concentrations in groundwaters from the Sahel. *Journal of Environmental Quality* 26, 1231–9.

Edmunds, W.M. and Smedley, P.L. (2000) Residence time indicators in groundwater: the East Midlands Triassic sandstone aquifer. *Applied Geochemistry* 15, 737–52.

Edmunds, W.M., Bath, A.H. and Miles, D.L. (1982) Hydrochemical evolution of the East Midlands Triassic sandstone aquifer, England. *Geochimica et Cosmochimica Acta* 46, 2069–81.

Edmunds, W.M., Miles, D.L. and Cook, J.M. (1984) A comparative study of sequential redox processes in three British aquifers. In: E. Eriksson (ed.) *Hydrochemical Balances of Freshwater Systems.* Proceedings Symposium, Uppsala, IAHS Publication 150, pp. 55–70.

Edmunds, W.M., Cook, J.M., Darling, W.G. et al. (1987) Baseline geochemical conditions in the chalk aquifer, Berkshire, UK: a basis for groundwater quality management. *Applied Geochemistry* 2, 251–74.

Edmunds, W.M., Cook, J.M, Kinniburgh, D.G. Miles, D.I. and Trafford, J.M. (1989) Trace element occurrence in British groundwaters. Research Report SD/89/3. British Geological Survey. Keyworth, 424 pp.

Edmunds, W.M., Kinniburgh, D.G. and Moss, P.D. (1992) Trace metals in interstitial waters from sandstones: acidic inputs to shallow groundwaters. *Environmental Pollution* 77, 129–41.

Edmunds, W.M., Smedley, P.L. and Spiro, B. (1995). Controls on the geochemistry of sulphur in the East Midlands Triassic aquifer, UK.

In: *Isotopes in Water Resources Management.* IAEA, Vienna 1996, Vol. 2, pp. 107–22.

Edmunds, W.M., Brewerton, L.J., Shand, P. and Smedley, P.L. (1997) The natural (baseline) quality of groundwater in England and Wales. Part 1. A guide to the natural baseline quality study. British Geological Survey Technical Report, WD/97/51, Keyworth.

Edmunds, W.M., Shand, P., Hart, P. and Ward, R.S. (2003) The natural (baseline) quality of groundwater in England and Wales: a UK pilot study. *The Science of the Total Environment* 310, 25–35.

Frengstad, B., Mitgård Skrede, A.K., Banks, D., Reidar Krog, J., and Siewers, V. (2000) The chemistry of Norwegian groundwaters. III. The distribution of trace elements in 476 crystalline bedrock groundwaters as analysed by ICP-MS techniques. *Science of the Total Environment* 246, 21–40.

Hawkes, H.E. and Webb, J.S. (1962) *Geochemistry in Mineral Exploration.* Harper, New York.

Hendry, J., Schwartz, F.W. and Robertson, C. (1991) Hydrogeology and hydrochemistry of the Milk River aquifer system, Alberta, Canada, a review. *Applied Geochemistry* 6, 369–80.

Kunkel, R., Hannapel, S., Schenk, R., Voigt, H.J., Wendland, F. and Wolter, R. (2004) A procedure to define the good chemical status of groundwater bodies in Germany. *Proceedings of the COST 629 Workshop: Integrated Methods for Assessing Water Quality.* Louvain-la.Neuve, Belgium, pp. 50–8.

Langmuir, D. (1971) The geochemistry of some carbonate groundwaters in central Pennsylvania. *Geochimica et Cosmochimica Acta* 35, 1023–45.

Langmuir, D. (1997) *Aqueous Environmental Chemistry.* Prentice-Hall, Englewood Cliffs, NJ.

Ledin, A., Petterson, C., Allard, B. and Aastrup, M. (1989). Background concentration ranges of heavy metals in Swedish groundwaters from crystalline rocks: a review. *Water, Air, and Soil Pollution* 47, 419–26.

Lee, L. and Helsel, D. (2005) Baseline models in trace elements in major aquifers of the United States. *Applied Geochemistry* 20, 1560–70.

Meinardi, C.R., et al. (2003). Basiswaarden voor spoorelementen in het zoete grondwater van Nederland: gegevens uit de landelijke en provinciale meetnetten (LMG, PMG, LMB, sprengen Veluwe), Rijksinstituut voor Volksgezondheid en Milieu (RIVM), 43 pp.

Parker, J.M., Young, C.P. and Chilton, J. (1991) Rural and agricultural pollution of groundwater. In: Downing, R.A. and Wilkinson, W.B. (eds) *Applied Groundwater Hydrology*. Oxford Science Publications, pp. 149–63.

Reimann, C. and Filimoser, P. (2000) Normal and log-normal data distribution in geochemistry; death of a myth; consequences for the statistical treatment of geochemical and environmental data. *Environmental Geology* 39,1001–14.

Reimann, C. and Garrett, R.G. (2005) Geochemical background – concept and reality. *Science of the Total Environment*, 350, 12–27.

Runnells, D.D., Dupon, D.P., Jones, R.L. and Cline, D.J. (1998) Determination of background chemistry of water at mining and milling sites, Salt Lake Valley, Utah, USA. In: Arehart, G.B.,

Hulston, J.R. (eds), *Water–Rock Interaction 9*. Ninth International Symposium on Water–Rock Interaction, Taupo, New Zealand, Balkema, Rotterdam, pp. 997–1000.

Shand, P. and Frengstad, B. (2001) Baseline groundwater quality: a comparison of selected British and Norwegian aquifers. British Geological Survey Report, IR/01/177. Keyworth.

Shand, P., Edmunds, W.M., Lawrence A.R., Smedley, P.L. and Burke, S. (2007) The natural (baseline) quality of groundwater in England and Wales. Hydrogeology Report of the British Geological Survey. Keyworth, Nottingham.

Smedley, P.L. and Edmunds, W.M. (2002) Redox patterns and trace element behaviour in the East Midlands Triassic sandstone aquifer, UK. *Groundwater* 40, 44–58.

Stumm, W. and Morgan, J.J. (1996) *Aquatic Chemistry*. 3rd edn. Wiley and Sons, New York.

Tukey, J.W. (1977) *Exploratory Data Analysis*. Addison-Wesley, Reading, MA.

Walraevens, K., Cardenal-Escarcena, J. and Van Camp, M. (2007). Reaction transport modelling of a freshening aquifer (Tertiary Ledo-Paniselian Aquifer, Flanders-Belgium). *Applied Geochemistry* 22, 289–305.

2 The Baseline Inorganic Chemistry of European Groundwaters

P. SHAND AND W. M. EDMUNDS

The baseline chemistry of groundwaters varies both spatially and with time due to many complex factors including climate, soil type, geology and residence time. The baseline is therefore represented by a range rather than a single concentration. The variations in baseline groundwater characteristics are presented for a variety of aquifers in Europe including different rock types (from carbonate to siliclastic) and ages (PreCambrian to Quaternary). Most aquifers have been affected to some degree by diffuse pollution and a set of guidelines has been developed to help in the determination of baseline.

More than 80 components were measured in these aquifers. Statistical data and plots are useful for comparison between areas and are of strategic value since they include data for elements for which there are no existing guidelines, but which may still be harmful (e.g. Be, Se, Sb, Tl, U). This represents a comprehensive and strategic database of the baseline concentrations for most inorganic substances and a reference to quantify pollution by such substances.

There is considerable overlap between the different aquifers for most solutes regardless of main aquifer type. The main hydrochemical characteristics in aquifers are determined, not specifically by rock type, but by a range of hydrochemical processes such as mineral dissolution, redox reactions, ion-exchange reactions, mixing, adsorption–desorption and rainfall inputs.

2.1 Introduction

The baseline chemistry of groundwater as defined in Chapter 1 in aquifers or groundwater bodies, is very variable, both spatially and with depth, as a function of the many complex geological, hydrogeological, geochemical and climatic factors which control hydrochemical evolution. It is well established that groundwater chemistry varies in response to these factors, as well as with time, related to solubility and kinetic controls (Hem 1985; Appelo and Postma 2005). However, the characterisation of the groundwater chemistry is difficult to establish for several reasons: sampling is often biased towards parts of the aquifer which are most productive (e.g. valley bottoms), samples may be limited to specific (shallow) horizons in the aquifer, the sample may represent a mixture of waters where long screen intervals are used to maximise groundwater abstraction, and the sample may not, therefore, be representative of groundwater within an aquifer. Nevertheless, where sufficient infrastructure exists, it is generally possible to ascertain the main controls on aquifer chemistry and to assess at least some of the spatial and depth variations within an aquifer. An assessment of the natural or baseline condition is more difficult because of the long history of anthropogenic influence both by the direct inputs of pollutants as well as the changes imposed by pumping, thus a reversal or change in flow regime often results in changes in water quality. Such an assessment is required in order to quantify the presence or absence of pollution (Shand and Frengstad 2001; Edmunds et al. 2003; Shand et al. 2007) as well as to provide limitations to remediation, that is, remediating to the baseline may be either prohibitively expensive or practically impossible.

This chapter presents an overview of major, minor and trace elements in groundwaters from a range of European aquifers. A total of 25 aquifers from 11 countries were investigated (Table 2.1). These aquifers were chosen so as to cover a wide range of aquifer lithologies (carbonate-poor siliciclastic to carbonate dominant) as well as of different age (Cambrian to Quaternary). The results across Europe are shown as statistical summaries and the overall trends in water quality are discussed. Detailed studies of many of the individual reference aquifers are presented in subsequent chapters of this book.

2.2 Database Compilation and Sampling

For the reference aquifers from each country, available groundwater chemical data were collated from various sources and the different databases were compiled to a common

framework. The quantity and quality of data were very variable both within and between the different countries. These were limited in part by a lack of historical data and partly due to data confidentiality, for example, in private company databases. The quality of the historical data was checked for individual countries using ionic balances as a means of detecting poor quality analyses which were then rejected. Owing to the lack of historical data in most European countries, the data were mainly from records over the last decade; within these datasets the range of measured components was often limited or varied over time.

In order to have a more consistent database and overcome data-poor aquifers, an additional dedicated hydrochemical sampling programme was completed for most of the aquifers studied. Both sampling methodology and analytical procedures may cause variations in data, hence the groundwaters were all sampled using an agreed protocol and using the same analytical techniques. In this exercise a wide range of major and minor elements as well as trace elements (*c.*63 trace elements) were analysed for each reference aquifer by a single laboratory. A protocol was established whereby unstable parameters such as pH, redox potential (Eh) and dissolved oxygen (DO) were measured in a flow-through cell where possible, and specific electrical conductance (SEC), alkalinity and temperature were measured at site. In addition, all samples were filtered through 0.45 µm membrane filters for the analysis of dissolved components. The number of analyses available for each aquifer and used in the summaries are shown in Table 2.2.

The compilation and amalgamation of data from each aquifer into a database

Table 2.1 European reference aquifers showing location, lithology and age of aquifer.

Country	Aquifer	Location	Lithology	Period
United Kingdom	Wessex Basin	Southern England	Chalk	Upper Cretaceous
	Lower Greensand	Southern England	Clay, silt and sands	Lower Cretaceous
	Sherwood Sandstone	Vale of York	Sandstone	Triassic
	Sherwood Sandstone	East Midlands	Sandstone	Triassic
Spain	Donana Aquifer	Southwest Spain	Sands/Gravels	Plio-Quaternary
	Madrid Basin	Central Spain	Clay, silt, sand and limestone	Tertiary
Portugal	Aveiro Aquifer Q	Northwest Portugal	Clay, silt, sand and gravel	Quaternary
	Aveiro Aquifer C	Northwest Portugal	Silt, sand and sands	Cretaceous
Belgium	Ledo–Paniselian	Northwest Flanders	Clay and Sand	Eocene
	Neogene Aquifer	Northeast Flanders	Sands	Neogene
France	Fontainbleau Sands	Southern Paris basin	Fine sands	Oligocene
	Valreas Aquifer	Southeast France	Clay, silt and sands	Miocene
	Lower Triassic	Lorraine	Sandstone	Triassic
Denmark	Miocene Aquifer	Jylland, west Denmark	Sands	Miocene
	Quaternary Aquifer	Odense, central Denmark	Sands	Quaternary
	Limestone & Chalk aquifers	Copenhagen, east Denmark	Limestone and Chalk	Cretaceous/Tertiary
Poland	Tertiary K-G Sands (MGWB 332)	Southern Poland	Sands	Tertiary
	Bogucice Sands (MGWB 451)	Krakow, South Poland	Sands	Tertiary
Estonia	Devonian Aquifer system	Southeast Estonia	Limestone (dolomitised)	Devonian
	Cambrian-Vendian System	Northern Estonia	Silt and sandstone	Cambrian-Vendian
Czech Republic	Cenomian	Northern Czech Republic	Sandstone, limestone	Upper Cretaceous
	Turonian	Northern Czech Republic	Sandstone, calcareous	Mid-upper Cretaceous
Bulgaria	Thracian	Eastern Bulgaria	Sandstone, conglomerate, clay	Plio-Quaternary
	Razlog	Southwest Bulgaria	Marble	PreCambrian
Malta	Mean sea-level aquifer	Malta and Gozo	Limestone	Oligocene

Table 2.2 Numbers of analyses for each of the reference aquifers.

Country	Aquifer	Number of Analyses
United Kingdom	Wessex Basin	59
	Lower Greensand	86
	Sherwood Sandstone York	43
	Sherwood Sandstone E Midlands	56
Spain	Donana Aquifer	98
Portugal	Aveiro Aquifer Q	72
	Aveiro Aquifer C	90
Belgium	Ledo–Paniselian	256
	Neogene Aquifer	84
France	Fontainbleau Sands	42
	Valreas Aquifer	14
	Lower Triassic	18
Denmark	Miocene Aquifer	64
	Quaternary Aquifer	2
	Limestone & Chalk aquifers	35
Poland	Tertiary K-G Sands	40
	Bogucice Sands	33
Estonia	Devonian Aquifer system	2394
	Cambrian-Vendian System	1551
Czech Republic	Cenomian	43
	Turonian	61
Bulgaria	Thracian	19
	Razlog	20
Malta	Mean sea-level aquifer	7

presented several problems. For example, many of the parameters had different detection limits as a consequence of different analytical techniques or changes in detection limit due to improvements in sensitivity over time. Where reasonable detection limits were available, a value of half of the detection limit was imposed for statistical and plotting purposes. However, if old data with very high detection limits were present, these were removed.

Although most laboratories have stringent quality control procedures and reference material checks for accuracy and precision, there was some concern as to how closely comparisons could be made between data from the different country laboratories. A set of samples (including international reference materials) were therefore sent to several laboratories as part of a 'round-robin' intercomparison exercise. The results for selected major and trace elements are shown in Fig. 2.1. The comparisons for major elements and the majority of trace elements are within the errors provided for the reference material, hence providing a reliable

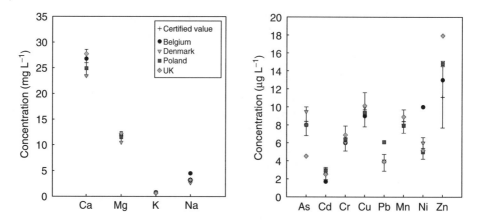

Fig. 2.1 Comparison of hydrochemical data from laboratories used routinely for partners from Belgium, Denmark, Poland and the UK completed as part of a round-robin exercise.

analytical framework for the comparison of data sets. However, for older data, quality control must remain an issue.

2.3 Data Presentation

Comparisons of datasets for major, minor and trace elements are presented in the following section using cumulative probability plots. These are considered the most effective means for displaying the range of data as well as potentially discriminating different populations of data. Cumulative probability (or cumulative frequency) plots display the range of data based on a percentile basis (Fig. 2.2). Different populations, represented by a change in slope on the plots may indicate different sources, for example, natural and contaminant, as well as major discontinuities within the groundwater body (e.g. reducing and oxidising environments). Cumulative probability plots have been used in the mineral exploration industry to discriminate anomalous concentrations from background (Runnells et al. 1998), and also

around polluted sites to discriminate point source pollution. Geochemical distributions are, however, often naturally complex and many common geochemical processes cause changes in simple distributions (Fig. 2.2). Therefore, these provide a useful first-order tool to help in assessing the internal hydrochemical functioning of an aquifer.

Several other problems need to be borne in mind with regard to interpreting ranges of data on the plots described above. For example, it could be argued that outliers, that is, data above the 97.7 percentile also represent the baseline. However, this criterion has been applied here for practical purposes, and it is a common practice to remove unusual or outlying data when applying statistical treatments.

The baseline range for an aquifer may typically show wide variations (Shand et al. 2007) and it is desirable to subdivide it into different sections for monitoring and interpretation. This allows, for example, the different properties of unconfined and confined parts of an aquifer to be compared and assessed: the different ages and flow

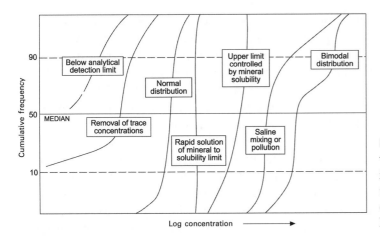

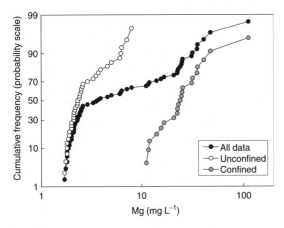

Fig. 2.2   Cumulative frequency versus log-concentration diagram illustrating geochemical processes that may give rise to a range of complex distributions of hydrochemical data.

Fig. 2.3 Cumulative frequency plot showing different populations of Mg concentrations in groundwater of the chalk aquifer of southern England. Two different systems can be defined: the unconfined and confined parts of the aquifer.

groundwaters are due to the dominance of congruent dissolution of low-Mg calcite, whereas in the older, confined groundwaters, incongruent dissolution of the calcite has led to an increase in Mg concentrations over time. When considering the water quality management of a groundwater body, anthropogenic influences around a local system of boreholes needs to be assessed with regard to local variations and not swamped by the wide ranges typically found across a large aquifer system. The initial statistical evaluation therefore allows the major components of the groundwater body to be examined and where necessary for the aquifer to be subdivided. In the plots used in this chapter, the systems are typically represented by the aquifer as a whole; more detailed interpretation is made in country chapters.

patterns make it likely that resource management, remediation strategies and protection controls will be different in each part of the aquifer. An example of differences between unconfined and confined parts of the same aquifer are shown in Fig. 2.3 for the major element Mg. The figure shows how Mg forms two distinct populations of data. The lower concentrations in the unconfined

2.4   How to Determine Baseline

The simplest way to determine the baseline of a specific substance (as defined in Chapter 1) is to analyse the range of concentrations in pristine waters. Hydrogeological evaluation and consideration of residence times will normally be able to pinpoint aquifer sections that are likely to be free of contamination. In addition, newly drilled boreholes may provide chemical data on initial conditions. As a result of groundwater abstraction, pristine groundwaters now remain at the present day mainly in areas where the aquifer is little developed, in confined or very deep sections. These groundwaters may also be very old (palaeowaters) and their quality may derive from chemical reactions which may not be representative of the parts of the aquifer which are actively developed. In confined aquifers, the conditions may be reducing and the chemical processes differ significantly from unconfined aquifers which are often oxidising. Groundwater dating (Chapter 5) is an invaluable tool for assessing such changes in chemistry with time.

Many unconfined aquifers are unprotected and subject to various forms of pollution from human activities. Groundwater sources or areas which have been grossly affected by point source pollution are often obvious and

may be avoided or discarded. However most of the important aquifers in Europe now show the effects of diffuse pollution, notably enhancement of nitrate concentrations, and a methodology needs to be developed to determine the baseline in the groundwater bodies in such aquifers.

One of the best methods to determine the natural baseline is the use of historical data. Unfortunately, few good-quality records remain which go back further than a few decades. Even where trends exist, they may be difficult to interpret unless accompanying information such as knowledge of changes in abstraction rates are available. A plot of nitrate trends over 100 years in groundwaters of a Permo–Triassic aquifer in England (Fig. 2.4) shows that gradients in nitrate concentration have increased with time and therefore extrapolation from the past few recent decades would not provide a realistic baseline value.

Baseline compositions may also be made by extrapolating from areas of similar geology, hydrogeology and landscape position.

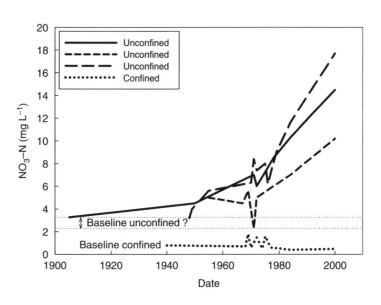

Fig. 2.4 Time series data for NO_3–N in groundwaters from three unconfined and one confined sites in the Sherwood sandstone aquifer in the UK.

However, care must be taken because it is unlikely that two different areas will have exactly the same baseline properties due to facies change, fracture/porosity development, recharge history, natural variability in flowpaths, aquifer mineralogy and geochemistry. Several of the reference aquifer studies emphasise the heterogeneity in groundwater chemistry *within* aquifers and it is advisable to use comparisons between countries or regions as a guideline only. Geochemical modelling (Chapter 4) may provide one way of examining natural trends, indicating the degree to which some parameters are modified and to test hypotheses regarding geochemical evolution.

The application of more advanced statistical tools such as hierarchical cluster analysis appears very useful where categories of data (rather than a single component) are used to recognise different populations (Daughney and Reeves 2005). All the abovementioned techniques (Table 2.3) should be used in conjunction since it is unlikely that any one will be wholly successful, and hence a combination is appropriate.

2.5 Geochemical Controls on Groundwater Chemistry in Europe

The chemical properties of groundwater are derived through reactions between water and the minerals and gases with which it comes in contact between the recharge and discharge areas. The dominant geochemical processes and sources involved in determining baseline groundwater chemistry include rainfall inputs, water–gas interaction especially in the soil and unsaturated zone, mineral dissolution, redox reactions, ion-exchange reactions (during freshening

Table 2.3 Summary of tools used to help in the determination of baseline in modified aquifers.

- Exclude known polluted samples
- Historical data
- Up-gradient and cross-gradient sampling
- Compare with pristine areas
- Graphical techniques
- Statistical techniques
- Geochemical modelling

and salinisation) and mixing (with old formation water or modern seawater). The timescales involved in these changes vary from seconds to millions of years. The dominant processes influencing each of the reference aquifers is shown in Table 2.4.

Most of the aquifers display a range of processes, however, these occur at different rates and in different hydrogeological environments, giving rise to large spatial and temporal variations in groundwater quality. Mineral dissolution rates vary considerably, for example, carbonates and sulphates dissolve comparatively rapidly whilst silicate minerals dissolve much more slowly. It is commonly found that the readily dissolved minerals have been removed from the shallow parts of the aquifer where recharge is occurring leading to the development of 'reaction fronts' (Appelo and Postma 2005).

There is commonly a zonation in redox processes as groundwaters evolve within an aquifer where oxidation of organic matter (or other reduced species) will cause reduction of species in a specific order: DO, NO_3, Fe, SO_4 (Stumm and Morgan 1996). A sharp redox boundary often exists close to the junction between unconfined and confined parts of the aquifer and this typically occurs where there is a change in residence time caused by

Table 2.4 Summary of the dominant processes controlling baseline water quality in the reference aquifers.

Aquifer	Dominant baseline processes
UK Wessex	Carbonate and feldspar dissolution, saline mixing
UK Greensand	Silicate dissolution, ion-exchange, redox
UK Vale of York	Carbonate, silicate, sulphate dissolution, redox, saline mixing
UK E Midlands	Carbonate, silicate, sulphate dissolution, redox, saline mixing
ES Donana	Evaporation, calcite and feldspar dissolution, redox, mixing, ion-exchange
ES Madrid	Carbonate and silicate dissolution, ion-exchange, clay neoformation
P Quaternary	Carbonate and silicate dissolution, redox, ion-exchange
P Cretaceous	Carbonate and silicate dissolution, redox, ion-exchange, saline mixing
B Ledo–Paniselian	Silicate dissolution, redox, ion-exchange
B Neogene	Carbonate and silicate dissolution, redox
F Fontainbleu	Silicate and carbonate dissolution
F Valreas	Carbonate dissolution, ion-exchange, saline mixing
F Triassic	Carbonate and sulphate dissolution
DK Miocene	Carbonate dissolution, redox, ion-exchange
DK Quaternary	Redox, saline mixing
DK Limestone	Carbonate dissolution, redox, ion-exchange, saline mixing
PL Kedzieryn	Carbonate and gypsum dissolution, redox, saline mixing/diffusion
PL Bogucice	Carbonate dissolution, redox, ion-exchange, diffusion
EE Devonian	Carbonate dissolution, redox, saline mixing
EE Cambrian-Vendian	Silicate dissolution, ion-exchange, saline mixing
CZ Cenomian	Calcite dissolution, ion-exchange, saline mixing
CZ Turonian	Calcite and silicate dissolution, saline mixing
BG Thracian	Calcite and silicate dissolution, redox
BG Razlog	Carbonate dissolution
M Sea-level aquifer	Carbonate and phosphate dissolution, saline mixing

the juxtaposition of different flow systems. This may give rise to very different baseline chemistries and it may be necessary to define these as different systems (Fig. 2.3).

Ion-exchange reactions are important in aquifers where the adsorbed 'exchangeable cations' were initially at equilibrium with, for example, original formation water, but where the water chemistry has changed as water of different composition is drawn in by pumping (Appelo and Postma 1995). The two most common examples of exchange processes are 'aquifer freshening' where the original water was marine or saline, and fresh water is now displacing the older more saline water; and where seawater (or deeper saline water) is introduced into a freshwater environment causing salinisation. In both cases, there is an exchange of cations between the exchange complex (e.g. clays) and the groundwater. Several of the reference aquifers (Table 2.4) have been influenced by mixing with saline water, which is reflected in the positive skewed distributions, and modification of the relative proportions of major cations. The origins of salinity include entrainment of formation waters from the matrix of dual porosity aquifers, up-coning of deeper saline water of different origins and saline intrusion in coastal regions from present day or palaeo-seawater.

2.6 Hydrochemical Characteristics of the European Reference Aquifers

The consistent approach to sampling and analysis in the 25 reference aquifers in this investigation allows a snapshot of European water quality emphasising the baseline quality. The ranges of chemical species within and between the selected groundwaters are summarised in Figs. 2.5–2.11 as cumulative probability plots. Since all data from the 25 aquifers are plotted, it is not always possible to follow trends in individual aquifers. These are presented in later chapters on selected individual aquifers. The intention is to give an overview of the range and variability of European water compositions from the shape of the envelope. The legend symbols are explained in Table 2.5. An overall statistical summary is shown in Tables 2.6 and 2.7.

2.6.1 Acidity and total mineralisation

The variations in pH are shown on Fig. 2.5(a). The average (median) pH varies from *c*.6.5 to 7.5 in most aquifer types regardless of whether they are carbonate or siliclastic. This highlights the fact that only small amounts (*c*.2–3%) of carbonate minerals (calcite, dolomite) in an aquifer are sufficient to buffer the pH to circum-neutral values. However, the overall acidity varies over several orders of magnitude (pH is the negative log of the hydrogen ion concentration, hence one unit increase is a ten fold decline in H^+ concentration) within many aquifers. A main exception is the Chalk of the UK, a fine-grained carbonate aquifer (UK2) which has a median value of 7.22 and with a narrow pH range. Several aquifers contain alkaline waters exceeding pH 9 where natural water softening to $Na–HCO_3$ compositions has taken place. The carbonate-poor Quaternary aquifer of Portugal (P2) has the lowest median pH of 6.1. Many aquifers contain a small percentage of low pH groundwaters, occurring in shallow siliciclastic aquifers or within decalcified parts of the aquifer in which mobilisation of trace metals such as aluminium may also be expected (Moss and Edmunds 1997; Shand et al. 2005).

The groundwaters also show a wide variation in SEC [Fig. 2.5(b)], a measure of the total mineralisation. The freshest groundwater bodies are represented by the siliclastic aquifers (PL2 and P1) where easily soluble components such as carbonate or evaporite minerals are often not present. The highest median SEC values are found in the Maltese aquifers where saline intrusion and mixing is a widespread problem. Most aquifers show distributions which are approximately log-normal (a linear distribution on log-scale probability plots) varying by around one order of magnitude. However, several show a positive skew corresponding to high SEC values due to the presence of older saline Na–Cl type formation waters in deeper parts of the aquifer, for example, Belgian Ledo–Paniselian aquifer (B1), or to seawater intrusion (Donana aquifer of Spain, ES2).

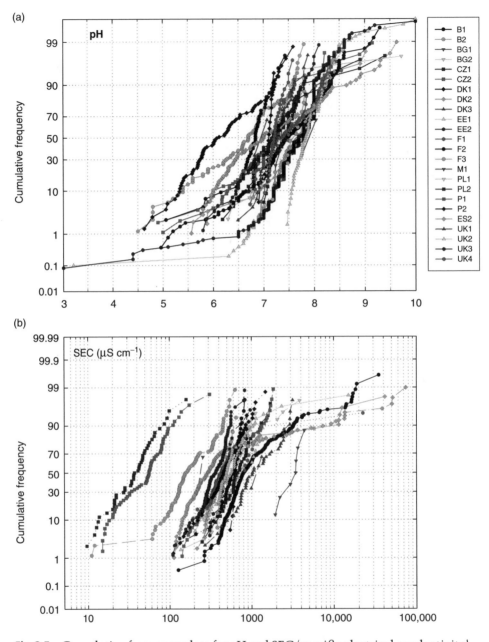

Fig. 2.5 Cumulative frequency plots for pH and SEC (specific electrical conductivity) in the European reference aquifer groundwaters. For aquifer key see Table 2.5.

Table 2.5 Aquifer symbols used for cumulative probability plots, Figs. 2.5 to 2.11.

Symbol	Country	Aquifer	Symbol	Country	Aquifer
					Water table
B1	Belgium	Ledo Paniselian	Malta	Malta	aquifer
B2		Neogene	PL1	Poland	Kedzieryn
BUL1	Bulgaria	Razlog	PL2		Bogucice
					Aveiro
BUL2		Thracian	P1	Portugal	Cretaceous
					Aveiro
CZ1	Czech	Cenomian	P2		Quaternary
CZ2		Turonian	E2	Spain	Donana
DK1	Denmark	Chalk	UK1	UK	Vale of York
DK2		Miocene	UK2		East Midlands
					Lower
DK3		Quaternary	UK3		Greensand
		Cambrian-			
EE1	Estonia	Vendian	UK4		Wessex
EE2		Devonian			
F1	France	Lorraine			
F2		Fontainbleu			
F3		Valreas			

2.6.2 *Major elements*

Major ions define the main characteristics of natural waters and reflect the main rock types and geochemical processes involved in water–rock interaction. Although contamination will add some major ions to groundwater, these effects are often negligible. Selected major element distributions in European aquifers are shown on Figs. 2.6 (cations) and 2.7 (anions and Si). Although the different aquifer units may show distinct distributions, sometimes approaching log-normal, there is considerable overlap between the aquifers, regardless of rock type, mineralogy or hydrogeological properties.

Calcium concentrations vary across European aquifers by over four orders of magnitude and even within individual aquifers, they are typically two to three orders of magnitude. Most of the Ca is derived from natural sources and represent baseline concentrations. The upper limiting factor in most freshwater systems is saturation with respect to calcite (Hem 1985) which will limit the maximum concentrations and the flattening-off in many of the distributions at around 70–120 mg L^{-1} is a reflection of this (Fig. 2.6). Aquifer freshening (ion-exchange mainly with Na) decreases Ca in solution and mixing with saline water may lead to an increase in concentrations. Nevertheless, it is clear that baseline concentration ranges in the system at the aquifer wide scale are large.

Magnesium distributions display many similar patterns to Ca (Fig. 2.6), but the median concentrations are lower and nota-

Table 2.6 Summary statistics for hydrochemical parameters, stable isotopes, and major and minor elements for European groundwaters studied.

Parameter	Units	Min.*	Max.	Median	Mean	95 %ile	97.7 %ile	n
T	°C	5.4	30.8	12.0	13.9	21.6	23.5	902
pH		3.00	9.76	7.60	7.53	8.25	8.40	3358
Eh	mV	2203	552	144	163	425	455	746
DO	mg L^{-1}	<0.1	13.0	1.1	2.8	9.5	10.0	715
SEC	µS cm^{-1}	10	74200	498	1150	2940	7150	1192
δ^2H	‰	−75.0	−4.0	−45.9	−42.7	−20.8	−19.6	397
δ^{18}O	‰	−10.7	−0.8	−7.1	−6.6	−3.9	−3.7	426
δ^{13}C	‰	−25.0	4.2	−13.5	−13.3	−6.7	−4.6	368
Ca	mg L^{-1}	0.30	2090	57.1	65.8	121	176	4820
Mg	mg L^{-1}	0.00	1213	19.6	24.4	45.6	57.8	4784
Na	mg L^{-1}	0.00	16691	21.1	146	364	859	2214
K	mg L^{-1}	<0.5	479	5.3	11.4	35.6	65.8	2047
Cl	mg L^{-1}	0.15	32000	19.0	160	369	663	5078
SO$_4$	mg L^{-1}	<0.02	2900	10.0	36.4	122	239	4820
Field HCO$_3$	mg L^{-1}	<0.1	1640	168	192	415	471	823
Lab. HCO$_3$	mg L^{-1}	<0.1	971	299	288	464	577	3823
NO$_3$–N	mg L^{-1}	<0.002	108	0.37	4.98	26.2	42.7	1137
NO$_2$–N	mg L^{-1}	<0.001	20.0	0.01	0.07	0.16	0.35	2641
NH$_4$–N	mg L^{-1}	<0.003	35.4	0.15	0.43	1.35	2.43	3421
P	mg L^{-1}	<0.02	8.3	0.39	0.19	0.6	1.2	780
TOC	mg L^{-1}	0.10	732	1.90	6.02	18.7	23.5	396
DOC	mg L^{-1}	0.15	130	1.50	3.02	9.03	14.0	367
F	mg L^{-1}	<0.05	5.60	0.15	0.30	1.18	2.06	712
Br	mg L^{-1}	<0.03	17.8	0.09	0.29	0.94	2.02	590
I	mg L^{-1}	<0.001	3.31	0.01	0.03	0.06	0.11	531
Si	mg L^{-1}	0.14	44.3	4.60	5.56	12.9	14.7	1684

* Typical detection limits shown, higher detection limits in some countries may have biased medians and means to slightly higher concentrations.

bly the concentration ranges in individual aquifers are small – often below an order of magnitude. This reflects the lower geochemical abundance of Mg, but also the slower dissolution kinetics of Mg carbonates. Mg is also released progressively from impure carbonates by incongruent dissolution, leading to higher Mg/Ca ratios in older, evolved groundwaters (see Chapter 1).

The large range of Na concentrations and tendency for a positive skew (Fig. 2.6) is controlled largely by mixing with saline formation waters (see also similar distribution for Cl). However, Ca–Na exchange during aqui-

fer freshening has led to different patterns of Na from Cl in some aquifers, reflected by an increase in Na/Cl ratio. Potassium concentrations also vary over several orders of magnitude and are dominated initially by reactions between groundwater and K-bearing mineral phases such as K-feldspar and clays. The high concentrations are probably dominated by mixing with old formation waters or with seawater during seawater intrusion. In some of the reference aquifers, K is enhanced due to anthropogenic influences locally, as described in the subsequent chapters, but this is masked by the greater

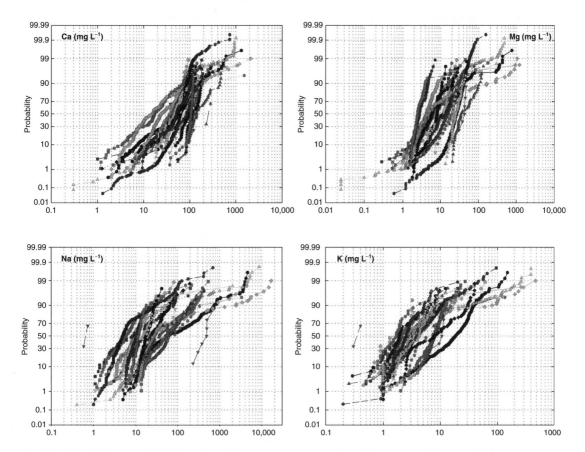

Fig. 2.6 Cumulative probability plots for major cations in the reference aquifer groundwaters. Symbols as in Figure 2.5.

Table 2.7 Summary statistics for trace elements for European groundwaters studied.

Parameter	Units	Min.*	Max.	Median	Mean	95 %ile	97.7 %ile	n
Ag	µg L^{-1}	<0.05	0.63	<0.05	<0.05	<0.05	0.10	571
Al	µg L^{-1}	<1	4707	4.00	23.9	68.2	132	750
As	µg L^{-1}	<0.05	79.0	0.50	2.97	13.1	20.6	813
Au	µg L^{-1}	<0.05	0.10	<0.05	<0.05	<0.05	<0.05	565
B	µg L^{-1}	<20	5971	34.2	129	562	868	758
Ba	µg L^{-1}	0.90	4848	46.9	122	454	787	808
Be	µg L^{-1}	<0.05	17.4	0.05	0.15	0.42	0.98	722
Bi	µg L^{-1}	<0.05	0.25	<0.05	<0.05	0.09	0.25	576
Cd	µg L^{-1}	<0.05	8.79	<0.05	0.16	0.59	1.13	784
Ce	µg L^{-1}	<0.01	65.0	<0.01	0.-1	0.27	0.90	579
Co	µg L^{-1}	<0.02	52.5	0.08	0.65	2.61	4.74	728
Cr	µg L^{-1}	<0.5	28.4	<0.5	1.34	5.99	9.98	783
Cs	µg L^{-1}	<0.01	205	0.02	0.52	0.71	1.90	623
Cu	µg L^{-1}	<0.1	134	1.20	3.11	12.3	17.4	786
Dy	µg L^{-1}	<0.01	2.46	<0.01	0.02	0.03	0.14	570
Er	µg L^{-1}	<0.01	1.18	<0.01	0.02	0.03	0.10	573
Eu	µg L^{-1}	<0.01	1.10	<0.01	0.01	0.03	0.07	578
Fe	µg L^{-1}	<0.005	120	0.35	1.35	4.94	9.70	4423
Ga	µg L^{-1}	<0.05	1.06	<0.05	<0.05	0.06	0.11	576
Gd	µg L^{-1}	<0.01	3.33	<0.01	0.03	0.04	0.17	578
Ge	µg L^{-1}	<0.05	5.12	<0.05	0.08	0.26	0.35	577
Hf	µg L^{-1}	<0.02	1.97	<0.02	<0.02	<0.02	<0.02	578
Hg	µg L^{-1}	<0.2	6.00	<0.2	<0.2	1.00	1.60	746
Ho	µg L^{-1}	<0.01	0.40	<0.01	<0.01	0.03	0.05	579
In	µg L^{-1}	<0.01	0.05	<0.01	<0.01	<0.01	<0.01	579
Ir	µg L^{-1}	<0.05	0.72	<0.05	<0.05	<0.05	<0.05	576
La	µg L^{-1}	<0.01	906	<0.01	1.77	0.26	0.85	578
Li	µg L^{-1}	<0.03	7073	11.3	34.9	80.0	142	756
Lu	µg L^{-1}	<0.01	0.50	<0.01	0.03	0.17	0.26	573
Mn	µg L^{-1}	<0.05	2.55	<0.05	0.11	0.44	0.75	1066
Mo	µg L^{-1}	<0.1	13.2	0.40	0.88	3.14	4.96	728

Table 2.7 (*Continued*)

Parameter	Units	Min.*	Max.	Median	Mean	95 %ile	97.7 %ile	*n*
Nb	µg L^{-1}	<0.01	0.16	<0.01	<0.01	<0.01	0.02	578
Nd	µg L^{-1}	<0.01	37.0	<0.01	0.22	0.21	1.03	578
Ni	µg L^{-1}	<0.2	114	1.00	3.25	14.8	25.1	786
Os	µg L^{-1}	<0.05	<0.05	<0.05	<0.05	<0.05	<0.05	576
Pb	µg L^{-1}	<0.1	30.0	0.39	0.81	2.00	4.99	785
Pd	µg L^{-1}	<0.2	5.10	<0.2	<0.2	<0.2	0.30	578
Pr	µg L^{-1}	<0.01	9.74	<0.01	0.04	0.04	0.16	578
Pt	µg L^{-1}	<0.01	0.06	<0.01	<0.01	<0.01	0.02	578
Rb	µg L^{-1}	0.06	445	3.13	6.83	23.9	36.8	627
Re	µg L^{-1}	<0.01	2.47	<0.01	0.02	0.02	0.06	578
Rh	µg L^{-1}	<0.01	0.12	<0.01	<0.01	0.01	0.01	578
Ru	µg L^{-1}	<0.05	0.07	<0.05	<0.05	<0.05	<0.05	576
Sb	µg L^{-1}	<0.05	6.00	<0.05	0.73	6.00	6.00	681
Sc	µg L^{-1}	0.34	83.2	3.16	3.77	7.72	9.68	578
Se	µg L^{-1}	<0.015	247	0.50	1.83	5.16	9.64	757
Sm	µg L^{-1}	<0.05	5.07	<0.05	0.05	0.04	0.22	577
Sn	µg L^{-1}	<0.05	3.38	0.05	0.10	0.37	0.51	577
Sr	µg L^{-1}	8.21	33764	240	645	2083	3115	836
Ta	µg L^{-1}	<0.05	<0.05	<0.05	<0.05	<0.05	<0.05	576
Tb	µg L^{-1}	<0.01	0.61	<0.01	0.01	0.03	0.03	578
Te	µg L^{-1}	<0.05	3.11	<0.05	<0.05	<0.05	0.09	576
Th	µg L^{-1}	<0.05	1.85	<0.05	<0.05	0.07	0.15	580
Ti	µg L^{-1}	<10	488	<10	<10	<10	<10	578
Tl	µg L^{-1}	<0.01	0.74	<0.01	0.03	0.20	0.20	613
Tm	µg L^{-1}	<0.01	0.15	<0.01	<0.01	0.03	0.03	578
U	µg L^{-1}	<0.05	56.3	0.07	1.20	5.70	9.89	626
V	µg L^{-1}	<1	169	<1	1.64	3.00	4.00	578
W	µg L^{-1}	<0.1	414	<0.1	0.85	0.30	0.50	578
Y	µg L^{-1}	<0.01	41.6	0.01	0.51	1.67	5.33	578
Yb	µg L^{-1}	<0.01	2.08	<0.01	0.02	0.03	0.10	578
Zn	µg L^{-1}	<0.2	2010	11.5	40.5	112	224	773
Zr	µg L^{-1}	<0.5	179	<0.5	1.25	3.00	5.92	578

overall natural range. The effect of increases in K (a nutrient) at a local scale may be significant; hence the need to define the baseline locally with regard to local receptors such as streams or stream-based ecosystems fed by natural groundwater discharge.

The distributions of Cl (Fig. 2.7) are dominated by the regional influence of saline water inputs to the groundwater. Initial inputs from rainfall are unlikely to exceed 10 mg L^{-1} for most aquifers, although concentrations may increase on recharge by a factor of two to five depending on evapotranspiration. Local increases have been observed in many of the studied aquifers due to agricultural and industrial activities or being close to urban centres, but this, as for Na, is swamped by the contribution from natural sources (saline water). Trend analysis in the reference aquifer reports has highlighted that increases in salinity are often locally observed due to up-coning of deeper saline formation water as a consequence of abstraction. The source of solutes in this case is natural, and a baseline property, but indirect anthropogenic influences have caused the change, hence the term anthropogenically modified baseline may be used to deal with such changes.

Sulphate concentrations approach normal distributions for most European aquifers although there is often a slight negative skew (Fig. 2.7). Sulphate baseline concentrations partly reflect inputs from rainfall and in many aquifers the initial inputs may have been augmented by industrial aerosols in the past half century (Edmunds and Kinniburgh 1986). At a regional scale, the ranges may be swamped by the influence of saline water. In aquifers with SO_4 concentrations above 100 mg L^{-1} another major influence is the presence of gypsum or anhydrite.

This is clearly highlighted in the UK Triassic Sandstone aquifer (UK1; Fig. 2.7) and is attributed to the dissolution of sulphate-bearing minerals in older groundwaters. As with Cl, sulphate in this aquifer and others has also been introduced into aquifers by anthropogenic influences related to land use near to outcrop. A more detailed discussion of sulphate is made in Chapters 4 and 5 using modelling approaches and a consideration of residence time indicators.

Alkalinity (as HCO_3) displays similar distributions to Ca (Fig. 2.7), but with a negative skew related to groundwater from carbonate-deficient aquifers. The alkalinity of groundwater is an excellent indicator of the buffering capacity of the system and distributions of low alkalinity groundwaters are a better measure than pH of the extent of carbonate-deficient aquifers. This is exemplified by the UK Lower Greensand aquifer where low alkalinities predominate. In some of the older deeper groundwaters, for example, in the Belgian Ledo–Paniselian aquifer, ion-exchange and development of Na–HCO_3 waters has allowed alkalinities to exceed 900 mg L^{-1}. Large variations in alkalinity are apparent in most European aquifers, the lowest concentrations representing carbonate-deficient systems or shallow clastic aquifers which have been decalcified.

2.6.3 Silicon

The presence of silicon, up to a few mg L^{-1}, in nearly all natural waters testifies to its natural abundance so that an undersaturation with respect to silicate minerals denotes either an absence of Si minerals (a rare occurrence) or very short residence times (Davis 1964), as found in many surface waters and some of the European groundwa-

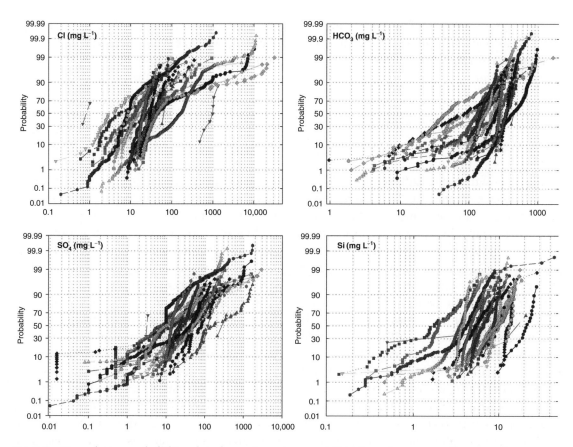

Fig. 2.7 Cumulative probability plots for major anions and Si in the reference aquifer groundwaters. Symbols as in Fig. 2.5.

ters. Silicon concentrations for individual aquifers in most European groundwaters (Fig. 2.7) lie within quite narrow ranges although with median concentrations ranging from 1.8 to 21 mg L^{-1}. Quartz is relatively insoluble at normal groundwater temperatures with saturation being around 3 mg L^{-1}. The dissolved silicon concentrations in most ground waters are derived from silicate weathering reactions (Garrels and Mackenzie 1967) and concentrations are likely to reflect the type and rate of silicate reactions (Appelo and Postma 2005). The silicate hydrolysis reactions, such as

for the weathering of K-feldspar, are non-reversible and dissolved silica is produced:

$$2KAlSi_3O_8 + 2H^+ + 9H_2O \rightarrow$$
$$Al_2Si_2O_5(OH)_4 + 2K^+ + 4H_4SiO_4$$

At the temperatures of most groundwaters, upper limits to the Si concentrations at neutral pH are likely to be controlled by chalcedony, an amorphous silica phase or secondary clay minerals (Paces 1978; Appelo and Postma 2005) rather than pure quartz at the ambient pH temperatures and the lack of equilibrium conditions with quartz reached in most groundwaters. The wide range of median values may indicate the different mineral assemblages and silicate reactions in each aquifer.

2.6.4 Nitrogen species

The distributions of nutrient species NO_3–N and NH_4–N in the European reference aquifers are shown in Fig. 2.8 along with dissolved organic carbon (DOC). Nitrogen may be present in water as a number of dissolved species: nitrate (NO_3), ammonium (NH_4), nitrite (NO_2) which is metastable, nitrogen gas (N_2) as well as organic nitrogen. Organic nitrogen was not analysed as part of this study, since it is only anticipated to be a significant component of polluted groundwaters.

The occurrence and mobility of the inorganic N-species are largely dependent on the redox condition of the water, with NO_3 being stable under oxidising conditions and NH_4 under reducing conditions. Under reducing conditions, denitrification is generally mediated by heterotrophic and autotrophic bacteria, which oxidise organic material (e.g. DOC, SOC) or inorganic material (e.g. Fe^{2+}, pyrite). The reduction of nitrate is dependent on other nutrients and especially reactive organic carbon (Ottley et al. 1997). The main product of denitrification in most aquifers is N_2 gas, as demonstrated by measurement of N_2/Ar ratios (Wilson et al. 1990).

Nitrate frequently occurs above natural baseline concentrations as a result of land use and other human impacts. Over the past century, there have been large changes to natural N and P cycles, mainly due to the application of fertilisers in agriculture. Although point sources of pollution have been avoided, diffuse pollution by fertiliser addition has led to significant increases in

nutrient inputs with time (Fig. 2.4) over large areas. As well as the extensive but variable applications of fertiliser, denitrification reactions also make the determination of baseline nitrate values very difficult. Atmospheric inputs have also significantly increased in the past due to emissions from the burning of fossil fuels; hence human impacts are present at the regional scale from recharge inputs. The variations in NO_3–N concentrations are large, mainly due to anthropogenic influences, but also since the populations include many data below detection limits corresponding to reducing environments where nitrate has reacted to N_2 gas. Some aquifers have median concentrations greater than the existing EU maximum admissible concentration (MAC) of 10.3 mg L^{-1}. Evidence from a number of the reference aquifers (see later chapters) indicates that baseline concentrations are unlikely to be more than c.1–3 mg L^{-1} NO_3–N, although this has been difficult to establish in some areas due to a lack of good historical data. Modern N-inputs in rainfall cannot be used as a guide since these have increased nitrogen species due to atmospheric pollution.

The concentrations of NH_4–N are also variable both within and between aquifers (Fig. 2.8), with an overall global median value of 0.15 mg L^{-1} NH_4–N. In many cases, the highest concentrations are considered to be naturally-derived as they are present in reducing groundwaters at depth or under confined conditions. Increasing ammonium has been found in the anaerobic Chalk aquifer in Berkshire (UK) in older groundwaters along the flowline attributed to exchange with minor amounts of marine clays, especially illite, reaching concentrations in excess of 0.5 mg L^{-1}. Organic nitrogen was not analysed as part of this study, but may

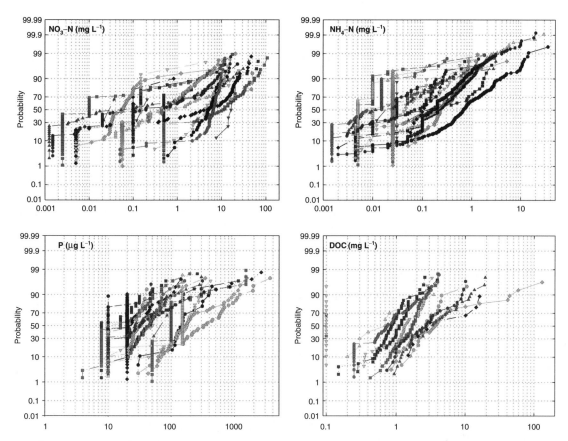

Fig. 2.8 Cumulative probability plots for selected dissolved nutrients in the reference aquifer groundwaters. Symbols as in Fig. 2.5. Note that TOC is plotted for the Bulgarian Razlog aquifer to indicate that DOC concentrations are very low.

form a significant component of total nitrogen present in polluted groundwaters.

2.6.5 Phosphorus

Dissolved phosphorus concentrations in groundwater are derived from a variety of natural and anthropogenic sources, occurring as inorganic orthophosphate, inorganic polyphosphates and organic phosphorus (P). The dominant natural source of P is apatite, especially fluorapatite, and exchangeable P on iron oxides (either desorbed or dissolved during reductive dissolution). Anthropogenic sources include inorganic and organic fertilisers, water treatment works and farmyard slurry. Under alkaline conditions P is easily sorbed to carbonates and iron oxide minerals, hence its mobility is likely to be low. A large proportion of total phosphorus in waters occurs in particulate form, and the total dissolved phosphorus (TDP) analysed in this study may contain a colloidal component. Phosphorus concentrations show log-normal distributions in many aquifers, but observed concentrations are often below the limits of detection used (indicated by vertical lines on Fig. 2.8). Although it is commonly assumed that P inputs from groundwater to rivers are low, especially in carbonate terrains, the data show that groundwater inputs may often be significant.

2.6.6   Organic carbon (DOC)

The occurrence of organic carbon in European groundwaters is described in more detail in Chapter 3. The overall data are shown on Fig. 2.8 to highlight the large ranges in concentration. Many of the high concentrations are derived from natural sources (e.g. Danish Miocene and UK Triassic sandstones). Although no DOC data were available for the Razlog aquifer in Bulgaria, total organic carbon (TOC, plotted on Fig. 2.8) was below the limit of detection. The nature of most DOC in groundwater is poorly documented, and its ability to modify redox status and solute transport is determined by the reactivity of the organic matter. A few aquifers may contain naturally high organic carbon (e.g. in Quaternary aquifers rich in organic matter such as the Ribe formation, Denmark) and contribute significantly to the DOC of groundwater. However the natural baseline values are typically low (median values ranging from 0.7 to 3.7 mg L^{-1}) and elevated concentrations may sometimes act as an indicator of contamination. Point source pollution may also enhance groundwater concentrations, for example, beneath slurry pits and around landfill sites. The baseline must therefore be assessed on a local scale.

2.6.7   Trace elements

Regional studies of trace element concentrations in the groundwaters of Europe are relatively scarce, although some national assessments are available (Edmunds et al. 1989; Banks et al. 1998; Lee and Helsel 2005). A total of 63 trace elements were analysed in most of the reference aquifers by ICP-MS (inductively coupled plasma mass spectrometry) to provide a common reference and improve the number of aquifers with good trace element data. Selected elements are shown in Figs. 2.9–2.11, mainly those that are present at concentrations above detection limit in most aquifers. Additional elements were analysed for the individual reference aquifers but not plotted. They include (brackets denote detection limits in µg L^{-1}):

B (20), Cr (0.5), Cs (0.01), Cu (0.1), Ge (0.05), Li (4), Pb (2), Rb (0.01), Sb (0.05), Sc (0.05), Se (0.5), V (1), Y (0.01), REEs (0.01–0.05).

Many metals were below the detection limit including (values in brackets are typical detection limits in µg L^{-1}):

Ag (0.05), Au (0.05), Be (0.05), Bi (0.05), Ga (0.05), Hf (0.02), Hg (0.1), In (0.01), Ir (0.05), Nb (0.01), Os (0.05), Pd (0.2), Pt (0.01), Re (0.01), Rh (0.01), Ru (0.05), Sn (0.05), Ta (0.05), Te (0.05), Th (0.05), Ti (10), Tl (0.01) W (0.1), Zr (0.5).

A number of elements are mobile under specific pH conditions, for example, detectable quantities of the rare earth elements are generally present only under acidic conditions. Trace elements often show concentration ranges of over three to four orders of magnitude and there is considerable overlap between the aquifers regardless of rock type or host mineralogy. The particular behaviour and areal distribution depend very much on whether they are controlled by redox processes (e.g. Fe, Mn, As), saturation with respect a specific mineral phase (e.g. Ba) or by geochemical abundance in the particular aquifer lithology.

2.6.7.1 Strontium

Strontium is derived mainly from Ca-rich minerals such as calcite, Ca-rich feldspars and/or gypsum/anhydrite. Concentrations in individual aquifers often vary over two to three orders of magnitude (Fig. 2.9), reflecting both sources and residence time. Although geochemically Sr behaves similarly to Ca in groundwaters (Hem 1985), it is not limited by mineral saturation at the concentrations

usually found in groundwater, in contrast to Ca which is limited by calcite saturation. Strontium, like magnesium is released by incongruent reactions and may increase with residence time. The populations of Sr therefore approach a more log-normal distribution than Ca. The highest baseline Sr concentrations are typically found in carbonate aquifers (e.g. Danish Chalk, Malta and Gozo aquifers) or sandstone aquifers containing gypsum (e.g. UK Vale of York). High concentrations may also be present in silicate aquifers, especially where mixing with seawater or formation water has occurred.

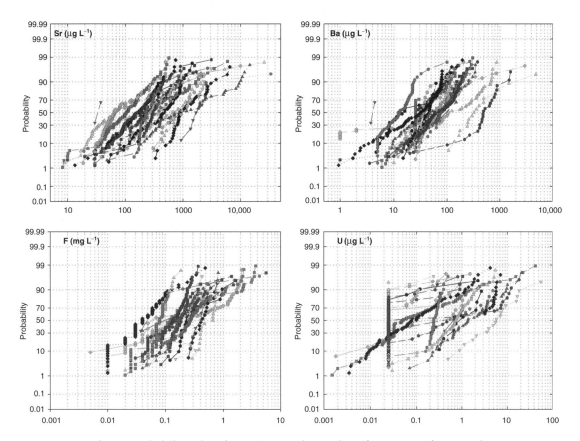

Fig. 2.9 Cumulative probability plots for Sr, Ba, F and U in the reference aquifer groundwaters. Symbols as in Figure 2.5.

Since strontium is typically derived from water–rock interaction, it is a good indicator of processes taking place in the aquifer and is unlikely to be impacted seriously by pollution.

2.6.7.2 Barium

Barium also shows a wide range of concentrations within and between aquifers (Fig. 2.9). Sources include K-rich minerals such as K-feldspar or biotite as well as barite ($BaSO_4$). Solubility is largely controlled by saturation with respect to barite and where SO_4 concentrations are high, Ba is held at low concentrations. Thus it is uncommon to find Ba in groundwaters at concentrations higher than a few hundred $\mu g\ L^{-1}$. Although Ba is a toxic element, relatively few European groundwaters contain waters higher than the MAC. Median concentrations are similar in most aquifers, the exceptions being much higher concentrations in the Estonian aquifers and to a lesser degree in the Danish Chalk where groundwaters have relatively low dissolved SO_4 (Fig. 2.7).

2.6.7.3 Fluoride

Fluoride, like Sr, is an element which owes its occurrence in groundwater almost entirely to water–rock interaction, apart from an initial rainwater input concentration of approximately $0.1\ mg\ L^{-1}$. Sources include the mineral fluorite (CaF_2) and phosphate minerals, such as fluorapatite; fluoride is also released from weathering of primary silicates such as biotite and hornblende (Edmunds and Smedley 2005). There are significant variations between the European aquifers (Fig. 2.9), with some displaying relatively restricted ranges whilst others display varia-

tions over three orders of magnitude. High concentrations are normally associated with granites or sediments derived from granitic rocks or with alkaline groundwaters where exchange of Ca for Na allows higher F solubility once Ca has been removed. Fluorine is an essential element for humans at low concentrations but harmful at higher concentrations. Most groundwaters have median values which fall in the optimum range for potability, although a small number of individual groundwaters have concentrations above current European drinking water standards ($1.5\ mg\ L^{-1}$).

2.6.7.4 Uranium

Uranium occurs as the mineral uraninite (UO_2) and its oxidised or partly oxidised massive form, pitchblende (U_3O_8), as well as a number of secondary minerals. However, discreet uranium minerals are comparatively rare. Uranium is redox sensitive and in water, U occurs dominantly as the hexavalent, U(VI), species. U(IV) concentrations are usually low (less than $0.06\ \mu g\ L^{-1}$) in reducing waters because of the low solubility of the mineral uraninite. Higher uranium concentrations are nearly always found in oxidising groundwaters where the uranium may be stabilised as oxy-anions (Langmuir 1997). Uranium varies over a large range, with solute concentrations sometimes spanning three to four orders of magnitude in some aquifers (Fig. 2.9). Median concentrations are greater than $1\ \mu g\ L^{-1}$ in a number of aquifers while maximum concentrations reach a few tens of $\mu g\ L^{-1}$. These are natural baseline concentrations and are derived from primary or secondary minerals along flow pathways, being especially enriched in sandstone aquifers with increasing residence time.

2.6.7.5   Fe, Mn, Co, Ni

Iron and manganese are common in aquifers both in primary minerals and as secondary oxides and oxy-hydroxides in most rock types. Primary sources of Fe and Mn include carbonates and silicates which release Fe^{2+} on weathering (Hem 1985). Both Fe and Mn are redox-sensitive elements in groundwater, being soluble in their reduced form but forming insoluble oxy-hydroxides under oxidising conditions. This is reflected in these elements having low baseline concentrations in unconfined parts of the aquifers where oxygen is present, but increasing significantly across redox boundaries (as described in detail in later chapters). Concentration ranges of these elements vary over four to five orders of magnitude (Fig. 2.10) in individual aquifers reflecting the importance of redox controls. Where aquifers are oxidising, such as the Razlog aquifer in Bulgaria, Fe remains low, but most groundwaters become reducing at depth or where confined, leading to reductive dissolution of oxides and oxy-hydroxides.

Fe hydroxides/oxides have a high sorption capacity for other metals such as Ni, Co, Mn and As (Dzombak and Morel 1990). Nickel and Co are shown on Fig. 2.10 and both display large ranges in the majority of aquifers. Ni and Co are primarily derived from weathering of ferro-magnesian minerals (e.g. olivine, pyroxene, biotite) but the often close association between high Fe and/or Mn and trace metals indicates a control largely by desorption/dissolution reactions involving oxide/oxy-hydroxide phases. The concentrations of many trace metals such as Ni and Co form distinct, if low, concentration populations in most aquifers related to lithology and prevailing geochemical processes and

the concentrations reported here represent the natural baseline against which the anthropogenic pollution can be measured.

2.6.7.6   Aluminium

Aluminium is a major constituent of many rock-forming silicate minerals. Gibbsite (Al hydroxide) is also a relatively abundant mineral and may form the main upper limit on aluminium concentrations in groundwater. However, complexes of Al with F, SO_4 and organic matter are also important (Driscoll and Postek 1996). Despite its relative geo chemical abundance, Al occurs at low concentrations, in the $\mu g\ L^{-1}$ range, because of the low solubility of the Al-bearing minerals at the near-neutral pH values of most natural waters (Nordstrom and Ball 1986). Aluminium is most soluble under acidic conditions as Al^{3+}, its most toxic form, and to a lesser degree at high pH as hydrolysed species. It is possible that some Al may also exist in colloidal form, which can pass through the 0.45 μm filters used as a standard to filter samples.

Median concentrations in the European aquifers are typically less than 10 $\mu g\ L^{-1}$ (Fig. 2.11) reflecting the circum-neutral pH values of most of the groundwaters. Maximum concentrations in some, mainly siliciclastic aquifers, reach a few hundred $\mu g\ L^{-1}$ in more acidic groundwaters.

2.6.7.7   Arsenic

Arsenic occurs naturally, mainly in minerals such as arsenopyrite (FeAsS) and arsenic-rich pyrite. It also occurs in significant concentrations in pyrite or sorbed onto iron oxides. Its presence in silicate minerals and carbonates is usually much lower (Smedley and

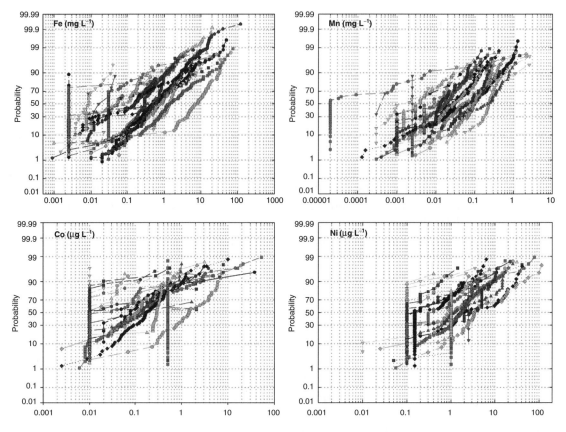

Fig. 2.10 Cumulative probability plots for Fe, Mn, Co and Ni in the reference aquifer groundwaters. Symbols as in Fig. 2.5.

Kinniburgh 2002). Sulphide minerals and iron oxides are therefore the most common sources and sinks of As in groundwaters. Release from sulphide minerals can occur through oxidation reactions while release from iron oxides can be via reductive dissolution or desorption reactions. Arsenic is redox sensitive having two main oxidation states in groundwater, As(III) and As(V). The As(III) form is the dominant form in reducing conditions and As(V) dominates under oxidising conditions. The spatial variations in total As are illustrated (Fig. 2.12) for the UK East Midlands aquifer. The cross section shows a profile of As concentrations relative to Eh along a flow line of 30 km from outcrop where the groundwater temperature is shown as a proxy for distance and residence time. In this aquifer, anthropogenic impacts are limited to the near outcrop and for As are negligible. The timescale of flow is around 30 kyr. Arsenic concentrations build up with time within the oxic section of the aquifer probably due to desorption as a consequence of increasing pH (Smedley and Edmunds 2002). Higher concentrations in the deeper parts of the aquifer are most likely controlled by reductive dissolution of Fe oxides since Fe concentrations are high.

The different As species have different toxicity with the reduced form being more harmful than the oxidised form as well as slightly more soluble. The current EC limit for As in drinking water is 10 µg L^{-1} and natural arsenic presents a problem in sections of some European aquifers. Concentrations in the reference aquifers typically vary by one to two orders of magnitude and median concentrations range from <0.1 to 6 µg L^{-1}.

2.6.7.8 Cadmium and zinc

Cadmium is an uncommon element and occurs as a trace constituent of sulphide minerals, particularly those containing Zn (e.g. sphalerite ZnS). It forms a divalent cation in solution under acidic oxidising conditions and may form a range of complex ions at higher pH. Natural concentrations in most groundwaters are limited, most likely by adsorption onto Mn or Fe oxide surfaces, although binding by organic substances may also be significant at high pH. Cadmium may also be adsorbed by calcite surfaces (Davis et al. 1987) or become incorporated into calcite via chemisorption. High concentrations, above 0.5 ppm, in soils are considered to be due to pollution including industrial, high-Cd phosphate fertiliser or

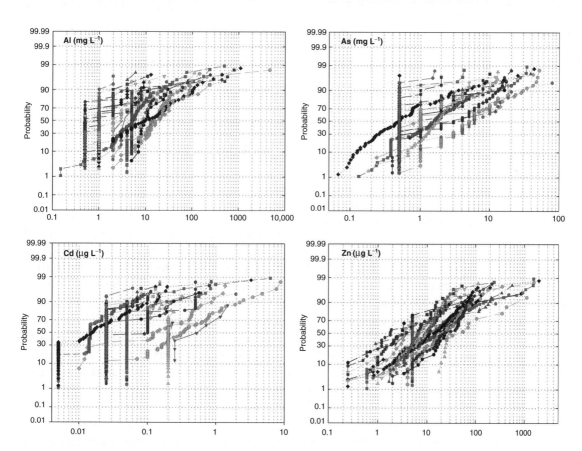

Fig. 2.11 Cumulative probability plots for Al, As, Cd and Zn in the reference aquifer groundwaters. Symbols as in Fig. 2.5.

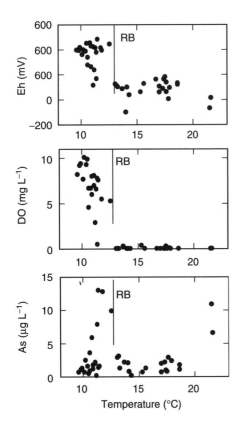

Fig. 2.12 Arsenic distribution in the East Midlands aquifer (UK) showing dissolved oxygen and Eh redox controls (modified from Smedley and Edmunds 2002). Scale is expressed as temperature as proxy for distance down-gradient.

sewage sludge applications (McBride 1994). Cadmium concentrations in Europe are generally low (below the detection limit of 0.2 µg L^{-1} or less) in most groundwaters (Fig. 2.11), the highest median and maxima were found in the Belgian Neogene aquifer, but did not exceed 1 µg L^{-1}.

In contrast to cadmium, Zn is present at measurable concentrations in almost all groundwaters. It is associated with sulphide minerals, typically those of Cu and Pb and is a trace constituent in most rocks. The most important sources of Zn in the environment are in ore deposits where it forms primary sulphide minerals such as sphalerite (ZnS) and is also released from secondary sulphate and carbonate minerals. It is relatively common in groundwaters but is strongly adsorbed on ferric hydroxides under alkaline conditions and may also bind with organic matter. The median concentrations of the individual aquifers vary over an order of magnitude from 1.8 to 60 µg L^{-1} (Fig. 2.11) and typical ranges within aquifers are typically over 2–3 orders of magnitude. For Zn, it is likely, as for other trace metals that groundwater concentrations largely represent the result of water–rock interaction. However, there are many sources of contamination including, for example, galvanised components of pumps, screens etc.

2.7 Discussion

The previous sections have described the wide range and complex distributions of a wide range of substances in the selected European reference aquifers, and have attempted to characterise these variations mainly using a simple statistical approach. The timescales involved in these geochemical changes vary from minutes to millions of years and groundwater compositions may be variable even where the bedrock geology and mineralogy are uniform. Variations in source minerals and groundwater flowpaths give rise to the observed range in the natural baseline. The degree of spatial heterogeneity is, therefore, likely to vary in response to the above factors and may be predictable or appear random. The characterisation of such variations requires careful sampling and care must also be taken with the interpretation of pumped, potentially mixed, samples from boreholes.

The groundwaters of Europe overall display a wide range of solute concentrations and statistical summaries are given in Tables 2.7 and 2.8. The ranges in water quality **within** all of the reference aquifers, may also vary over several orders of magnitude. The ranges can be described statistically using the median (more robust than the mean) together with an upper limit (97.7 percentile) to remove outlying data (Shand et al. 2007). The ranges in concentration highlight the order of magnitude variations in the aquifer or groundwater body as well as showing distinctions between groundwater bodies having different lithologies. These summary data must be placed in the context of spatial variability to be of practical value for management. Median concentrations for individual aquifers are also shown in Table 2.8. The statistical summaries mask information on the spatial and depth variations within each groundwater body and further hydrogeological and geochemical characterisation is essential for the protection as well as management of water resources. Emphasis should especially be placed on understanding depth variations within aquifers particularly since groundwater bodies are likely to be vertically zoned or in indirect contact with other groundwater bodies having different quality. Interpretation of quality differences should be linked to physical characterisation to assess any changes that might be caused by change in flow regimes induced by abstraction. Examples of spatial variations in individual aquifers are presented in later chapters.

The reference aquifer studies indicate the uniqueness of each groundwater system from a geochemical perspective, yet with similar processes taking place in each. Characterisation of such heterogeneity is a pre-requisite for establishing an effective and efficient monitoring network. It is also commonly the case that pollution inputs are masked within the overall large ranges of concentration of the individual solutes, therefore quantification of pollution needs to be carried out on a local smaller system to test whether concentrations are other than baseline. Owing to subtle differences in geological structure, lithological facies changes and mineralogy, care must be taken when extrapolating the results between and within aquifers; other factors such as the cover of glacial drift in northern Europe may lead to heterogeneity in water quality. It is therefore recommended that data for baseline assessment be sought from each groundwater body rather than making assumptions based on data from remote areas with similar geology.

Knowledge of groundwater residence time is important, since older groundwater is less likely to be impacted by pollution and because residence time plays an important role in reactions controlling the baseline quality. Conversely, significant pollution has occurred in most aquifers in water younger than the second half of the twentieth century. The timescales of groundwater movement are typically of the order of meters per year and it is therefore likely that the influence of pollution from the past few decades will continue to be present in the aquifers for a considerable time period in the future.

The natural age distribution and original baseline properties have been modified in many of Europe's aquifers by groundwater abstraction. This has been most evident in the unconfined parts of aquifers but may also

Table 2.8 Median concentrations in studies European groundwaters. See Table 2.5 for aquifer key.

Parameter	Units	B1	B2	BUL 1	BUL 2	CZ 1	CZ 2	DK 1	DK 2	DK 3	EE 1	EE 2
T	°C	10.0	12.4	14.6	14.0	10.3	9.8	9.8	10.4	10.8	9.1	8.5
pH		6.70	7.48	7.31	7.00	7.29	7.20	7.11	7.70	6.95	7.80	7.70
Eh	mV	271	148	320				−77	−80	−47		
DO	mg L^{-1}	0.2	1.9	6.5	5.3	1	6	0.01	0.02	0.015	0.1	6.8
SEC	µS cm^{-1}	158	799	637		37	56	633	421	1157	695	458
δ^2H	‰		−50									
$\delta^{18}O$	‰		−7.3						−7.7			
$\delta^{13}C$	‰	−13.3	−10.4						−12.9			
Ca	mg L^{-1}	24	62	87	64	43	100	114	37	257	43	64
Mg	mg L^{-1}	4.3	12	14	25	5.8	6.1	19	7.0	20	15	27
Na	mg L^{-1}	9.1	52	17		8.4	5.9	20	28	41	60	5.0
K	mg L^{-1}	3.8	18	2.0		3.1	2.1	3.7	2.9	7.1	8.0	3.8
Cl	mg L^{-1}	13	43	12	18	5.2	13	47	32	64	160	8.2
SO$_4$	mg L^{-1}	10	54	64	55	16	47	82	7.9	282	7.5	9.0
HCO$_3$	mg L^{-1}			252		159	260	320	137	439	156	357
NO$_3$–N	mg L^{-1}	0.150	0.282	4.22	1.56	0.500	11	0.100	0.057	0.029		
NO$_2$–N	mg L^{-1}	0.005	0.020	0.003	<0.001	0.003	0.003	0.002	0.799		0.006	0.010
NH$_4$–N	mg L^{-1}	0.160	0.680	0.039	0.094	0.070	0.025	0.104	0.226	0.188	0.200	0.158
P	mg L^{-1}	0.324	0.145	<0.02	<0.02	<0.02	<0.02	0.021	0.166	0.070	<0.02	<0.02
TOC	mg L^{-1}	21.9		0.1				1.9	5.7	2.6		
DOC	mg L^{-1}	2.7				1.1	1.7	2.9	3.7	2.8		
F	mg L^{-1}	0.205	0.235	0.205		0.150	0.120	0.425	0.060	0.140	0.660	0.186
Br	mg L^{-1}	0.030	0.090					0.060	0.165	0.170	1.770	<0.03
I	mg L^{-1}	0.003	0.014					0.009	0.004	0.004	0.052	0.003
Si	mg L^{-1}	9.9	21			1.8	2.0	13	8.3	17	3.5	4.2
Ag	µg L^{-1}	<0.05	<0.05	<0.05				<0.05	<0.05	<0.05	<0.05	<0.05
Al	µg L^{-1}	14	10	<1		5	5	1	9	4	3	<1
As	µg L^{-1}	5	1	1	1	<1	1	1	1	6	1	1
Au	µg L^{-1}	0.025	0.025	0.025				<0.05	<0.05	<0.05	<0.05	<0.05
B	µg L^{-1}	22	514	<20		<20	<20	45	198	63	167	41
Ba	µg L^{-1}	42	15	54		25	25	46	35	105	325	488
Be	µg L^{-1}	<0.05	<0.05	<0.05		0.05	0.05	<0.05	<0.05	<0.05	<0.05	<0.05
Bi	µg L^{-1}	<0.05	<0.05	<0.05				<0.05	<0.05	<0.05	<0.05	<0.05
Cd	µg L^{-1}	0.32	<0.05	<0.05	1.50	0.05	0.05	<0.05	<0.05	<0.05	<0.05	<0.05
Ce	µg L^{-1}	0.13	0.04	<0.01				<0.01	0.150	<0.01	<0.01	<0.01
Co	µg L^{-1}	1.45	0.08	<0.02		0.05	0.50	<0.02	<0.02	0.84	<0.02	<0.02
Cr	µg L^{-1}	0.8	<0.5	6.6	0.5	1.0	1.0	0.8	<0.5	<0.5	<0.5	<0.5
Cs	µg L^{-1}	0.01	<0.01	<0.01				<0.01	<0.01	<0.01	0.02	<0.01
Cu	µg L^{-1}	1.75	3.95	0.90	15.00	1.00	1.00	0.60	1.10	0.60	5.70	1.40

Aquifer*											
EE 2	F 1	F 2	F 3	P 1	P 2	PL 1	PL 2	UK 1	UK 2	UK 3	UK 4
19.6		11.9	17.2	20.7	17.3	12.0	11.2	11.4	12.6	11.0	11.4
7.00	7.50	6.84	7.48	7.34	6.10	7.24	7.43	7.26	7.84	7.18	7.22
	140		54	91	369	54	94	202	240	124	517
	4.21	6.44		0.96	0.715	0.02	0.07	0.75	0.5	0.3	7.61
295	735	601	496	431	456	568	671	1025	492	366	586
−29		−52	−53	−26	−23	−71	−70	−54	−58	−46	−39
−5.0		−7.2	−7.5	−4.7	−4.3	−10.0	−9.9	−7.9	−8.3	−7.1	−6.4
−12.6			−8.0	−12.9	−16.8	−13.5	−13.3	−14.6	−12.8	−14.2	−14.4
15	38	93	101	18	43	82	86	140	46	52	106
4.8	19	8.2	22	6.1	6.0	12	14	35	25	2.8	2.5
28	73	18	27	54	26	15	23	36	10	10	11
2.5	6.6	1.9	2.0	7.9	7.8	3.4	2.6	4.0	4.3	2.5	1.7
45		31	13	41	37	3.4	25	37	15	20	21
10	48	46	40	46	41	50	27	170	48	22	14
45	155	238	284	131	85	209	330	333	172	148	269
0.452		22.150	20	0.010	4.43	0.150	0.100	0.053	0.570	0.196	6.60
		<0.001	0.005	0.003	0.010	0.002	0.002	0.003	0.005	0.002	0.002
0.025			0.030	0.005		0.010	0.136	0.011	0.005	0.030	0.011
0.033	<0.02	<0.02	0.034	0.100	0.020	0.178	0.044	0.027	0.150	0.032	0.044
0.8			0.4	1.5	4.2	1.7	0.9	3.1	0.9	1.0	1.2
						1.4	0.9	3.0	0.7	2.7	0.7
			0.412	0.270	0.040	0.182	0.180	0.170	0.080	0.133	0.080
			0.500	0.154	0.135	<0.03	<0.03	0.113	0.072	0.078	0.084
				0.006	0.019	0.007	0.004	0.014	0.004	0.004	0.003
9.6	6.3	9.8	12	5.7	5.4	9.3	8.6	6.7	3.7	6.7	4.7
<0.05	<0.05	<0.05	<0.05	<0.05	0.100	<0.05	<0.05	<0.05		<0.05	<0.05
16	6	<1	5	5	10	<1	1	1	1	<1	1
1	6	1	1	2	<1	1	1	1	2	3	1
<0.05	<0.05	<0.05	<0.05		<0.05	<0.05	<0.05	<0.05		<0.05	<0.05
20	<20	<20	30	117	78	48	42	61	<20	<20	<20
58	78	30	38	75	31	64	62	48	105	59	12
<0.05	<0.05	<0.05	<0.05	0.18	0.05	<0.05	<0.05	<0.05	0.05	<0.05	<0.05
<0.05	<0.05	<0.05	<0.05	<0.05	<0.05	<0.05	<0.05	<0.05	<0.05	<0.05	<0.05
0.12	<0.05	<0.05	<0.05	<0.05	<0.05	<0.05	<0.05	<0.05	0.20	0.10	<0.05
0.03	<0.01	<0.01	0.02	0.02	<0.01	<0.01	<0.01	<0.01	<0.01	<0.01	<0.01
<0.02	<0.02	0.05	<0.02	0.12	0.17	<0.02	<0.02	0.15	0.29	<0.02	0.02
3.3	<0.5	19	<0.5	0.1	<0.5	<0.5	<0.5	<0.5	1.0	0.6	<0.5
0.02	2.1	0.02	<0.01	0.20	0.11	<0.01	<0.01	0.09	0.17	0.01	<0.01
3.80	1.60	1.45	3.10	0.97	3.27	0.10	0.30	1.90	2.00	1.40	2.20

Table 2.8 (*Continued*)

Parameter	Units	B1	B2	BUL 1	BUL 2	CZ 1	CZ 2	DK 1	DK 2	DK 3	EE 1	EE 2
Dy	μg L^{-1}	0.02	<0.01	<0.01				<0.01	<0.01	<0.01	<0.01	<0.01
Er	μg L^{-1}	0.01	<0.01	<0.01				<0.01	<0.01	<0.01	<0.01	<0.01
Eu	μg L^{-1}	<0.01	<0.01	<0.01				<0.01	<0.01	<0.01	0.01	<0.01
Fe	μg L^{-1}	7598	500	8	110	30	30	663	890	2650	300	500
Ga	μg L^{-1}	<0.05	<0.05	<0.05				<0.05	<0.05	<0.05	<0.05	<0.05
Gd	μg L^{-1}	0.02	<0.01	<0.01				<0.01	<0.01	<0.01	<0.01	<0.01
Ge	μg L^{-1}	<0.05	<0.05	<0.05				0.07	<0.05	<0.05	0.27	<0.05
Hf	μg L^{-1}	<0.02	<0.02	<0.02				<0.02	<0.02	<0.02	<0.02	<0.02
Hg	μg L^{-1}	<0.2	<0.2	<0.2		<0.2	<0.2	<0.2	0.95	<0.2	<0.2	<0.2
Ho	μg L^{-1}	<0.01	<0.01	<0.01				<0.01	<0.01	<0.01	<0.01	<0.01
In	μg L^{-1}	<0.01	<0.01	<0.01				<0.01	<0.01	<0.01	<0.01	<0.01
Ir	μg L^{-1}	<0.05	<0.05	<0.05				<0.05	<0.05	<0.05	<0.05	<0.05
La	μg L^{-1}	0.090	0.07	<0.01				<0.01	0.04	<0.01	0.01	<0.01
Li	μg L^{-1}	6.500	48	3.5		25	25	10	8.0	30	10	14
Lu	μg L^{-1}	<0.01	<0.01	<0.01				<0.01	<0.01	<0.01	<0.01	<0.01
Mn	μg L^{-1}	53	60	<2	<2	99	12	18	92	407	76	25
Mo	μg L^{-1}	<0.1	0.15	0.75		1.0	1.0	1.4	0.45	1.6	2.5	0.70
Nb	μg L^{-1}	<0.01	<0.01	<0.01				<0.01	<0.01	0.01	<0.01	<0.01
Nd	μg L^{-1}	0.11	0.03	<0.01				<0.01	0.02	<0.01	<0.01	<0.01
Ni	μg L^{-1}	1.10	<0.2	<0.2	5.0	1.0	1.0	2.0	0.85	3.85	<0.2	<0.2
Os	μg L^{-1}	<0.05	<0.05	<0.05				<0.05	<0.05	<0.05	<0.05	<0.05
Pb	μg L^{-1}	1.30	0.55	<0.1	15.5	1.0	1.0	<0.1	1.0	3.0	<0.1	<0.1
Pd	μg L^{-1}	<0.2	<0.2	<0.2				<0.2	<0.2	<0.2	<0.2	<0.2
Pr	μg L^{-1}	0.02	<0.01	<0.01				<0.01	<0.01	<0.01	<0.01	<0.01
Pt	μg L^{-1}	<0.01	<0.01	<0.01				<0.01	<0.01	<0.01	<0.01	<0.01
Rb	μg L^{-1}	3.4	7.2	0.30				1.6	1.1	1.7	4.3	1.8
Re	μg L^{-1}	<0.01	<0.01	0.04				0.010	<0.01	0.02	<0.01	<0.01
Rh	μg L^{-1}	<0.01	<0.01	<0.01				<0.01	<0.01	<0.01	<0.01	<0.01
Ru	μg L^{-1}	<0.05	<0.05	<0.05				<0.05	<0.05	<0.05	<0.05	<0.05
Sb	μg L^{-1}	<0.05	<0.05	0.06		0.25	0.25	<0.05	<0.05	<0.05	<0.05	<0.05
Sc	μg L^{-1}	4.7	9.0	2.4				6.4	4.1	5.5	1.8	2.6
Se	μg L^{-1}	<0.5	<0.5	1.0		<0.5	0.50	1.1	<0.5	1.0	12	<0.5
Sm	μg L^{-1}	<0.05	<0.05	<0.05				<0.05	<0.05	<0.05	<0.05	<0.05
Sn	μg L^{-1}	<0.05	0.10	<0.05				<0.05	0.11	0.14	<0.05	<0.05
Sr	μg L^{-1}	67	477	331		200	240	906	279	994	742	241
Ta	μg L^{-1}	<0.05	<0.05	<0.05				<0.05	<0.05	<0.05	<0.05	<0.05
Tb	μg L^{-1}	<0.01	<0.01	<0.01				<0.01	<0.01	<0.01	<0.01	<0.01
Te	μg L^{-1}	<0.05	<0.05	<0.05				<0.05	<0.05	<0.05	<0.05	<0.05
Th	μg L^{-1}	<0.05	<0.05	<0.05				<0.05	<0.05	<0.05	<0.05	<0.05
Ti	μg L^{-1}	<10	<10	<10				<10	<10	<10	<10	<10
Tl	μg L^{-1}	<0.01	<0.01	<0.01				<0.01	<0.01	<0.01	<0.01	<0.01

Aquifer*											
EE 2	F 1	F 2	F 3	P 1	P 2	PL 1	PL 2	UK 1	UK 2	UK 3	UK 4
<0.01	<0.01	<0.01	<0.01	<0.01	<0.01	<0.01	<0.01	<0.01	<0.01	<0.01	<0.01
<0.01	<0.01	<0.01	<0.01	<0.01	<0.01	<0.01	<0.01	<0.01	<0.01	<0.01	<0.01
<0.01	<0.01	<0.01	<0.01	<0.01	<0.01	<0.01	<0.01	<0.01	<0.01	<0.01	<0.01
96	55	<5	31	1355	164	1425	605	361	80	160	<5
<0.05	<0.05	<0.05	<0.05	<0.05	<0.05	<0.05	<0.05	<0.05	<0.05	<0.05	<0.05
<0.01	<0.01	<0.01	<0.01	0.04	<0.01	<0.01	<0.01	<0.01	<0.01	<0.01	<0.01
<0.05	<0.05	<0.05	<0.05	0.14	<0.05	<0.05	<0.05	<0.05	<0.05	<0.05	<0.05
<0.02	<0.02	<0.02	<0.02		<0.02	<0.02	<0.02	<0.02	<0.02	<0.02	<0.02
0.90	<0.2	<0.2	1.20		<0.2	<0.2	<0.2	<0.2	<0.2	<0.2	<0.2
<0.01	<0.01	<0.01	<0.01	0.03	<0.01	<0.01	<0.01	<0.01	<0.01	<0.01	<0.01
<0.01	<0.01	<0.01	<0.01		<0.01	<0.01	<0.01	<0.01	<0.01	<0.01	<0.01
<0.05	<0.05	<0.05	<0.05		<0.05	<0.05	<0.05	<0.05	<0.05	<0.05	<0.05
0.01	<0.01	<0.01	0.02	<0.01	<0.01	<0.01	<0.01	<0.01	<0.01	<0.01	<0.01
2.0	33	3.5	27	14	2.2	22	22	34	14	5.4	0.85
<0.01	<0.01	<0.01	<0.01	0.03	<0.01	0.23	0.06	<0.01	<0.01	<0.01	<0.01
29	7	4	4	17	30	230	62	214	4	13	<2
0.30	0.40	0.35	0.40	0.22	0.12	0.30	0.30	0.80	0.59	<0.1	<0.1
0.01	<0.01	0.01	<0.01		<0.01	<0.01	<0.01	0.01	<0.01	<0.01	<0.01
0.02	<0.01	<0.01	0.01	0.03	<0.01	<0.01	<0.01	<0.01	<0.01	<0.01	<0.01
3.40	<0.2	<0.2	0.30	1.67	<0.2	<0.2	<0.2	1.20	1.00	2.80	0.35
<0.05	<0.05	<0.05	<0.05		<0.05	<0.05	<0.05	<0.05	<0.05	<0.05	<0.05
1.00	0.30	<0.1	1.00	0.21	0.25	<0.1	<0.1	1.00	<0.1	0.40	1.00
<0.2	<0.2	<0.2	<0.2		<0.2	<0.2	<0.2	<0.2	<0.2	<0.2	<0.2
<0.01	<0.01	<0.01	<0.01	0.03	<0.01	<0.01	<0.01	<0.01	<0.01	<0.01	<0.01
0.01	<0.01	<0.01	<0.01		<0.01	<0.01	<0.01	<0.01	<0.01	<0.01	<0.01
3.4	13	1.7	1.2	8.2	9.7	1.8	1.4	2.5	2.8	2.2	1.3
<0.01	<0.01	<0.01	<0.01		<0.01	<0.01	<0.01	<0.01	<0.01	<0.01	<0.01
<0.01	<0.01	<0.01	<0.01		<0.01	<0.01	<0.01	<0.01	<0.01	<0.01	<0.01
<0.05	<0.05	<0.05	<0.05		<0.05	<0.05	<0.05	<0.05	<0.05	<0.05	<0.05
0.68	<0.05	<0.05	<0.05	6.0	0.05	<0.05	<0.05	<0.05	<0.05	<0.05	<0.05
4.3	4.6	3.1	4.1	0.50	3.6	5.8	4.5	2.1	1.0	3.3	1.6
0.60	0.50	3.250	0.50	1.1	0.52	<0.5	<0.5	0.90	0.15	<0.5	0.65
<0.05	<0.05	<0.05	<0.05	<0.05	<0.05	<0.05	<0.05	<0.05	<0.05	<0.05	<0.05
0.10	0.38	<0.05	<0.05		0.05	0.07	<0.05	0.18	0.08	0.06	0.07
65	359	222	1222	104	171	672	299	853	271	162	233
<0.05	<0.05	<0.05	<0.05		<0.05	<0.05	<0.05	<0.05	<0.05	<0.05	<0.05
<0.01	<0.01	<0.01	<0.01	0.03	<0.01	<0.01	<0.01	<0.01	<0.01	<0.01	<0.01
<0.05	<0.05	<0.05	<0.05		<0.05	<0.05	<0.05	<0.05	<0.05	<0.05	<0.05
<0.05	<0.05	<0.05	<0.05		<0.05	<0.05	<0.05	<0.05	<0.05	<0.05	<0.05
<10	<10	<10	<10		<10	<10	<10	<10	<10	<10	<10
<0.01	<0.01	0.01	<0.01		0.04	<0.01	<0.01	<0.01	0.20	<0.01	0.01

Table 2.8 (*Continued*)

Parameter	Units	B1	B2	BUL 1	BUL 2	CZ 1	CZ 2	DK 1	DK 2	DK 3	EE 1	EE 2
Tm	µg L^{-1}	<0.01	<0.01	<0.01				<0.01	<0.01	<0.01	<0.01	<0.01
U	µg L^{-1}	<0.05	<0.05	5.26				<0.05	<0.05	3.01	<0.05	0.56
V	µg L^{-1}	<1	<1	<1				<1	1.0	<1	1.0	<1
W	µg L^{-1}	0.20	<0.1	<0.1				<0.1	<0.1	<0.1	<0.1	<0.1
Y	µg L^{-1}	0.16	0.04	<0.01				<0.01	0.02	0.02	0.02	<0.01
Yb	µg L^{-1}	<0.01	<0.01	<0.01				<0.01	<0.01	<0.01	<0.01	<0.01
Zn	µg L^{-1}	20	24	3.3	60	5.0	5.0	20	7.7	11	29	1.8
Zr	µg L^{-1}	<0.5	<0.5	<0.5				<0.5	<0.5	<0.5	<0.5	<0.5

* see Table 2.5 for aquifer nomenclature.

affect the confined aquifer where decreases in piezometric level have led to loss of artesian conditions. Such abstraction may cause the mixing of different stratified layers, for example, by pulling down younger, often polluted groundwater in the cone of depression around a borehole or up-coning of older, deeper groundwater. In these cases it is likely that the local baseline will be modified. In some cases, this can lead to derogation of the groundwater due to salinity or changes in redox status. Although the sources of individual solutes may be natural, the cause is anthropogenic. The changes in flow systems with time often make it difficult to interpret hydrochemical data within the present day modified flow regime. However, since the natural flow rates in most aquifers are low (of the order of m yr^{-1}, except for example in karstic areas) the kinetic dependence of many chemical reactions means that groundwaters often retain a memory of previous flow regimes.

The data plots (Figs. 2.5–2.11) show that it is often difficult to distinguish between the groundwater chemistry in different aquifer types, for example, carbonate vs. silicate types. This is due to the fact that many silicate aquifers contain secondary calcite as cements: carbonate minerals dissolve easily and they will often dominate the groundwater chemistry. Only in cases where the carbonate cement has been dissolved away (e.g. in shallower parts of recharge zones) will differences become apparent. However, differences in trace elements may be characteristic of each type. As well as lithological control, other processes (mixing, redox reactions and ion-exchange reactions) are common to all aquifers and in combination with residence time will generally dominate the evolution of groundwater chemistry in aquifers.

Diffusional exchange between groundwaters in aquifers with dual porosity characteristics (where differences in quality between matrix pore water and water moving much more rapidly in fractures) has been considered in several reference aquifers. This is the case for the European Chalk for example; saline or geochemically modified pore water is released very slowly to the more active part of the flow system. Exchange may also occur between aquifers and adjacent aquitards.

Many solutes found naturally in groundwater exceed existing regulatory standards as shown relative to MAC concentrations. In such cases, trend reversal or remediation

Aquifer*											
EE 2	F 1	F 2	F 3	P 1	P 2	PL 1	PL 2	UK 1	UK 2	UK 3	UK 4
<0.01	<0.01	<0.01	<0.01	0.03	<0.01	<0.01	<0.01	<0.01	<0.01	<0.01	<0.01
<0.05	0.41	0.90	0.34	0.05	0.09	<0.05	<0.05	1.4	0.73	<0.05	0.24
<1	<1	<1	<1	3.0	<1	<1	<1	<1	<1	<1	<1
<0.1	<0.1	0.10	<0.1		<0.1	<0.1	<0.1	<0.1	<0.1	<0.1	<0.1
0.02	<0.01	0.02	0.01	0.01	0.59	0.02	<0.01	0.02	<0.01	0.01	<0.01
<0.01	<0.01	<0.01	<0.01	0.03	<0.01	<0.01	<0.01	<0.01	<0.01	<0.01	<0.01
20	9.6	2.8	44	15	13	6.9	3.5	9.2	9.3	23	6.4
<0.5	<0.5	<0.5	<0.5	3.0	<0.5	<0.5	<0.5	<0.5	<0.5	<0.5	<0.5

may be difficult, impossible or prohibitively expensive, so special regulations may be required for these. Most solutes considered as pollutants (i.e. of anthropogenic origin) also have a natural baseline and careful assessment of the baseline (natural) component needs to be undertaken. A summary of both anthropogenic and natural components which may exceed existing standards for the reference aquifers is provided in Table 2.9. Of the anthropogenic substances, most are derived from the agricultural application of fertiliser although some cases have been noted where mining activities (e.g. BG Thracian) have affected large areas. Of the baseline components, Fe, Mn, F, Cl and SO_4 are elements often present at excedence levels, but noteworthy are the number of aquifers with naturally high concentrations of the trace elements As, Ba and Ni.

It is inappropriate to cite individual values for baseline concentrations and a range of values must therefore be considered; narrowing down the size of the groundwater body or system or separating geochemically defined units (e.g. reducing groundwaters) may limit the range. For many elements, pristine areas or time periods must be identified. The location of historical data can be important to determine the pre-anthropogenic concentrations, especially for nitrate.

The statistical data and plots are useful for comparison of baseline between areas and may be useful in the future to assess changes from the present day. In addition they are of strategic value since they include new data for elements for which there are no existing statutory guidelines, but which may still be harmful for human consumption (e.g. Be, Se, Sb, Tl, U). The present definition of these baseline ranges for a system can be used to recognise future contamination incidents involving such uncommon elements.

2.8 Conclusions

There is a considerable range in water quality within all of the reference aquifers, with concentrations typically varying over several orders of magnitude. The ranges can be described statistically using the median (more robust than the mean) and an upper limit (97.7 percentile) to remove outlying data.

Most aquifers have been affected to some degree by diffuse pollution. Baseline is therefore unlikely to be modern-day status: the

Table 2.9 Summary of solutes (natural and anthropogenic) which may exceed drinking water standards in the studied reference aquifers (see Table 2.1 for further details of aquifers).

Aquifer	Natural (baseline) components	Anthropogenic (non-baseline) components identified
UK Wessex	F, Fe, Mn	NO_3, K
UK Greensand	Fe, As	NO_3, K, Cl, SO_4
UK Vale of York	SO_4, As, Ba, Fe, Mn, U	NO_3, K, Cl, SO_4
UK E Midlands	Mn, As	NO_3, K, Na, Cl, SO_4
ES Donana	Mg, K, Na, Cl, SO_4, As, Ni, Se	NO_3, SO_4, Br, Co, Cu, Zn
ES Madrid	As	NO_3, organic C
P Quaternary	Fe, Mn, CH_4	NO_3, SO_4, K
P Cretaceous	Fe, Mn, F, As, Ba	NO_3
B Ledo–Paniselian	Na, K, Mg, Ca, Fe, Mn, NH_4, Cl, SO_4, F	NO_3, NO_2, K, Cl
B Neogene	As, Ba	NO_3, K, Cl, P, Cu, Cr, Pb, Zn
F Fontainbleu	Fe, Mn	NO_3, SO_4, Cl, Na, K
F Valreas	Fe, Mn, As	NO_3, SO_4, Cl, K, Zn
F Triassic	SO_4, Fe, Mn, As, Ba, U	NO_3
DK Miocene	Cl, DOC, As, Al, Se	–
DK Quaternary	Al, As, Fe, Ni, DOC	–
DK Limestone	SO_4, Cl, F, As, Ni, Zn	–
PL Kedzieryn	SO_4, NH_4, Fe, Mn	SO_4, Cl
PL Bogucice	Fe, Mn, NH_4	SO_4, Cl
EE Devonian	Fe, Cd, Li, Pb	NO_3
EE C-Vendian	Na, Cl, F, Fe, Mn, Ba, Cd, Li, Pb	–
CZ Cenomian	Fe, Mn, F	NO_3, SO_4, Cl
CZ Turonian	NH_4, Fe, Mn, F, Al	NO_3, NH_4
BG Thracian	SO_4, Mn, Ba	NO_3, SO_4, PO_4, Na, Cl, heavy metals
BG Razlog	Cl, SO_4, Na, K, NH_4	–
M Sea-level aquifer	Na, Cl, F, P, Ba	NO_3

data from aquifers require data manipulation (removal of outliers and anthropogenic influences) making determination of the baseline difficult. It is commonly the case that pollution inputs are masked within the overall large range of baseline, therefore, quantification of pollution needs to be done on a local scale, that is, within a smaller system.

Historical records provide the most important approach for determining the baseline for a number of key elements (major ions and

nitrate for example). However, old records are often difficult to find, and many long-term data may have been lost as a result of organisational changes and the shift to digital records. In some countries, the data are confidential and not available for public use.

The statistical data and plots given in this chapter and in the reference aquifer studies are useful for comparison between areas, and are of strategic value since they include data for elements for which there are no existing guidelines, but which may still be harmful (e.g. Be, Se, Sb, Tl, U). The definition of these baseline ranges can be used to recognise contamination incidents by uncommon elements for which no guidelines exist. More than 80 components have been measured in these aquifers with overall summaries given in Tables 2.6–2.8. This represents a comprehensive and strategic database of the baseline concentrations for most inorganic substances, and a reference to quantify pollution.

There is considerable overlap in concentration for most solutes between the different aquifers regardless of lithology. The main hydrochemical characteristics in most aquifers are determined, not specifically by rock type, but by a range of hydrochemical processes such as mineral dissolution, redox reactions, ion-exchange reactions, mixing, adsorption–desorption and rainfall inputs. Owing to subtle differences in geological structure, facies changes, mineralogy, etc. care must be taken when extrapolating the results between and within aquifers. This is particularly the case for trace elements where sources tend to be much more heterogeneous. It is recommended that data be sought from the groundwater body of interest rather than relying on the applicability of data from remote areas. A protocol for baseline definition is given in Chapter 21.

A number of solutes such as As, Ba, F and Fe have concentrations above standards for which regulatory standards are in operation, but the source of these is natural. Trend reversal or remediation may be either impossible or prohibitively expensive in areas where this occurs and special regulations may therefore be required for these. Many of these solutes, in addition, are also pollutants and a careful assessment of the baseline

(natural component) needs to be undertaken to distinguish natural baseline from anthropogenic origin.

The ranges in concentration highlight the order of magnitude variations in the aquifer or groundwater body. These must be placed in the context of spatial variability to be of practical use for groundwater management. Knowledge of the spatial and depth variations is then also essential for the protection as well as management of water resources. The baseline may also be modified by groundwater abstraction without the introduction of pollutants.

References

Appelo, C.A.J. and Postma, D. (2005) *Geochemistry, Groundwater and Pollution.* 2nd edn. Balkema, Leiden.

Banks, D., Midtgard, A.K., Morland, G. et al. (1998) Is pure groundwater safe to drink? Natural contamination of groundwater in Norway. *Geology Today* 14, 105–13.

Daughney, C.J. and Reeves, R.R. (2005) Definition of hydrochemical facies in the New Zealand National Groundwater Monitoring Programme. *Journal of Hydrology (NZ)* 44, 105–30.

Davis, J.A., Fuller, C.C. and Cook, A.D. (1987) A model for trace element sorption processes at the calcite surface: adsorption of Cd^{2+} and subsequent solid solution formation. *Geochimica et Cosmochimica Acta* 51, 1477–90.

Davis, S.N. (1964) Silica in streams and groundwaters. *American Journal of Science* 262, 870–91.

Driscoll, C.T. and Postek, K.M. (1996) The chemistry of aluminium in surface waters. In: Sposito, G. (ed.) *The Environmental Chemistry of Aluminium.* Lewis Publishers, Boca Raton, pp. 363–417.

Dzombak, D.A. and Morel, F.M.M. (1990) *Surface Complexation Modeling: Hydrous Ferric Oxide.* Wiley & Sons, New York.

Edmunds, W.M. and Kinniburgh, D.G. (1986) The susceptibility of UK groundwaters to acidic deposition. *Journal of the Geological Society of London* 143, 707–20.

Edmunds, W.M. and Smedley, P.L. (2005) Fluorine in natural waters – occurrence, controls and health aspects. In: Selnius, O., Alloway, B., Centeno, J.A. et al. (eds) *Essentials of Medical Geology*. Academic Press, pp. 301–29.

Edmunds, W.M., Cook, J.M., Kinniburgh, D.G. et al. (1989) Trace element occurrence in British groundwaters. Research Report British Geological Survey SD/89/3. 420 pp.

Edmunds, W.M., Shand, P., Hart, P. et al. (2003) The natural (baseline) quality of groundwater in England and Wales: a UK pilot study. *Science of the Total Environment* 310, 25–35.

Garrels, R.M. and Mackenzie, F.T. (1967) Origin and composition of some springs and lakes. In: Stumm, W. (ed.) *Equilibrium Concepts in Natural Water Systems*. Advances in Water Chemistry Series no. 67. American Chemical Society, Washington, DC, pp. 222–42.

Hem, J.D. (1985) Study and interpretation of the chemical characteristics of natural water. US Geological Survey Water Supply Paper, 2254.

Langmuir, D. (1997) *Aqueous Environmental Geochemistry*. Prentice-Hall, New Jersey.

Lee, L. and Helsel, D. (2005) Baseline models of trace elements in major aquifers of the United States. *Applied Geochemistry* 20, 1560–70.

McBride, M.C. (1994) *Environmental Chemistry of Soils*. Oxford University Press, New York.

Moss, P.D. and Edmunds, W.M. (1997) Processes controlling acid attenuation in the unsaturated zone of a Triassic sandstone aquifer (UK) in the absence of carbonate minerals. *Applied Geochemistry* 7, 573–83.

Nordstrom, D.K. and Ball, J.W. (1986) The geochemical behavior of aluminium in acidified surface waters. *Science* 232, 54–6.

Ottley, C. J., Davidson, W. and Edmunds, W.M. (1997) Chemical catalysis of nitrate reduction by Fe (II). *Geochimica et Cosmochimica Acta* 61, 1819–28.

Paces, T. (1978) Reversible control of aqueous aluminum and silica during irreversible evolution of natural-waters. *Geochimica et Cosmochimica Acta* 42 (10), 1487–93.

Runnells, D.D., Dupon, D.P., Jones, R.L. et al. (1998) Determination of background chemistry of water at mining and milling sites, Salt Lake Valley, Utah, USA. In: *Proceedings of the 9th International Symposium on Water–Rock Interaction*. Taupo, New Zealand. A.A. Balkema, Rotterdam, pp. 997–1000.

Shand, P. and Frengstad, B. (2001) Baseline groundwater quality: A comparison of selected British and Norwegian aquifers. British Geological Survey Internal Report IR/01/177.

Shand, P., Haria, A.H., Neal, C. et al. (2005) Hydrochemical heterogeneity in an upland catchment: further characterization of the spatial, temporal and depth variations in soils, streams and groundwaters of the Plynlimon forested catchment, Wales. *Hydrology and Earth System Sciences* 9, 621–34.

Shand, P., Edmunds, W.M., Lawrence, A.R., Smedley, P. L. and Burke, S. (2007) *The Natural (Baseline) Quality of Aquifers in England and Wales*. British Geological Survey Research Report RR/07/06. 72 pp.

Smedley, P.L. and Edmunds, W.M. (2002) Redox patterns and trace-element behavior in the East Midlands Triassic sandstone aquifer. *Ground Water* 40, 44–58.

Smedley, P.L. and Kinniburgh, D.G. (2002) A review of the source, behavior and distribution of arsenic in natural waters. *Applied Geochemistry* 17, 517–68.

Stumm, W. and Morgan, J.J. (1996) *Aquatic Chemistry*. Wiley & Sons, New York.

Wilson, G. B., Andrews, J.N. and Bath, A.H. (1990) Dissolved gas evidence for denitrification in the Lincolnshire Limestone groundwaters, eastern England. *Journal of Hydrology* 113, 51–60.

3 Organic Quality of Groundwaters

D. C. GOODDY AND K. HINSBY

Summary

Organic carbon is present in all natural waters and plays an important role in many geochemical processes. Total organic carbon (TOC), measured on unfiltered samples, has been analysed on more than 400 groundwater samples from eight European Union countries, and the operationally defined dissolved organic carbon (DOC), measured on filtered samples, was analysed on roughly 250 groundwater samples from four of these countries. TOC was found at a median concentration of 2.69 mg L^{-1} with a range from 0.12 to 59.4 mg L^{-1} and DOC had a median concentration of 2.16 mg L^{-1} with a range from 0.18 to 58.9 mg L^{-1}, demonstrating very high natural organic carbon values can be found locally in some pristine aquifers. A relationship between the organic carbon utilised by bacteria and TOC is observed although the correlation is not clear. Generally, there is a linear correlation between the TOC, DOC and the chemical oxygen demand (COD) values. Organic contaminants derived from anthropogenic activities are generally not identified in the TOC/DOC analysis as the contaminant concentrations are typically several orders of magnitude lower than the bulk aqueous carbon measurement. TOC/DOC analysis can be an important indicator of pollution at landfills and similar settings with high loads of organic carbon, but in many other situations the TOC or DOC concentration is not a good tracer or indicator of contamination. The TOC or DOC is though a very important component in the biogeochemical cycling of elements and consequently it is recommended as a component to be measured on all groundwater samples. The measurement of a filtered organic carbon fraction is recommended where groundwaters contain significant amounts of particulate material. Measurements of TOC and DOC on the same samples in this study show comparable levels, however, the TOC/DOC ratio varied and TOC was not generally found to be above DOC as may have been expected. Further investigations are needed to find the reasons for this. Further research is also needed to evaluate what part of the TOC is easily available for biogeochemical processes. Increased knowledge on this issue would help to better understand the development of the different redox environments and bacterial regrowth in water supply systems.

3.1   Introduction

All natural waters contain dissolved organic compounds. Geochemists have tended to ignore them in the past because they are complex and difficult to analyse, but it is now recognised that they play a major role in weathering processes (Heyes and Moore

1992; Shand et al. 2005), in diagenesis (Bottrell 1996) and in the transport of trace metals (Benedetti et al. 1996). Organic carbon concentrations can vary considerably both spatially and temporally so a large number of analyses are frequently required to produce a representative estimate of concentrations. Natural organic carbon in shallow groundwaters is derived mostly from the overlying soils and consists of a mixture of macro molecules normally classified as humic and fulvic substances. For example, it has been shown that humic substances constitute the dominant fraction of the organic carbon (up to 90%) in Canadian groundwater systems (Thurman 1985). The remainder is a complex mixture of lower molecular weight compounds. Some of the more hydrophobic compounds present in surface and recharge waters will be sorbed onto the aquifer solids. Deeper and older waters may contain organic material which has been mobilised from the matrix (e.g. Grøn 1989; Grøn et al. 1996; Orem et al. 1999) such as lower molecular weight compounds resulting from kerogen degradation or bitumen dissolution. The natural organic carbon content of groundwater is generally not a health concern in itself, but exceptions have been reported (e.g. Orem et al. 1999). There's however no general guideline value for the total contents of organic carbon in groundwater. Organic carbon in groundwater can also be derived from anthropogenic sources. The anthropogenic carbon occurs at much lower concentrations (generally 2–4 orders of magnitude), but the organic pollutants such as chlorinated solvents and pesticides are more toxic and a health concern at these lower concentration levels. Finally chlorination may create carcinogenic by-products in groundwater (Stuart et al. 2001).

3.2   Measured Organic Fractions

TOC (total organic carbon) and DOC (dissolved organic carbon) measurements are standard methods globally used for measuring the amount of organic carbon in water samples (e.g. ASTM 1988). TOC measurements include all organic compounds in the analysed water sample, that is, TOC measurements for instance also include suspended particles and bacteria in addition to a large number of dissolved organic molecules of different sizes (Fig. 3.1). DOC measurements are performed on filtered samples (conventionally 0.45 μm silver filters although polycarbonate filters are increasingly being used), that is, most suspended/ 'particulate' carbon is removed, but some viruses and small bacteria may still pass through the filter.

Organic carbon is generally very important for the evolution of different redox environments in aqueous environments, and the amount and reactivity of the organic carbon in the aquifers is therefore an important parameter when evaluating the state and trends of groundwater quality. However, the amount of TOC/DOC itself does not provide information on how reactive or reducing the environment is. Generally, the biologically viable TOC/DOC in European groundwaters is likely to originate from the soil, as any organic matter present in the rock matrix is mostly unavailable to microorganisms. Exceptions do exist though in shallow aquifers where the aquifers may contain young reactive organic matter (Jakobsen and Postma 1999; Hansen et al. 2001). The majority of DOC that travels from the soil zone to deep within the aquifer will be old and highly recalcitrant, and therefore not prone to further microbial breakdown over the timescales of

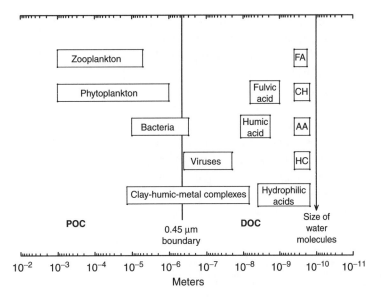

Fig. 3.1 Continuum of particulate and dissolved organic carbon in natural waters. FA – fatty acids; CH – carbohydrates; AA – amino acids; HC – hydrocarbons. TOC measurements include all the organic 'compounds' shown above, DOC values include only the dissolved organic 'compounds' which are to the right of the 0.45 micron boundary (<0.45 μm).

more active groundwater circulation (Darling and Gooddy 2006). Groundwater with a small but very reactive amount of 'young' organic carbon may be much more reactive than groundwater with a much larger but 'old' non-degradable amount of organic carbon.

The degradability of organic carbon is not only important for the 'health' of the aquifers (Grøn et al. 1992), but also for the water supply systems (DEPA 2002). The amount of highly reactive organic carbon, which are especially relevant for the water supply systems may be estimated by the method proposed by Kooij et al. (1982). This method estimates the easily assimilable organic carbon (AOC), that is, the part of the TOC in the water, which is easily available to aerobic bacteria. High AOC contents may cause bacterial growths in the aquifers or a problem with bacterial regrowth in water pipelines if the water is not desinfected/chlorinated (Kooij 1992; DEPA 2002). Water chlorination on the other hand is known to cause carcinogenic disinfection by-products when chlorine

reacts with the DOC, and it is therefore often avoided if possible (Stuart et al. 2001; DEPA 2002). AOC concentrations are generally much lower than TOC/DOC concentrations typically 2–3 orders of magnitude in Danish groundwater (DEPA 2002), however, in some cases they may be close to the TOC/DOC values. Large AOC values indicate young potential contaminated groundwater or relatively young and highly reactive organic carbon in the sediments. Large AOC values furthermore show that the groundwater has a significant potential for reducing oxygen and nitrate and hence that reduced geochemical environments in the aquifers should develop relatively fast. Groundwater generally has AOC values, which are much lower than surface waters (Jørgensen 2000).

Increased understanding regarding the evolution and variation of natural organic and inorganic groundwater quality is strongly needed as a basis for the recognition of human impacts on aquifers and for aquifer management. We present baseline values for the non-volatile organic carbon (NVOC) fraction expressed as total dissolved organic carbon (TOC) and dissolved (passed through a 0.45 µm filter organic carbon (DOC) in groundwaters, which are comparable to earlier studies. These were sampled from 20 European reference aquifers. TOC and DOC were analysed in groundwater from more than 300 wells. The number of bacteria, the concentration of easily AOC and the COD were also determined in selected samples from Denmark.

3.3   Description of Investigated Aquifers

Fourteen different aquifers were sampled for TOC analysis in groundwater in eight EU countries (Fig. 3.2, Tables 3.1 and 3.2). The sampled aquifers can be grouped in three major aquifer types: (1) sand (including gravel), (2) carbonates (chalk/limestone) and (3) sandstone. The aquifer sediments were deposited in different geological periods from the Devonian sandstone in Estonia, which is more than 350 million years old, to the relatively recent unconsolidated Quaternary sand aquifers in Portugal and Spain less than 2 million years old. Hence the dataset contain analyses from aquifers representing deposits from the last three geological eras: the Palaeozoic, Mesozoic and Cenozoic. The aquifer sediments represent several transitional, terrestrial and marine depositional environments, which at the time of deposition have contained varying amount of organic matter.

Most of the investigated aquifers are unconsolidated sands (5) or sandstone (6), illustrating that these are the most common and important aquifer types in Europe. The

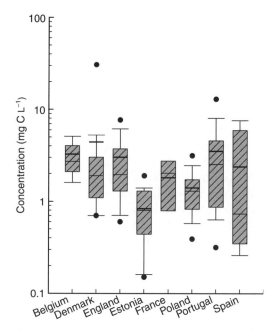

Fig. 3.2   Box–Whisker plot of TOC concentrations in all European aquifers sampled.

sands and sandstones range from unconsolidated mature very pure quartz sands to arkosic immature sandstones. Four of the investigated aquifer or aquifer systems are composed of various kinds of carbonates or include a carbonate, which likewise is among the most common and important aquifer types in Europe. The carbonates range from finegrained chalk to bryozoan limestone with large fragments of Bryozoa and other marine animals. The distribution of TOC concentrations in the selected aquifers is believed to illustrate the most common distribution of TOC values in European aquifers. The investigated aquifers are generally semi-confined and recharged either directly or through Quaternary deposits of varying thickness. Table 3.1 lists the name and some basic geological information about the investigated aquifers.

Table 3.1 Summary of aquifer types used in this baseline study.

Country	Aquifer name	Type	Geological period	Age (million years)*	Depositional environment
Belgium	Ledo–Paniselian	Sand	Neogene (Miocene)	>5	Marine
Denmark	Ribe Formationen	Sand	Neogene (Miocene)	>5	(Fluvio)-deltaic
	Bryozokalk	Bryozoan limestone	Paleogene	>55	Marine
England	Cheshire	Sandstone	Permo-Triassic	>200	Aeolian/fluvial
	Dorset	Chalk	Cretaceous	>65	Marine
	Thames	Chalk	Cretaceous	>65	Marine
	Vale of York	Sandstone	Permo-Triassic	>200	Aeolian/fluvial
Estonia	Middle Devonian (D2)	Sandstone	Devonian	>350	Marine/Deltaic
France	Valréas	Sandstone	Neogene (Miocene)	>5	Marine
Poland	Bogucice	Sandstone	Neogene (Miocene)	>5	Marine
	Kedzierzyn-Glubczyce	Sand	Neogene (Miocene)	>5	Barrier/shallow Marine
Portugal	Aveiro Quaternary	Sand	Neogene (Pleistocene)	>0.01	Fluvial/Aeolian
	Aveiro Cretaceous	Sandstone/ limestone	Cretaceous	>65	Transitional Fluvial/Deltaic/ Marine
Spain	Doñana aquifer system	Sand	Neogene (Plio-Pleistocene)	>1	Eolian, fluvial, fluvio-deltaic

* The listed ages are not exact ages for the aquifer sediments but the approximate minimum age of the geological period in which the aquifer sediment was deposited.

Table 3.2 Statistical summary for TOC in investigated European groundwater.

Country	*n*	Minimum	Maximum	Median	Mean	Standard deviation
Belgium	18	1.22	10.5	2.72	3.26	2.05
Denmark	46	0.70	59.4	1.99	4.50	10.4
England	101	0.49	19.2	1.96	3.02	3.06
Estonia	19	0.15	1.90	0.80	0.83	0.46
France	5	0.60	3.33	2.03	1.82	1.08
Poland	120	0.31	5.69	1.31	1.43	0.85
Portugal	120	0.12	19.3	3.25	4.26	3.22
Spain	10	0.26	7.75	0.74	2.40	2.99
Mean		0.48	15.9	1.85	2.69	3.01
Denmark NMP	9883	0.10	71	1.4	2.0	2.5

All measurements are in mg C L^{-1}. Data from the Danish National Monitoring Programme (Denmark NMP) is included for comparison.

3.4   Results

3.4.1   TOC concentrations

Results from the study countries for TOC are statistically summarised in Table 3.2 and shown graphically in Fig. 3.2. For most of the countries there is a fairly large discrepancy between the Mean and Median values which reflects the extreme concentration of the outliers (see black circles on Fig. 3.2) especially for Denmark. For the two countries with the lowest TOC concentration in groundwater, Poland and Estonia, the mean and median concentrations are very similar. The sample set for France is too small to be statistically significant and is included for comparison.

3.4.2   DOC concentrations

A limited number of samples from Denmark, England, France and Poland were also analysed for the DOC fraction. Table 3.3 shows the range of values obtained for these four countries. The high overall mean concentration is probably a reflection of a few very high organic carbon concentration waters from Denmark. These samples bias the statistical evaluation since they were specifically collected from wells known to have high TOC values in order to demonstrate the possible large range of natural organic carbon in groundwater.

3.4.3   Comparing TOC and DOC data

A cumulative frequency plot for the TOC (based on 439 analyses) and for the DOC (based on 246 analyses) is shown in Fig. 3.3. The curve shape is typical of a lognormal distribution and indicates that both populations resemble this kind of statistical function. The very high concentration outliers from Denmark which exceed 50 mg C L^{-1} occur in the top 99.5 percentile and so are not shown on the plot.

A direct comparison can be made between DOC and TOC samples collected in the UK is shown in Fig. 3.4(a). A fairly weak positive correlation is observed between the two parameters ($R^2 = 0.32$) with TOC generally lower in concentration that DOC, suggesting some degradation of the organic carbon in the unfiltered samples between sampling and analysis which may be due to microbial action.

In comparison, Fig. 3.4(b) shows a cross plot for DOC and TOC samples collected

Table 3.3 Statistical summary for DOC investigated European groundwater. All measurements are in mg C L^{-1}.

Country	n	Minimum	Maximum	Median	Mean	Standard deviation
Denmark	12	1.67	58.9	3.98	11.98	18.8
England	113	0.18	18.6	2.19	3.09	3.04
France	3	1.28	3.16	1.32	1.92	1.07
Poland	118	0.28	3.93	1.14	1.32	0.75
Mean		0.85	21.1	2.16	4.58	5.92

in Poland. The range of values is much smaller but also the two parameters are much more closely correlated ($R^2 = 0.81$) with TOC values being roughly 8% higher than the DOC values, more likely reflecting a high degree of particulate matter in the unfiltered fraction.

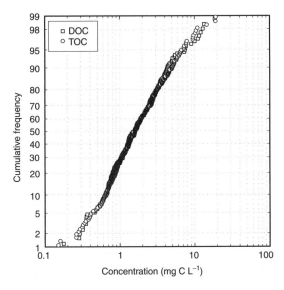

Fig. 3.3 Cumulative frequency plot for both DOC and TOC from all European aquifers sampled.

3.4.4 Comparing TOC between aquifers

Fig. 3.5(a) shows an example of TOC data for two Permo-Triassic sandstone aquifers and two Chalk aquifers both from the UK. The data suggest a much broader range of concentrations for the Permo-Triassic sandstones and additionally that the Permo-Triassic sandstones have generally higher TOCs.

In contrast, Fig. 3.5(b) shows the same data but with each plot based on aquifer type and region. Permo-Triassic sandstones from Cheshire and the Vale of York show quite different properties, with the Vale of York encompassing a greater range of values. Similarly there are considerable differences between Chalk from the Dorset region and Chalk from the Thames region of the UK. This highlights the difficulties encountered when only considering aquifer types.

This is also illustrated by a recent study performed on data from the National Danish groundwater monitoring database. This study demonstrates TOC values comparable to the British values illustrated in Fig. 3.5. However, here the different types of silicious rocks (unconsolidated Tertiary and Quaternary sands) show slightly lower average TOC concentrations than most Danish carbonaceous

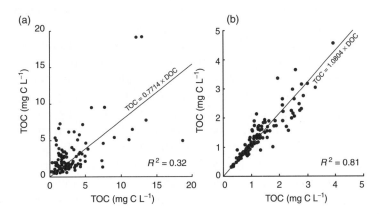

Fig. 3.4 Cross plot of TOC and DOC from (a) all UK aquifers sampled and (b) all Polish aquifers sampled.

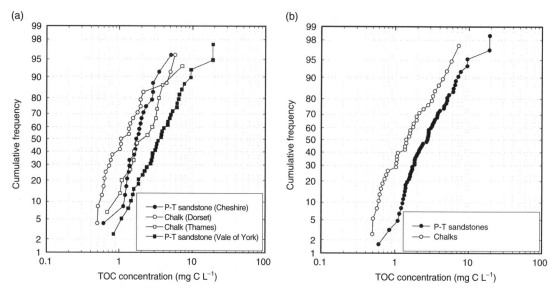

Fig. 3.5 Cumulative frequency plot for (a) TOC concentration in two Permo-Triassic sandstone and two Chalk aquifers from the UK and (b) showing data as split by aquifer type and region.

rock types, for example, the Bryozoan limestones (Ernstsen et al. 2005).

3.4.5 Comparing TOC with other parameters

Measurement of TOC can be problematic and in itself not necessarily a meaningful parameter accordingly, TOC values measured as part of the Danish National Monitoring Programme (Czakó 1994; Henriksen and Stockmarr 2000) have been compared with the COD in Fig. 3.6(a). A good positive correlation can be seen between the two parameters ($R^2 = 0.86$) with COD values approximately 2.7 times greater than TOC values. The value of 2.7 corresponds to the ratio of the atomic mass between oxygen ($O_2 = 32$) and Carbon (12). As the reduced carbon compounds are oxidised to CO_2 this indicate that organic carbon is quantitatively the only important element, which has been

oxidised in the investigated groundwaters. This will probably be the case for many natural groundwaters globally, although other reducing elements like, for example, Fe^{2+} will be quantitatively important in some natural groundwater types. COD often requires a considerably larger volume of water for determination than a TOC measurement so a simple factor difference makes for easy conversion between the two parameters.

TOC has also been compared with the amount of AOC in Fig. 3.6(b). The data are from this study and from a study performed by the Water Resources Division of Copenhagen Energy. No strong correlation is observed although generally, higher concentrations of TOC are concomitant with higher AOC. This is likely to relate to the structure of the TOC in the groundwater and would take considerably more time to characterise although this is a relationship

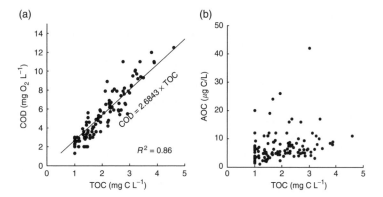

Fig. 3.6 Samples taken from the Danish National Monitoring Programme showing (a) the relationship between COD and TOC and (b) the relationship between AOC and TOC.

that should be investigated further in the future.

3.5 Discussion

3.5.1 Contaminated and natural environments

By understanding the range of baseline organic carbon concentrations in groundwater it is possible to use this parameter as an indicator of groundwater contamination at sites with high loads of organic carbon (e.g. Christensen et al. 2001; Gooddy et al. 2002) simply by comparing concentrations of DOC. In a review of landfill studies Christensen et al. (2001) have shown that there is decline in both the volatile and non-volatile fraction of the organic carbon concentration with increasing distance from mixed landfill sites. The concentrations found in a specific study conducted by Christensen et al. (2001), which are fairly typical at mixed landfills, are considerably above any of the maximum detected natural background concentrations found in our study for at least the first 50 m downgradient of the landfill, and they remain above

the overall median groundwater TOC concentration for more than 150 m. This indicates a considerable increase in the load of microbiological available carbon originating in the landfill, which can lead to the formation of distinct redox zones and mobilisation of, for example, natural or leachate derived heavy metals at the landfill.

A number of studies in the UK carried out by Gooddy et al. (2002) show the DOC concentrations beneath an unlined waste lagoons used for storing cattle manure. Concentrations up to 200 mg L^{-1} can be observed, nearly 100 times greater than a typical groundwater value observed in the reference aquifers. These lagoons also form distinct redox zones beneath them in a similar manner to the landfill example cited above.

Not all detections of a high concentration of DOC are, however, indicative of contamination. Some of the samples found in the reference aquifer from Denmark have organic carbon concentrations up to more than 100 mg L^{-1}. The high carbon concentration manifests itself in the form of a dark brown coloured liquid rich in naturally derived fulvic and humic acids, which occur in these particular aquifer sediments (Hinsby et al. 2001). Such occurrences are, however, extremely rare, the 95 percentile is below 10 mg L^{-1} TOC for all Danish aquifer types. Organic carbon rich waters may be found both in deep aquifer systems (>100 m below surface) like in the Danish cases (Grøn et al. 1996; Jørgensen et al. 1999; Hinsby et al. 2001) and in shallow soil, bog and aquifer sediments (e.g. Weis et al. 1989; Ernstsen et al. 2005). In case of doubts on the origin of the organic carbon further characterisation of the composition is needed (e.g. Weis et al. 1989; Grøn et al. 1996).

Where groundwaters contain DOC they are also susceptible to forming trihalomethanes, known carcinogens, during the chlorination processes. Stuart et al. (2001) demonstrated a relationship between the potential to form trihalomethanes following chlorination against the amount of DOC present in the water. The relationship was well correlated ($R^2 > 0.9$) and showed increasing formation of trihalomethanes with increasing concentrations of DOC. This is a concern for water managers and water authorities from both uncontaminated and contaminated groundwater with relatively high concentrations of organic carbon.

The use of TOC and DOC as an indicator of groundwater contamination also has its limitations. Johnson et al. (2001) examined the changes in concentration that occurred over a year of two pesticides in groundwater following their application. Peaks in concentration pesticide concentration were found to vary by nearly an order of magnitude. However, it was not possible to obtain this level of sensitivity (roughly 3 orders of magnitude lower) with organic carbon analysis. Consequently these concentrations, which exceeded the EC MAC, would go undetected if trying to observe contamination by DOC

analysis alone. Similarly, other micropollutants such as chlorinated solvents will generally not be identified as groundwater contaminants by organic carbon analysis.

3.6 Conclusions and Recommendations

The results show that the baseline values for TOC and DOC are comparable. The mean and median concentrations were around 2 mg C L^{-1}, while minimum and maximum values ranged from <0.5 mg C L^{-1} to >50 mg C L^{-1}. The AOC seem to correlate with TOC although the correlation is not clear and needs to be investigated further. Generally, there is a linear correlation between the TOC, DOC and the COD values. Very high natural organic carbon values were found locally in some pristine aquifers. Higher concentrations of DOC can lead to greater formation of trihalomethanes following chlorination. Organic contaminants derived from anthropogenic activities (e.g. pesticides and chlorinated solvents) are generally not identified in the TOC/DOC analysis as the contaminant concentrations are 3–4 orders of magnitude lower than the natural levels. In many situations organic carbon concentrations is not a good tracer or indicator of contamination, however it is a very important component in the biogeochemical cycling of elements and consequently it is recommended as a component to be measured on all groundwater samples. The measurement of a filtered organic carbon fraction (DOC) is to be recommended where groundwaters contain significant amounts of particulate material, otherwise, the more operational simple and cheaper

TOC measurement can be made. Measurements of the reactive part of the organic carbon readily available for microbiological processes provide valuable additional information and should be developed further.

References

ASTM. (1988) Standard D-4129-88: Test method for total organic carbon in water, pp. 85–90.

Baun, A., Jensen, S.D., Berg, P.L., Christensen, T.H. and Nyholm, N. (2000) Toxicity of organic chemical pollution in groundwater downgradient of a landfill (Grindsted, Denmark). *Environmental Science and Technology* 34, 1647–52.

Benedetti, M.F., Van Riemsdijk, W.H., Koopal, L.K., Kinniburgh, D.G., Gooddy, D.C. and Milne, C.J. (1996) Metal ion binding of natural organic matter; from the laboratory to the field. *Geochimica et Cosmochimica Acta* 60 (14), 2503–13.

Bottrell, S.H. (1996) Organic carbon concentration profiles in recent cave sediments: Records of agricultural pollution or diagenesis? *Environmental Pollution* 91 (3), 325–32.

Christensen, T.H., Kjeldsen, P., Bjerg, P.L. et al. (2001) Biogeochemistry of landfill leachate plumes. *Applied Geochemistry* 16, 659–718.

Czakó, T. 1994. Groundwater monitoring network in Denmark: example of results in the Nyborg Area. *Hydrological Sciences Journal* 39, 1–17.

Darling, W.G. and Gooddy, D.C. (2006) The hydrogeochemistry of methane: evidence from English groundwaters. *Chemical Geology* 229, 293–312.

Ernstsen, V., Larsen, C.L. and Tougaard, L. (2005) NVOC krav til drikkevand. Arbejdsrappport fra Miljøstyrelsen Nr. 18 2005, Report for the Danish EPA, Geological Survey of Denmark and Greenland (in Danish).

DEPA. (2002) Investigation of number of bacterial cells and regrowth potential in drinking water. Danish Environmental Protection Agency, Miljøstyrelsen, Report: Miljøprojekt 719. In Danish (English summary), on-line report at: http://www.mst.dk/udgiv/publikationer/2002/87-7972-246-6/html/indhold.htm.

Gooddy, D.C., Clay, J.W. and Bottrell, S.H. (2002) Redox-driven changes in pore-water chemistry of the Chalk unsaturated zone beneath unlined cattle slurry lagoons. *Applied Geochemistry* 17, 903–21.

Grøn, C. (1989) Organic halogens in Danish ground waters. PhD Thesis, Tehnical University of Denmark, 239 pp.

Grøn, C., Tørslev, J., Albrechtsen, H.-J. and Jensen, H.M. (1992) Biodegradability of dissolved organic carbon from an unconfined aquifer. *The Science of the Total Environment* 117/118, 241–51.

Grøn, C., Wassenaar, L. and Krog, M. (1996) Origin and structures of groundwater humic substances from three Danish aquifers. *Environment International* 22, 519–34.

Hansen, L.K., Jakobsen, R. and Postma, D. (2001) Methanogenesis in a shallow sandy aquifer, Romo, Denmark. *Geochimica et Cosmochimica Acta* 65, 2925–35.

Heyes, A. and Moore, T.R. (1992) The influence of dissolved organic-carbon and anaerobic conditions of mineral weathering. *Soil Science* 154, 226–36.

Henriksen, H.J. and Stockmarr, J. (2000) Groundwater resources in Denmark – modelling and monitoring. *Water Supply* 18, 550–7.

Hinsby, K., Harrrar, W.G., Nyegaard, P. et al. (2001) The Ribe Formation in western Denmark: Holocene and Pleistocene groundwaters in a coastal Miocene sand aquifer. In: Edmunds and Milne (eds) *Palaeowaters in Coastal Europe: Evolution of Groundwater Since the Late Pleistocene.* Geological Society, London, Special Publication, no. 189.

Jakobsen, R. and Postma, D. (1999) Redox zoning, rates of sulfate reduction and interactions with

Fe-reduction and methanogenesis in a shallow sandy aquifer, Romo, Denmark. *Geochimica et Cosmochimica Acta* 63, 137–51.

Johnson, A.C., Besien, T.J., Bhardwaj, C.L. et al. (2001) Penetration of herbicides to groundwater in an unconfined chalk aquifer following normal soil applications. *Journal of Contaminant Hydrology* 53 (1–2), 101–17.

Jørgensen, C. (2000) Removal of AOC and NVOC during artificial recharge – Investigation at the Arrenæs site. Final report of EU project ENV4-CT95-0071.

Jørgensen, N.O., Morthorst, J. and Holm, P.M. (1999) Strontium-isotope studies of "brown water" (organic-rich groundwater) from Denmark. *Hydrogeology Journal* 7, 533–9.

Kooij, D. (1992) Assimilable organic carbon as an indicator of bacterial regrowth. *Journal of American Water Works Association* 84 (2), 57–65.

Kooij, D., Visser, K.A. and Hijnen, W.A.M. (1982) Determining the concentration of easily assimilable organic carbon in drinking water. *Journal of American Water Works Association,* 74(10), 540–5.

Orem, W.H., Feder, G.L. and Finkelman, R.B. (1999) A possible link between Balkan endemic nephropathy and the leaching of toxic organic compounds from Pliocene lignite by groundwater; preliminary investigation. *International Journal of Coal Geology* 40, 237–52.

Shand, P., Haria, A.H., Neal, C. et al. (2005) Hydrochemical heterogeneity in an upland catchment: further characterisation of the spatial, temporal and depth variations in soils, streams and groundwaters of the Plynlimon forested catchment, Wales. *Hydrology and Earth System Sciences* 9 (6), 611–34.

Stuart, M.E., Gooddy, D.C., Kinniburgh, D.G. and Klinck, B.A. (2001) Trihalomethane formation potential: a tool for detecting non-specific organic groundwater contamination. *Urban Water* 3 (3), 197–208.

Thurman, E.M. (1985) *Organic Geochemistry of Natural Waters.* Martinus Nijhoff/Dr W. Junk Publishers, Dordrecht, 497 pp.

Weis, M., Abbt-Braun, G. and Frimmel, F.H. (1989) Humic-like substances from landfill leachates-characterization and comparison with terrestrial and aquatic humic substances. *Science of the Total Environment* 81/82, 343–52.

4 Geochemical Modelling of Processes Controlling Baseline Compositions of Groundwater

D. POSTMA, C. KJØLLER, M. SØGAARD ANDERSEN,
M. T. CONDESSO DE MELO AND I. GAUS

Reactive transport models were developed to explore the evolution in groundwater chemistry along the flow path in three aquifers; the Triassic East Midlands aquifer (UK), the Miocene aquifer at Valréas (France) and the Cretaceous aquifer near Aveiro (Portugal). All three aquifers contain very old groundwaters and variations in water chemistry that are caused by large-scale geochemical processes taking place at the timescale of thousands of years. The most important geochemical processes are ion exchange (Valréas and Aveiro) where freshwater solutes are displacing marine ions from the sediment surface, and carbonate dissolution (East Midlands, Valréas and Aveiro). Reactive transport models, employing the modelling code PHREEQC (Parkhurst and Appelo 1999; Appelo and Postma 2005), which included these geochemical processes and one-dimensional solute transport were able to reproduce the observed patterns in water quality. These models may provide a quantitative understanding of the evolution in natural baseline properties in groundwater.

4.1 Introduction

Geochemical modelling can be used to help predict the baseline concentration levels for various dissolved inorganic substances in aquifers over space and time. A geochemical model quantifies the main geochemical processes that occur in the aquifer and may relate these to solute transport processes. A model contains assumptions concerning the original conditions in the aquifer with respect to water and sediment composition and a set of constraints defining the geochemical processes that may occur. If one defines the composition of the water recharging the aquifer and the transport parameters, then predictions for the development of water composition over space and time can be made. The model can be tested against observed water and sediment composition data. Good agreement between model and observations may validate the model, although it never proves what has actually happened. Once the model is considered credible it may also become a tool for predicting how the groundwater chemistry will develop in the future. Predictions will be most reliable in an undisturbed system but, with some caution, they can also be used to predict the effect on the water quality of changing flow conditions such as abstraction of groundwater from the aquifer. Likewise, a geochemical model can also be used in attempting to predict baseline concentration trends and analysing their susceptibility to change.

The modelling code used in this chapter is PHREEQC (Parkhurst and Appelo 1999; Appelo and Postma 2005). This programme combines a geochemical model, handling aqueous speciation, mineral equilibria, ion exchange, adsorption and reaction kinetics, with a one-dimensional transport model, specifying flow, diffusion and dispersion. The application of a reactive transport model requires a set of field data showing an evolution in water composition either in space or in time. Given the long residence time of groundwater in aquifers, the dataset will typically consist of the evolution in water chemistry along a flow line through the aquifer. A meaningful geochemical model

application requires that the evolution in water chemistry is determined by geochemical processes rather than being merely historical in origin. This chapter presents geochemical models for three European aquifers: the Triassic East Midlands aquifer (UK), the Miocene aquifer at Valréas (F) and the Cretaceous aquifer near Aveiro (P) to explore processes controlling the natural baseline properties of groundwater and their evolution.

4.2   Triassic East Midlands Sandstone Aquifer, UK

4.2.1   Geology and hydrogeology

The East Midlands Triassic sandstone aquifer is located in Nottinghamshire in the east of England. The aquifer is contained in the Sherwood Sandstone Group and has a thickness ranging between 120 and 300 m. It is dipping at about 1° towards the east (Fig. 4.1). The Sherwood Sandstone is underlain by a sequence of Permian mudstones,

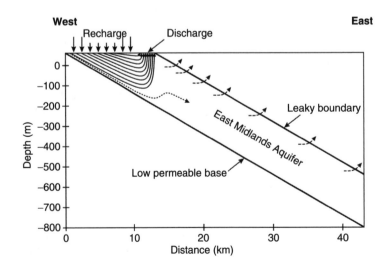

Fig. 4.1   Vertical transect of the flow system in the Triassic East Midlands aquifer, prior to groundwater abstraction.

marls and limestones with a low permeability, which form the base of the aquifer. At its western end, the aquifer is unconfined and the sandstone outcrops in places where it is not covered by glacial drift. Towards the east, the Sherwood Sandstone is overlain first by the gypsum bearing Colwick Formation and further eastwards by the Mercia Mudstone Group, which forms a confining unit. The groundwater geochemistry of the East Midlands aquifer has previously been studied by Edmunds et al. (1982), Edmunds and Smedley (2000) and Smedley and Edmunds (2002).

The Sherwood Sandstone is dominated by quartz with minor amounts of feldspars. Alteration products of the feldspars such as illite and kaolinite are present at the top of the sandstone (Moss and Edmunds 1992). Near the surface, the Sherwood Sandstone has been decalcified. However, in the deeper parts, 1–4% (w/w) of the carbonate minerals calcite and dolomite are present (Edmunds et al. 1982). The eastern and deeper part of the aquifer also contains gypsum, especially towards the confining unit. The aquifer sediments contain little organic carbon or pyrite.

Modelling of the groundwater flow prior to groundwater abstraction suggests that only 10% of the recharged precipitation entered the confined part of the aquifer. The remaining 90% discharged to springs and rivers at the boundary between the unconfined and the confined part of the aquifer (Fig. 4.1). Therefore groundwater flow is fast in the unconfined and slow in the confined section of the aquifer. Because of the slow flow velocity, the confined part of the aquifer contains very old groundwater of more than 10,000 years in age. In the confined part of the aquifer, some groundwater discharge is considered to occur through faults or through the overlying Mercia Mudstone.

4.2.2 Hydrochemistry and modelling

Fig. 4.2 displays the water chemistry in the East Midlands aquifer based on the 1991–92 data-set reported by Edmunds and Smedley (2000). The groundwater residence time is based on ^{14}C dating, and the correlation of the ^{14}C age with the groundwater temperature (Edmunds and Smedley 2000). Waters that contain nitrate, derived from agriculture, must have recharged during the last 50 years. These waters also contain enhanced concentrations of sulphate, calcium, magnesium and bicarbonate, resulting from soil treatment and industrial activity. Based on the nitrate content, the groundwaters in the East Midlands aquifer are accordingly subdivided into recent and old waters. Apparently, some mixing between recent and older waters must occur around the boreholes, because ^{14}C ages suggest that some of these anthropogenically influenced water samples are up to 5000 years old (Fig. 4.2[a]).

The groundwaters around the town of Gainsborough, indicated by triangles in Fig. 4.2, are somewhat off the main groundwaters flow path. They are characterised by high sulphate, calcium and magnesium concentrations and will be discussed separately.

The groundwaters along the main flow path from Worksop to Lincoln show increases in calcium, bicarbonate and $P(CO)_2$ and a decrease in pH along the flow lines (Fig. 4.2). In addition, these waters are close to saturation with calcite $(CaCO_3)$ (Fig. 4.2[g]) and dolomite $(CaMg[CO_3]_2)$ (data not shown). This behaviour could correspond to a gradual depletion of the small amount of calcite and dolomite, initially present in the

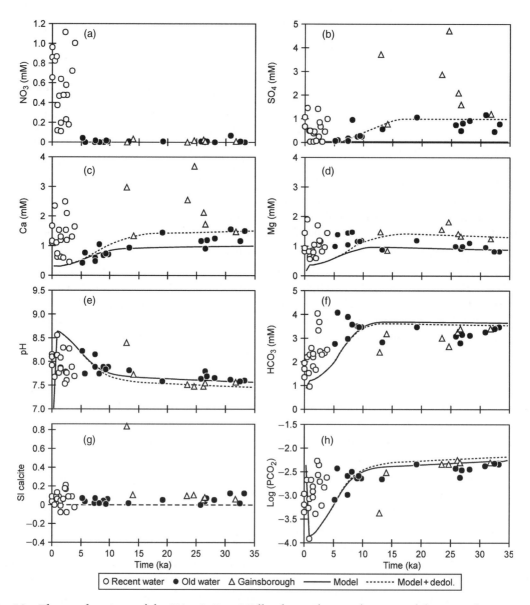

Fig. 4.2 The geochemistry of the Triassic East Midlands aquifer, as a function of the groundwater residence time expressed in thousands of years (ka). All field data (symbols) are from Edmunds and Smedley (2000). Circle symbols indicate data situated approximately along a flow path from Worksop to Lincoln. The subdivision between recent (open circles) and old water (filled circles) is based on the presence or absence of nitrate. Triangle symbols are for data from wells around Gainsborough. The solid line represents a PHREEQC model with carbonate (calcite and dolomite) dissolution where the up-flow part of the system becomes depleted in carbonate minerals and where dissolution conditions change from being open to closed with respect to CO_2. The dotted line is for a model that in addition dissolves 1 mM gypsum along the flow line.

Sherwood Sandstone. In support of this, Moss and Edmunds (1992) found that near the surface in the unconfined part of the aquifer the Sherwood Sandstone has been decalcified. In this scenario, carbonate dissolution will begin in the unsaturated zone in the presence of soil CO_2 (the system is open with respect to CO_2), resulting in a high $P(CO)_2$, high calcium and bicarbonate concentrations and a low pH. As the rock above the groundwater table becomes depleted in carbonate minerals, the system becomes closed with respect to CO_2. Carbonate mineral dissolution below the water table will result in a lower $P(CO)_2$, lower calcium and bicarbonate concentration and a higher pH. The details of such a reaction scheme have been outlined by Langmuir (1971) and Appelo and Postma (2005).

These processes have been modelled with PHREEQC using a model column of 30 km, subdivided into 60 cells, and an average groundwater flow velocity of 0.86 m yr^{-1}. This gives a residence time of up to 35,000 years in the aquifer. Longitudinal dispersivity was taken as 600 m or 2% of the total flow path. The groundwater temperature increases with residence time (depth) (Edmunds and Smedley 2000). In the model, the temperature increases linearly from 10.6°C in the first cell to 19.6°C in the last cell. Initially, all cells contain trace amounts of calcite and dolomite and the water is in equilibrium with these minerals. A soil $P(CO)_2$ of $10^{-2.35}$ is imposed on the first cell and the water is then transported through the model column where the carbonate minerals will dissolve if present. The results of this reactive transport model are shown in Fig. 4.2 as the solid lines. Initially calcite and dolomite dissolution occurs close to the source of CO_2 and lead to high concentrations of calcium and bicarbonate

accompanied by a low pH. As the first cells become depleted in carbonate minerals, dissolved CO_2 will be transported downstream until it reaches a cell containing carbonate minerals. Since the source of CO_2 is now limited to the amount dissolved in the water, the resulting concentrations of calcium and bicarbonate become lower and the pH higher. Comparison of model results with field data (Fig. 4.2) along the Worksop–Lincoln flow path (filled circle symbols) show good agreement in the overall trends even though the field data contains some scatter. The increase in temperature along the flow path affects the solubilities of dolomite and calcite differently. The solubility of dolomite decreases more rapidly with increasing temperature than for calcite and causes some calcite to recrystallise into dolomite. The result is (from 10 ka onward) a slight decrease in pH, magnesium and bicarbonate concentration and an increase in calcium concentration, trends that might be present in the field data as well. Overall, the model illustrates how carbonate mineral equilibria in combination with an advancing mineral dissolution front may affect the baseline concentrations on a timescale of thousands of years.

The groundwaters near Gainsborough (Fig. 4.2) have distinctly higher calcium and sulphate concentrations which have been attributed to the dissolution of gypsum ($CaSO_4 \cdot 2H_2O$), a mineral that is known to be present in the confining unit. In an aquifer containing both dolomite and calcite, the dissolution of gypsum will cause the dissolution of dolomite and the precipitation of calcite. The process is termed de-dolomitisation (Plummer et al. 1990; Appelo and Postma 2005). By dissolution of gypsum, the calcium concentration increases, causing calcite to precipitate; this again lowers the aqueous carbonate concentration and causes dolomite to dissolve. The overall mass transfer (at 17°C and $SI_{calcite}$ = 0.1) is described by the equation:

$$1.6\ CaSO_4 + 0.6\ CaMg(CO_3)_2 \rightarrow$$
$$1.2\ CaCO_3 + Ca^{2+} + 0.6\ Mg^{2+} + 1.6\ SO_4^{2-}$$

For the East Midlands aquifer, de-dolomitisation has already been suggested by Edmunds et al. (1982) as an important geochemical process in the aquifer. The result is an increase in the groundwater sulphate concentration which is coupled to increases in magnesium and calcium concentration.

As shown in Fig. 4.2, there is no simple increase in the sulphate content of the groundwater near Gainsborough with increasing groundwater residence time. Probably, the de-dolomitisation process has

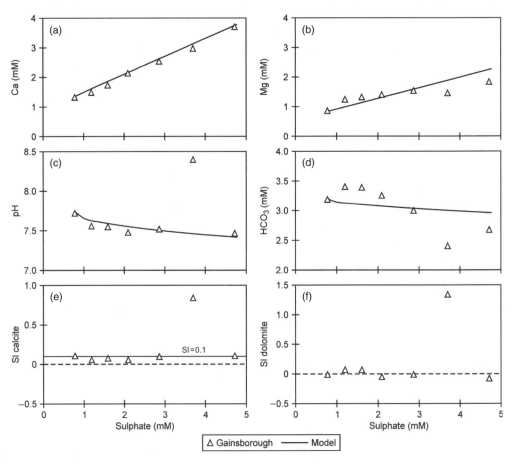

Fig. 4.3 Groundwater composition of the wells around Gainsborough plotted as a function of their sulphate concentration. Lines are for a PHREEQC model simulating de-dolomitisation, where gypsum dissolution ($CaSO_4 \cdot 2H_2O$) causes dissolution of dolomite [$CaMg(CO_3)_2$] and precipitation of calcite ($CaCO_3$).

been affected by groundwater abstraction, introducing water from gypsum bearing layers into the Sherwood Sandstone. Instead, the de-dolomitisation reaction is traced by ordering the Gainsborough groundwater samples according to their sulphate content (Fig. 4.3). De-dolomitisation is then modelled with PHREEQC by successively adding gypsum to a groundwater in equilibrium with calcite and dolomite. The model predicted lines are shown in Fig. 4.3, and are in good agreement with the field data trends, suggesting de-dolomitisation to be the main cause for the variations in water chemistry.

The increase in the groundwater sulphate concentration along the flow path going from Worksop to Lincoln, is a modest 1 mM (Fig. 4.2). De-dolomitisation is included in the PHREEQC carbonate dissolution model by linearly adding 1 mM of gypsum between 6.3 and 14.3 km (corresponding to 7.3–16.6 ka in Fig. 4.2). The model result is included in Fig. 4.2 as the dotted line. Generally, the effect of de-dolomitisation on the model results remains quite small although increases in the concentrations of calcium, magnesium and sulphate can be noted.

4.3 Miocene Valréas aquifer, France

4.3.1 *Geology and hydrogeology*

The Miocene Valréas aquifer is located in the southeastern part of France, just east of the Rhône Valley. The basin in which the aquifer sediments have been deposited is surrounded by mountains towards the west, north and east, while towards south the Valréas basin extends into the Carpentras basin. The size of the Valréas basin is approximately 10 by 20 km.

The aquifer is located in a Miocene sandstone formation, deposited as outwash material from the surrounding mountains (Huneau 2000). At the margins of the Valréas basin, the sandstone deposits are approximately 400 m in thickness, increasing to up to 600 m in the central part of the basin (Fig. 4.4). The sandstone is cemented by carbonate minerals and X-ray diffraction analyses carried out on sediment samples from a deep borehole near Richerenches in the central part of the basin indicates the presence of both aragonite and calcite in the upper approximately 100 m of the sandstone. Deeper in the aquifer, the carbonates mainly

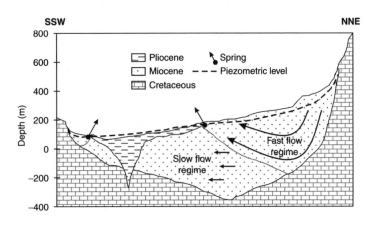

Fig. 4.4 Conceptual model showing the palaeo-flow in the Miocene Valréas aquifer (Huneau 2000). The approximate length of the cross section is 30 km.

consist of calcite. The sediment samples contain up to 30% (w/w) carbonates and have a cation-exchange capacity (CEC) between 40 and 60 meq kg^{-1}. Due to cementation with carbonate, the porosity of the sandstone is only 5–10%, and the hydraulic conductivity ranges between 1×10^{-7} and 1×10^{-6} m s^{-1} (Huneau 2000). The base of the aquifer is formed by the Cretaceous bedrock (Fig. 4.4). In the southern part of the Valréas basin, the Miocene sandstone is covered by a sequence of Pliocene clay and marl up to 200 m thick (Fig. 4.4). These deposits are presumably of marine origin and were deposited during a marine transgression into the Rhône Valley and its tributaries (Huneau 2000). The hydraulic properties of the Pliocene deposits are not well known, but considering their lithology, the hydraulic conductivity is probably less than 1×10^{-9} m s^{-1}.

The groundwater flow direction in the Miocene aquifer is from northeast towards southwest, and due to the presence of the Pliocene cover in the southern part of the Valréas basin, recharge to the aquifer takes place only in the northern part of the basin. The Pliocene cover thus separates the aquifer into an unconfined part towards north and a confined part towards the south (Fig. 4.4). The palaeo-flow conditions in the aquifer were presumably controlled by the presence of the Pliocene cover, causing a more rapid flow in the unconfined part of the aquifer as compared with the confined part of the aquifer. Previously, several springs were present at the boundary between the confined and unconfined parts of the aquifer, (Fig. 4.4; Huneau 2000).

Flow velocities calculated from ^{14}C measurements are 0.2 and 2.5 m yr^{-1} for the confined and unconfined part of the aquifer,

respectively (Huneau 2000). This suggests that under pristine conditions approximately 90% of the recharge to the unconfined part of the aquifer was discharged through these springs, leaving about 10% of the recharge to flow in the confined part of the aquifer. Since the 1950s, extensive abstraction of groundwater has been carried out from the confined part of the aquifer for irrigation purposes. Due to this, the springs at the confined/ unconfined boundary have dried up and now groundwater flow takes place at a more uniform velocity of around 2 m yr^{-1} throughout the entire aquifer. The hydrogeology of the Valréas aquifer is discussed in more detail in Chapter 13.

4.3.2 Hydrochemistry and modelling

The Valréas aquifer displays a Ca–HCO$_3$ type of water at its up-flow end, changing to groundwater enriched in Mg–HCO$_3$ further down the flow path, which again is replaced by a Na–HCO$_3$ type water in the downflow part of the aquifer (Fig. 4.5). There are also increases in pH and alkalinity along the flow path. The observed pattern is typical for freshening of an aquifer that originally has been filled with saline water (e.g. Chapelle and Knobel 1983; Appelo 1994; Walraevens and Cardenal 1994; Andersen et al. 2005). The salt water had been flushed from the aquifer a long time ago and present day chloride concentrations are less than 50 mg L^{-1}. However, the sediment ion exchanger still contains the saline cations Na$^+$, K$^+$ and Mg^{2+}, and these are at present being displaced by Ca^{2+} in the recharging water. The marine influence probably dates back to the time of deposition of the Pliocene cover in the southern part of the Valréas basin, and the

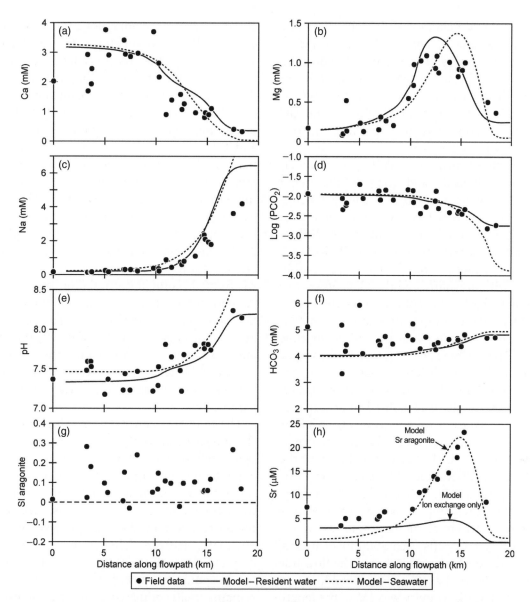

Fig. 4.5 Modelled hydrochemistry along a flow path in the Valréas aquifer together with field data from Huneau (2000). (a)–(f) show a model where the initial solution was assumed to be the solution now present at the end of the flow line (full line), and a model with seawater as the initial solution and $K_{Na\backslash Mg} = 0.6$ (dotted line). (h) shows a model where Sr is derived from the seawater (full line) and a model where Sr comes from the dissolution of aragonite containing 2 mol% Sr (dotted line).

observed hydrochemistry along the flow line in the aquifer is apparently the result of millions of years of geochemical evolution. The groundwater is close to equilibrium with the mineral aragonite ($CaCO_3$) and the downflow increases in alkalinity and pH suggest ongoing carbonate dissolution along the flow path (Fig. 4.5).

Reactive transport modelling was undertaken to identify and quantify the major controlling hydrogeological and geochemical processes that have been responsible for the present day hydrochemistry. A 20-km-long flow path was modelled with PHREEQC. The groundwater was brought into equilibrium with aragonite and the CEC was set to 45 meq kg^{-1}. The column was initialised with a water composition similar to that resident at the downstream end of the flow line. The flushing solution was water with a chemical composition as observed in the area of recharge. The transport of groundwater in the model was defined in accordance with the palaeo-flow conditions, with a flow velocity of 0.2 m yr^{-1} in the confined part and 2 m yr^{-1} in the unconfined part of the aquifer. In PHREEQC this was achieved by specifying variable cell lengths, whilst keeping the time step constant, so that the upstream 10 km of the flow line was represented by four cells each 2500 m in length, while the downstream 10 km of the flow line was modelled by 40 cells each 250 m in length. The dispersivity was 1% of the length of the flow path. The reactive transport model was allowed to run until 10 pore volumes of the model column were replaced by the flushing solution, corresponding to 0.5–0.6 million years.

The model results are given as the full lines, together with the field data in Fig. 4.5. Generally, the model is able to predict the

changes in water quality along the flow path quite well. Approximately 10 km downstream the flow line, there is an ion-exchange front where Ca^{2+} from the flushing solution displaces Mg^{2+} from the ion-exchange complex. Further downstream, near 16 km, there is another ion-exchange front where Mg^{2+} and Ca^{2+} displace K^+ and Na^+. This chromatographic pattern of different water qualities arises from the differences in binding strength of cations to the sediment exchanger in combination with their differing concentrations (Appelo and Postma 2005). Each ion-exchange front has its own characteristic velocity and, as time passes, the fronts become more and more separated. At both ion-exchange fronts, the uptake of Ca^{2+} on the ion-exchange complex causes aragonite to dissolve and as a result, the pH and alkalinity increase while the $P(CO)_2$ decreases. There is also good agreement between model and field data for these parameters.

The field data could only be simulated by the model when the palaeo-flow conditions were used, with the flow rate in the confined zone being ten times lower than in the unconfined zone. When the present day uniform flow velocity of 2 m yr^{-1} is used, the model fails to reproduce the observed water chemistry pattern. The amount of water extracted from the confined part of the aquifer by pumping during the last 50 years remains small compared with the total water content so that the water chemistry patterns still reflect the palaeo-flow conditions rather than the present day flow conditions. This is supported by ^{14}C dating, which indicates the presence of very old waters in the confined part of the aquifer (Huneau 2000).

The model used the present day water composition in the downstream end of the aquifer as the water composition initially present in the whole aquifer. However, this water is apparently already influenced by ion exchange as is indicated by its Na–HCO$_3$ type of composition. If our hypothesis of a freshening aquifer is correct, then the water originally present in the aquifer must have been seawater. To test this, a new model was set up with seawater as the initial solution in the flow tube. The results after modelling 40 pore volumes of flushing, corresponding to 2–3 million years, are included in Fig. 4.5(a)–(f) as the dotted lines and are generally also in good agreement with the field data. The critical point seems to be the behaviour of magnesium. If the default exchange coefficient of $K_{Na\backslash Mg}$ = 0.5 is used, the model tends to produce a sharper peak than observed in the field data (not shown in Fig. 4.5). However, exchange coefficients are not constants in a strict sense since they depend on the sediment characteristics. Appelo and Postma (2005) list a range of $K_{Na\backslash Mg}$ = 0.4–0.6 for the exchange of Na$^+$ for Mg^{2+}. If instead of the median value of 0.5 a value of 0.6 is used, indicating weaker binding of Mg^{2+} to the sediment exchanger, then the result shown by the dotted line in Fig. 4.5(b) is obtained. An almost similar result could have been obtained if the water originally in the aquifer was more saline than 35‰ seawater. For example, twice concentrated seawater would also decrease the binding of Mg^{2+} to the surface. This follows from the general relationship between the concentration of mono- and divalent ions on the ion exchanger and in the aqueous phase (e.g. Appelo 1996):

$$\frac{[j^{2+}]^{1/2}}{[i^+]} = \frac{K_{i\backslash j}\, \beta_j^{1/2}}{\beta_i}$$

where [ ] denotes aqueous concentrations, β denotes equivalent fractions on the exchanger and $K_{i\backslash j}$ denotes the selectivity coefficient. In this equation, a proportional increase in the concentration of both mono- and divalent ions in the aqueous phase, corresponding to evaporation, causes a decrease in the concentration of divalent cations relative to monovalent ions on the ion-exchange complex.

The modelled distribution shown with the dotted line in Fig. 4.5 is the result of 2–3 million years of geochemical evolution of the aquifer. This timescale corresponds approximately to the transition between Miocene and Pliocene. Perhaps more saline water was initially present in the aquifer in relation to the 'evaporite period' of the Late Miocene (Messinian) (e.g. Schreiber et al. 1976; Troelstra et al. 1980).

The field data also show a distinct maximum for strontium at 15 km down the flow line (Fig. 4.5[h]). Strontium is preferentially accumulated in aragonite, as compared with calcite, due to the large ionic radius (1.18 Å) of Sr^{2+} (Morse and Mackenzie 1990). The dissolution of aragonite, in relation to the ion-exchange fronts, could therefore be a source of Sr^{2+} to the groundwater (Plummer et al. 1976). However, strontium is also present in trace concentrations in seawater and as such, the increase in concentration along a flow line may also be a result of the freshening of a previously saline aquifer. In order to clarify the origin of Sr^{2+} in the Valréas aquifer, two geochemical models were set up. One model in which Sr^{2+} was included in the seawater composition according to the seawater concentration suggested by Goldberg (1963), and another model, in which the aragonite present in the aquifer sediments contained 2 mol% of Sr. This is only slightly higher than the average

Sr content of 1 mol% found in modern arago-
nitic organisms (Morse and Mackenzie 1990).
The results of modelling 2.2 million years of
infiltration in the aquifer with these two
models are also shown in Fig. 4.5. The contribution of Sr^{2+} from the seawater appears
to be negligible, thereby suggesting that dissolution of aragonite in combination with
ion exchange are the major processes controlling the distribution of Sr^{2+} in the aquifer.

4.4 Cretaceous Aveiro aquifer, Portugal

4.4.1 Geology and hydrogeology

The Cretaceous Aveiro aquifer system covers over 1800 km^2 of the northwest part of
the Portuguese mainland and adjacent continental shelf. The aquifer is part of a thick
sequence of mainly siliciclastic sediments
deposited in the northernmost part of the
Lusitanian basin. Sedimentation took place
in fluvial, deltaic or shallow marine environments. The Cretaceous sandstone units constitute the main aquifer which is confined
over two thirds of its extent by a low permeability marly clay formation of Upper
Cretaceous age. The confining unit limits
modern recharge to the aquifer (Fig. 4.6). The
average thickness of the confining unit
increases towards the coast and may be over
150 m thick above the deepest parts of the
aquifer. Uplift and erosion of the confining
marly clay unit in the east allowed the sandstone formation to outcrop and receive direct
recharge from rainfall infiltration. Calculated
apparent flow velocities for the aquifer,
based on ^{14}C dating, indicate a gradual
decrease of flow velocities along the flow
path. Flow velocities decrease from a few
metres per year in the areas closest to the
recharge area to less than half a metre per
year downstream in the confined part of the
aquifer. The hydrogeology of the Aveiro
Cretaceous aquifer is described in more
detail in Chapter 11.

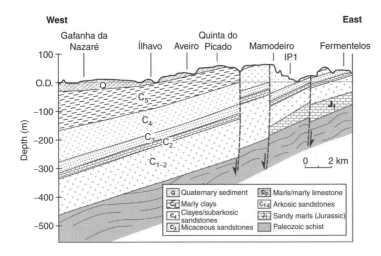

Fig. 4.6 The Cretaceous Aveiro
multilayer aquifer in cross
section. The principal aquifer
layers are the top of unit C_1, units
C_2 and C_3, the bottom of unit C_4.

4.4.2 Hydrochemistry and modelling

The hydrochemistry of the aquifer can be divided into three main facies (Condesso de Melo 2002). In the area of recharge, the infiltrating rainwater percolates through the unsaturated zone to reach the underlying groundwater body, reacting with carbonate minerals in a system open to CO_2. The resulting groundwater is dominated by Ca^{2+} and HCO_3^- with pH values less than 7.0 and water temperatures less than 20°C. Down the flow path, the aquifer gradually becomes confined and contains very limited modern recharge. Holocene to pre-industrial groundwater occurs in the intermediate, confined part of the aquifer. Here redox potential and dissolved oxygen progressively decrease along the flow path with groundwaters changing from oxidising to reducing conditions. These are groundwaters with either a Ca–HCO_3 or Na–HCO_3 hydrochemical facies and pH values greater than 6.2 but less than 8.3 and groundwater temperatures around 21°C (Fig. 4.7). The $P(CO)_2$ decreases in this part of the aquifer from 0.04 to 0.001 down the flow path. Finally, Late Pleistocene-Early Holocene groundwaters have been identified in the deeper and confined part of the aquifer. They are Na–HCO_3 or Na–Cl type waters, with pH in the range 7.1–8.9, and average groundwater temperatures higher than 22°C.

Although the isotopic and chloride data show that the aquifer has been fresh for more than 18 ka, the freshening chemical pattern of the aquifer suggests that the aquifer has previously contained seawater. This seawater has been almost completely flushed from the aquifer and occurs now only as traces of formation water in the low permeability parts of the aquifer. A conceptual model was derived for the geochemical evolution of the water in the Aveiro Cretaceous aquifer (Fig. 4.7). The model assumes that calcium bicarbonate water, which originated from rainwater infiltration and dissolution of carbonates by soil CO_2, recharges the aquifer. This water moves along the flow path and because of ion exchange eventually evolves to a Na–HCO_3 type water in the areas close to the coast. In the downstream end of the aquifer, cation-exchange reactions occur and calcium and magnesium are adsorbed on the sediment while sodium is released into solution. Measurements of the exchanger composition on clay samples are consistent with this cation-exchange reaction (Oliveira 1997).

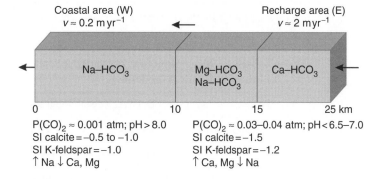

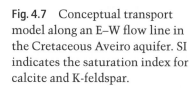

Fig. 4.7 Conceptual transport model along an E–W flow line in the Cretaceous Aveiro aquifer. SI indicates the saturation index for calcite and K-feldspar.

The loss of calcium and magnesium from solution, owing to the ion-exchange reaction, causes some dissolution of carbonate minerals in order to maintain equilibrium.

Based on the palaeohydrology, the aquifer was assumed to be initially filled with seawater. The measured composition of a groundwater sample from well #154 in the aquifer outcrop area was chosen as the flushing water composition. This sample was considered to be representative of the local recharge and is substantially undersaturated with dolomite and calcite. Its chloride content is about 20 ppm and the sample chemical composition corresponds approximately to five-fold evaporated rainwater. The hydrochemical data of the two water samples used in the geochemical modelling are summarised in Table 4.1.

The potential reactants and products in the model were selected based on petrographic and mineralogical information, the

measured saturation indices and the previous results of inverse mass-balance modelling for the aquifer (Condesso de Melo et al. 2001). The phases identified as major reactants include carbon dioxide, calcite, potassium feldspar and an ion-exchange complex. The evaluation of petrographic, mineralogical and saturation indices data also helped to determine which of these mineral phases were dissolving or precipitating within the aquifer system. According to Parkhurst et al. (1996), the reaction-path models need to be consistent with these determinations, or otherwise, they cannot be considered plausible. The phases identified as reactants and the dissolution constraints used in the modelling are summarised in Table 4.2.

The exchangeable cations and the CEC of the aquifer sediments were determined in four samples of the principal aquifer layer by ion displacement using both a 1 M NaCl solution and a 1 M NH$_4$Cl solution (Appelo

Table 4.1 Flushing and initial water compositions used for geochemical modelling of the Aveiro aquifer.

Well Ref.	pH	T (°C)	Na (mM)	K (mM)	Ca (mM)	Mg (mM)	Si (mM)	Al (mM)	HCO$_3$ (mM)	Cl (mM)	SO$_4$ (mM)	δ^{13}C (‰)
154	5.34	18.5	0.65	0.05	0.12	0.18	0.15	$1.0\,10^{-3}$	0.12	0.58	0.13	−17.99
Seawater	8.22	16	468.4	10.21	10.29	53.14	0.15	$3.7\,10^{-5}$	2.32	546	28.23	1.5

The hydrochemical composition for the seawater is from Goldberg (1963).

Table 4.2 Mineral phases, exchange species and constraints used in the geochemical modelling.

Phases	CO$_2$ (+), calcite (+), K-feldspar (+)
Exchange species	CaX$_2$, MgX$_2$, NaX
Balance	Na, K, Ca, Mg, Al, Si, SO$_4$, Cl, TIC, Alk

(+) = enters solution/dissolves; TIC = total inorganic carbon; Alk = alkalinity.

and Postma 2005). The results ranged from 2.5 to 5.6 meq kg^{-1} sediment. The CEC of the aquifer sediments was recalculated as CEC per litre of contacting groundwater by assuming a total aquifer porosity of 0.20 and a sediment bulk density of 1.8 kg L^{-1}, yielding 0.020 and 0.050 mol L^{-1}.

A 25-km-long flow tube was modelled using two different discretisation approaches: (1) dividing the column into 25 cells of 1000 m length, corresponding to an average flow velocity of 1 m yr^{-1}; and (2) dividing the upstream 10 km of the flow path into 10 cells of 1000 m in length while the downstream 15 km were divided into 150 cells of 100 m length. This second model simulated the varying flow conditions within the aquifer system, with flow velocities in the confined part of the aquifer about ten times lower than the flow velocities in the aquifer outcrop area. A dispersivity of 4% of the total flow length was used in both models, and was determined by the chloride distribution in the aquifer. All cells in the column were assigned a CEC (either 0.020 or 0.050 mol L^{-1}) and an SI of -1.0 for both calcite and K-feldspar. The flushing solution was allowed to equilibrate initially with CO_2, calcite (SI = -1.5) and K-feldspar (SI = -1.2).

The best model fit for the field results was obtained by trial and error, using different combinations of parameters to simulate the field data. The model with the variable discretisation completely failed to describe the observed hydrogeochemistry, since it could neither reproduce the evolution of ion-exchange fronts nor the pH or the bicarbonate concentration in the aquifer. The best model results are shown in Fig. 4.8 and are for the uniform cell model using a CEC of 0.020 mol L^{-1}. The model reproduces the ion-exchange fronts with calcium from the flushing solution initially displacing magnesium from the ion-exchange complex, and further downstream, both calcium and magnesium displacing sodium. Generally, the model describes the observed hydrochemical evolution satisfactorily confirming that the dominant reactions in the aquifer are calcite and K-feldspar dissolution, and cation exchange. Varying the P(CO)$_2$ in the initial solution influenced the estimated calcium, pH and bicarbonate distribution in the aquifer. When the saturation index of calcite was increased in the initial solution, the P(CO)$_2$ decreases in the column since more calcite is dissolved in the first cell of the column. The pH also increased due to the lower P(CO)$_2$.

According to the model, the present day hydrogeochemical pattern is the result of about 50 ka of geochemical evolution in the aquifer, which means that the aquifer has been fresh since the Late Pleistocene. This is in agreement with ^{14}C and chloride data for the aquifer that revealed fresh water for at least the last 35 ka, and also with the geological data. The correspondence in timescales furthermore indicates that the palaeohydrology is still dominating the aquifer geochemistry, and is not yet affected by the extensive groundwater abstraction that has taken place since the late 1960s. Thus, the reactive transport modelling of the Cretaceous Aveiro aquifer has helped to confirm the principal processes contributing to the hydrogeochemical evolution of the aquifer and has provided additional evidence about the timescale for aquifer freshening. At the detailed level, however, this interpretation may be complicated by the mixing that is inevitably caused by modern pumping in this multilayer aquifer.

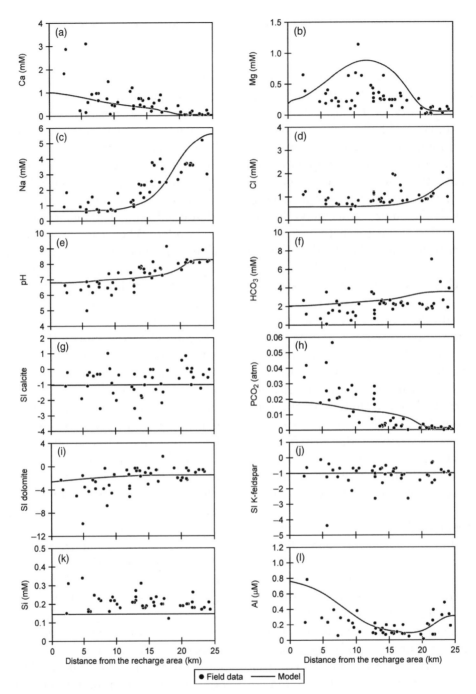

Fig. 4.8 Results of modelling the hydrogeochemical evolution of the Cretaceous Aveiro aquifer using a reaction-path geochemical model (time = 50 ka; CEC = 0.020 mol L^{-1}). Field data is indicated by symbols and model results by lines.

4.5 Discussion

From the modelling carried out on the various aquifers, several factors emerge that dominate the present groundwater composition and its susceptibility to change. These are the hydrogeology and continuity of the flow system, the progression of mineral dissolution fronts and the progression of ion-exchange fronts.

The discontinuity of the flow velocity in the aquifers is illustrated very well in the Valréas and the East Midlands aquifers, and to a certain extent also in the Aveiro aquifer. These aquifers are partly confined and drainage from their distal parts is poor (e.g. Figs. 4.1, 4.4 and 4.6). As a result, there are two flow regimes, one close to the area of recharge where the rate of groundwater flow is fast and second a distal confined part where the flow velocity is very low. Under pristine conditions considerable discharge from springs and rivers, occurred at the boundary between the confined and the unconfined parts of these aquifers. Both in the East Midlands and Valréas aquifers shallow groundwater discharge from the mainly unconfined sections was estimated by modelling to amount to 90% of the recharge under pristine conditions. In the Valréas aquifer, radiocarbon dating shows a discontinuity in age distribution corresponding to the two flow regimes. In recent years, groundwater abstraction has occurred from the confined parts of the aquifers, and as a result the flow field has become more homogeneous and the shallow groundwater discharge has ceased. However, since the abstracted volume of groundwater is still small compared to the total volume present in the aquifers, the groundwater in the confined parts of the aquifers remains old and the water chemistry largely reflects pristine conditions, as does radiocarbon dating. The very slow groundwater flow rates in the confined parts of the aquifers also imply that changes in water chemistry may develop over extremely long time spans. While the residence time of the groundwater in the confined part of the Valréas aquifer is around 30,000 years, modelling suggests that, given the present day input and flow conditions, 2–3 million years are required to develop the present day patterns in groundwater chemistry. This time span is approaching the age of the aquifer sediments!

Mineral fronts may control the groundwater chemical composition at almost any timescale. Thus at one end of the scale $CaCO_3$ dissolution fronts progress at the timescale of hundreds of years (Jakobsen and Postma 1999). At the other end of the scale, the East Midlands aquifer shows the effects of $CaCO_3$ dissolution on a time scale of thousands of years. When $CaCO_3$ dissolution by CO_2 occurs in an aquifer above the groundwater table, then the consumed CO_2 is replenished through the gas phase. The system is open for CO_2 and the Ca^{2+} and HCO_3^- concentrations will be high while the pH will be low. When the calcite dissolution front progresses to below the groundwater table, replenishment of CO_2 gas becomes impossible and the system is closed for CO_2. The resulting Ca^{2+} and HCO_3^- concentrations are lower and the pH is high. Any aquifer, where the original rock contains a limited amount of $CaCO_3$, is expected to go through a development from open system $CaCO_3$ dissolution to closed system $CaCO_3$ dissolution with corresponding changes in water chemistry. In the East Midlands aquifer this change in dissolution mode must have taken place somewhere between 5,000 and 10,000 years ago and the resulting variation is still moving through the aquifer.

Ion-exchange fronts are typically found in freshening aquifers, that is, aquifer systems that originally contained seawater, which is replaced by freshwater. In addition to flushing the chloride contained in the initial water, the Ca^{2+} in freshwater must also replace Mg^{2+} and Na^+ on the cation-exchange sites on the sediment grains. Since the amounts of adsorbed Mg^{2+} and Na^+ are much larger than the Ca^{2+} concentration in the flushing solutions, the ion-exchange fronts will move much slower through the aquifer than the chloride front. In addition, ion-exchange fronts may be coupled to mineral dissolution since binding of Ca^{2+} ions can cause calcite dissolution, raising the pH and the bicarbonate concentration. The Aveiro aquifer seems to be a classical freshening aquifer, displaying the theoretically expected chromatographic patterns along the flow line. The modelled timescale for changing the aquifer from the initial seawater conditions to the present day conditions is in the order of 50,000–100,000 years and the distal part of the aquifer is still not completely flushed. The Valréas aquifer is also a freshening type aquifer but with different flow conditions and therefore a different ion distribution along the flow path. As described above, the flow rates in the confined part decrease by a factor of 10 and the ion distribution patterns therefore become compressed. In this case, the model suggests a geochemical evolution covering more than 2 million years.

4.6 Conclusions

The investigated aquifers are all large hydraulic systems that show distinct variations in their pristine chemical composition and with groundwater residence times at a time scale of thousands of years. The geochemical processes occurring in these aquifers are strongly retarding some of the solutes as compared with groundwater flow and thereby control the variations in groundwater composition that operate on timescales of up to millions of years. For all three aquifers, the changes in water chemistry are well described by relatively simple reactive transport models which add confidence to our quantitative understanding of the systems. Therefore, geochemical models may be used to predict the changes in water quality over time in the pristine aquifers. Likewise, but more cautiously, they can be used to predict changes in groundwater composition when the flow rates are accelerated by orders of magnitude due to groundwater abstraction.

The overall conclusion is that many large aquifers should be considered as dynamic systems where the baseline concentrations of many solutes may vary both in time and space. From this perspective it is useful to consider and understand the **baseline processes** leading to the baseline concentrations.

References

Andersen, M.S., Nyvang, V., Jakobsen, R. et al. (2005) Geochemical processes and solute transport at the seawater/freshwater interface of a sandy aquifer. *Geochimica et Cosmochimica Acta* 69, 3979–94.

Appelo, C.A.J. (1994) Cation and proton exchange, pH variations, and carbonate reactions in a freshening aquifer. *Water Resources Research* 30, 2793–805.

Appelo, C.A.J. (1996) Multicomponent ion exchange and chromatography in natural systems. In: Lichtner, P.C., Steefel, C.I. and Oelkers, E.H. (eds) *Reactive Transport in Porous Media, Reviews in Mineralogy*, vol. 34. Mineralogical Society of America, Washington, DC, USA, pp. 193–227.

Appelo, C.A.J. and Postma, D. (2005) *Geochemistry, Groundwater and Pollution.* 2nd edn. A.A. Balkema, Rotterdam, The Netherlands.

Chapelle, F.H. and Knobel, L.L. (1983) Aqueous geochemistry and the exchangeable cation composition of glauconite in the Aquia aquifer, Maryland. *Ground Water* 21, 343–52.

Condesso de Melo, M.T. (2002) Flow and hydro-geochemical mass transport model of the Aveiro Cretaceous multilayer aquifer (Portugal). PhD thesis, Universidade de Aveiro, Portugal.

Condesso de Melo, M.T., Carreira Paquete, P.M.M. and Marques da Silva, M.A. (2001) Evolution of the Aveiro Cretaceous aquifer (NW Portugal) during the Late Pleistocene and present day: evidence from chemical and isotopic data. In: Edmunds, W.M. and Milne, C.J. (eds) *Palaeowaters in Coastal Europe: Evolution of Groundwater since the Late Pleistocene.* Special Publications 189, Geological Society, London, UK, pp. 139–54.

Edmunds, W.M. and Smedley, P.L. (2000) Residence time indicators in groundwater: the East Midlands Triassic sandstone aquifer. *Applied Geochemistry* 15, 737–52.

Edmunds, W.M., Bath, A.H. and Miles, D.L. (1982) Hydrochemical evolution of the East Midlands Triassic sandstone aquifer, England. *Geochimica et Cosmochimica Acta* 46, 2069–81.

Goldberg, E.D. (1963) The oceans as a chemical system. In: Hill, M.N. (ed.) *The Sea – Ideas and Observations on Progress in the Study of the Seas – Vol. 2: The Composition of Seawater.* John Wiley & Sons, New York, USA, pp. 3–25.

Huneau, F. (2000) Fonctionnement hydrogeologique et archives paleoclimatiques d'un aquifere profond mediterraneen – Etude géochimique et isotopique du bassin Miocène de Valréas (Sud-Est de la France). PhD thesis, Université d'Avignon et des Pays de Vaucluse, France.

Jakobsen, R. and Postma, D. (1999) Redox zoning, rates of sulphate reduction and interactions with Fe-reduction and methanogenesis in a shallow sandy aquifer, Rømø, Denmark. *Geochimica et Cosmochimica Acta* 63, 137–51.

Langmuir, D. (1971) The geochemistry of some carbonate ground waters in central Pennsylvania. *Geochimica et Cosmochimica Acta* 35, 1023–45.

Morse, J.W. and Mackenzie F.T. (1990) *Geochemistry of Sedimentary Carbonates.* Elsevier, Amsterdam.

Moss, P.D. and Edmunds, W.M. (1992) Processes controlling acid attenuation in the unsaturated zone of a Triassic sandstone aquifer (U.K.), in the absence of carbonate minerals. *Applied Geochemistry* 7, 573–83.

Oliveira, T.I.F. (1997) Capacidade de troca cat-iónica no Cretácico de Aveiro e sua influência no quimismo da água. M.Sc. thesis, Universidade de Aveiro, Portugal.

Parkhurst, D.L., Christenson, S. and Breit, G.N. (1996) Ground-water quality assessment of the Central Oklahoma aquifer, Oklahoma. Geochemical and geohydrologic investigations. Chapter C. United States. US Geological Survey Water-Supply Paper 2357.

Parkhurst, D.L. and Appelo, C.A.J. (1999) User's guide to PHREEQC (version 2) – A computer program for speciation, reaction-path, 1D-transport, and inverse geochemical calculations. US Geological Survey Water Resources Investigations Report no. 99–4259.

Plummer, L.N., Vacher, H.L., Mackenzie, F.T. et al. (1976) Hydrogeochemistry of Bermuda: A case history of ground-water diagenesis of biocalcarenites. *Geological Society of America Bulletin* 87, 1301–16.

Plummer, L.N., Busby, J.F., Lee. R.W. et al. (1990) Geochemical modeling of the Madison Aquifer in parts of Montana, Wyoming, and South Dakota. *Water Resources Research* 26, 1981–2014.

Schreiber, C., Friedman, G.M., Decima, A. et al. (1976) Depositional environments of Upper

Miocene (Messinian) evaporite deposits of the Sicilian Basin. *Sedimentology* 23, 729–60.

Smedley, P.L. and Edmunds, W.M. (2002) Redox patterns and trace-element behaviour in the East Midlands Triassic sandstone aquifer, UK *Ground Water* 40, 44–58.

Troelstra, S.R., van de Poel, H.M., Huisman, C.H.A. et al. (1980) Paleoecological changes in the latest Miocene of the Sorbas basin, S.E. Spain. *Geologie Méditerranéenne* VII, 115–26.

Walraevens, K. and Cardenal, J. (1994) Aquifer recharge and exchangeable cations in a Tertiary clay layer (Bartonian clay, Flanders-Belgium). *Mineralogical Magazine* 68A, 955–6.

5A   Timescales and Tracers

R. PURTSCHERT

The hydrochemical variations of any substance dissolved in groundwater are a consequence of environmental changes to the initial composition of the water entering the saturated zone and changes due to time-dependent water–rock interactions within the aquifer. Apart from natural processes on groundwater quality, in recent years, the effects of human impact also influence the groundwater chemistry. Hence, groundwater quality is closely linked to the point in **time of recharge** and the **residence time** of groundwater in the subsurface. For the definition and understanding of baseline water quality, some knowledge of residence time is therefore required. Environmental tracers provide an important tool for the estimation of groundwater residence times over a large range of timescales. The range of potential tracers that can be used to measure residence times in modern waters is reviewed in the context of defining natural groundwaters and recognising pollution.

5A.1   Introduction

5A.1.1   Groundwater quality as a function of residence time

The chemistry of groundwater varies as a result of a wide range of processes and factors. The heterogeneity of aquifers with regard to mineralogy and hydraulic properties is reflected in a statistical scatter of concentrations. Superimposed on these irregularly distributed variations, groundwater chemistry may also change systematically in space and time. The controlling factors of this systematic evolution are changing input concentrations in the recharge and modifications of the groundwater composition within the aquifer. Both factors vary as a function of time. Climatic and environmental changes affect the composition of rainwater, for example, due to changing atmospheric circulation patterns. The effects of changing recharge conditions are even more pronounced within the unsaturated zone. Increasing evaporation due to higher air temperatures leads to an increase in concentrations of solutes concentrations in the recharging water. Changing vegetation cover is another factor affecting quality and quantity of recharge. In the saturated zone groundwater naturally acquires solutes through interaction with the aquifer material. Because the rates of geochemical processes such as redox reactions and the dissolution of carbonate and other minerals are limited by kinetics, the baseline groundwater composition evolves in time and space. In general, young groundwater is found at shallow depth, in an oxidising environment and in unconfined aquifers whereas old groundwater is found in deeper layers, in a reducing environment and in confined parts of aquifers. A subdivision of groundwater bodies according to groundwater residence time may therefore be sensible in order to characterise and describe the baseline composition of groundwater.

A primary (but not exclusive) criterion for baseline conditions is the absence of groundwater recharged after the commencement of intensive agricultural and industrial activities about 50 years ago. Most aquifers in Europe contain younger groundwater components and are therefore potentially (but not necessarily) affected by human impact. The reliable identification and quantification of young groundwater components is therefore a key issue in a baseline study. Moreover, in many cases, knowledge of residence time is the only way to distinguish between anthropogenically and naturally elevated concentrations of any parameter.

5A.1.2 Groundwater residence time in context with other quality relevant issues

The investigations considered in this chapter involved several work packages dealing with (1) the description of the chemical status as function of groundwater residence time in different reference aquifers; (2) an understanding of baseline processes including chemical modelling and (3) the monitoring and analysis of geochemical trends. All these issues are mutually linked with groundwater residence time:

1. As deduced in this and the following chapter, the definition of baseline quality is related to the age of the groundwater. In some cases the chemical evolution itself can be used as a dating tool. However, it is important to note that each aquifer is unique and

therefore the chemical clocks have to be calibrated and adjusted for each groundwater body separately.

2. Geochemical modelling, tantamount to an understanding of geochemical processes in an aquifer, is inseparably linked with time. A model includes the identification and quantification of geochemical reactions and the consideration of the rates at which they take place. If a model is used to predict future trends, a reliable temporal calibration of the time-dependent parameters of the model is particularly needed. Once the processes and their timescales are understood they can be related to timescales of the water movement or the evolution of groundwater within an aquifer system.

3. The identification of significant and sustained geochemical trends can in principle be based on statistical procedures, without considering the dynamics of the groundwater body under investigation (Grath et al. 2001). However, as derived in Chapter 6 of this volume, a direct link exists between the flow velocity, which can be determined by dating techniques, and the temporal evolution of a chemical trend at a given location. For conservative water quality indicators, the spatial variation of concentrations will be transformed in a corresponding variation of concentrations as a function of the observation time. Or in other words, when groundwater is moving the spatial pattern will pass through a point fixed in space and will lead to a trend as function of time at that given point. For non-conservative parameters, the amplitude and frequency of the variations will be shifted or attenuated. This is the case for ion-exchange reactions. An understanding of the geochemical processes (inclusive of modelling as discussed above) is

therefore crucial for the interpretation of trends. For example, the presence of seasonal trends indicates that the groundwater must contain very young groundwater components. Consequently, the age of the water also helps to establish a sensible sampling frequency in monitoring programmes. If modern water components are absent, a yearly sampling programme seems appropriate. However, care needs to be taken because changes in mixing ratios can occur much faster than the mean groundwater residence time suggests. Last but not least, it is also important not only to monitor the concentrations of contaminants, but also the age or more precisely the age structure of the investigated groundwater body.

Tracers with suitable half-lives, such as ^{39}Ar with 269 years, are unique tools for recognising trends before the aquifer is contaminated (the keywords being 'early warning'). For example, only with such a tracer it is possible to recognise a shift of the mean residence time from 200 to 100 years. In the unfortunate but common case that a groundwater body is already contaminated, the replenishment rate (based on the mean age) of the water defines the reaction time for remediation measures. If the mean residence time is only a few years, artificially forced trend reversals can be expected within the same time frame. However, for contaminated aquifers with groundwater residence times of 20–30 years it is not possible to achieve a good status within, for example, 10 years. In any case remediation of pollution is extremely difficult, costly or may even be technically unfeasible. It is therefore essential that old, pristine groundwaters are identified and given extra protection.

5A.1.3 Main objectives

The main objectives of groundwater dating in the context of baseline determination are to
1. identify, date and quantify the proportion of young groundwater components;
2. establish timescales for old groundwater components.
The large range in groundwater residence times, between years and millions of years, requires a corresponding set of dating tools in order to derive the age structure in aquifers. Young groundwater, which is most vulnerable to contamination, can be dated with several tracers. Tritium (^{3}H), CFCs, SF$_6$ and ^{85}Kr are all applied in an increasing number of studies for the investigation of modern water. Older waters can be dated with naturally produced isotopes with corresponding longer half-lives (e.g. ^{39}Ar, ^{14}C). All these tracers have been applied and tested within the context of the present studies. For very old water, for instance in large sedimentary basins with residence times of hundreds of thousand of years, suitable dating methods are available (e.g. ^{81}Kr, ^{36}Cl or ^{4}He), although not applied within the present studies because such old waters were not investigated.

A few hydrogeologically well characterised reference aquifers have been selected to cross check the different tracers, to test their limitations for general use and to demonstrate the dating methodology and the link between groundwater quality and age. In this chapter, a brief introduction to the groundwater dating approach using environmental tracers is presented. Some applications in the reference aquifers are described in the following chapter.

5A.2   Dating Techniques

5A.2.1   *The meaning of groundwater residence time*

As a starting point we need to define the meaning of groundwater age and assess what we need to do to 'date' groundwater. In the context of groundwater hydrology, the age, or more correctly, the mean residence time of groundwater corresponds to the average time during which water remains in the sub-surface system before it reaches a designated point along its flow path, such as a catchment outlet or the wellhead of a borehole.

This interpretation of age rests on the assumption of minimal mixing of different water ages in a water sample. However, this is seldom the case. Spatially distributed recharge, hydrogeological heterogeneities, inter-aquifer mixing and, last but not least, mixing during sampling due to extended screened borehole intervals leads to age distributions rather than a single groundwater age. The age distribution or age structure of a water sample is described by the proportion and age of all contributing water components.

Groundwater dating using environmental tracers and radiogenic isotopes will provide a tracer residence time which may be different for different types of tracers (Bethke and Johnson 2002). For most gaseous tracers, the starting point of the dating clock is determined by the moment the water parcel is isolated from the atmosphere. This usually happens at the groundwater table. On the other hand, if tracers are part of the water molecule the time clock starts at the time of origin of precipitation. The time difference, or in other words the travel time of water through the unsaturated zone, is usually negligible compared to the mean travel time in the European aquifers considered in these studies. However, in areas with extended unsaturated zones or semi-confined aquifers, where the infiltration through the low permeability cover is slow, this delay has to be taken into account if the residence time of recent water components is estimated.

5A.2.2   *Hydraulic ages*

The driving force for groundwater flow is the gradient of hydraulic head. Water flows from high elevation to low elevation and from high pressure to low pressure. The specific discharge or Darcy velocity **q** can be calculated according to Darcy's law:

$$q = -K \cdot \mathrm{grad}(h)$$

where q is Darcy velocity ($\mathrm{ms^{-1}}$), K is hydraulic conductivity ($\mathrm{ms^{-1}}$) and h is pressure head (m).

The effective water velocity v is related to the total porosity $\varnothing$ according to $v = q/\varnothing$. This velocity can in principle be used for the calculation of groundwater flow times between two points along a flow path with known head difference. The main uncertainties of simulated age distributions in aquifers, using, for example, particle tracking methods, arise from the lack of information about head distributions in the past and the ubiquitous heterogeneity inherent in aquifer systems. The inability to describe and represent this heterogeneity adequately is a fundamental problem in groundwater hydrology and will continue, even with improved models, to place limits on the reliability of model predictions.

5A.2.3   *Tracer ages*

Environmental tracers provide a unique possibility for a spatially and temporally integrated view of aquifer dynamics. Thanks to their complementary sensitivity, tracer methods can reveal the conceptual weaknesses of groundwater models and help to adapt them accordingly. However, special care should be taken in interpretations of tracer concentrations, because the concentrations may be affected greatly by a number of processes including mixing, hydrodynamic dispersion and diffusion into the rock matrix. Diffusive exchange of ^{14}C dissolved in the water with aquitards could, for example, lead to increased tracer ages compared to the hydraulic water ages (Sudicky and Frind 1981).

As shown later, groundwater residence times in the investigated reference aquifers range from a few years up to several tens of thousands of years. For these timescales, tracer dating relies on substances, which were already present in the geosphere at the time of recharge (in contrast to artificially released tracers for short-term experiments). Environmental tracers for old groundwaters are of natural origin. For young groundwaters, which were recharged within the last 50 years, tracers used are often 'transient' and a result of human activities. The principle of dating groundwater using environmental tracers is based on variable concentrations of substances with known chronologies. Different processes can be responsible for such variations: (1) radioactive decay; (2) chemical or physical accumulation processes in the subsurface and (3) variable input concentrations during recharge. A tracer is suitable for groundwater dating if the temporal evolutions of these processes are well defined and no other, unquantifiable, processes cause concentration variation in the subsurface. Such processes will include local tracer contamination, tracer

degradation and tracer dilution due to pronounced water–rock interaction, which are difficult to predict. Neglecting these factors may, however, cause misinterpretation of tracer concentration in the means of groundwater residence time (Table 5A.1).

The dating range of a method (Fig. 5A.1) is given by the characteristic timescale on which the tracer concentration varies over time. Half-life, input function and accumulation rate are the main characteristics of a tracer that determine the residence times on which a tracer is most sensitive. For example, the radioactive isotope ^{39}Ar with a half-live of 269 years and a constant initial activity is an ideal tracer for groundwater dating in the range between 100 and 1000 years (0.3–3 half-lives). In order to cover a large range of groundwater residence times, multiple tracer measurements should be applied for the resolution of the age structure of an aquifer. Some commonly applied tracer methods with their corresponding dating range are shown in Fig. 5A.1; noble gases and inert chemical substances are the most appropriate tools used for groundwater dating.

The interpretation of measured tracer concentrations in the determination of mean residence time relies on several assumptions about the flow regime (advection, dispersion, diffusion, mixing) and the physical or chemical interaction with the aquifer matrix. A thorough knowledge of groundwater chemistry and geology is therefore required in order to minimise ambiguities. The remaining uncertainties can be further reduced with the simultaneous application of two or more complementary methods and the combination of tracer measurements and numerical flow models. The appropriate combination of methods (dating strategy) has to be adapted

Table 5A.1 Factors affecting tracer ages.

Factor	Tracer	Conditions	Effects	Age influence
Recharge temperature	CFCs, SF$_6$	All	Temperature at the water table -overestimated -underestimated $\pm 2°C$, < 1970: < 1 year $\pm 2°C$, 1970–1990: 1–3 years $\pm 2°C$, > 1990: > 3 years	 Too young Too old
Excess air	SF$_6$ (CFCs)	All	Overestimation	Too old
Thickness of unsaturated zone	All but in particular ^{3}H (and ^{3}He)		Retardation in the unsaturated zone 0–10 m, < 2 years 30 m, 8–10 years Depending on the water content of the soil and the recharge rate (^{3}H)	Too old
Local sources, contamination (e.g. landfills)	CFCs, SF$_6$	Urban areas, towns, industry	Global input functions are distorted. Measurements in local precipitation are necessary. If the soil or aquifer is contaminated no correction possible	Too young
	^{85}Kr	Reprocessing plants		Too young
	^{3}H	^{3}H processing industry		Variable
Reducing conditions	CFCs in particular CFC-11 (SF$_6$)	Anaerobic conditions	Microbiological degradation	Too old
Sorption	CFCs	Sediments with high content of organics		Too old
Subsurface production	^{3,4}He, ^{39}Ar, ^{36}Cl	Crystalline and volcanic rocks	Decay and neutron activation of isotopes in the rock matrix	Too old
Mixing of different water components	All		Mixing of old (tracer free) and recent water components. Age of young component Age of old water component	 Too old Too young
Hydrodynamic dispersion	All but in particular ^{3}H	All	Gases >1975: negligible <1975: considerable ^{3}H and ^{3}H–^{3}He >1965: negligible <1965: considerable	 Too old Too young Too old Too young

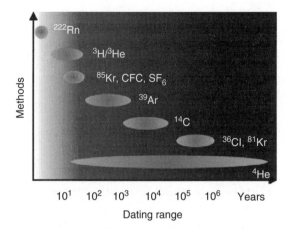

Fig. 5A.1 Tracer methods used for groundwater dating as a function of their dating ranges. Several methods can be used for recent waters (white) while the selection is sparse for ancient waters (grey) and palaeowaters (black). However, for each timescale, at least one tracer is available allowing investigation of the age structure of a whole aquifer.

to the individual aquifer, objectives, as well as available financial resources. It is recommendable to apply during a first phase, several methods for selected sampling locations. This provides the basis for a better understanding of the system and the relevant processes involved. After this evaluation or 'calibration phase', simple and relatively cheap methods might then be used to achieve a larger number of samples, for example, for monitoring programmes.

The quantification of modern (and potentially contaminated) water components is crucial for the definition of the baseline quality of a groundwater body. Special focus was therefore put on the investigation of waters of this age range, all of which are based on the identification of substances that are a consequence of activities during or following the industrial era. The increase of

the concentrations of these tracers in the environment went in parallel with the increasing groundwater pollution from intensive farming, which led to high concentrations of nitrate and agrichemicals.

5A.2.4 Short description of environmental tracers

5A.2.4.1 $^3H/^3He$

The application of the heavy radioactive hydrogen isotope tritium (^3H) ($T_{1/2} = 12.32$ yr) (Lucas and Unterweger 2000) and its decay product helium-3 (^3He) as tracers in hydrology started in the 1950s and early 1960s when large amounts of ^3H were released during the tests of thermonuclear bombs in the atmosphere. Concentrations of ^3H in Northern Hemisphere precipitation greater than 2000 TU were measured during this period of time (IAEA 1992) (1 TU [Tritium Unit] corresponds to a $^3H/^1H$ ratio of $1/10^{18}$ or 0.118 Bq kg^{-1} of water). Because almost all nuclear tests were conducted in the Northern Hemisphere, much lower concentrations were recorded in the Southern Hemisphere. In comparison, the natural background activity in precipitation produced by the interaction of cosmic rays with nitrogen and oxygen is about 5 TU (Kaufman and Libby 1954; Craig 1961; Roether 1967). Hence, a ^3H concentration above about 0.3 TU in a groundwater sample indicates the presence of post-bomb water. The International Atomic Energy Agency (IAEA) implemented a worldwide network of stations to observe tritium concentrations in precipitation (www-naweb.iaea.org/napc/ih/GNIP/IHS_GNIP.html). Nowadays, the limiting factors for the application of tritium are (1) the decreasing concentration of bomb tritium in modern precipitation, which converge towards the natural background level with small gradients and (2) local sources of tritium in industrial areas which disturb the known atmospheric signal. Tritium by itself only provides information on the presence of young water in a sample with a detection limit corresponding to about 0.2% of current concentration in precipitation. As a further development of the method, ^{3}H has been combined with its decay product ^{3}H$_{trit}$ (tritiogenic ^{3}He) to determine the so called ^{3}H/^{3}He age τ (Schlosser et al. 1988; Schlosser 1992; Solomon et al. 1993; Beyerle et al. 1999(b)):

$$\tau = \frac{1}{\lambda} \cdot \ln\left(1 + \frac{^3\mathrm{He}}{^3\mathrm{H}}\right)$$

The ^{3}H/^{3}He age τ (Tolstikhin and Kamenskiy 1969) corresponds to a piston-flow age and is relatively insensitive to the input history of ^{3}H. The concentration of dissolved ^{3}He increases as soon as the groundwater is isolated from the atmosphere. Therefore τ corresponds to the residence time of groundwater in the saturated zone. The achievable dating resolution by technical means is less than 60 days (Beyerle et al. 1999a). The uncertainty increases significantly for waters with elevated concentrations of radiogenic helium produced in the subsurface because ^{3}He$_{trit}$ has to be separated from other helium components (Schlosser 1992; Solomon et al. 1992).

5A.2.4.2 ^{85}Kr

^{85}Kr has been predominantly released from nuclear fuel reprocessing plants (e.g. Sellafield in northwest England and La Hague in northern France). The activity in the atmosphere has been steadily increasing over the last 45 years and in 2006 was about 90 dpm cc^{-1} Kr (decays per minute per cm^3 STP krypton) in the Northern Hemisphere (Fig. 5A-2). The lack of significant sources in the Southern Hemisphere results in about 15% lower concentrations than in the Northern Hemisphere (Weiss et al. 1989).

The half-life of 10.76 years and the input function (Fig. 5A.2) define the dating range of this isotope which is suitable for residence times in the range 5–50 years. Apart from the above-mentioned sources no others contribute to the release into the environment. Near reprocessing plants, elevated concentration may be observed during time periods when the gas is released to the atmosphere (Loosli 1983). As a noble gas, no chemical reactions or degradation have to be taken into account for the conversion into groundwater residence times. Calculated model ages are relatively insensitive to hydrodynamic dispersion due to the continuously increasing input function (Corcho et al. 2005; Engesgaard et al. 2004). Another important advantage of the method arises from the fact that an isotope ratio (^{85}Kr/Kr) is measured; therefore degassing or gas stripping do not affect ^{85}Kr-tracer ages, in contrast to other gas-based methods. For the same reason, recharge conditions such as recharge temperature, infiltration height and excess air do not have to be known. The main limitation of the ^{85}Kr method is the large sample volume required and the relatively high analytical demands. Two hundred liters of water has to be degassed in the field or in the laboratory. The ^{85}Kr activity is measured after Kr separation by gas chromatography in proportional counters ranging in volume from 10 to 20 cm^3 (Loosli and Purtschert 2005).

5A.2.4.3   CFCs

Chlorofluorocarbons CFC-11 ($CFCl_3$), CFC-12 (CF_2Cl_2) and CFC-113 ($C_2F_3Cl_3$) are anthropogenic organic compounds that have been produced since the 1930s for a number of industrial and domestic purposes ranging from aerosol propellants to refrigerants. Concentrations have been steadily increasing over the last 50 years (Fig. 5A.2). In contrast to the spatially variable nature of ^{3}H concentrations in rainfall, atmospheric concentrations of CFCs show little spatial variation. Because of various environmental regulations limiting the use of CFCs the release rate was reduced to about half of the peak values during the late 1980s. As a result of the decreasing gradient of the atmospheric input function, the sensitivity to date very young ground water has diminished. However, because CFCs are detectable in very low concentrations they are sensitive tools for the detection of recent water components where mixtures with older waters are present. Similar to every tracer that relies on absolute concentrations, the conversion into a residence time depends on the recharge conditions (Busenberg and Plummer 1992). The amount of tracer found in the water is not only determined by the atmospheric input concentration, but also by the recharge temperature and the amount of excess air dissolved in the water (Table 5A.1). The main limitations of the CFC dating method are contamination in industrial areas and degradation under anoxic conditions (Busenberg and Plummer 1992; Corcho et al. 2002). CFC-12 proved to be the most stable of the CFCs. As for all gas tracers any water contact or gas exchange with the atmosphere has to be avoided during sampling. CFCs are measured by gas chromatography with a detection limit less than 0.2% of the modern concentration in the atmosphere (Hofer and Imboden 1998). A comprehensive summary of the use of CFC in hydrology is provided by the IAEA (IAEA 2006).

5A.2.4.4 SF_6

Sulphur hexafluoride (SF_6) is primarily of anthropogenic origin, but also occurs naturally in minerals, rocks and magmatic fluids (Busenberg and Plummer 2000). Sulphur hexafluoride is very stable with an estimated life time of over 800 years (Morris et al. 1995). Concentrations of SF_6 in air have rapidly increased (Fig. 5A.2; Maiss and Levin 1994) from a steady-state value of about 0.02–0.05 to 4 parts per trillion (pptV) during the past 35 years and are expected to continue increasing while atmospheric concentrations of CFCs are steady or beginning to decrease. Contamination due to point sources seems, like CFCs, to be the main factor limiting SF_6 as a dating tool for shallow groundwaters (Ho and Schlosser 2000). Because of the low solubility of SF_6 in water, tracer ages are sensitive to excess air, so this needs to be accurately determined (Table 5A.1).

5A.2.4.5 ^{39}Ar

^{39}Ar is produced in the atmosphere mainly by the interaction of cosmic rays with argon. The resulting modern atmospheric equilibrium activity is constant at 1.67×10^{-2} Bq m^{-3} of air (Loosli 1983). With a half-life of 269 years, ^{39}Ar fills the dating gap between the recent groundwater indicators and ^{14}C (Fig. 5A.1). The interpretation of measured activities is in principle straightforward because, as a noble gas, [39]Ar undergoes no geochemical reactions. However, in aquifers with very high subsurface neutron fluxes (Loosli et al. 1989) the application of [39]Ar may have some limitations because of subsurface production. In such rarely occurring cases, more care has to be put into the interpretation of the data (Purtschert et al. 2001).

[39]Ar/Ar ratios in groundwater are in the order 10^{-15}. Although initial recent attempts were made to use the AMS technique (Collon et al. 2004a), the detection of [39]Ar is at the moment only possible by the radioactive counting technique (Loosli 1983). Because of the very low activities, a relatively large amount of water (more than 2000 L) has to be degassed in the field to obtain a sufficient volume of argon. The [39]Ar activity of the separated argon is measured in high pressure proportional counters in an ultra-low background environment (Collon et al. 2004b; Loosli and Purtschert 2005).

5A.2.4.6 Radiocarbon (^{14}C)

Radiocarbon dating of dissolved inorganic carbon (DIC) in groundwater started about 40 years ago and is one of the most widely used applications of radiocarbon dating (Clark and Fritz 1997; Kalin 2000). However, the interpretation of ^{14}C results from groundwater is complex because carbon in groundwater may be derived from several sources with different ^{14}C concentrations. Atmospheric ^{14}C is produced in the upper atmosphere by the reaction of thermal neutrons, which are produced by cosmic rays, with nitrogen:

$$^{14}N + n \rightarrow {}^{14}C + p$$

The ^{14}C is oxidised to CO_2 and mixes within the lower atmosphere where it is assimilated in the biosphere and hydro-sphere. [14]CO_2 of the soil zone dissolves in the infiltrating water and ^{14}C decays in the subsurface with a half-life of 5730 years. The problem is that the formation of carbonic acid drives the dissolution of carbonate minerals in the aquifer. Geochemical correction models (Ingerson and Pearson 1964; Fontes and Garnier 1979; Mook 1980; Eichinger 1983) which are either based on the chemical composition of the groundwater, the ^{13}C content of DIC, or both, are used to take into account the admixture of dead carbon (i.e. zero ^{14}C) from the aquifer material. Although much progress has been made in the last years to model the reactions occurring along groundwater flow lines (Plummer et al. 1991), it is questionable to calculate ^{14}C ages in cases when chemical processes dominate the decrease of ^{14}C in DIC. A common range of corrected input activities is 50–60 pmc and this corresponds to a chemical dilution factor of about two. Corrected input activities of less than 25 pmc, for example, indicated when δ^{13}C values are above –5‰, indicate that at least 75% of the total activity drop is due to chemical or isotopic dilution processes and not due to radioactive decay. In this situation reliable groundwater dating using radiocarbon is no longer feasible (Loosli et al. 2001).

Thermonuclear testing in the atmosphere resulted also in a significant increase of ^{14}C in the atmosphere (Levin et al. 1992). The presence of such 'excess' [14]C (Fig. 5A.2) indicates, in a similar way to bomb ^{3}H, the presence of young groundwater components (Kalin and Long 1994; Corcho et al. 2007).

5A.2.4.7 Radiogenic helium

Helium is produced in the subsurface by the decay of uranium and thorium in minerals (Tolstikhin et al. 1996). In most cases, it can

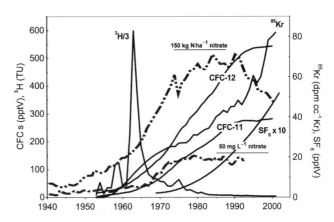

Fig. 5A.2 Input functions of ^{3}H, ^{85}Kr, SF$_6$, CFCs, ^{36}Cl and ^{14}C compared to nitrate concentrations in Danish groundwaters and fertilizer use in Denmark since 1940 (Danish groundwater monitoring programme). These groundwaters were dated using CFCs and contain at least 2 mg L^{-1} oxygen.

be assumed that 100% of the He produced escapes from the rock into the water-filled pore volume (Pearson et al. 1991). Under closed-system conditions, this in situ accumulation can be calculated and ^{4}He ages can be estimated:

$$Age = \frac{C \cdot \phi}{P_{\text{in situ}} \cdot (1 - \phi)}$$

where C is the measured He concentration in water, P is the in situ production rate in the rock and ϕ denotes the porosity.

However, this ideal case is rarely realised. Groundwater residence times estimated by this method are often much higher than hydraulic ages or ages derived from other tracer methods. This discrepancy is caused by additional He sources. Fundamentally, two He sources can be recognised. The first is He produced in situ in the aquifer matrix and which is then immediately transferred to the groundwater. This 'real time' production rate can easily be calculated using known U and Th concentrations of the aquifer rock. The second source, He that was produced before the groundwater entered the aquifer, or that was produced outside the aquifer, is much more difficult to quantify but is in many cases the dominant one. This 'offline'

produced He could be stored in mineral grains (Solomon et al. 1996) or in less permeable zones within the aquifer or in adjacent aquitards above or below the aquifer (Lehmann et al. 2002). Other authors postulate a steady-state helium flux from the whole upper crust beneath the aquifer (Torgersen and Clarke 1985; Torgersen and Ivey 1985). This variety of possible sources, whose relative contributions may change spatially and temporally, limits the use of He as a quantitative dating tool. Nevertheless, He concentrations in groundwater can be used as a qualitative residence time indicator. After calibration with other independent and absolute dating methods, He is also suitable as a quantitative dating tool over a large range of residence times up to millions of years.

5A.2.4.8 ^{81}Kr

^{81}Kr (T$_{1/2}$ = 230 kyr) has been proposed as the ideal tracer isotope for dating old water and ice in the age range of 100–1000 kyr (Oeschger 1987). ^{81}Kr is mainly produced in the upper atmosphere by cosmic-ray induced spallation and neutron activation (Lal and Peters 1962). Human activities producing nuclear fission have a negligible effect on the ^{81}Kr concentration because the stable ^{81}Br shields ^{81}Kr from the decay of fission products (Collon et al. 2004b). Subsurface production of ^{81}Kr can also be assumed to be negligible based on theoretical calculations and on first results in groundwater studies (Lehmann et al. 2003; Sturchio et al. 2004)

However, the task of analysing ^{81}Kr at or below the atmospheric level has always been an experimental challenge. In recent studies, accelerator mass spectrometry (AMS) has been used to analyse ^{81}Kr in old groundwater samples (Collon et al. 2000). With the relatively new atom trap trace analysis (ATTA) method (Du et al. 2003) the required sample size could be reduced from of 15 tons of water to less than 2 tons of water, similar to that for ^{39}Ar (Loosli and Purtschert 2005). The water is degassed in the field and Kr is extracted and separated by gas chromatographic methods.

5A.2.4.9 ^{36}Cl

^{36}Cl can be used for groundwater dating in two ways. High levels of ^{36}Cl in groundwater indicate, similarly to ^{3}H and ^{14}C, that recharge occurred since the bomb tests. Bomb ^{36}Cl in young groundwaters overwhelms the natural fallout by several orders of magnitude (Phillips 1999). However, quantitative groundwater dating using thermonuclear ^{36}Cl is complicated, lacking well constrained local input functions (Corcho et al. 2005).

The main application of ^{36}Cl is due to its long half-life ($T_{1/2}$ = 301 kyr), which makes this isotope a valuable tool for dating very old groundwater on timescales up to 1.5 million years. Problems arise from the fact that the input concentration (or ^{36}Cl/Cl ratio) varies spatially and temporally and that the Cl and ^{36}Cl concentration may be

modified after recharge in the subsurface (Torgersen et al. 1991; Andrews and Fontes 1993). However, this problem may be overcome by using supplementary chemical and isotopic data (Love et al. 2000; Lehmann et al. 2002). The required sample size is about one liter of water and the ^{36}Cl/Cl ratio is measured by AMS (Synal et al. 1990).

5A.2.5   *Box models and multiple tracer measurements*

Tracer concentrations representing young residence time indicators are interpreted with the aid of so-called lumped-parameter or box models (Maloszewski and Zuber 1996). It is assumed that the age distribution of the water can be approximated by a simple mathematical function. Examples are the dispersion model (DM) and the exponential model (EM) which are used in this chapter (Fig. 5A.3). The piston-flow model was neglected because (1) it turned out to be unrealistic in most cases; and (2) it can be approximated by a dispersion model with a small dispersion coefficient. Other age distributions such as the combination of a piston flow and an exponential model (EPM) can be approximated as a dispersion model with an adapted dispersion parameter. To take into account the possibility of an admixture of ancient and tracer-free water components, a mixing parameter m is included in the modelling. The mixing parameter quantifies the portion of water that is described by the above mentioned age distribution (Fig. 5A.3). This leads to mathematical expressions (Maloszewski and Zuber 1996) for the age distribution h (t, m, T_m, d) with model parameters p_j namely the proportion of young water (m), the mean

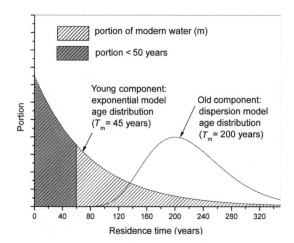

Fig. 5A.3 Hypothetical age structure of ground-water with two main components. The old component follows a dispersion model age distribution with a mean residence time of 200 years. This component is therefore free of transient tracers. One part of the young component, which is described by an exponential model age distribution, is younger than 50 years (grey area) and contains young residence time indicators (^{3}H, ^{85}Kr, CFC, etc.), while the rest is also free of transient tracers. However, it is defined that the portion of recent water (m) corresponds to the total amount of water that follows the younger age distribution (shaded area).

residence time (T_m) and a dispersion parameter (d) for the DM.

The convolution of tracer input c_{in} to tracer output c_{out} for a certain sampling time T_s can be calculated according to the formula:

$$C_{out}(T_s, p_j)$$
$$= \int_0^\infty c_{in}(T_s - t) \cdot \exp(-\lambda t) \cdot h(t, p_j) \cdot dt$$

The decay constant λ is equal to zero for stable isotopes. We can consider several tracers being applied at the same time. If each tracer, i, behaves like an ideal tracer and the

assumed age distribution reflects roughly the real natural situation, it should be possible to find a model and model parameters p_j which are consistent with all measured tracer concentrations. These model parameters can be determined by minimising the sum of the weighted squared deviations between the modelled (C_i^{mod}) and measured (C_i^{meas}) concentrations or activities:

$$\chi^2 = \sum_i^n \frac{(C_i^{mod}(pj) - C_i^{meas})^2}{\sigma_i^2}$$

where
i is the ^{85}Kr, ^{3}H, ^{3}He, SF$_6$ (total n tracers); σ_i the experimental 1σ-errors; and p_j the model parameters (T_m, d, m).

The weightings σ_i^2 give preference to the most accurate data. If the remaining best fit difference between the model values and the measured values is smaller than the uncertainties of measurements (small χ^2), the model seems to be appropriate and no additional processes have to be taken into account. In other words, within the uncertainty, all tracer concentrations are not contradictory with the assumed age distribution and the corresponding model parameters. In cases of large disagreements (large χ^2), the model is incomplete and has to be improved in order to explain all tracer concentrations. In this case, we can also reject the tracer which causes the disagreement. CFCs under reducing conditions are an example (as described in the case studies in the next chapter). Standard methods of least squares fitting can been applied to solve this minimisation problem (Corcho et al. 2007).

It is obvious that this procedure is only applicable if a sufficient number of tracer methods are applied at the same site. The identification of processes that affect the interpretation of mean residence time is only possible with the combination and comparison of several methods.

5A.3 Summary and Conclusions

A comprehensive set of tracers and methods to derive groundwater residence times has been developed and established over the last two decades. Although for each time scale at least one method is available, it is still an ambitious task to determine the age of groundwater because several processes other than ageing may change concentrations of dating tracers. Natural systems such as aquifers are very complex and therefore difficult to describe with simple models. We always have to deal with mixing of water components with different age. The age distribution of a water sample is very often the result of a com-

bination of natural mixing processes resulting from hydrodynamic dispersion, upwelling of deeper water and mixing induced from human impacts, for example, extraction of water from different depths in a large screen interval or changing gradients due to heavy pumping. It is not possible to determine details about the age structure of a groundwater body with only one tracer. Only the combination of several isotopic and chemical tracers at the same time provides a reliable basis for an understanding of a flow system. Furthermore, it is important to emphasise that dating is only one objective. The determined age structure has to be included into the general knowledge of the system, which may consist of knowing the recharge area, the flow path and flow velocities, the evolution of the chemical and isotopic components mainly due to water–rock interaction and the reservoir size as well as the balances of the water masses.

Tracer	Sample size	Half-life; Dating range	Pro; Contra	Assessment
Tritium	0.1–1 L	12.32 yr 5–45 yr	Simple, cheap; dating inaccurate local contamination	
T/He-3	40 mL	12.32 yr 1–45 yr	Very accurate degassing; analytical requirement	
CFC	0.5 L	– 5–45 yr	Simple, cheap; contamination degradation	
SF6	0.5 L	– 2–30 yr	Easy sampling, accurate; excess air sensitivity	
Kr-85	300 L	10.8 yr 2–40 yr	Very accurate sampling, analytical requirement	
Ar-39	2000 L	269 yr 50–1000 yr	Unique dating range sampling; analytical requirement	
C-14	1 L 80 L	5730 yr 1–30 yr	Routine method; interpretation complicated	

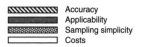

Accuracy
Applicability
Sampling simplicity
Costs

Fig. 5A.4 Short characterisation of a subselection of dating methods which have been applied in the case studies. Criteria of assessment are the dating accuracy, the limitations due to secondary effects such as degradation and local contamination, the simplicity of the sampling protocol and the costs per sample (high costs result in a low rating).

Several methods can be applied for dating and identifying modern water components. Each method has advantages and disadvantages. The selection of the appropriate tracers has to be adjusted to the individual aquifer and aims of the investigation. Factors that may lead to misinterpretation of measured tracer concentrations are summarised in Table 5A.1. As a consequence of these complications, the following procedure is recommended:

1. Start with an easy to handle method which can indicate the presence of young water components (e.g. tritium or CFCs).

2. If young waters are present select additional methods in order to identify factors which may affect tracer concentrations in terms of residence time (e.g. admixture of old water, CFCs degradation, local contamination).

3. Based on the investigations of step 2, select the simplest and cheapest method(s) for investigations with larger sample numbers.

This is also valid if only the presence or lack of modern water has to be demonstrated. Contamination and/or degradation also have to be excluded in this case.

A very brief assessment of some common dating methods is given in Fig. 5A.4. A series of applications of these tracers in selected reference aquifers are presented in the following chapter.

References

Andrews, J.N. and Fontes, J.C. (1993) Comment on Chlorine 36 dating of very old groundwater 3. Further studies in the Great Artesian Basin, Australia by Torgersen, T. et al. *Water Resources Research* 29, 1871–4.

Bethke, C.M. and Johnson, T.M. (2002) Paradox of groundwater age. *Geology (Boulder)* 30, 107–10.

Beyerle, U., Aeschbach Hertig, W., Hofer, M. et al. (1999a) Infiltration of river water to a shallow aquifer investigated with ^{3}H/^{3}He, noble gases and CFCs. *Journal of Hydrology* 220, 169–85.

Beyerle, U., Aeschbach-Hertig, W., Hofer, M. et al. (1999b) Infiltration of river water to a shallow aquifer investigated with 3H/3He, noble gases and CFCs. *Journal of Hydrology* 220, 169–85.

Busenberg, E. and Plummer L.N. (1992) Use of chlorofluorocarbons (CCl_3F and CCl_2F_2) as hydrologic tracers and age-dating tools: The alluvium and terrace system of central Oklahoma. *Water Resources Research* 28, 2257–84.

Busenberg, E. and Plummer, L.N. (2000) Dating young groundwater with sulfur hexafluoride; natural and anthropogenic sources of sulfur hexafluoride. *Water Resources Research* 36, 3011–30.

Clark, I.D. and Fritz, P. (1997) *Environmental Isotopes in Hydrogeology*. Lewis. Boca Raton, FL.

Collon, P., Kutschera, W., Loosli, H.H. et al. (2000) ^{81}Kr in the Great Artesian Basin, Australia; A new method for dating very old groundwater. *Earth and Planetary Science Letters* 182, 103–13.

Collon, P., Bichler, M. Caggiano, J. et al. (2004a) Developing an AMS method to trace the oceans with ^{39}Ar. *Nuclear Instrumentas and Methods in Physics Research* 223–4, 428–34.

Collon, P., Kutschera, W. and Lu, Z.T. (2004b) Tracing noble gas radionuclides in the environment. *Annual Review of Nuclear and Particle Science* 54, 39–67.

Corcho, A., Purtschert, R., Hofer, M. et al. (2002) Comparison of residence time indicators (^{3}H/^{3}He, SF$_6$, CFC-12, and ^{85}Kr) in shallow groundwater: A case study in the Odense aquifer, Denmark. In: Goldschmidt Conference, Davos. *Geochimica et Cosmochimica Acta* 66, A152.

Corcho, J.A., Purtschert, R., Hinsby, K. et al. (2005) ^{36}Cl in modern groundwater dated by a multi tracer approach (^{3}H/^{3}He, SF$_6$, CFC-12 and ^{85}Kr): A case study in quaternary sand aquifers in the Odense Pilot River Basin, Denmark. *Applied Geochemistry* 20, 599–609.

Corcho, J.A., Purtschert, R., Barbecot, F. et al. (2007) Constraining groundwater age distribution using [39]Ar: A multiple environmental tracer (^{3}H, [3]He, [85]Kr, [39]Ar and ^{14}C) study in the semi-confined Fontainebleau Sands aquifer (France). *Water Resources Research*, in press.

Craig, H. (1961) Isotopic variations in meteoric waters. *Science*, 133, 1702–3.

Du, X., Purtschert, R., Bailey, K. et al. (2003) A new method of measuring [81]Kr and [85]Kr abundances in environmental samples. *Geophysical Research Letters* 30, 2068.

Eichinger, L. (1983) A contribution to the interpretation of ^{14}C-groundwater ages considering the example of a partially confined sandstone aquifer. *Radiocarbon* 25, 347–56.

Engesgaard, P., Hoiberg, A. L., Hinsby, K. et al. (2004) Transport and time lag of chlorofluorcarbon gases in the unsaturated zone, Rapis Creek, Denmark. *Vadose Zone Journal* 3, 1249–61.

Fontes, J.-C. and Garnier J.-M. (1979) Determination of the initial ^{14}C activity of the total dissolved carbon: A review of the existing models and a new approach. *Water Resources Research* 15, 399–413.

Grath, J., Scheidleder, A., Uhlig, S. et al. (2001) The EU Framework Directive: Statistical aspects of the identification of groundwater pollution trends, and aggregation of monitoring results. Final Report. In Austrian Federal Ministry of Agriculture and Forestry.

Ho, D.T. and Schlosser, P. (2000) Atmospheric SF_6 near a large urban area. *Geophysical Research Letters* 27, 1679–782.

Hofer, M. and Imboden, D.M. (1998) Simultaneous determination of CFC-11, CFC-12, N_2 and Ar in water. *Analytical Chemistry* 70, 24–729.

IAEA (1992) Statistical treatment of data on environmental isotopes in precipitation. Technical Reports Series 331, International Atomic Energy Agency, IAEA.

IAEA (2006) *Use of Chlorofluorocarbons in Hydrology a Guidebook*. IAEA, Vienna.

Ingerson, E. and Pearson, F.J. Jr. (1964) Estimation of age and rate of motion of groundwater by the ^{14}C-method. In: Y. Miyake and Koyama, T. (eds.) *Recent Researches in the Fields of Hydrosphere, Atmosphere and Nuclear Geochemistry*. Maruzen, Tokyo. pp. 263–83.

Kalin, R.M. (2000) Radiocarbon dating of groundwater systems. In: Cook, P.G. and Herczeg, A.L. (eds.) *Environmental Tracers in Subsurface Hydrology*. Kluwer Academic, Boston, MA, pp. 111–44.

Kalin, R.M. and Long, A. (1994) *Application of Hydrogeochemical Modelling for Validation of Hydrologic Flow Modelling in the Tucson Basin Aquifer, Arizona, USA*. Tecdoc-777, IAEA, Vienna.

Kaufman, S. and Libby, W.F. (1954) The natural distribution of tritium. *Physical Reviews* 93, 1337–44.

Lal, D. and Peters, B. (1962) Cosmic ray produced isotopes and their application to problems in geophysics. *Elementary Particle and Cosmic Ray Physics* 6, 1–74.

Lehmann, B.E., Love, A., Purtschert, R. et al. (2003) A comparison of groundwater dating with [81]Kr, [36]Cl and [4]He in 4 wells of the Great Artesian Basin, Australia. *Earth and Planetary Science Letters* 211, 237–50.

Lehmann, B.E., Purtschert, R., Loosli, H.H. et al. (2002) 81Kr-calibration of 36Cl- and 4He-evolution in the western Great Artesian Basin, Australia. *Geochimica et Cosmochimica Acta* Davos, p. A445.

Levin, I., Bösiger, R., Bonani, G. et al. (1992) Radiocarbon in atmospheric carbon dioxide and methane; global distribution and trends. In: Taylor, R.E., Long, A. and Kra, R.S. (eds) *Radiocarbon after Four Decades. An Interdisciplinary Perspective* (Co-publication with Radiocarbon) Springer-Verlag, New York.

Loosli, H.H. (1983) A dating method with [39]Ar. *Earth and Planetary Science Letters* 63, 51–62.

Loosli, H.H., Blaser, P., Darling, G. et al. (2001) Isotopic methods and their hydrogeochemical context in the investigation of palaeowaters. In: Edmunds W.M. and Milne, C.J. (eds.)

Palaeowaters in Coastal Europe: Evolution of Groundwater since the Late Pleistocene. Special Publication 189, Geological Society, London. pp. 193–212.

Loosli, H.H., Lehmann, B.E. and Balderer W. (1989) Argon-39, argon-37 and krypton-85 isotopes in Stripa groundwaters. *Geochimica et Cosmochimica Acta* 53,1825–29.

Loosli, H.H. and Purtschert, R. (2005) Rare Gases. In: Aggarwal, P., Gat, J.R. and Froehlich, K. (eds) *Isotopes in the Water Cycle: Past, Present and Future of a Developing Science.* IAEA, Vienna. pp. 91–95.

Love, A.J., Herczeg, A.L., Samson, L. et al. (2000) Sources of chloride and implications for 36Cl dating of old groundwater, South Western Great Artesian basin, Australia. *Water Resources Research* 36, 1561–74.

Lucas, L. and Unterweger M.P. (2000) Comprehensive review and critical evaluation of the half-life of tritium. *Journal of Research of the National Institute of Standards and Technology* 105, 541–49.

Maiss, M. and Levin, I. (1994) Global increase of SF_6 observed in the atmosphere. *Geophysical Research Letters* 21, 569–72.

Maloszewski, P. and Zuber, A. (1996) Lumped parameter models for the interpretation of environmental tracer data. In: *Manual on Mathematical Models in Isotope Hydrology.* IAEA, Vienna. pp. 9–58.

Mook, W.G. (1980) Cabon-14 in hydrogeological studies. In: Fritz, P. and Fontes, J.C. (eds) *Handbook of Environmental Isotope Geochemistry,* Vol. 1, Elsevier Science Publishing Company, Amsterdam, pp. 49–74.

Morris, R.A., Miller, T.M, Viggiano, A.A. et al. (1995) Effect of electron and ion reactions on atmospheric lifetimes of fully fluorinated compounds. *Journal of Geophysical Research - Atmosphere* 100(D1), 1287–94.

Oeschger, H. (1987) Accelerator mass spectrometry and ice core research. *Nuclear Instruments Methods in Physics Research* B29, 196–201.

Pearson, F.J. Jr, W. Balderer, H.H. Loosli, et al. (1991) Applied isotope hydrogeology: a case study in Northern Switzerland. Technical Report 88–01, Elsevier, Amsterdam.

Phillips, F. (1999) Chlorine-36. In: Cook, P.G. and Herczeg, A.L (eds.) *Environmental Tracers in Subsurface Hydrology.* Kluwer, Boston, MA, pp. 379–96.

Plummer, L.N., Prestemon, E.C. and Parkhurst D.L. (1991) An interactive code (NETPATH) for modeling net geochemical reactions along a flow path. Water-Resources Investigations Report 91-4078, US Geological Survey.

Purtschert, R., Lehmann, B.E. and Loosli H.H. (2001) Groundwater dating and subsurface processes investigated by noble gas isotopes (^{37}Ar, ^{39}Ar, ^{85}Kr, ^{222}Rn, ^{4}He). In: Cidu, R. (ed.) *Water Rock Interaction,* WRI-10, Villasimus, Italy, pp. 1569–73.

Roether, W. (1967) Estimating the tritium input to groundwater from wine samples: Groundwater and direct run-off contribution to Central European surface waters. In: *Isotopes in Hydrology.* IAEA, Vienna, pp. 73–9.

Schlosser, P. (1992) Tritium/^{3}He dating of waters in natural systems. In: *Isotopes of Noble Gases as Tracers in Environmental Studies.* IAEA, Vienna, pp. 123–45.

Schlosser, P., Stute, M., Doerr, H. et al. (1988) Tritium/^{3}He dating of shallow groundwater. *Earth and Planetary Science Letters* 89, 353–62.

Solomon, D.K., Poreda, R.J., Schiff, S.L. et al. (1992) Tritium and helium 3 as groundwater age tracers in the Borden aquifer. *Water Resources Research* 28, 741–55.

Solomon, D.K., Schiff, S.L., Poreda, R.J. et al. (1993) A validation of the ^{3}H/^{3}He-method for determining groundwater recharge. *Water Resources Research* 29, 2591–962.

Solomon, D.K., Hunt, A. and Poreda, R.J. (1996). Source of radiogenic helium 4 in shallow aquifers: Implications for dating younggroundwater. *Water Resources Research* 32, 1805–13.

Sturchio, N.C., Lu, Z.T., Purtschert, R. et al. (2004) One million year old groundwater in the Sahara revealed by krypton-81 and chlorine-36. *Geophysical Research Letters* 31, L05503. DOI: 10.1029/2003GL019234, 2004, pp. 1–4.

Sudicky, E.A. and E.O. Frind. (1981) Carbon 14 dating of groundwater in confined aquifers: Implications of aquitard diffusion. *Water Resources Research* 17, 1060–4.

Synal, H.A., Beer, J., Bonani, G. et al. (1990) Atmospheric transport of bomb-produced ^{36}Cl. *Nuclear Instruments and Methods in Physics Research* B 52, 483–8.

Tolstikhin, I.N. and Kamenskiy, I.L. (1969) Determination of ground-water ages by the T-^{3}He method. *Geochemistry International* 6, 810–11.

Tolstikhin, I., Lehmann, B.E. Loosli, H.H. et al. (1996) Helium and argon isotopes in rocks, minerals, and related groundwaters: A case study in northern Switzerland. *Geochimica et Cosmochimica Acta* 60, 1497–514.

Torgersen, T. and Clarke, W.B. (1985) Helium accumulation in groundwater, I: An evaluation of sources and the continental flux of crustal ^{4}He in the Great Artesian Basin, Australia. *Geochimica et Cosmochimica Acta* 49, 1211–18.

Torgersen, T. and Ivey, G.N. (1985) Helium accumulation in groundwater, II: A model for the accumulation of the crustal ^{4}He degassing flux. *Geochimica et Cosmochimica Acta* 49, 2445–52.

Torgersen, T., Habermehl, M.A., Phillips, F.M. et al. (1991) Chlorine 36 dating of very old groundwater 3. Further studies in the Great Arthesian Basin. *Water Resources Research* 27, 3201–13.

Weiss, W., Sartorius, H. and Stockburger, H. (1989) Global distribution of atmospheric ^{81}Kr; a database for the verification of transport and mixing models. In: *Isotopes of Noble Gases as Tracers in Environmental Studies.* IAEA, Vienna. pp. 29–92.

5B   Dating Examples in European Reference Aquifers

R. PURTSCHERT, J.A. CORCHO ALVARADO AND H. H. LOOSLI

In the previous chapter a brief summary of dating tracers and methods applied to the European reference aquifers were presented. In this chapter, five reference aquifers with a wide range of groundwater residence times and hydrogeological conditions were selected as exemplary case studies. It is shown how multiple tracer measurements can be used to resolve the age structure of a groundwater body, how disruptive effects for tracer dating can be identified and how groundwater quality and residence time are related.

5B.1   Age Classification

Estimated ranges of groundwater residence times in the investigated reference aquifers are summarised in this chapter. In most cases, groundwaters with ages up to many thousand of years can be found. In the context of baseline groundwater quality assessment, three classes of residence times are distinguished:

1 modern waters: younger than 50–60 years;
2 ancient waters: older than 60 years but infiltrated in Holocene;
3 palaeowaters: recharged during or prior to the last glacial period.

1 These waters are part of the 'active' groundwater flow system of the water cycle (Seiler and Lindner 1994; Seiler et al. 1999) and provide the main groundwater resource in Europe. Young waters are typically found at shallow depths and in unconfined parts of the aquifers. Because of the fast turnover time and the lack of confinement, these waters are most vulnerable to contamination. The investigation of recent waters is a crucial task and somehow a contradiction in terms. On the one hand the natural quality of shallow groundwater is of particular interest, and on the other pristine conditions are no longer present because of probable contamination. Dating methods with environmental tracers provide an instrument to extrapolate back from contaminated conditions to pristine conditions. Fortunately, several methods are available to date recent groundwaters (Figs. 5A.1 and 5A.2, Chapter 5A).

2 Groundwater pollution increased in parallel with industrial and agricultural development in Europe and elsewhere. Nitrate concentrations started to increase since the 1950s (Fig. 5A.2, Chapter 5A). Water recharged prior to this time is more likely to be in a pristine condition and free of contaminants. The composition of 'ancient' groundwater is often determined by time-dependent geochemical reactions between the waters and the aquifer material. The longer the groundwater residence time, the longer the contact time between water and aquifer matrix. The large age range of European groundwaters produces a correspondingly large range of solute concentrations in pristine waters. However, groundwaters with residence times slightly above 50 year are very interesting because they best represent the pristine condition of young waters. Only few tracers are available for this important age range. ^{39}Ar was used in several reference aquifer investigations.

3 Palaeowaters recharged more than 10,000 years ago, originated during a time of different climatic and under different hydrogeological conditions than today (Edmunds and Milne 2001). Although the quality of palaeowaters is often very high, they have played a minor role for European drinking water supply. However, as a consequence of increasing pollution of shallow aquifers, the pressure to exploit such high-quality reserves increases. Different (drier) recharge regimes in the past and increasing abstraction rates today lead to a misbalance of renewal and abstraction rate. Exploiting palaeowaters in many parts of the world corresponds therefore to groundwater mining, although in wetter countries with adequate recharge the older pristine waters may be gradually replaced by modern waters. The most important dating methods for palaeowaters are ^{14}C in combination with radiogenic helium.

5B.2   Case Studies in Selected Aquifers

5B.2.1   UK Triassic Sandstone aquifer

5B.2.1.1   Hydrogeology

The East Midlands (Sherwood) Sandstone is situated on the western rim of the North Sea Basin. The aquifer consists of coarse- to very fine-grained sandstone and thickens northwards from 120 to 300 m. The formation dips eastwards with a gradient of about 1/50. The modern hydraulic gradient (1/250) and hydraulic conductivity indicate a flow velocity of about 0.2 m yr^{-1}. Earlier isotope data (Bath et al. 1987) indicate a higher flow velocity (~ 0.6 m yr^{-1}). The aquifer is confined by a thick sequence of mud rocks (Mercia Mudstone) (see cross-section in Chapter 1 and Fig. 5B.1. From the piezometric gradient, it can be concluded that any water flow through the Mudstone would be upward and vertical infiltration through the confining layer is very unlikely. Therefore, recharge occurs exclusively in the western part, where the Sherwood Sandstone outcrops (Chapter 4). Prior to development, much of the recharge water was probably discharged via springs and rivers with a lesser portion entering the deep aquifer. This resulted in a shallow fast circulating groundwater system near the outcrop and a much slower flowing movement of groundwater in the deep part of the aquifer. Discharge in the deep aquifer is believed to occur by leakage through faults and also the confining Mercia Mudstone.

5B.2.1.2   Dating results

The residence time of groundwater in the Sherwood Sandstone has been established in a number of studies using different isotopic and chemical indicators (Bath et al. 1979, 1987; Andrews et al. 1983; Edmunds and Smedley 2000). In the present studies, nine wells were re-sampled for an extensive set of modern tracers in order to investigate in more detail the application of young residence time indicators and to increase knowledge

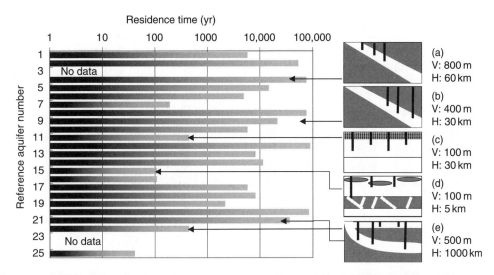

Fig. 5B.1 Estimated age spectra of reference aquifers investigated in the reference aquifers. Aquifer numbers refer to Table 2.1 in Chapter 2 in this volume. Arrows indicate aquifers which are discussed in this chapter. The hydrological situation, aquifer dimensions (V: vertical; H: horizontal), and approximate locations of sampling wells are indicated in the simplified cross sections. (a) Triassic sandstone aquifer, UK; confined aquifer sampling locations in the transition unconfined–confined. (b) Ledo–Paniselian aquifer, Belgium; confined aquifer, sampling locations mainly in confined part. (c) Fontainebleau sands aquifer, France; unconfined aquifer with thick unsaturated zone. (d) Pleistocene sand aquifers around Odense, Denmark: semi-confined aquifer with separated deeper layers. (e) Turonian (unconfined) and Cenomanian aquifer (confined), Czech Republic.

in particular in the age range between modern water and the dating range of the ^{14}C method. Samples were taken for ^{85}Kr, $^{3}H/^{3}He$, CFC-12, SF_6, ^{37}Ar, ^{39}Ar, ^{14}C and stable noble gases including helium isotopes.

It turned out that all SF_6 concentrations measured in the sampled groundwaters exceed the values in modern air-equilibrated water samples. This was most probable caused by an analytical interference with H_2S (Hofer personal communication). Therefore, in this case, SF_6 could not be used for dating. Young tracer concentrations above the detection limit were found in five wells (Figs. 5B.2 and 5B.3). In the diagrams, groundwater temperature was used as an evolution index because water temperatures

increase with depth and along the flow paths (see also Fig. 5B.4). Waters warmer than about 12.5°C are free of modern tracer and hence are older than 50 years. The estimate of age and mixing ratio of modern water components, applying the fitting procedure explained in the previous chapter, resulted in maximum 50% modern water at Far Baulker in the recharge area (Fig. 5B.2). The results of the dispersion model and the exponential model do not differ significantly. The calculated residence times of recent water components range from 12 to 60 years. The consistency of all tracers in the means of this simple box model approach is quite good for those wells with more than 10% modern water (low χ^2). For lower concentrations, the

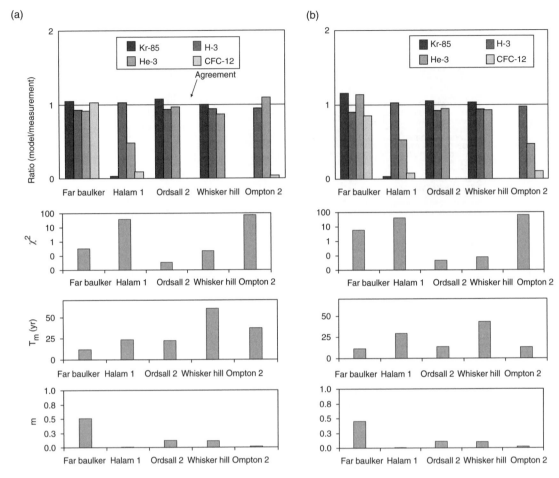

Fig. 5B.2 Estimation of residence time and mixing ratio of recent water components in the UK Triassic Sandstone aquifer. (a) Dispersion model, (b) exponential model. From top to bottom: Ratios of modelled versus measured tracer concentrations. Values close to unity for all tracers indicate consistency between the tracers in the frame of the applied model. χ^2 values: Large numbers indicate inconsistency of tracers; Tm : Mean residence time; m: Estimated portion of recent water components (see Fig. 5A.3).

different tracer ages and mixing portions do not agree (high χ^2). In particular the gaseous tracers [85]Kr and CFC-12 indicate a higher amount of recent water compared to the tritium results. This disagreement is most likely the result of a weak contamination with modern air during or shortly before sampling. It is concluded that a reliable

and simultaneous determination of mixing ratio and residence time of recent water components is only possible when at least more than 5% young water is present. However, the identification of recent water by at least one tracer is already valuable information, and is also important for the future monitoring of these wells. Pumping in

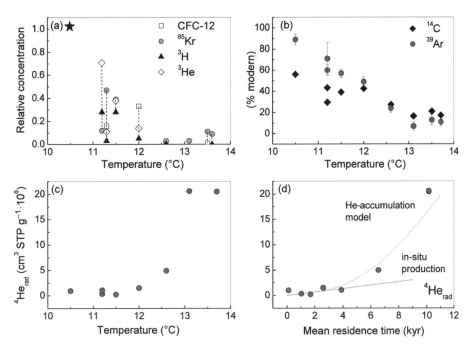

Fig. 5B.3 (a) Relative concentrations (reference well χ: Far Baulker) of young residence time indicators as function of groundwater temperature. (b) 14C and 39Ar activities decrease linearly with groundwater temperature. (c) Radiogenic helium versus groundwater temperature. (d) Radiogenic helium as function of mean residence time. The solid line indicates the in situ accumulation rate. The dotted line is based on model calculations assuming an external helium flux entering the aquifer from below.

large-screen boreholes disturbs the flow system and produces highly mixed waters with modern components but also containing much older components from deeper parts of the aquifer.

For a more detailed understanding of the age structure of the aquifer ^{39}Ar, ^{14}C and ^{4}He were measured. A general increase of residence time in the direction of groundwater flow or increasing depth is demonstrated by decreasing ^{39}Ar and ^{14}C activities and an increase of the concentration of radiogenic ^{4}He (Fig. 5B.3[b]–[d]). Calculated ^{14}C model ages, which take into account the dilution of atmospheric ^{14}C with dead carbon of the dissolving carbonate, increase up to 8000 years,

whereas for the same waters, ^{39}Ar ages do not exceed 1400 years which has to be interpreted as a discrepancy of the two methods if piston flow is assumed. However, the linear correlation with temperature, and thus between the ^{14}C and ^{39}Ar concentrations, is a strong indication of mixing. In the long screened intervals of the water abstraction wells, a mixture of relatively young water (ages of a few hundred years) from shallow depths and older waters at the bottom of the aquifer (ages up to 13,000 years) is extracted.

The mean residence time of this mixture increases further downgradient. The temporal evolution of helium (Fig. 5B.3[d]) dissolved in groundwater in excess of the atmospheric equilibrium concentration depends on the subsurface accumulation rate. In most cases, it is assumed that accumulation from in situ produced helium results in a linear He increase with time. The accumulation rate originating from 'external' sources is very often diffusion controlled and therefore spatially not constant. If we assume an external helium flux enters the aquifer from below and superimposes on helium produced in situ at a rate P, then, because the screen, intervals of boreholes in deep confined aquifers normally do not reach the bottom of the aquifer, it takes some time until the external helium flux 'can be seen' in the well (Torgersen and Clarke 1985; Torgersen and Ivey 1985). The resulting helium evolution as a function of the mean groundwater residence time was calculated and is compared with the data on Fig. 5B.3(d). The mean groundwater ages of the wells were calculated using ^{39}Ar and ^{14}C concentrations. The model and data show a very similar pattern. With such a calibration of the 'helium clock' it is possible to date groundwaters beyond the dating range of ^{14}C. But one has to keep in mind that the helium accumulation rate may be different for other aquifers and that the clock has to be calibrated for each individual case.

5B.2.1.3   Summary and implications

The groundwaters in the investigated part of the UK Triassic Sandstone aquifer show pronounced age stratification with depth. Wells that are in or very close to the recharge area extract modern water components with a mixing portion between 1 and 50%. Further along the flow path, in the confined part, water components in the age range of 50–300 years dominate. Possibly a part of this water advanced deeper into the confined part of the aquifer due to high extraction rates over the last 30 years. The portion of this intermediate water decreases linearly with depth. Near the base of the aquifer, water residence times up to 12–13 kyr can be found (wells Grove and Hayton). Older Pleistocene waters are recorded further downstream (Bath et al. 1979; Edmunds and Smedley 2000).

For groundwater management and monitoring, it is crucial to realise that mean residence times calculated, for example, from ^{14}C activities have to be interpreted as the result of mixing of younger (less than 300 years) and much older waters (even if no ^{3}H or other indicators for modern waters are detectable). This has consequences for the assessment of the future evolution of groundwater quality. A shift of mixing ratios between young and old water, disturbing the baseline conditions, occurs possibly within a relatively short time interval, although ^{14}C ages are of the order of several thousands of years.

The evolution of a set of chemical parameters as a function of the mean residence time is depicted in Fig. 5B.4. A detailed discussion of the relevant geochemical processes is given in Chapters 1 and 4. It must be borne in mind that the trends shown in the Fig. 5B.4 are the combined result of time-dependent water–rock interactions and gradual mixing of waters with different ages (from different depths). Wells plotted on the left of the dashed line extract varying amounts of recent water components based on tracer dating. It is considered that groundwater temperature is indeed a good indicator for the age of the water (Fig. 5B.4[a]). The decrease in nitrate and sulphate concentrations can be attributed to the decreasing amount of modern water and less

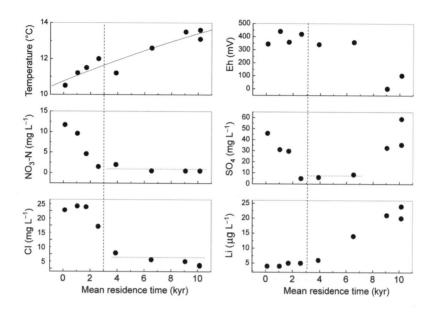

Fig. 5B.4 Downgradient trends for various chemical parameters in the Triassic sandstone aquifer plotted as a function of mean groundwater residence time calculated based on mixing ratios and age of the different water components. Samples to the right of the dotted vertical lines are free of modern water components.

to diminution under reducing conditions; Eh values decrease further downgradient indicating the redox boundary (Smedley and Edmunds 2002). Concentrations of NO_3, SO_4 and Cl at the right side of the modern water line (in age range 4000–6000 years) can be regarded as best representatives for **baseline** conditions for waters in the shallow part of the aquifer (dotted lines in the graphs). The progressive dissolution of gypsum and anhydrite leads to a notable increase in SO_4 further downgradient (Edmunds and Smedley 2000).

Lithium concentrations are linearly correlated with the mean groundwater residence time solely as a result of time-dependent rock–water interaction. Two conclusions can be drawn: (1) the definition of a **baseline** concentration of Li without any specification of the age of the groundwater is

meaningless; and (2) provided that the Li-age relationship is identified, Li can in principle be used as a (low cost) residence time indicator (Edmunds and Smedley 2000). However, the calibration of such a chemical dating tool has to be performed for each aquifer by using absolute dating tools. Caution has also to be placed with extrapolations beyond the calibrated dating range.

5B.2.2 Ledo–Paniselian aquifer, Belgium

5B.2.2.1 Hydrogeology

The Ledo–Paniselian aquifer (LPA) in Belgium forms part of an alternating sequence of marine Tertiary clay and sand deposits gently dipping towards the north (Walraevens 1990, 1998). Piezometric levels demonstrate that recharge to the Ledo–Paniselian aquifer takes place in topographically higher regions where the aquifer is covered by Bartonian Clay and therefore is semi-confined. Discharge occurs to the south – in the unconfined area – where the Ledo–Paniselian crops out from beneath the Quaternary sediments. The remaining part of the recharged groundwater flows to the north, where the depth of the aquifer increases (Fig. 5B.1). Here, upward leakage through the Bartonian Clay occurs, causing flow velocities in the Ledo–Paniselian to diminish gradually.

5B.2.2.2 Dating results

Extensive geochemical, hydrodynamic and modelling studies have allowed a general understanding of this freshening ground-water system of a sandy aquifer in Belgium (e.g. Walraevens et al. 1990). From 1996 to 2001, samples were collected (partly repeatedly) from 39 wells for ^{14}C, δ^{13}C and noble

gas isotopes measurements as well as for the hydrochemistry. These were collected from two infiltrating areas along two assumed flow lines about 20 km to the north and from greater depths.

The two main objectives for the extensive isotope study were
1 to directly determine the timescales of the chemical evolution of the groundwaters and to compare them with the hydraulic calculations;
2 to compare various dating tools for old waters (^{14}C, ^{4}He, climatic signals).

^{14}C ages are model ages because for each sample an initial ^{14}C-value (as starting activity for radioactive decay) has to be determined. In the groundwater flow direction of the Ledo–Paniselian aquifer, the following general trends have been established: the ^{14}C activity decreases, scattering considerably up to about 10 km from the recharge area and showing low values below 5 pmc at longer distances and mainly in the eastern

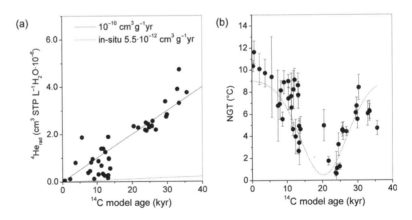

Fig. 5B.5 (a) Accumulation of radiogenic helium as a function of corrected 14C age in the Ledo–Paniselian aquifer. The in situ accumulation rate was calculated based on measured U and Th concentrations of the aquifer rock and the porosity. The observed line indicates that helium from an external source contributes to the overall helium budget. (b) Noble gas recharge temperatures of the Ledo–Paniselian groundwaters plotted against distance from the recharge area. The (hand draw) broken fitted line indicates a transition of Holocene temperatures to colder, Pleistocene climatic conditions.

flow path. The $\delta^{13}C$ values increase regularly from about −13‰ at 5 km distance, to about −4‰ at 20 km; values above −6‰ are more frequent along the eastern flow path. There, also the HCO_3 content is higher (between 600 and 700 mg L^{-1}) whereas it is around 400 mg L^{-1} in the western path up to about 12 km distance. This demonstrates that in the eastern (and partly in the northern) part, the carbonate chemistry and the exchange processes occurred under different conditions than in the western flow direction. In general, reliable ^{14}C model ages can be calculated if the $\delta^{13}C$ values are below about −6‰. Such ages between modern and about 20,000 years could be calculated for the Ledo–Paniselian aquifer and are plotted in Fig. 5B.5(a). In addition, waters must be very old if very low ^{14}C activities (up to a few pmc) and 'normal' $\delta^{13}C$ values have been measured. However, in this case a quantitative conversion into model ages is questionable and lower age limits are preferable because of a small amount of contamination during sampling; a small admixture of young water or underground production may have introduced the small ^{14}C activity. If $\delta^{13}C$ values are more positive (measured values are up to +4‰ in the Ledo–Paniselian aquifer) calculated ^{14}C model ages are highly questionable because additional processes should be considered.

As in the UK Triassic aquifer, a systematic increase of the He content with the age of the water was measured in the Ledo–Paniselian groundwater (Aeschbach-Hertig et al. 2003). On Fig. 5B.5(a), this increase of radiogenic He is plotted as a function of calculated ^{14}C model ages. Two straight lines are added, representing a linear increase with time: the lower line corresponds, within about a factor of two, to the in situ production rate calculated with assumed U and Th concentrations of the rocks. Apparently, some waters in the Ledo–Paniselian aquifer accumulate He at a rate which is typical for in situ production. The line with a slope, which is about a factor 20 higher, is a fit through the measured higher concentrations. These higher values are explained by additional He, for example, transferred from the aquitards above and below into the aquifer. Some waters in the Ledo–Paniselian aquifer contain even higher concentrations, but these locations are often not along the main flow paths or are at larger distances from the recharge area. The linear increase of the He concentration with time supports, in a qualitative way, the calculated ^{14}C model ages and may even be used for a quantitative extrapolation of ^{14}C ages beyond the dating range of this method.

Preliminary estimations of hydraulic water ages for the Ledo–Paniselian agree up to now only by an observed linear correlation, however, hydraulic ages seem to be considerably lower. A relatively large scatter in the correlation is also observed. One reason may be that the concept of flow lines is too simplified and that mixing processes influence the two dating methods differently. By following the evolution of chemical and isotopic concentrations along the assumed flow lines, areas of intensive mixing of different water bodies could be determined at about 10 km distance from the recharge area for the Ledo–Paniselian groundwater. Such areas may in general be good locations to collect monitoring samples and to follow time series, because natural and man-made induced trends may be detected. Within a few years changes of the ^{14}C activity were even measured for samples older than a few kyr, mainly at and to the north of the mixing zones. Mixing and age structures are connected and therefore several parameters have to be determined. Geochemical and isotopic information is supplementary and has to be combined, especially to allow predictions of future water quality by time-calibrated flow models.

In Fig. 5B.5(b), Noble gas recharge temperatures (NGRT) are plotted against distance from the recharge area (Aeschbach-Hertig et al. 2003). Rarely have so many NGRT been determined in an aquifer system. To increase the reliability, sampling from the area of cooler climatic conditions at about 12 km distance from recharge has been repeated giving consistent results. However, it was found that from the whole set of measured concentrations of Ne, Ar, Kr and Xe, about 20% could not be converted into reliable NGRT or showed large errors (Aeschbach-Hertig et al. 2003). Fig. 5B.5(b) demonstrates that waters with ages up to about 10 kyr yield NGRT of 8–10°C, in agreement with modern air temperatures, increasing confidence in the applied calculation model (Aeschbach-Hertig et al. 1999). Some samples, however, showed lower temperatures (4–6°C), which may be an effect of mixing. Samples in the age range 11–25 kyr recharged clearly under cooler climatic conditions with a maximal temperature difference of about 9°C to the Holocene. Before 25 kyr NGRT increase again gradually. The stable isotope results, as additional climate indicators, show that the Ledo–Paniselian paleowaters do not have significantly lower values, as already shown in the discussion of the continental dependence of $\delta^{18}O$ in Europe (Edmunds and Milne 2001).

5B.2.2.3   Summary and implications

With respect to the dating methods, the following conclusions can be drawn from the measurements in the Ledo–Paniselian aquifer: (1) ^{14}C is the main dating method for waters older than about 1000 years; (2) ^{14}C ages may be supported in a semi-quantitative way by ^4He results; (3) ^{14}C and hydraulic ages agree only qualitatively; (4) the combination of geochemical with isotopic data allows the determination of areas of intensive mixing and the time calibration of flow models and (5) based on ^{14}C results and on NGRT, some waters in the Ledo–Paniselian aquifer infiltrated during the Pleistocene.

Because of water–rock interaction, a large range of concentrations of chemical components has to be expected. This range can be lowered by adding isotope results, which allow dividing the waters into subgroups along their evolution along flow lines. In this way, the evolution of **baseline** chemistry can be understood as time-dependent processes. In the Ledo–Paniselian groundwater, for example, the Ca concentration decrease and the HCO_3 and Na concentrations increase (due to cation exchange) with time on both flow paths (Fig. 5B.6). The change of water types corresponds to age scales of 10–30 kyr and longer. This conclusion demonstrates that the combination of geochemical and isotopic data improves our quantitative understanding of a flow system.

5B.2.3   Fontainebleau Sands aquifer, France

5B.2.3.1   Hydrogeology

The Fontainebleau Sands aquifer forms part of the shallow sequence of sediments in the Paris Basin (France), which is the largest sedimentary basin in Western Europe. The Oligocene sandy aquifer is enclosed between two clayey layers: above is the Beauce

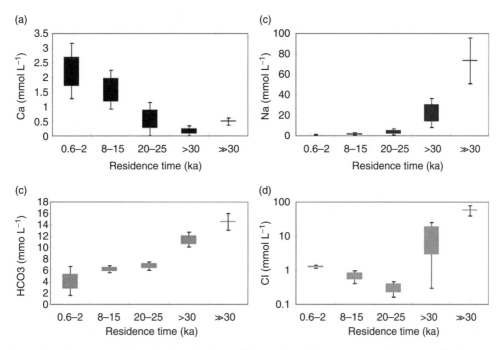

Fig. 5B.6 Hydrochemical evolution along the main flow path of the Ledo–Paniselian Aquifer.

formation which was altered by diagenesis from limestone to millstone and clay (Thiry 1988); and below are Oligocene and Eocene marls which separate the Fontainebleau Sands from the underlying Eocene multi-layered aquifer.

The sediments comprise very fine, well-sorted quartz grains with an average diameter of 100 µm. The Fontainebleau Sands formation has a thickness of 50–70 m, a hydraulic transmissivity of $1 \cdot 10^{-3}$–$5 \cdot 10^{-3}$ m^2 s^{-1} and a mean total porosity of about 25%. The upper part of the formation comprises up to 99% pure quartz sands (white facies), while the content of organic matter, carbonates, sulphides, feldspar and clays (dark facies) increases with depth (Bariteau 1996).

The mean precipitation in the area is about 700 mm yr^{-1} (measured at Trappes climatic station of Meteo France; observation period from 1991 to 2000). The estimated recharge rate varies between 80 and 210 mm yr^{-1} based on hydrograph data (Mercier 1981; Bariteau 1996). Water table depths in the Fontainebleau Sandstone range between 20 and 40 m below ground surface (Fig. 5B.1[c]). This relatively thick unsaturated zone plays a key role in the transport of substances into the saturated zone; therefore it has to be considered in the interpretation of young residence time indicators (^{85}Kr, ^{3}H/^{3}He, CFC, SF$_6$).

5B.2.3.2 Dating results

As a first step, young groundwater components were characterised using the fitting approach presented in the previous chapter (Section 5A.2.5). Unconfined conditions, homogeneous recharge and aquifer properties

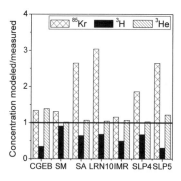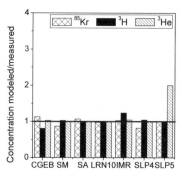

Fig. 5B.7 Ratios of modelled versus measured tracer concentrations. Values close to unity for all tracers indicate consistency between the dating methods and that the applied box-model is adequate. Only the results of the most plausible exponential model (EM) are shown: (left) Without considering infiltration through a 35 m thick unsaturated zone; (right) results with corrected input function.

lead to an exponential age distribution if the borehole is screened over the whole aquifer thickness (Vogel 1967). This situation is fairly well justified in this area. The exponential model (EM) was therefore applied in this study. In addition, the thick unsaturated zone will shift the input concentrations of ^{3}H, ^{3}He and ^{85}Kr at the water table compared to atmospheric values. It is known that gaseous tracers, and tracers that are part of the water molecule such as tritium, are delayed by different rates during transport though the unsaturated zone (Cook and Solomon 1995). Furthermore, the volatile ^{3}He is completely lost to the atmosphere as long as ^{3}H is transported through the unsaturated zone. In other words, the $^{3}H/^{3}He$ clock starts at the water table. The interpretation of ^{3}H, ^{3}He and ^{85}Kr with the EM model, using atmospheric input values, reveals large discrepancies for the 'best-fit' calculations (Fig. 5B.7). The modelled ^{3}H concentrations seem to be too low and ^{85}Kr (and ^{3}He) too high compared to the measurements. Because ^{3}H concentrations in precipitation have decreased since the bomb peak, higher residence times in the saturated or unsaturated

zone have to be assumed in the modelling for tritium in order to diminish the discrepancy. Thus, a one-dimensional model of the unsaturated zone was implemented to calculate the different shifts of the input functions for ^{85}Kr and ^{3}H. The resulting travel times of ^{3}H and ^{85}Kr through the unsaturated zone are 11–25 years and 1–10 years, respectively (Corcho et al. 2007). Using the calculated input functions at the water table, a much better agreement between the modelled and measured values could be attained for all of the wells except for SLP 5 where degassing of ^{3}He within the aquifer or during sampling is most likely the reason for this deviation (Fig. 5B.7).

The best-fit model parameters for mean residence time (T_m) and mixing portion of young water components (m) range between 1–13 years and 7–40%, respectively. This means that a large fraction of older water, which cannot be detected using young residence time indicators, contributes to the overall age spectra. In some cases, a strong correlation between T_m and m was observed (Corcho et al. 2007). In other words, an increased mean residence time T_m (less water

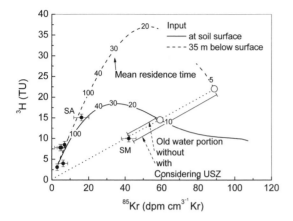

Fig. 5B.8 Measured 3H versus 85Kr concentrations compared with exponential model calculations. The lines are labelled with the mean residence time Tm. Points of intersection (open circles) indicate age of the modern component of sample SM. Length of arrows relative to total distance to the origin of coordinate axes defines the mixing proportion of old water.

younger than 50 years) can be compensated by a higher mixing portion of this water (more water younger than 50 years). This problem becomes clearer if only the two tracers ^{85}Kr and ^{3}H are considered (Fig. 5B.8). The lines on Fig. 5B.8 represent the EM with input functions at the soil surface and at the water table (35 m depth) respectively. With two tracers, it is possible to determine two parameters. The mean residence time of the young component is defined by the intercept of the mixing line with the model curve. The portion of young water is given by the relative distance of the data point to the origin of the coordinate axes. This can be done for sample SM. It becomes evident that the selection of the input function is crucial for dating and mixing estimates. With the atmospheric input, a mean residence time T_m of about 12 years

and a mixing portion m of 70% can be estimated from Fig. 5B.8. With the input at the water table, T_m of 4 years and a mixing portion m of 40% results. However, most data points lie in the area where the model curves have a strong gradient towards the origin of the coordinate axes. In this area, for high mean EM residence times, it is not possible to distinguish between ageing and admixture of old water. This ambiguity can only be resolved with a tracer which is sensitive to residence times above the 50 years limit. ^{39}Ar with a half-life of 269 years is such a tracer. Ages calculated with the exponential model based on ^{39}Ar (Table 5B.1) range between 100 and 400 years and confirm the findings that the main weighting of the age distribution is older than 50 years. Within uncertainties, most samples can be explained consistently with a one component EM. However, at least for sample SA, clearly a two-component mixing scenario with a young component and an old component of about 400 years has to be assumed.

5B.2.3.3   Summary and implications

For the interpretation of environmental tracer data in terms of groundwater residence times for this aquifer, it is crucial to take into account several observations:
1 A thick unsaturated soil zone overlies the aquifer.
2 Samples were taken from large-screened borehole intervals intercepting the water table in an area with spatially distributed recharge. Each water sample is therefore a mixture of waters of different age. This situation is not uncommon, because wells for water supply are screened over large parts of the aquifer in order to maximise well yield. The distribution of groundwater ages in this case

Table 5B.1 Radioactive and stable isotope measurements and calculated apparent tracer ages for groundwaters from the Fontainebleau aquifer in 2001.

Well	Data					Piston flow ages (yr)			Exponential ages (yr)	
	^{85}Kr (dpm/ cm$^{3-1}$ Kr)	^{3}H (TU)	^{39}Ar (%mod)	^{14}C (pmC)	^{13}C (‰)	^{3}H/^{3}He	^{85}Kr	^{39}Ar	^{85}Kr	^{39}Ar
SM	43.0 ± 5.0	10.0*	79 ± 7	80.2° 6	−16.3	8 ± 1	11 ± 1	91 ± 35	11 ± 2	103 ± 45
CGEB	6.8 ± 0.7	8.5	73 ± 5	75.1	−14	9 ± 1	29 ± 1	122 ± 27	122 ± 3	144 ± 37
SA	16.1 ± 4.1	15.1	69 ± 5	84.2	−14.3	11 ± 1	20 ± 2	144 ± 28	46 ± 4	174 ± 41
LRN10	6.1 ± 4.8	7.8	77 ± 5	73.7	−14.1	15 ± 1	30 ± 5	101 ± 25	137 ± 10	116 ± 33
IMR	2.9 ± 0.4	3.1	55 ± 5	69.8	−13.8	9 ± 2	35 ± 1	232 ± 36	299 ± 2	318 ± 65
SLP4	6.2 ± 2.5	7.8	59 ± 5	75.5	−13.9	2 ± 1	30 ± 3	205 ± 33	335 ± 30	270 ± 56
SLP5	5.6 ± 2.8	4.0	51 ± 5	73.8	−13.7	13 ± 2	31 ± 5	261 ± 38	150 ± 25	373 ± 76

*Error 0.2 TU; ° 0.6 pmC.

is approximated by an exponential function (Zuber 1986).

3 A large fraction of water is older than 50 years. The sensitivity of young residence time indicators (e.g. ^{85}Kr, ^{3}H/^{3}He, SF$_6$) to characterise the overall age distribution is limited and a tracer with extended dating range is required. For that reason ^{39}Ar was applied in this study.

The fact that all wells contain various amounts of recent, and probably anthropogenically influenced water components, makes the evaluation of baseline chemistry difficult. However, knowing the portion of recent water components helps to derive conclusions about the natural groundwater chemistry. First of all it may be concluded that the portion of young water decreases with depth (Fig. 5B.9). As a first approximation, any anthropogenically derived substances should decrease with decreasing proportion of recent water components. The intercept with the coordinate on the graph represents the composition of water that is free of recent water components. A correlation between mixing ratio and concentration was found, for example, for Si, Eh and Cl

(Fig. 5B.9). No clear trend could be observed for NO_3 and SO_4.

In shallow aquifers, elevated chloride concentrations are very often attributed to anthropogenic influence such as road salting in winter. In the Fontainebleau aquifer, it seems that the chloride concentration increases with decreasing amount of young water, making it likely that it is most probably natural, and the increase in silica is clearly the result of prolonged water–rock interaction with the sandstone.

5B.2.3.4 Pleistocene sand aquifers around Odense, Denmark

Hydrogeology

The investigated area is located on the Island of Funen, close to the city of Odense, Denmark. Geologically, the site is situated within a complex setting of Quaternary glacio-fluvial sand aquifers with confining clay tills. The semi-confined glacio-fluvial sands, which constitute the main aquifer on the island, overlie a sequence of mainly Palaeocene marls and clays of varying thickness (typically

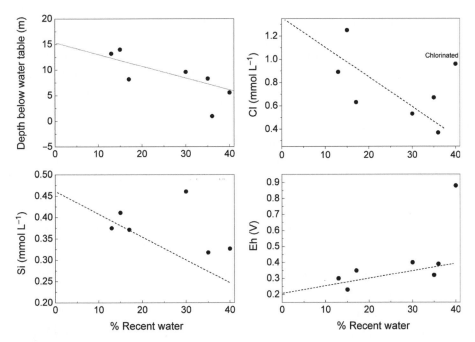

Fig. 5B.9 Mean screen depth and selected chemical parameters as a function of proportion of modern water components.

10–20 m) which form the lower boundary of the Quaternary aquifer system. Recharge occurs preferentially through sand 'windows' (Hinsby et al., this volume). The potentiometric head is at depths of between 4 and 10 m below surface level (b.s.l.) Screened intervals (5–14 m) within the sands are at depths ranging from 18 to 56 m b.s.l. A Chalk aquifer underlies the Odense shallow aquifer and in some areas deep wells also extract water from this aquifer.

Dating results

Three wells (34h, 74e and OdJ) were sampled during the first half of 2001 for the analysis of $^3H/^3He$, ^{85}Kr, SF_6 and CFC-12. Two depths were sampled from OdJ, one mixed sample from the complete screened section, and one sample from the bottom of the screen. ^{85}Kr, ^3H and CFC-12 concentrations above the detection limit (DL) were observed in all of the samples, in accordance with the hydrogeological situation which points to modern recharge. A CFC-12 concentration in excess of air-equilibrated water was measured in one sample and attributed to contamination. The SF_6 concentration in one sample (74e) was lower than the DL.

Measured concentrations of gases were converted into corresponding atmospheric input concentrations based on the in situ water temperature, elevation of the recharge area and the excess air content (Schlosser et al. 1989; Busenberg and Plummer 2000). No corrections are necessary for ^{85}Kr because the isotope ratio ($^{85}Kr/Kr$) is unaffected by details of the infiltration process. As a first approximation, the tracer data were interpreted assuming piston flow (PF) and neglecting hydrodynamic dispersion and mixing of different water components. Using the local input functions, residence times between 14 and 48 years were calculated (Corcho et al. 2002, 2005). Low SF_6 concentrations, close to the detection limit, allowed only the estimation of a minimum age of about 30 years for wells 74e OdJ. The comparisons of tracer ages depict two obvious discrepancies (Fig. 5B.10): CFC-12 ages are generally too high compared to the other tracers and $^3H/^3He$ ages for well OdJ are too low compared to ^{85}Kr ages. Neither dispersion nor mixing can explain the disagreements between the ^{85}Kr and CFC-12 ages. It is known from similar studies that CFCs can be degraded under reducing conditions (Busenberg and Plummer 1992). Because of the increasing input function this results in an overestimation of residence times. Thus, degradation stays as the only explanation

for the deviations. A degradation rate of 0.8–3.6 × 10^{-4} day^{-1} could be estimated in this case. $^3H/^3He$ ages correspond to the residence time of the originally tritium bearing water component while the other tracers indicate a mean age which is the combined result of the age of the young water component and dilution with an old tracer-free water component (Corcho et al. 2002).

^3H, ^3He and ^{85}Kr were interpreted applying the fitting method described in Chapter 5(a). In this case CFC-12 was excluded because an additional process (degradation), which is not included in the dispersion model, has modified the concentration in groundwater. The resulting mean residence times of the young component (T_m) range from 17 to 27 years. In the deepest well OdJ, an admixture of between 30% and 40% of old water was calculated. This is in perfect agreement with the fact that the screen of the well OdJ is located in the deepest part of the aquifer where hydraulic contact with the underlying chalk aquifer is likely. The simulated age distribution using a regional numerical flow model (Troldborg 2004) depicts a similar conclusion (Fig. 5B.10, inset).

Summary and implications

The combined application of a set of different residence time indicators for young groundwater allowed a very reliable dating of four samples from three boreholes in the Odense aquifer. It was possible to identify processes that erroneously affect residence times obtained by using only a single method. Under reducing conditions, groundwater residence times obtained from CFC measurements tend to be too high due to degradation. Considering the admixture of an old, and tracer-free, water component and dispersive mixing, it was possible to consistently

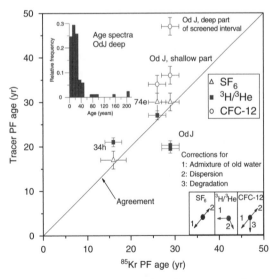

Fig. 5B.10 SF6, CFC-12 and 3H/3He piston flow ages (PF) versus 85Kr PF age at three wells of the Odense sands aquifer. Arrows indicate direction of age correction if the simple PF model is complemented with additional processes.

interpret ^{3}H/^{3}He, ^{85}Kr and SF$_6$ concentrations in the means of groundwater residence time.

This case study emphasises the importance of combining several tracer methods if detailed information about the age structure of a groundwater is needed, and also how the intercomparison of different tracers helps to identify disagreements.

5B.2.3.5 Turonian and Cenomanian aquifer, Czech Republic

Hydrogeology

The Bohemian Cretaceous Basin hosts the semi-confined Turonian sands aquifer and the underlying confined Cenomanian sandstone aquifer (Paces et al., Chapter 17, this volume). Both aquifers are an important source of high-quality groundwater in the Czech Republic. The sandstone layers are made of up to 99% of quartz sands. The two aquifers are separated by low-permeability marlstone and claystone aquitards (Paces et al., Chapter 17, this volume). The thickness of the semi-confined Turonian sands aquifer varies between less than 10 m in the south and east, to about 190 m in the western part. The main source of recharge to the aquifer is precipitation. The artesian Cenomanian sands aquifer is typically 30–80 m thick in the studied area, with an average transmissivity of 48 m^2 d^{-1} (with values ranging from 0.02 to 148 m^2 d^{-1}) and a mean porosity of about 0.20 (Herčík et al. 2003). The Cenomanian aquifer is recharged by infiltrating rainwater in a 1–2 km wide zone along the Luzice fault where the sands crop out. Groundwater flows from north to south. The wells for the present investigation are located in the western section of the

Basin, which is most exploited for water supply and is where earlier intensive uranium mining took place. The Stráz tectonic block, with an area of 240 km^2, is located at the north-western margin of the Bohemian Basin (Paces et al., Chapter 17, this volume). In this block, both aquifers are fully developed and contaminated due to uranium mining by underground acid leaching.

Dating results

The results of the tracer measurement in the Turonian and Cenomanian aquifer are summarised in Table 5B.2. In two wells sampled from the Turonian aquifer, young water components could be identified on the basis of ^{85}Kr and/or ^{3}H concentrations. ^{3}He/^{3}He ages are slightly lower than ^{85}Kr ages, which possibly indicates the presence of post-bomb water and dilutes absolute concentrations (^{85}Kr), but does not affect isotope ratios (^{3}H/^{3}He). ^{39}Ar concentrations below 100% modern confirm this finding. Waters from two wells (VP7523, VP7512) do not have ^{85}Kr or ^{3}H concentrations above the detection limit and are therefore older than about 50 years. The lowest ^{39}Ar activity of 60% modern corresponds to the mean residence times of Turonian aquifer samples of less than 200 years. ^{14}C ages calculated with the Pearson correction model (Ingerson and Pearson 1964) are, within uncertainty, consistent with residence times less than 1000 years for all of the wells. In summary, it can be concluded that residence times of the groundwater of the Turonian aquifer have an age between modern and a few hundred years.

Waters of the underlying, mostly confined Cenomanian aquifer are free of ^{85}Kr and ^{3}H and contain consequently no modern water components. Potential tools for further

Table 5B.2   Results of tracer measurements of the Turonian (upper 4 wells) and Cenomanian aquifers (sampling year 2003).

Well	Depth (mbs)	^{3}H (TU)	^{85}Kr (dpm/ cm^3 Kr)	^{3}He$_{trit}$ (TU)	^{39}Ar (%mod)	^{14}C (pmC)	δ^{13}C (‰)	^{4}He$_{rad}$ (10^{-8}) (cm^{3}g^{-1})	Piston flow ages (yr)			
									^{3}H/^{3}He	^{85}K	^{39}Ar	^{14}C
VP7523	200	≤1.2 TU	<0.8		85±5	73.4	−12.7		−	>50	63	−1,761
VP7512	201	≤1.2 TU	<1.1		60±5	63.8	−13.0		−	>50	198	−1,094
VP7524	29	8.7	39.6±3.2	1.5	82±5	78.8	−12.9	0.04	3	9	77	−2,262
VP7520	131	10.5	12.0±1.3	11	92±5	63.9	−13.5	0.63	13	20	32	−762
VP7502	219	≤1.2 TU	<1	11	196±6	16.3	−8.4	10	>50	>50	−	10,630
VP7506	284	≤1.2 TU	<1		21±4	53.8	−12.0		>50	>50	600	600
VP7500	411	≤1.2 TU	<0.5	469	<8	45.8	−10.6	152	>50	>50	>1200	965
VP7515	410	≤1.2 TU	<2	1383	<15	24.0	−14.1	328	>50	>50	>1200	7,080
VP7517	386	≤1.2 TU				14.2	−13.0		>40			11,565
VP7519	247	≤1.2 TU		28305		6.3	−11.9	921	>50			17,725
Karany B	< 200 (?)	≤1.2 TU	<0.3	4836	38±3	40.8	−12.7	15706	>50	>50	376	25,600

constraining the age spectra are ^{39}Ar, ^{14}C and radiogenic helium. ^{14}C activities decrease in the direction of groundwater flow from 54 to 6 pmc (Table 5B.2) The two samples VP7502 and Karany B do not follow this general trend (Fig. 5B.11[a]). Well VP7502 is situated close to the recharge area and is anomalous in regard to several tracer concentrations compared to the other samples: low ^{14}C, high δ^{13}C, ^{39}Ar above 100% modern and containing a high concentration of total dissolved inorganic carbon (TDIC). These features can possibly be attributed to gas intrusion (^{39}Ar rich argon, ^{14}C free CO$_2$) through the Luzice fault (Paces et al., Chapter 17, this volume; Weinlich et al. 1999). Karany B is the most distant well from the recharge area but significantly shallower than the wells further upstream. The elevated ^{14}C and ^{39}Ar activities are therefore most probably the result of an admixture of younger waters in the age range of a few

hundred years. The residence times of the other wells can be calculated based on ^{39}Ar and ^{14}C activities. ^{14}C ages were calculated based on geochemical reaction path modelling using NETPATH (Plummer et al. 1991). The main geochemical reactions controlling the ^{14}C evolution are calcite dissolution and reprecipitation, ion exchange and sulphate and dolomite dissolution. Well VP7506 with an ^{39}Ar activity of about 20% modern, corresponding to an age of 600 years, was selected as an initial well. The calculated ^{14}C ages are plotted as a function of distance to the recharge area in Fig. 5B.11(b). Groundwater ages increase linearly up to about 18 kyr corresponding to a flow velocity of 2.3 m yr^{-1}. These ages are in agreement with ^{39}Ar concentration less than 15% modern, which indicates, on one hand, a lower limit of residence times of 800 years, and on the other hand, that underground production (Lehmann et al. 1993) must be

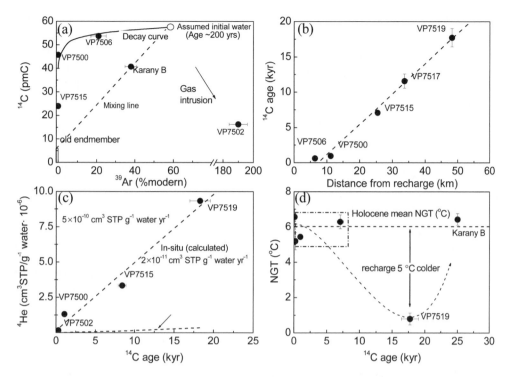

Fig. 5B.11 (a) Relationship between 39Ar and 14C in the Cenomanian aquifer. The composition of initial water was estimated based on results from the overlaying Turonian aquifer. Well VP7502 is situated close to the recharge area. (b) Increase of 14C age as function of distance from recharge area. (c) Accumulation of radiogenic helium as a function of age. The dashed line indicates the expected increase if helium originated solely from in situ production. (d) Noble gas temperature versus 14C age. Samples younger than 10,000 years reflect the mean annual soil temperature during the Holocene. Sample VP7519 recharged during the last glacial period with 5°C cooler temperatures than today.

very low in this part of the aquifer. The age of 18,000 years means that this water recharged during the last ice age when the climate was cooler than today. Indeed the sample VP7519 is depleted in $\delta^{18}O$ and the noble gas recharge temperature is about 5°C lower than for the rest of the wells (Fig. 5B.11[d]). The downgradient ageing is also reflected by the increasing ^4He concentrations (Fig. 5B.11). The helium accumulation rate based on ^{14}C ages is about 5×10^{-10} cm^3 STP g^{-1} yr^{-1}, a value which is an order of magnitude higher than the in situ accumulation rate calculated from measured U and Th concentration in the aquifer.

Summary and implications

Groundwater residence times in the Turonian aquifer range up to a few hundred years with recent groundwater components present in some of the wells. The corresponding vulnerability to pollution is also demonstrated by elevated nitrate concentrations at these locations (Chapter 17, this volume).

In contrast, groundwater in the investigated parts of the confined Cenomanian aquifer is not yet affected by anthropogenic activities. Baseline groundwater quality is influenced mainly by time-dependent water–rock interactions (Chapter 17, this volume). The proximity of one well to a tectonically active fault zone impedes the application of ^{14}C and ^{39}Ar as dating tools. Gas from external sources with elevated ^{39}Ar and CO_2 production rates interferes with the atmospheric ^{39}Ar and ^{14}C, respectively. For the other wells, ^{39}Ar and ^{14}C ages agree well and a remarkably good correlation between ^{14}C ages and distance and ^{4}He concentrations can be observed.

References

Aeschbach-Hertig, W., Peeters, F., Beyerle, U. et al. (1999) Interpretation of dissolved atmospheric noble gases in natural waters. *Water Resources Research* 35, 2779–92.

Aeschbach-Hertig, W., Kipfer, R., Blaser, P. et al. (2003) A noble gas palaeotemperature record from the Ledo–Paniselian aquifer in Belgium. In: Society, E.G. (ed.) *EGS 28th General Assembly*. Geophysical Research Abstracts, Nice.

Andrews, J.N., Balderer, W., Bath, A.H. et al. (1983) Environmental isotope studies in two aquifer systems: A comparison of groundwater dating methods. *IAEA, Isotope Hydrology 1983*. IAEA, Vienna. pp. 535–77.

Bariteau, A. (1996) Modélisation géochemique d'un aquifer: la nappe de l'Oligocène en Beauce et l'altération des Sables de Fontainebleau. Thèse. University Paris Sud.

Bath, A.H., Edmunds, W.M. and Andrews, J.N. (1979) Palaeoclimatic trends deduced from the hydrochemistry of a Triassic sandstone aquifer, United Kingdom. *IAEA, Isotope Hydrology 1978*. IAEA, Neuherberg. pp. 545–68.

Bath, A.H., Milodowski, A.E. and Strong, G.E. (1987) Fluid flow and diagenesis in the East Midlands Triassic sandstone aquifer. In: Goff, J.C. (ed.) *Fluid Flow in Sedimentary Basins and Aquifers*. Geological Society, London, pp. 127–40.

Busenberg, E. and Plummer, L.N. (1992) Use of chlorofluorocarbons (CCl_3F and CCl_2F_2) as hydrologic tracers and age-dating tools: The alluvium and terrace system of central Oklahoma. *Water Resources Research* 28, 2257–84.

Busenberg, E. and Plummer, L.N. (2000) Dating young groundwater with sulphur hexafluoride; natural and anthropogenic sources of sulfur hexafluoride. *Water Resources Research* 36, 3011–30.

Collon, P., Kutschera, W. and Lu, Z.T. (2004) Tracing noble gas radionuclides in the environment. *Annual Review of Nuclear and Particle Science* 54, 39–67.

Cook, P.G. and Solomon, D.K. (1995) Transport of atmospheric trace gases to the water table: Implications for groundwater dating with chlorofluorocarbons and krypton 85. *Water Resources Research* 31, 263–70.

Corcho, J.A., Purtschert, R. and Hofer, M. (2002) Comparison of residence time indicators (^{3}H/^{3}He, SF$_6$, CFC-12, and ^{85}Kr) in shallow groundwater: A case study in the Odense aquifer, Denmark. In: Goldschmidt Conference. *Geochimica et Cosmochimica Acta* Davos, pp. A152.

Corcho, J.A., Purtschert, R., Hinsby, K. et al. (2005) 36Cl in modern groundwater dated by a multi tracer approach (H/3He, SF6, CFC-12 and 85Kr): A case study in Quaternary sand aquifers in the Odense Pilot River Basin, Denmark. *Applied Geochemistry* 20, 599–609.

Corcho, J.A., Purtschert, R., Barbecot, F. et al. (2007) Constraining the age distribution of highly mixed groundwater using 39Ar: A multiple environmental tracer (3H, 3He, 85Kr, 39Ar and 14C) study in the semiconfined Fontainebleau Sands aquifer (France). *Water Resources Research* 43, W03427.

Edmunds, W.M. and Milne, C.J. (eds). (2001) Palaeowaters in Costal Europe: Evolution of groundwater since the late Pleistocene. Geological Society Special Publication 189.

Edmunds, W.M. and Smedley, P.L. (2000) Residence time indicators in groundwater: The East Midland Triassic sandstone aquifer. *Applied Geochemistry*, 15, 737–52.

Grath, J., Scheidleder, A., Uhlig, S. et al. (2001) The EU Framework Directive: Statistical aspects of the identification of groundwater pollution trends, and aggregation of monitoring results. Final Report. Austrian Federal Ministry of Agriculture and Forestry.

Herčík, F., Herrman, Z. and Velečka, J. (2003) *Hydrogeology of the Bohemian Cretaceous Basin.* Edited by AQUATEST a.s. and KAP Ltd., Czech Geological Survey, Prague.

Ingerson, E. and Pearson, F.J. Jr. (1964) Estimation of age and rate of motion of ground-water by the ^{14}C-method. In: Miyake Y. and Koyama, T. (eds.) *Recent Researches in the Fields of Hydrosphere, Atmosphere and Nuclear Geochemistry.* Maruzen, Tokyo. pp. 263–83.

Lehmann, B.E., Davis, S.N. and Fabryka Martin, J.T. (1993) Atmospheric and subsurface sources of stable and radioactive nuclides used for groundwater dating. *Water Resources Research*, 29, 2027–40.

Mercier (1981) Inventaire des ressources aquifèrs et vulnérabilité des nappes du départemen des Yvelines. 81SGN3481DF, B.R.G.M., Service géologique régional Ile de France.

Plummer, L.N., Prestemon,. E.C and Parkhurst, D.L. (1991) An interactive code (NETPATH) for modeling net geochemical reactions along a flow path. Water Resources Investigations Report 91-4078, US Geological Survey.

Schlosser, P., Stute, M., Sonntag, C. et al. (1989) Tritiogenic He3 in shallow groundwater. *Earth and Planetary Science Letters* 94, 243–53.

Seiler, K.P. and Lindner, W. (1995) Near-surface and deep groundwater. *Journal of Hydrology* 165, 33–44.

Seiler, K. P., Maloszewski, P., Weise, S.M. et al. (1999) Environmental isotopes as early warning tools to control the abstraction of deep groundwaters. International Symposium on Isotope Techniques in Water Resources Development and Managment. IAEA, Vienna. pp. 258–59.

Smedley, P.L. and Edmunds W.M. (2002) Redox patterns and trace element behaviour in the East Midlands Triassic sandstone aquifer, UK. *Ground Water* 40, 44–58.

Thiry, M., Bertrand Ayrault, M., Grisoni, J.-C, Ménilley, F., Schmitt, J.M. (1998) Les Grès de Fountainebleau: silicifications de nappes liées à l'éolution géomorphologique du bassin de Paris durant le Plio-Quaternaire. *Bulletin de la Société Géologique de France* 8(4), 419–30.

Torgersen, T. and Clarke, W.B. (1985) Helium accumulation in groundwater, I: an evaluation of sources and the continental flux of crustal ^{4}He in the Great Artesian Basin, Australia. *Geochimica et Cosmochimica Acta* 49, 1211–18.

Torgersen, T. and Ivey, G.N. (1985) Helium accumulation in groundwater, II: A model for the accumulation of the crustal ^{4}He degassing flux. *Geochimica et Cosmochimica Acta* 49, 2445–52.

Troldborg, L. (2004) The influence of conceptual geological models on the simulation of flow and transport in Quaternary aquifer systems. PhD. Technical University of Denmark, Denmark.

Vogel, J.C. (1967) *Investigation of Groundwater Flow with Radiocarbon.* IAEA Isotopes in Hydrology. IAEA, Vienna. pp. 355–69.

Walraevens, K. (1990) Hydrogeology and hydrochemistry of the Ledo–Paniselian semi-confined aquifer in east and west Flanders. *Academiae Analecta* 52, 12–66.

Walraevens, K. (1998) Natural Isotopes and Noble Gases in Groundwater of the Tertiary Ledo–Paniselian Aquifer in East and west Flanders. *Natuurwet. Tijdschr* 78, 246–60.

Weinlich, F.H., Braeuer, K., Kaempf, H. et al. (1999) An active subcontinental mantle viotile system in the western Eger rift, Central

Europe: Gas flux, isotopic composition (He, C and N) and compositional fingerprints. *Geochimica et Cosmochimica Acta* 63, 3653–71.

Zuber, A. (1986) On the interpretation of tracer data in variable flow systems. *Journal of Hydrology* 86, 45–57.

6 Identifying and Interpreting Baseline Trends

M. VAN CAMP AND K. WALRAEVENS

Baseline trends are recognised and identified by investigating time series of chemical analyses of groundwater samples. While systematic variations in chemical parameters can easily be recognised, interpretation of time series in terms of hydrochemical trends should be based on the understanding of chemical processes affecting groundwater quality and not just on the results of descriptive statistical procedures. Interpretation should consider the relationship between groundwater hydrodynamics, by which groundwater masses can move, and the heterogeneity within the groundwater masses themselves. The timescale of the chemical variations is important and is coupled to the flow velocity in the system.

A number of examples from different European countries and different geological settings have been collected and are presented. They illustrate typical trends encountered in many aquifer systems. Sometimes trends are induced by altered hydrodynamics caused by exploitation. This results in shifting of groundwater bodies with different quality. Salinisation and increase of redox sensitive components can be explained in this way.

6.1 Significance of baseline trends

The chemical quality of groundwater evolves over long timescales during flow from source to discharge area under natural hydrogeological conditions. Under these conditions characteristic geochemical trends or gradients are established along flow lines resulting from water–rock interaction or changing recharge composition in the course of time. Exploitation of groundwater by pumping over just a few decades disturbs the natural baseline quality and may accelerate the natural processes. For water quality management it is important both to recognise the original spatial geochemical patterns and to find evidence of how these patterns have responded to development of the aquifer.

In Europe, the understanding of natural baseline trends is important in view of the emphasis in the Water Framework Directive (WFD, European Union 2000) of the need to identify 'long term anthropogenically induced upward trends in pollutant concentrations and the reversal of such trends': it is first necessary to identify whether anthropogenic or strictly natural processes are involved. In order to distinguish anthropogenically induced trends from baseline trends, process interpretation is required. If restricted to a purely statistical approach, an upward baseline trend will be considered as pollution.

Baseline has been defined in Chapter 1 as the range of concentrations of a given element, species or chemical substance present in solution, being derived from natural geological, biological or atmospheric sources. Baseline trends are here defined as a systematic and continued change in time in the concentration of a chemical component or value of a physico-chemical property in the groundwater, which originates from natural processes or changes in the hydrodynamics of aquifer systems. Not included are changes due to input in the groundwater body of anthropogenic components, thus eliminating contaminated waters and groundwater bodies.

The objectives of this chapter are to consider how hydrochemical data may be used to identify natural groundwater baseline trends and discriminate their rates. Because baseline trends refer to an evolution in time, the occurrence of trends can only be recognised when time series data are available. Most time series show oscillations, but only when systematic patterns occur, can clear baseline trends be identified. First of all some of the classical methods of identifying spatial and temporal trends are reviewed, emphasising the need to consider the type and scale of groundwater bodies. Second the advantages and limitations of a geostatistical approach are considered and a new approach is proposed to couple the hydrodynamic and hydrochemical information. Finally, trends obtained from a number of European reference aquifers are considered to illustrate several features of changing baseline chemistry.

6.2   Baseline trends related to hydrodynamic and hydrochemical system analysis

6.2.1   Relation of hydrodynamics and timescales to baseline trends

Spatial variations of concentrations and chemo-physical parameters can be transformed into time series by flow of the groundwater. This transformation from the space to the time domain is conditioned by the flow velocity, described by Darcy's Law, for the given hydrodynamic conditions (hydraulic gradient and hydraulic conductivity):

$$v_{\text{eff}} = \frac{k*\text{grad}\,(h)}{n_{\text{e}}}$$

where v_{eff} = effective groundwater flow velocity (L/T); k = hydraulic conductivity (L/T); grad(h) = hydraulic gradient; n_{e} = effective porosity.

Groundwater flow velocities are usually of the scale of a few metres per year up to hundreds of metres per year (around pumping wells). Most of the data series found in European aquifers span time lengths of at most a few decades. This limits the size of spatial concentration gradients (spatial scale) that can be converted to observable trends in time series, depending on the local groundwater flow velocities. In fractured rock aquifers, flow velocities can be much higher if groundwater moves through open fractures or fissures.

Concentration gradients due to hydrochemical processes which act on the scale of an aquifer (e.g. the freshening of an aquifer) have a spatial size of the order of at least few km to tens of km (10^3–10^4 m). Under natural flow conditions, it will require too much time (10^2–10^4 years) to convert these gradients into observable changes. In most aquifer systems, it can be expected that changes of groundwater quality during a period of a few years are unlikely to be caused by the lateral shifting of groundwater bodies with different composition. The intrinsic scale difference between the lateral extension of an aquifer system (usually 10^4–10^5 m) and its thickness (usually 10–100 m) indicates that if vertical concentration gradients exist, upward or downward flow components can cause changes in concentration levels on a much shorter timescale, for example, years to decades. However, under natural hydrodynamic conditions, these vertical flow components are rather small in most aquifer systems and anthropogenically induced changes in the flow situation (groundwater exploitation) are the main reason for upconing of deeper waters.

The hydrochemistry and the hydrodynamics of an aquifer can be in steady or transient state. After changing the boundary conditions, it will require some time before the hydrodynamics and the hydrochemistry have adapted to the new conditions. There exists, however, a difference in timescale to reach a steady state for hydrodynamics and hydrochemistry, whereby hydrochemistry will require a much longer time. As long as a hydrochemical steady state is not reached, at least in some parts of the aquifer system, concentrations can be expected to change. However, this will often be so slow that it will not be noticeable within a few years.

6.2.2 Relation of groundwater body heterogeneity to baseline trends

Groundwater composition will usually be variable in space in an aquifer system. Based on major ion concentrations and their relative occurrence, different water types can be distinguished (Stuyfzand 1986), which reflect the genesis and ongoing hydrochemical processes. A groundwater body can be divided into more homogeneous sub-bodies based on these types, but even within a single water type or sub-body, variations in concentrations are likely to occur. These variations can be related to the heterogeneity of the aquifer sediment characteristics and composition if water–rock interactions are involved, to spatial and temporal variations in recharge water composition at the start of the flow cycle, and probably to many other reasons. As an example, we can consider $CaHCO_3$ type waters, originating from dissolution of calcite in equilibrium with CO_2, in which calcium concentrations can vary because of fluctuations of CO_2 partial pressure in the unsaturated zone through which the water percolates, or due to heterogeneous distribution of calcite in the soils. Mixing and the effects of dispersion can introduce some homogenisation but a perfectly uniform composition will never be reached. The size of these minor variations will depend on water types and water bodies, and may need to be quantified for each case.

These small-scale spatial variations have nothing to do with processes or mechanisms which act on a larger scale such as freshening of an aquifer or intrusion of salt water. The presence and recognition of these variations and their size are important because they can lead to trends in observed time series of the related parameters. When groundwater is moving, these lower and higher concentration 'islands' will pass through a point fixed in space (an observation well, for example) and will lead to a fluctuating pattern in the observed concentrations around an average value. Over the longer term, no trend will occur as periods of increasing and decreasing concentrations will alternate. How long these periods will last depends on two factors: the size of the high and low concentration islands and the flow velocity of the groundwater. The former can be quantified as the 'spatial correlation distance'.

As an example, two fields were generated using a geostatistical simulation technique (Fig. 6.1), both with a log-normal distribution of parameter values (with an average of 1.00 and a standard deviation of 0.25) but with two different correlation distances: one for 10 m and one for 100 m. Shown are maps of the logarithmic values and a cross section in which the values are back-transformed to a linear scale. As can be seen in the profiles, concentrations can double between low and high peaks for the chosen distribution characteristics.

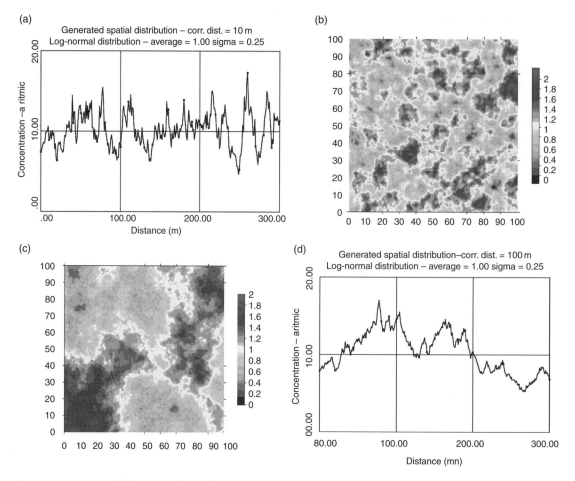

Fig. 6.1 Generated log-normally distributed parameter fields with correlation distances of 10 and 100 m: (a) cross section; (b) unconditional simulation (a and b: spatial correlation = 10 m); (c) unconditional simulation; (d) cross section (c and d: spatial correlation = 100m).

6.2.3 Hydrochemical characterisation

In order to characterise natural trends, groundwaters can be presented graphically and classified according to their composition into groups with more or less similar chemical characteristics, often termed hydrochemical facies or water types. The interpretation of time trends requires the aquifer to be subdivided into fairly homogeneous hydrochemical facies. Ideally, the groundwater classification should be able to allow identification and recognition, besides the composition, of the genesis and evolution of the groundwater in aquifer systems. Besides hydrochemical classification systems, legal and regulatory classifications exist in different countries and regions, which are highly dependent on local hydrogeological and regulatory situations. A simple differentiation between fresh, brackish, saline and possibly hypersaline (brines) waters can be made based on salinity or chloride concentrations, indicating their potential use as potable or consumable water. For industrial use, waters are often classified according to their hardness in grades of soft and hard water.

Piper (1944) used a graphical presentation to help classify groundwaters according to the relative composition of their main ions. The presentation consisting of two ternary diagrams (one for the cations and one for the anions) and one binary diagram is still widely used and known as the Piper diagram. The lower left ternary triangle, or cation ternary, compares the cation composition as an equivalent fraction of calcium, magnesium and the sum of sodium and potassium. Waters with similar fractions of cations (and anions) may have different absolute amounts of total dissolved solids (TDS). Similarly, the lower right ternary triangle or anion ternary, contrasts the anion composition in terms of fraction of equivalents of the sulphate ion, chloride ion and the sum of bicarbonate and carbonate ions. The central diamond subdiagram is a combination of the cation and anion fractions (Freeze and Cherry 1979). The Piper diagram is considered as the first step in chemical classification of water facies for the purpose of studying the evolution of groundwater (Manharawy and Hafez, 2003). Regions within the diagrams may be delineated and can correspond with different hydrochemical facies. Classifications based on Piper diagrams only account for the relative occurrences of the main ions and do not reflect absolute concentrations. Ongoing chemical processes which affect major ion composition in an aquifer are seen as a shifting in the location of points on the diagram. Mixing between two different water types gives a composition, which lies on the connecting line between the two points. Other graphical presentations often used in groundwater classification are by Stiff (1951) and Schoeller (1955).

A number of variations on the ternary diagram approach have also been suggested. Furtak and Langguth (1967) used the Piper graphical procedure and classified the waters into ten different types (Table 6.1).

The Kansas Geological Survey (1989) adopted six types of water chemistry that are predominant in groundwaters in sedimentary aquifers:

Class 1: $Ca(HCO_3)_2$ to $Mg(HCO_3)_2$
Class 2: $CaSO_4$ to $MgSO_4$
Class 3: $CaCl_2$
Class 4: $NaHCO_3$
Class 5: Na_2SO_4
Class 6: $NaCl$

Table 6.1 Water types recognised in the classification of Furtak and Langguth (1967).

Group	Water type
1	Earthalcalic carbonate
2	Earthalcalic carbonate-sulphate
3	Earthalcalic sulphate
	Earthalcalic chloride
	Earthalcalic nitrate
4	Earthalcalic-alcalic carbonate
5	Earthalcalic-alcalic sulphate
	Earthalcalic-alcalic chloride
	Earthalcalic-alcalic nitrate
6	Alcalic-carbonate
7	Alcalic-sulphate
	Alcalic-chloride
	Alcalic-nitrate
8	Calcium carbonate
9	Calcium magnesium carbonate
10	Magnesium carbonate

Stuyfzand (1986) based his classification on a modified version of the cation ternary diagram and used a distinct separation of groups (Fig. 6.2). Each water type is represented by a symbolic name consisting of the dominant cation and anion (or 'mix' if there is none) name and supplementary indicators for salinity, hardness and cation exchange. Therefore, this classification is best suited for salinising or freshening aquifers where cation exchange is important.

To include the redox status as a criterion in groundwater classification, a complementary classification system was proposed by Pannatier et al. (2000) which is based on local redox conditions in an aquifer system and the advance of oxidation and reduction processes. Modified ternary diagrams have also been proposed by Derron (1999) for non-clastic aquifer environments such as crystalline rocks. In this example, the ternary diagram has Ca, Mg and Si poles to help distinguish waters from mafic and ultra-mafic rocks and granito-gneiss complexes. In a study of natural tracers, in recent groundwaters from different Alpine aquifers (Kilchmann et al. 2004) a further subdivision of water types was made based on the trace element compositions.

6.2.4 Baseline trends related to hydrochemical processes

Groundwater quality in an aquifer system reflects the main processes which have occurred and have been described in Chapter 1. These processes depend on (1) the lithology and initial water composition; (2) the hydrodynamic boundary conditions which impose the flow cycles; and (3) the groundwater flow velocity and reaction kinetics. Aquifers can be subdivided according to the main processes which control the groundwater composition. Different processes may occur in the same part of the aquifer, but the main aquifer type is determined by the dominant process. Four main aquifer types may be distinguished:

Freshening and salinising aquifers. Here both saline and fresh waters are found, mixing and cation exchange will be the main processes determining groundwater quality. Where evaporites occur, brines can be generated and salinisation will be induced.

Carbonate aquifers. If the sediments have a marine origin, calcite will be present. After the old connate seawater has been flushed out and the cation exchange complex has reached a freshwater equilibrium, calcite dissolution may become the main process.

Siliceous aquifers. Many sandstones and other continental sediments may be entirely carbonate free and the weathering of silicates becomes the main quality determining process.

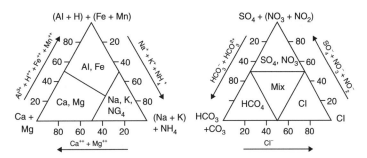

Fig. 6.2 Triangular diagrams used in the Stuyfzand (1986) classification.

Carbonate cements may also be removed over long time periods so that shallow aquifers formerly with carbonate may also be dominated by silicate reactions.

Anoxic aquifers. Along flow lines, redox conditions will change from oxidising to reducing and a sequence of reactions will occur.

6.2.4.1 Salinising and freshening aquifers

When both fresh and saline waters are present in an aquifer, changes in the distribution of these waters can lead to a change in the baseline concentrations. Changes in the location and extension of the fresh and salt groundwater bodies will be caused by the hydrodynamics of the aquifer system. As long as a hydrochemical steady state is not achieved (this will take much longer than reaching a hydrodynamic steady state, Walraevens and Van Camp 2005), the freshwater/saltwater distribution will be in a transient situation, leading to an increase or decrease of the fresh- and saltwater components with time. Under natural hydrodynamic conditions, this evolution is usually so slow that no significant impact can be expected on the timescale of typical monitoring series (say a decade). However, if the natural flow situation is altered, the movement of groundwater bodies may be faster. Considering the fact that the lateral extension

of aquifers is much larger than the thickness, it is more likely that observable changes in salinity will be caused by vertical flow components, provided that a vertical stratification of the salinity exists. Local groundwater extraction is the most common reason for the introduction of vertical flow components, inducing an upconing effect of deeper waters to the pumping wells.

The two main processes which occur in freshening and salinising aquifers are mixing of a saline and freshwater end-member and cation exchange (Walraevens and Van Camp 2005). Diffusion is only important on long timescales but can influence water quality near underlying or overlying aquitards in semi-confined layers. Fresh water is often of the $CaHCO_3$ type and originates from the dissolution of calcite in water containing atmospheric CO_2. Saline water has often a comparable composition to sea water and is of the NaCl type.

Recognising freshening in time series (Table 6.2)

As the freshwater replaces saltwater, chloride concentrations drop. The calcium of the freshwater will be replaced during the ion-exchange process initially by sodium, and later by magnesium. These become the main cations, resulting in $NaHCO_3$ and $MgHCO_3$ type waters. This process can be recognised by an increase in the ratio Na/Cl.

Recognising salinisation in time series (Table 6.2)

The most direct indicator of salinisation is the increase in chloride concentration as the saltwater component increases. Since Cl is a conservative constituent, the seawater component can be calculated from the increasing Cl concentration. For the saline end-member, average seawater composition can be assumed.

Sodium will increase initially, but when cation exchange occurs, it will partly be replaced by calcium, giving a deficit compared to the chloride concentration. Calcium concentrations will increase because of cation exchange and will be higher than could be expected from the HCO_3 concentrations based on calcite dissolution. The resulting water is of the CaCl type, which is a direct indicator of salinisation.

6.2.4.2 Carbonate aquifers

In carbonate aquifers, calcite and/or dolomite dissolution is the main process that determines groundwater quality. In general, the main groundwater type is $CaHCO_3$ or $CaMgHCO_3$ water.

In karstic aquifers, flow velocities can be very high and a fast response of wells and springs to rainfall events is possible. In the case of springs, a correlation of discharge rates with meteorological conditions can indicate this. In such cases, changes in spring flow rates may rapidly be reflected in the groundwater quality.

Related trends in time series

Calcite dissolution will proceed congruently provided the groundwater is undersaturated with respect to this mineral giving rise to stoichiometric ratios of elements in the groundwater. The calcite in most carbonate aquifers is impure and at saturation, incongruent reaction produces a less soluble but purer calcite and releases impurities such as Mg and Sr to the groundwater, recognisable in time series. This process is observed clearly in the European Chalk aquifer (Edmunds et al. 1987). Where dolomite is

Table 6.2 Main hydrochemical processes and related parameter trends in freshening and salinising aquifers.

Process(es)	Indicators	Indicators
Salinisation by salt water intrusion + cation exchange	Cl increasing Na increasing Na/Cl < 1 Ca increasing Ca/HCO_3 > 1	
Freshening by fresh water intrusion + cation exchange		Cl decreasing Na decreasing Na/Cl > 1 Ca decreasing Ca/HCO_3 < 1
Dissolution of evaporites (halite, anhydrite, gypsum)	Na, Cl, Ca, SO4 increasing	

Table 6.3 Main hydrochemical processes and related parameter trends in carbonate aquifers.

Process(es)	Indicators (decreasing)	Indicators (increasing)
Calcite congruent dissolution	–	Ca
Calcite incongruent dissolution	Ca	Mg
Dolomite congruent dissolution	–	Ca, Mg
Dolomite incongruent dissolution	Ca	Mg

Table 6.4 Experimentally determined lifetime of 1 mm crystals of some silicate minerals and related parameter trends (Lasaga 1984).

Mineral	Mean lifetime (years)	Increasing parameters
Quartz	34,000,000	Si
K-feldspar	520,000	K, Si
Na-feldspar (albite)	80,000	Na, Si
Ca-feldspar (anorthite)	112	Ca, Na, Si

present, Mg will be released in roughly equal proportions under conditions of congruent reaction but also will increase further by incongruent dissolution (Table 6.3):

$$Ca_{1-x}Mg_xCO_3 \rightarrow aCa_{1-y}Mg_yCO_3 + (x - ay)$$
$$Mg^{2+} + (1 - x - a + ay)Ca^{2+}$$
$$+ (1 - a)CO_3^{2-}$$

where $y < x$.

6.2.4.3 Siliceous aquifers

In non-carbonate lithologies or in aquifers that have been decalcified, silicate weathering is the main mechanism that determines groundwater quality (Appelo and Postma 2005). Silicate weathering is a relatively slow process compared with carbonate dissolution that adds silica and cations (Na, K, Ca) to the groundwater and results in secondary mineral formation, mainly clays; iron that is released by the weathering can form oxides and/or hydroxides. Sodium may increase relative to chloride in siliceous aquifers following mineral weathering since it is not balanced by chloride derived from a seawater component.

Related trends in time series

Because the silicate weathering progresses very slowly compared to groundwater flow, no significant impact of these processes on the timescales of observation series (at most decades) can be expected. (Typical lifetimes calculated for 1 mm crystals at 25°C and pH 5 are listed in Table 6.6, Lasaga 1984). The relative rates of the different weathering processes causes a sequential disappearance of the different silicate minerals (Goldich weathering sequence). Concentration gradients caused by silicate weathering have to be seen on a regional scale, often the scale of the whole aquifer. The main dissolution processes and their related parameter trends are listed in Table 6.4. The most rapid reactions such as plagioclase dissolution, will have the largest impact on groundwater composition and reaction rates should be compared to percolation times of typical flow cycles. The occurrence of the different processes can be checked by calculating the saturation indices of the different mineral phases. Undersaturation means that dissolution of the mineral can continue.

6.2.4.4 Anoxic aquifers

Anoxic groundwaters are common at depth in most aquifers once oxygen has been consumed by geochemical processes. A theoretical redox sequence (Table 6.5) can be defined and this is commonly observed in many anoxic aquifers (Berner 1981). Vertical stratification can be expected where the flow components are mainly downward as in recharge areas. A complete series of redox reactions will necessarily be found in an aquifer due to the relative abundances of solutes and electron acceptors/donors. In particular, sulphate reduction and methanogenesis require strongly reducing conditions and the presence of reactive organic carbon.

Related trends in time series

Changes in the concentrations of a redox species in a time series can be triggered by different mechanisms. Shifting of the redox zone stratification is induced by vertical flow components, especially by prolonged pumping in aquifer exploitation. Changes of recharge rates of the aquifer will modify the amount of oxidised water that enters the system (e.g. the impact of dry and wet years). Changes in vegetation and land use can also alter the amount of organic matter in the soils available for oxidation.

6.3 Recognising baseline trends in time series data

First, classic statistical methods will be reviewed. In addition a geostatistical approach to evaluate time-trend analysis as a function of the geostatistical charactersitics of groundwater sub-bodies will be described.

Table 6.5 Main hydrochemical processes and related parameter trends in anoxic aquifers.

Process(es)	Redox zone	Increasing parameters	Decreasing parameters
Oxidation organic matter (by oxygen)	Oxic	HCO_3	OC
Oxidation of pyrite (by oxygen)	Oxic	SO_4, H^+	
Denitrification	Post-oxic		NO_3
Reduction of Mn^{+4} to Mn^{+2}	Post-oxic	Mn^{+2}	
Reduction of Fe^{+3} to Fe^{+2}	Post-oxic	Fe^{+2}	
Surface complexation (release by dissolution of iron hydroxides)	Post-oxic	As	
Sulphate reduction	Sulphidic	HS^-, HCO_3	SO_4
Methanogenesis	Methanogenic	CH_4	HCO_3

Source: After Berner (1981).

6.3.1 Statistical approach

As a first exploration of trends, traditional statistical techniques can be used. However, applying a purely mathematical approach, which is not based on the actual hydrochemical context will never yield an understanding of the occurring trends in terms of the hydrogeochemical system that is driving the processes involved. Extrapolation of past trends into the future on a purely mathematical basis without a process-based scheme is not recommended.

6.3.1.1 Distribution analysis

Histograms can be constructed to help identify the distribution pattern of the visualised parameter. Multimodal distributions will be recognised as multiple maxima will occur. It is advised to transform concentration levels to logarithmic values before processing. If the distribution analysis reveals a multimodal distribution for one or more parameters (as described in Chapter 2), this will probably be reflected in the occurrence of different types of groundwater (e.g. both fresh and saline waters in a freshening or salinising aquifer).

Cumulative distribution function (CDF) curves can be used to check if a parameter follows a normal distribution. If that is the case the points will fit on a straight line (the CDF-axis has to use a probability scale). Log transformation of the parameter (concentration) can reveal a log-normal distribution. If the distribution is multimodal, the data will lie on different line segments. For the freshening Ledo–Paniselian aquifer in Flanders (Belgium), the aquifer was subdivided into different parts, each showing a characteristic water type. On the CDF-curves of the Ledo–Paniselian $CaHCO_3$ waters (Fig. 6.3[a]) it can be seen that Ca, HCO_3 and Cl approach a log-normal distribution. The curve for Na shows higher values than Cl for the upper curve segment, indicating that in the waters with the highest Na values ion exchange is likely to have increased their concentration. These waters are evolving towards the $NaHCO_3$ type (but Ca is still the main cation). The curve for sulphate follows a log-normal distribution in the upper curve segment, but a significant deviation for the lower concentrations. These have originated by sulphate reduction. In the Ledo–Paniselian NaCl waters (Fig. 6.3[b]) the old connate seawater (upper segments) can be distinguished from diluted seawater with lower concentrations (lower segments). The curves here have a typical S-shape.

If the distribution analysis shows that a parameter follows a multimodal distribution, likely because different water types occur in the groundwater body, it is advised to break up the whole dataset into subsets according to the different water types, before attempting any characterisation. For example, in a freshening or salinising aquifer, joining both fresh $CaHCO_3$ with saline NaCl waters will give a tremendous range of concentrations which is not relevant to the recharge area where only freshwater occurs. The start of a salinising process will initially give chloride concentrations far below the average level for the aquifer but clearly above the average of the $CaHCO_3$ water subset. Applying the descriptive statistics for the whole aquifer will only show progressing salinisation, when water quality is very close to pure seawater, far too late for the start of prevention strategies.

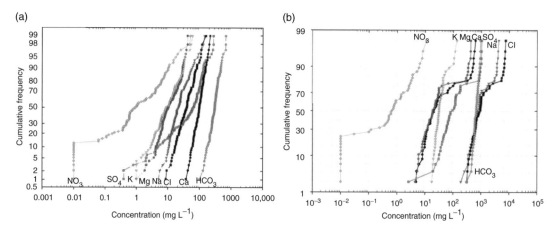

Fig. 6.3 CDF curves for the main elements and species in two groups (CaHCO$_3$ and NaCl) of Ledo–Paniselian groundwaters: (a) CaHCO$_3$; (b) NaCl.

6.3.1.2 Cross plots

Cross plots of parameters can help in understanding the hydrochemical system and identify the processes which act in the aquifer system. These cross plots should always be interpreted taking into account the hydrodynamic and hydrochemical system. Not only concentrations can be plotted, but also ratios of parameters can give meaningful graphs.

As an example (Fig. 6.4) a cross-plot of the sodium to chloride ratio (on the y-axis) versus the chloride concentration (on the x-axis) is plotted for the waters in the freshening Ledo–Paniselian aquifer (Walraevens and Van Camp 2005). On this cross-plot, the location of the data points is related to the groundwater type (as indicated) and the points shift according to the impact of cation exchange and mixing with old saline connate waters.

CaHCO$_3$ waters from the recharge area have low Cl-levels and a Na/Cl ratio ≤ 1. In NaHCO$_3$ waters, which are found downgradient, the Na/Cl ratio increases strongly due

to cation exchange (calcium being replaced by sodium) while chloride concentrations remain constant. Further downstream a transition zone with old saline waters (NaCl type) exists and Cl concentrations are higher.

6.3.1.3 Regression analysis

The classical approach for trend estimation is to fit a linear or quadratic regression model. This approach is not flexible with regard to the shape of the trend. The LOESS smoother (Cleveland 1979; Cleveland and Devlin 1988) applies the linear regression method only locally, and is much more flexible with regard to the shape of the trend (Grath et al. 2001). Next, a method for testing the trend must be applied. The test of Mann-Kendall and the ANOVA ('analysis of variance') test based on the LOESS ('local regression') smoother are well-known.

Trend analysis should be performed on the arithmetic mean of the concentration data in order to obtain data that are representative for a certain period.

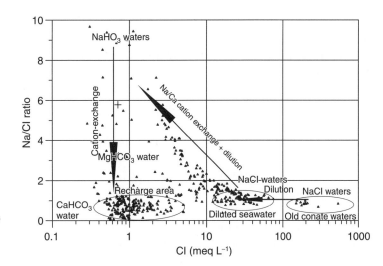

Fig. 6.4 Cross plot of Na/Cl ratio versus Cl-concentration in the Ledo–Paniselian groundwaters.

6.3.2 *Geostatistical approach*

A methodology was developed to evaluate time-trend analysis as a function of the geostatistical characteristics of groundwater sub-bodies. The mathematical description of trends in time series in itself does not give an insight and understanding into the underlying processes and possible causes of the occurring trends. Yet the geostatistical approach presented here allows a clear link to be made between time-trend analysis and the hydrodynamic and hydrogeochemical background of the sampled aquifer systems.

6.3.2.1 Methodology for assessing the relationship between time trends and small-scale spatial variations

The geostatistical characterisation may be viewed here as a quantitative way to express the natural variability of groundwater composition in time series and its relationship with spatial occurrence and distribution. The question with every time series of measured groundwater quality is whether the observed variations in an observation well

can be related to fluctuations due to the small-scale natural variability of the groundwater sub-body (introduced by heterogeneities of the aquifer and/or of occurring physico-chemical processes), or if they are the result of recently introduced processes affecting groundwater composition, such as contamination or other anthropogenic influences (e.g. caused by lowered water tables because of exploitation of the aquifer).

The procedure is defined on a step-by-step basis for easy application to real world cases. It combines traditional statistical trend recognition techniques with geostatistical analysis in the form of variogram construction. The variogram is used to check if an existing trend can be contributed by the small-scale spatial variability of the groundwater sub-body.

The proposed steps are as follows:

1. Regionalisation of groundwater quality using a geochemical process-based approach

Since the chemical composition can be heterogeneous due to the influence of several distinct physico-chemical processes many parameters are characterised by multimodal distribution curves. A separation into different fairly homogeneous groundwater sub-bodies is a useful and necessary step in reducing the diversity of the analytical data and to simplify the complex chemical relationships between the observed compositions. The definition of homogeneous groundwater sub-bodies can be based on the groundwater type as described earlier. Another approach instead of using water types is applying a cluster analysis on the sample data, which will divide the set of chemical compositions into subsets, by minimising the variance within these subsets, and maximising the variance between them.

The spatial distribution of the groundwater types can be visualised on a map. This map can then be used to derive the geographical extension of each one of the hydrochemical sub-bodies (e.g. with $CaHCO_3$ or NaCl type water). If necessary, more than one water type can be grouped together in a single sub-body. Each defined sub-body can be characterised by descriptive statistics on each of the parameters (e.g. mean, median, standard deviation) and the distribution function can be found by constructing probability plots to check if the values are normally or log-normally distributed. Each sub-body will have its own geostatistical characterisation which determines the spatial variability of the different parameters and their spatial correlation.

2. Assessment of the effective groundwater flow velocity

The average effective groundwater flow velocity in the vicinity of the observation wells should first be retrieved or estimated. This can be done by using piezometric level measurements in neighbouring wells, or by estimation using a regional map of hydraulic heads and groundwater flow. Applying Darcy's Law with known or estimated values for hydraulic conductivity and porosity ($v_{eff} = k \cdot i$/porosity) will give the effective groundwater flow velocity (e.g. expressed in $m\ yr^{-1}$). More specifically, the upstream gradient should be used in the formula. When flow regimes are transient (e.g. due to seasonal fluctuations or an increase in pumping rate over the years) an average value for the gradient must be chosen.

3. Assessment of the small-scale spatial variability

For each available time series, the hydrochemical domain of the observation well must be identified (as defined in step 1) and, for each parameter in the time series, a variogram must be constructed using the data points in the corresponding region. In a variogram, the variance of the parameter is plotted as a function of distance. The variance is calculated as half of the squared difference between all possible pairs of data points $[\gamma = (z_i - z_j)^2/2]$ and is plotted versus the distance as derived from their spatial position (coordinates in a geographical reference system). The variogram should be constructed for small distances (up to a few hundreds of metres) to reveal the small-scale spatial variability. The use of software is advised when there are more than a few data points in the close neighbourhood of the observation well with the time series.

The next step is to determine a 'characteristic length' for the observation well time series. This characteristic length corresponds with the distance the groundwater has travelled during the length of the observation series. It is found by simply multiplying the length of the time series with the average effective groundwater flow velocity (obtained from step 2). The characteristic length is a distance and thus expressed in length units. The time series and the groundwater flow velocity should be expressed using the same time units.

On the variogram, the expected variance for the characteristic length can be found. This value gives half of the squared difference of the parameter value that can be expected over the distance the groundwater has moved since the start of the time series. The possible spatial variation of the parameter $(z_i - z_j)$ can be calculated by doubling the value on the vertical axis of the variogram for the characteristic length and taking the square root (criterion = $[2 \cdot \gamma]^{1/2}$). This value (criterion) can be compared with the observed change of the parameter during the observation period. If the observed changes are smaller than the calculated value, they can possibly be contributed by the local spatial variability of the hydrochemical sub-body. Normal statistical regression procedures can also be applied to the time series to detect a trend and the change in parameter value due to the trend component can be compared with the criterion. If it is smaller, the trend can probably only reflect the small-scale variability and not the existence of external stressing factors or processes.

6.3.2.2 *Thematic example*

As an idealised example to illustrate this geostatistical approach, a hypothetical time series was generated (Fig. 6.5[a]). It assumes pumping in a coastal freshwater body, originally consisting of $CaHCO_3$ type waters. Risk for salinisation is monitored by mea-

surements of Cl concentrations. The generated observation series is 50 years long and has an increasing saline component which reaches nearly 0.6% after 50 years. The $CaHCO_3$ water has a low Cl concentration with log-normally distributed concentrations with an average of 30 ppm and a standard deviation of 0.10. The 30 ppm concentration is consistent with rainwater composition (around 10 ppm Cl) and an evaporation ratio of 3. The spatial correlation distance is 100 m, derived from variogram analyses of the Cl concentration of the freshwater body. Variations in Cl concentration may be due to fluctuations in evapotranspiration rates. The first 25 years of fluctuations in chloride concentration are due only to the heterogeneity of the freshwater body, while in the second half of the series, values increase because of salinisation. Because the $CaHCO_3$ water, which is formed by calcite dissolution CO_2 equilibrium, contains 3.62 mmol bicarbonate (for a pCO_2 of 10^{-2} atm), chloride will become the dominant anion when its concentration reaches $c.128$ ppm. The water type changes from $CaHCO_3$ to $CaCl_2$, which is a typical indicator for the salinising process. This water type is found at the end of the time series.

If observation series of limited length are used to identify trends on a purely mathematical ground (e.g. by linear fitting), without knowledge of the ongoing hydrochemical processes, a 10-year graph of the years 12–22 (Fig. 6.5[b]) could easily be interpreted as an indication of salinisation, although it is induced by small-scale variations of the Cl level before a seawater component is present in the water, as opposed in the graph between 30 and 40 years, which is related to an increased seawater component (Fig. 6.5[c]).

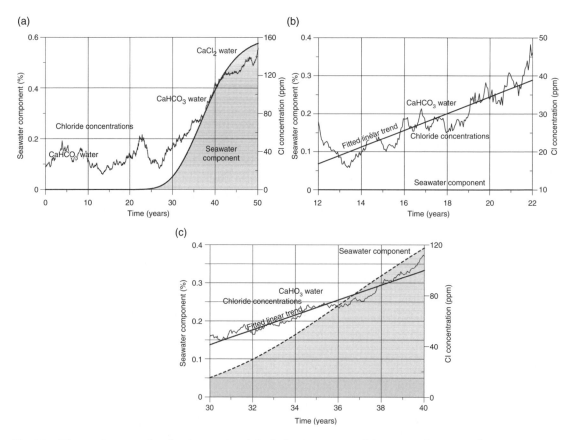

Fig. 6.5 Thematic example of a time series: (a) whole time series; (b) extracted 10-year observation series between 12 and 22 years; (c) extracted 10-year observation series between 30 and 40 years.

6.4 Practical procedure for identifying and explaining baseline trends

Based on the concept of hydrochemical aquifer types with their corresponding dominating hydrochemical processes and typical water types, a worksheet was developed to help recognise and explain observed baseline trends. The worksheet is implemented as a list of multiple-choice questions; based on the responses a possible explanation of the observed trends of parameters in the observation wells is given.

6.5 Application to European reference aquifers

Long-term time series of chemical data, extending over decades are uncommon in Europe and in many cases records have been lost. The European aquifers studied here contain some illustrative examples of trends in baseline conditions. Trends are then further explored in some other of the reference aquifers.

6.5.1 Belgium: the Ledo–Paniselian aquifer

The Ledo–Paniselian aquifer is not well characterised in terms of its temporal evolution; time series data do not exist, but samples were taken and analysed in many observation wells, giving a detailed image of the spatial variability of chemical parameters (Walraevens 1987). In this aquifer, both fresh- and saltwater are present as the end-members of a mixing situation, and concentration ranges of the main ions vary widely from weakly mineralised $CaHCO_3$ waters in the recharge area to compositions approaching seawater in the deepest part of the aquifer. To study the spatial variability in a meaningful way, the whole aquifer must be divided into smaller subregions of a more homogeneous composition. Distribution functions of the whole set of analyses show for some elements bimodal or even trimodal distributions, clearly indicating the occurrence of distinct water compositions in the aquifer. Zonation can be defined/characterised using the groundwater classification types and within each group or sub-region the spatial correlation can be investigated using variogram analyses.

As an example, the variogram for Ca content in the $CaHCO_3$ group (this is the largest sub-region in terms of number of analyses) is shown in Fig. 6.6(a). As can be seen on the variogram, a sill effect can be recognised, meaning that even over short distances (less than 500 m) some differences in the Ca concentrations occur, due to small-scale variations

in the composition of the groundwater body. The map distribution of the calcium content (Fig. 6.6[b]) shows that, over a distance of 2 km, Ca concentrations can double. Calcite dissolution is the main process which determines the Ca concentrations in the $CaHCO_3$ waters. The average of the log-normally distributed calcium concentrations is $c.96$ ppm, with 68 and 135 ppm for one standard deviation (less and more than the average). As the groundwater flow passes across an observation well these small-scale variations can be converted into time variations.

6.5.2 Belgium: the Neogene aquifer

Very few long time series exist in the Neogene aquifer system (see also Chapter 12), but an interesting one shows the evolution of arsenic in a production well of an important water exploitation over a period of nearly 10 years (Fig. 6.7). Although the As values show rather noisy, short time fluctuations, a linear fit of the series reveals an increase of around 20% over 10 years. Because arsenic is a redox bound species, occurring in natural waters as arsenate (oxidised form) or arsenite (reduced form) with different mobility, the shifting of redox zones by long-term pumping can slowly increase the arsenic level. Also a correlation with SO_4 reduction in the deeper parts of the aquifer system and ad/desorption on hydroxides can be a possible cause of the arsenic mobility.

In a neighbouring pumping station in the Neogene aquifer three observation wells at different depths (30, 50 and 80 m) show a vertical stratification of SO_4, Fe and As concentrations (Figs. 6.8 and 6.9): SO_4 decreases with depth (as a result of sulphate reduction), while Fe and As are increasing with

(a)

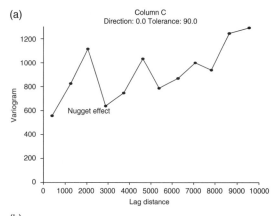

(b)

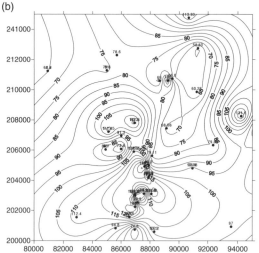

Fig. 6.6 Spatial distribution of calcium content in the $CaHCO_3$ waters in the recharge area of the Ledo–Paniselian aquifer: (a) variogram of the calcium content; (b) map of calcium concentration (in mg L^{-1}).

depth, indicating hydrochemical processes going on as redox conditions change with increasing depths. It is here likely that the reduction of iron hydroxides has released adsorbed arsenic which was before bounded by surface complexation. Although systematic changes in concentration of these parameters with time have not (yet) been observed,

it is clear that in aquifers with a strong vertical stratification, changes in the hydrodynamics (upward vs. downward flow components) can cause baseline trends for the involved elements and species.

6.5.3 *Bulgaria: the Razlog karst aquifer*

The Razlog mountain karst aquifer (see Chapter 2) is situated in the southwest part of Bulgaria (Machkova, personal communication). The basin is made up of metamorphic rocks and marbles with thicknesses up to 1000 m. Any anthropogenic influence can be neglected because of the high mountain topography and the state environmental protection measures (a UNESCO recognised natural park).

Aquifer discharge takes place via several springs, some of them being measured (discharge rate and composition) since 1962 (Machkova, personal communication). The largest spring has a long-term average discharge rate of 1100 L s^{-1}. The aquifer recharge area is in the Pirin mountains, close to the Greek border. Snow accumulation and melting has a strong influence on recharge and discharge variations, with cyclic seasonal fluctuations in karst spring discharge rates. The precipitation maximum in late autumn is totally transferred to the late spring and early summer recharge of the following year due to the permanent snow accumulation at the spring recharge zone.

The mountain rainfall water which feeds the karst system is quite pure. Spring waters are of the $CaHCO_3$ type. Long-term calcium and bicarbonate concentrations are around 35–40 and 150–170 ppm, respectively (Fig. 6.10). They have low Cl (*c*.5 ppm), Na + K around 10 ppm and SO_4 is 15–20 ppm. Nitrate is generally below 1 mg L^{-1}.

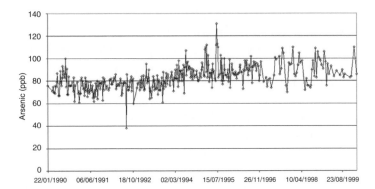

Fig. 6.7 Arsenic evolution in a production well in the Neogene aquifer in Flanders (Belgium).

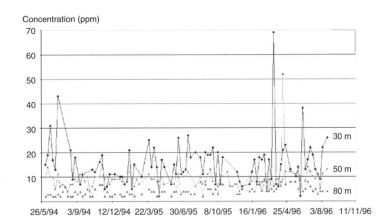

Fig. 6.8 Sulphate concentrations in the Neogene aquifer in Oud-Turnhout.

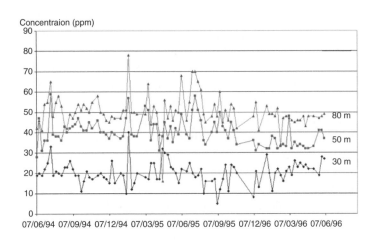

Fig. 6.9 Arsenic concentrations in the Neogene aquifer in Oud-Turnhout.

The time series includes 109 samples taken between 1962 and 2002. Evolution of spring discharge over the last 40 years of one large source (Fig. 6.10) shows a decreasing trend. A linear fit shows a decrease with 345 L s^{-1} over 4 decades. Chloride and sulphate concentrations show an increasing trend and bicarbonate decreases with time.

The gradual changes in composition can probably be related to a change in hydrodynamics of the aquifer system as suggested by the slow decrease in discharge rate. A change in aquifer recharge and/or a shift of the discharge area may be correlated with changes in climatological and meteorological conditions such as altered precipitation rates or shifting in time of the main precipitation period, changes in winter snow cover and snow accumulation. Percolation times through the karst system (relative changes in the diffuse and conduit flow components) can have been changed affecting calcite dissolution rates. Changes in temperature and precipitation may also have altered vegetation cover and impacted the composition of the infiltrating rainwater.

6.5.4   Czech Republic: the Turonian aquifer

Observation well records in the Turonian aquifer in Czech Republic (Paces, personal communication) span more than 15 years of measurements, starting around 1985 (Fig. 6.11). In series STPT-33 a sudden increase in TDS occurs in the early 1990s correlated with higher sulphate and calcium levels, shifting the water type from $CaHCO_3$ to $CaSO_4$ type. Nitrate increases sharply over this period. The last years concentrations have decreased to pre-1990 levels, indicating a trend reversal. Because the

hydrogeological and hydrochemical context of this observation well is not known, an anthropogenic source of the observed evolution cannot be excluded.

6.5.5   Denmark

Evolution of groundwater quality in the old well of the Carlsberg brewery (Hinsby, personal communication) provides a long time series between 1897 and 1945 (Fig. 6.12), which shows a sudden rise in Cl, Na, Ca and SO_4 in 1940, due to increased salinisation by upconing of deeper more saline water after at least 40 years of pumping. After this change in water quality the well was not used any more. It demonstrates that in case of upconing of more saline waters, the baseline quality deterioration can be quite sudden.

6.5.6   Estonia: the Basement aquifer

Time series from the Cambro–Vendian Basement aquifer in Estonia provide a 20 years record and have been sampled regularly and show an increase in a saline component (Karro et al. 2004). In well 613 (Fig. 6.13), a steady increase in Cl concentration over the 20-year monitoring period is observed, from less than 200 ppm in 1978, to over 700 ppm in 1997. The last years' concentrations have decreased. The sodium concentrations increase, but they are depleted by around 50% compared to the Cl values. Together with the sodium and chloride changes, calcium is also increasing, while bicarbonate shows no systematic trend. The increase in Ca concentration is not due to increased calcite dissolution, but to the replacement of sodium by calcium by cation exchange. The exchanged moles of Na were calculated based on the chloride

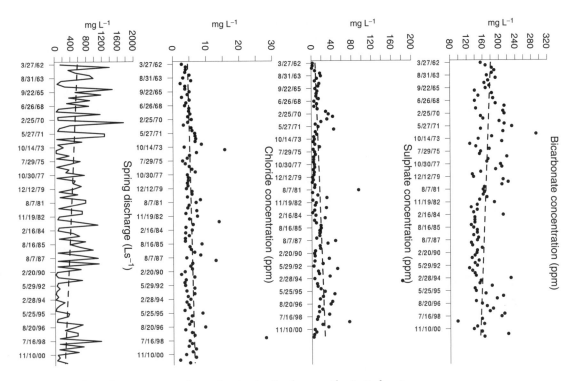

Fig. 6.10 Selected parameters of a spring in the Razlog aquifer in Bulgaria.

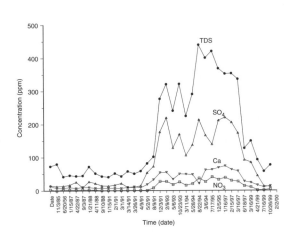

Fig. 6.11 Observation well STPT-33 in the Turonian aquifer in the Czech Republic.

concentrations and the seawater sodium to chloride ratio. The difference between the expected sodium content and measured values is the amount which is exchanged with calcium. This exchanged amount increases as the saline component becomes more prominent.

The relative occurrence of the main ions (Fig. 6.14) shows that calcium and magnesium form around 60% of the cations, with the sodium contribution increasing to the end of the series, and a dominance of Cl over HCO_3. The water is of the CaCl type, typically found in salinising situations where ion exchange (replacement of Na by Ca) occurs. Sulphate is virtually absent in the initial water, most likely because of sulphate reduction, but starts to increase as the marine component becomes more important. So while the reducing conditions were able to reduce sulphate in the initial CaCl water, the upconing deeper saline waters still have a high SO_4 level.

On a cross-plot of Na and HCO_3 as a function of Ca it is seen that bicarbonate concentrations are quite constant, between 2.5 and 3 meq L^{-1}, and not proportionally increasing with Ca. The excess of Ca above 2.5–3 meq L^{-1} is due to cation exchange, and not to

intensified calcite dissolution, and is induced by an increasing saline component as the increase of Na indicates.

It is possible that this is still the result of the history of the aquifer during glacial times, when less mineralised glacial waters have filled the reservoir. The cause of the increased salinisation is most likely linked to groundwater exploitation of the aquifer, attracting deeper more saline waters. The origin of the saline water at depth is described by Karro et al. (2004).

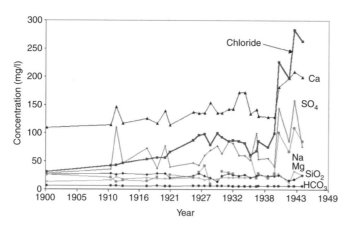

Fig 6.12   Evolution of water quality in the Carlsberg well in Denmark (1897–1945).

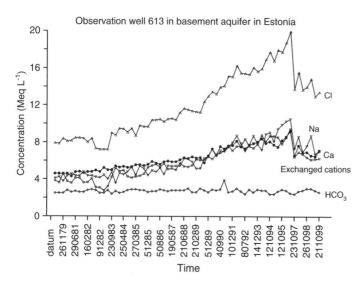

Fig. 6.13   Evolution of Na, Cl, Ca, HCO_3, and calculated exchanged cations in well 613 from Estonia.

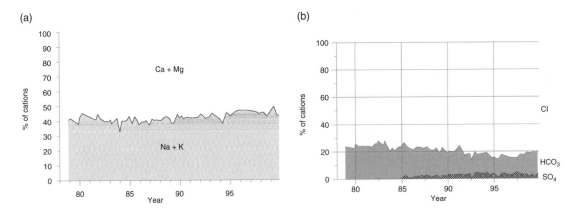

Fig. 6.14 Relative occurrence of main (a) cations, and (b) anions in well 613 from Estonia.

6.6 Conclusions

Identification of baseline trends in time series should not exclusively be based on statistical methods, but should also consider the hydrochemical context in which these trends are found. Interpretation of trends requires a profound understanding of the ongoing geochemical processes and should take into account the intrinsic heterogeneity of groundwater bodies. Trends can be caused not only by triggered chemical reactions but also by altered input concentrations into the aquifer system. Changes in the hydrodynamics and flow cycles of aquifers, due to exploitation or changes in land use, hydrography or climate, can move groundwater masses and cause changes in groundwater quality at a particular place and in pumping and observation wells. Salinisation originating from inflow or upconing of saltwater in coastal or salinised aquifers are among the most frequently found trends deteriorating groundwater quality. In systems where high flow velocities occur, like in karst aquifers, a short response time between changes in aquifer input and groundwater quality can be expected. Some hydrochemical processes act on timescales which are much longer than even the longest observation series. Examples of trends have been found in locations, ranging from Scandinavian to Balkan countries.

Acknowledgement

The authors of this chapter like to thank the project participants who provided usable time series data: the teams from Bulgaria, the Czech Republic, Denmark and Estonia.

References

Appelo, T. and Postma, D. (2005) *Geochemistry, Groundwater and Pollution*. A.A. Balkema Publishers, Leiden, The Netherlands, 649 pp.

Berner, R.A. (1981) A new geochemical classification of sedimentary environments. *Journal of Sedimentary Petrology* 51, 359–65.

Cleveland, W.S. (1979) Robust locally weighted regression and smoothing scatterplots. *Journal of the American Statistical Association* 74, 829–36.

Cleveland, W.S. and Devlin, S.J. (1988) Locally weighted regression: An approach to regression analysis by local fitting. *Journal of the American Statistical Association* 83, 596–610.

Derron, M.-H. (1999) Low temperature water–rock interaction: Geochemistry of metals in the weathering of alpine mafic rocks (in French). PhD-Université de Lausanne, 239 pp.

Edmunds, W.M., Cook, J.M., Darling, W.G. et al. (1987) Baseline geochemical conditions in the Chalk Aquifer, Berkshire, U.K. a basis for groundwater quality management. *Applied Geochemistry*, 2, 251–74.

El Manharawy, S. and Hafez, A. (2003) A new chemical classification system of natural waters for desalination and other industrial uses. *Desalinisation* 156: 153–80.

European Union. (2000) Directive 2000/60/EC of the European parliament and of the council of 23 October 2000 establishing a framework for Community action in the field of water policy. *Official Journal of the European Union* L 327.

Freeze, J.R. and Cherry, J.A. (1979) *Groundwater*. Prentice-Hall, Englewood Cliffs, NJ, 604 pp.

Furtak, H. and Langguth, H.R. (1967) Zur hydrochemischen Kennzeichnung von Grundwässern und Grundwassertypen mittels Kennzahlen. Mem. IAH-Congress, 1965, VII: 86–96, 5 Fig.; Hannover.

Grath, J., Scheidleder, A., Uhlig, S. et al. (2001) The EU Framework Directive: Statistical aspects of the identification of groundwater pollution trends, and aggregation of monitoring results. Final Report. Austrian Federal Ministry of Agriculture and Forestry, Environment and Water Management (Ref: 41.046/01-IV1/00 and GZ 16 2500/2–1/6/00), European Comission (Grant Agreement Ref.: Subv 99/130794). Vienna.

Kansas Geological Survey (1989) Hydrochemistry of the Dakota aquifer. Annual Report FY89, Open-File Report #90–27.

Karro, E., Marandi, A. and Vaikmae, R. (2004) The origin of increased salinity in the Cambrian-Vendian aquifer system on the Kopli Peninsula, northern Estonia. *Hydrogeology Journal* 12, 424–35.

Kilchmann, S., Waber, H.N. Parriaux, A. and Bensimon, M. (2004) Natural tracers in recent groundwaters from different Alpine aquifers. *Hydrogeology Journal* 12, 643–61.

Lasaga, A.C. (1984) Chemical kinetics of water–rock interactions. *Journal of Geophysical Research* 89 (B6), 4009–25.

Pannatier, E.G., Broers, H.P., Venema P. et al. (2000) A new process-based hydro-geochemical classification of groundwater. TNO-report NITG 00–143-B.

Piper, A.M. (1944) A graphic procedure in the geochemical interpretation of water analysis. *Transactions American Geophysical Union* 25, 914–28.

Schoeller, H. (1955) *Géochemie des eaux souterraines. Applications aux eaux de gîsements de pétrole*. Institut Français du Pétrole, France.

Stiff, H.A., Jr. (1951) The interpretation of chemical water analysis by means of patterns. *Journal of Petroleum Technology* 3(10), 15–17.

Stumm, W. and Morgan, J.J. (1995) *Aquatic Chemistry: Chemical Equilibria and Rates in Natural Waters*. 3rd edn. John Wiley & Sons, New York, USA, 1022 pp.

Stuyfzand, P.J. (1986) A new hydrochemical classification of watertypes: Principles and application to the coastal dunes aquifer system of the Netherlands. In: *Proceedings of the 9th Salt Water Intrusion Meeting*, Delft, pp. 641–55.

Walraevens, K. (1987) Hydrogeology and hydrochemistry of the Ledo–Paniselian Aquifer in East- and West-Flanders (in Dutch). PhD thesis. University Ghent, Belgium.

Walraevens, K. and Van Camp, M. (2005) Advances in understanding natural groundwater quality controls in coastal aquifers. In: Araguas, L., Custodio, E. and Manzano, M. (eds) *Groundwater and Saline Intrusion*. Series: Hidrogeologia y aguas subterraneas no. 15. Instituto Geologico y Minero de Espana, pp. 449–63.

7 Monitoring and Characterisation of Natural Groundwater Quality

M. T. CONDESSO DE MELO, E. CUSTODIO,
W. M. EDMUNDS AND H. LOOSLI

The groundwater monitoring requirements described are focused on unpolluted, pristine waters. These form the basis for defining the natural properties of groundwater derived from atmospheric or geological sources and for distinguishing natural phenomena from human impacts. Baseline monitoring networks imply long-term observation often at low frequencies and standardised measurements and data that contribute to an understanding of the flow system, identification of controlling geochemical processes and trends. Results provide key information for aquifer management and protection of high-quality resources so as to safeguard the sustainability of groundwater resources. Above all, the monitoring should lead to the long-term stability and sustainability of water quality in the groundwater body with the preservation of the 'good chemical status' for present and future generations as defined in the EU (European Union) Water Framework Directive (WFD). The requirements for optimum design and sampling of baseline groundwater quality are discussed. The current practice for monitoring in 13 European countries are then considered, showing the diversity of conditions that exist prior to the implementation of the EU Groundwater Directive, and how fast monitoring conditions may change upon policy modifications.

7.1 Introduction

Groundwater is a key natural resource for present and future generations, and the preservation of its quality is a crucial step towards the overall sustainability of resources (Custodio 2006). Natural groundwater-quality may evolve due to water–rock interaction over time periods ranging from a few months to many thousands of years. Monitoring the changes in natural groundwater quality requires a different approach towards the setting up, maintenance and operation of routine groundwater monitoring networks, being something more than compliance monitoring to enforce regulations. The emphasis is placed on the definition of the natural background status (baseline conditions) of groundwater bodies and aquifer systems and how these may vary with time. Accordingly the results may then be used as a reference quality for future monitoring purposes and remediation actions. This groundwater reference quality may be the baseline water quality, defined in Chapter 1 as 'the concentration of a given element, species or chemical substance present in solution which is derived from natural geological, biological, or atmospheric sources'; or, the 'good groundwater chemical status' following the EU Groundwater Directive recommendations (EU 2006).

Groundwater monitoring forms an essential part of water-quality assessment in most country programmes. For example, the United States Geological Survey (USGS) implemented the National Water-Quality Assessment (NAWQA) programme in 1991 to develop long-term consistent and comparable information on groundwater, and other water bodies in support of national, regional, state and local information needs and decisions related to water-quality management and policy (NAWQA 2004). The NAWQA programme is comprehensive and focuses on the overall condition of water, how it changes with time and how natural and human activities affect quality. As in many other national monitoring programmes, the NAWQA studies are currently focused on pollution (fate of agricultural chemicals, urbanisation impacts and nutrient impacts, for example). Rather than a holistic approach, the present study emphasises how the natural, unpolluted groundwater background quality may be recognised and, through monitoring, interprets how natural properties may vary with time and how superimposed human impacts may be identified.

The setting up of a monitoring network of natural groundwater quality requires the prior detailed three-dimensional characterisation of the physical framework of the groundwater body. Otherwise the groundwater-quality network maybe unable to fulfil its goal and may be too expensive when we compare the results and the importance of the information provided. From the reference aquifer studies in this book it is clear that each groundwater body has an individual pattern of physical and chemical characteristics which are determined by climatic, geochemical, hydrological and ecological conditions prevailing in the catchment area. A complete characterisation of the groundwater body requires an understanding of all natural processes that may contribute towards water-quality variation and change within the groundwater system.

A baseline monitoring network should be designed on the basis of an understanding of the groundwater system. Once established, it should also provide appropriate data to validate and update the conceptual model of the studied groundwater body and to act as the basis for interpretation of the main observed spatial and temporal water-quality trends, so distinguishing natural phenomena from human impacts.

As with all monitoring, baseline monitoring requires the interpretation of a specific set of physico-chemical data, collected at a preselected geographical location (borehole, spring, piezometer, etc.) and representative of a certain aquifer depth and area. Tracers may be included to calculate the residence times, to indicate past recharge regimes or mixing processes, and to provide early signs of water-quality degradation (not necessarily due to anthropogenic contamination). Sampling frequency needs to reflect monitoring objectives and will depend on the type of groundwater body; a deep confined aquifer, for example, requires lower sampling frequencies than an unconfined aquifer, but often would benefit from the use of a specific tracer to indicate the onset of any modern recharge.

The interpretation of the groundwater baseline monitoring data should enable the identification of natural variations in the groundwater body and how robust these are during abstraction. Spatial and temporal trends may be induced by natural processes (baseline changes, for example, increase in Cl due to aquitard diffusion, high Fe and Mn values induced by redox conditions, high F or As resulting from water–rock interaction) and those which are exacerbated as a result of human impact (e.g. increase in Cl in response to saltwater intrusion or depletion in water levels, salt recycling by irrigation return flows, high NO_3 due to land use change and agriculture practices).

The main outputs of groundwater baseline monitoring therefore are baseline trends, both in space and time, which may be used as early warning tools for regulators, for defining policy and for end-users' awareness. Baseline quality data may then be used as a control for comparison in areas suffering from anthropogenic impacts.

In this chapter we consider the basic requirements for the monitoring of natural groundwater quality and review the current practice in the main countries considered in this book, with the addition of others (Slovakia, Germany, Holland, Austria, Lithuania) who have provided up-to-date information to illustrate the very contrasting and complex situation prior to the implementation of the WFD.

7.2   Purpose of monitoring systems

Groundwater baseline monitoring programmes start from an understanding of the hydrogeological system and of all the related biogeochemical factors that may affect the groundwater baseline quality. The assessment of groundwater usually involves setting up a programme collecting successive groundwater samples at individual sites. Prior to this, regional characterisation of groundwaters will have established the main spatial variations (age, salinity, zonation in redox and major ion ratios, for example). The combined monitoring programme at many sites will therefore contribute to a spatial analysis of the regional baseline patterns over time.

Monitoring networks for groundwater baseline quality imply long-term observation, often at low frequencies and standardised measurement of groundwater-quality data that should contribute to:

1 An understanding of the groundwater flow system patterns (identification of groundwater recharge and discharge areas, main flow paths, residence times, water balance);

2 identification of the principal geochemical processes controlling water quality and active within the aquifer;

3 detecting trends in the concentrations of key naturally derived constituents that may have implications for potability, human health or the health of the dependent ecosystems that receive groundwater;

4 an early warning of the superimposed human impact and land management practices on groundwater quality;

5 providing enough geochemical data to establish cause–effect relationships;

6 providing advice on aquifer management and/or remediation actions in order to safeguard the sustainability of groundwater resources.

Perhaps the most important objective is to guarantee the long-term stability and sustainability of the water quality in the groundwater body with the preservation of the good chemical status for present and future generations.

A further important and often overlooked area for monitoring is the unsaturated zone (see also Chapter 1). In porous media, the main natural properties of recharging groundwater are determined in the uppermost few metres through evaporation of rainfall, soil biogeochemical processes and shallow water–rock interaction. Vertical movement of moisture proceeds slowly, typically of rates between 0.5 and 1.0 m yr^{-1}. The attenuation of acidity takes place through silicate weathering especially the depletion of base cations. Chloride which remains inert is concentrated by evapotranspiration and may be used as the measure recharge rates and also in favourable circumstances as a record of recharge history (Edmunds and Gaye 1994). Monitoring the unsaturated zone has been used extensively to demonstrate the slow movement of contaminants towards the water table creating the persistent pollution problems due especially to agrochemicals (Parker et al. 1991), farmyard slurry (Gooddy et al. 2002) and from leachates beneath landfill sites (Williams et al. 1992). However several studies also emphasise the natural attenuation of acidity and the extent of mobilisation of aluminium and other metals (Ohse et al. 1984; Stuyfzand 1984; Moss and Edmunds 1992).

7.3   Design of Groundwater Baseline Quality Networks

The basic steps in designing a groundwater baseline quality monitoring network are to:
1 conduct a detailed study of the regional hydrogeology;
2 build a realistic and appropriate conceptual model of the groundwater system;
3 select and/or install the groundwater sampling sites;
4 define the range of variables and tracers to monitor;
5 verify the laboratory analytical protocols and detection limits;
6 choose the sampling protocol and any preservation methods in agreement with the laboratories;
7 define the sampling frequency;
8 decide on which type of data analysis and interpretative approaches should be used so that early warning is provided in time for decision making and sustainable aquifer management;
9 check periodically the validity of the conceptual model.

The initial survey of a groundwater body may need to involve a wide range of chemical and isotopic tracers to establish the geochemical controls on water quality. However, once a conceptual model has been built using detailed studies, the main monitoring programme may involve only a few key indicator elements, those most sensitive to change.

7.3.1   Location and density of the monitoring sites

The density of monitoring wells should depend on the size and heterogeneity of the groundwater body, the geological and hydrogeological characteristics of the groundwater system, the aquifer vulnerability and the risk posed to the aquifer from potential contaminant sources. A guideline value for natural baseline monitoring has been suggested (Vrba 1989):
•   10–100 km^2/sampling point for regional monitoring;
•   100–10,000 km^2/sampling point for national monitoring.

Each aquifer however is unique and selection of sites is ideally based on local knowledge and experience, and sampling point density may change according to local conditions. Within the available budget, sampling sites should be representative, and should be able to monitor geochemical evolution along the same flow path. Given the long turnover time of most aquifers, water-quality distribution is the result of former flow patterns and may not follow current ones, especially when aquifer development is intense. Then, flow paths to be considered depend on the new circumstances and the pumping regime.

The location of monitoring sites requires experience and is not a task for newcomers as has been the practice in many organisations.

7.3.2 *Measurement and sampling frequency*

The sampling frequency is a function of the nature of the groundwater body and also a compromise between the logistical and financial resources for sample collection and laboratory analyses. Optimal frequencies have also to be adjusted to hydrogeological aspects (residence time, flow velocity) and seasonal influences (climate, atmospheric inputs) and statistical considerations. The nature and response-time of the aquifer and of the target element or substance guide the choice of sampling frequency.

Groundwater velocity in porous media is usually slow, in the order of a few centimetres to some tens of metres per year and, consequently, the variability of natural groundwater quality is also slow. In an unconfined aquifer the measurement and sampling frequency needs to be higher than in a confined aquifer,

due to the different conditions and temporal changes in water quality. Operational sampling and analysis frequencies could therefore vary from once every 6 years for a confined, slow-flow aquifer and for a parameter which is not related to any anthropogenic influence, to several times a year for the outcrop area with rapid groundwater flow (or fluctuating water table) and a parameter which is a rapidly responding indicator of groundwater-quality change. The sampling frequency should therefore be a function of the nature of the groundwater body and also a compromise between the logistical and financial resources for sample collection, including enough pumping time to get a reproducible sample under comparable conditions, and laboratory analyses.

7.3.3 *Parameter selection*

Baseline quality changes described elsewhere in this text are likely to be recognisable to a large extent by major ion ratios (e.g. for solution phenomena, ion exchange and salinity). Increase and decrease in major anions will be particularly indicative of salinity (Cl, SO_4) and of water softening associated with cation exchange (HCO_3). Redox processes can be followed using reduction-oxidation potential (Eh), dissolved oxygen (DO), nitrate (NO_3) and total dissolved iron (Fe_T). A number of trace elements may also be indicative of increased residence times, since the build up of key trace elements are related to time dependent geochemical processes (Edmunds and Smedley 2000). Key isotopic indicators such as tritium or organics such as pesticides should also be included to indicate the presence or absence of modern water (Gaye 2001; Loosli and Purtschert 2002). In practice, very few situations will be monitored at the minimum requirement and the complex pattern of land use of a relatively densely populated country means that many sampling sites will require monitoring for a wide range of elements or species.

Specific parameters for monitoring natural properties of groundwater are summarised in Table 7.1 and include field measurements such as pH, groundwater temperature (T), specific electrical conductance (SEC), reduction–oxidation potential (Eh), dissolved oxygen (DO) and alkalinity. For laboratory analysis Na, Ca, K, Mg, Cl, SO_4, Si, NO_3, NH_4, B, Ba, Sr, F, Br and certain other trace metals that may have significant mobility in specific geological environments such as Fe,

Mn, As and Cr, should be considered. Thus a good knowledge of the aquifer lithology (carbonate versus non-carbonate) and geological setting (e.g. presence of secondary minerals) may be used to guide decisions about which specific parameters need to be measured and in which areas.

Isotope indicators should also be considered for groundwater baseline monitoring purpose in association with other chemical tracers. Some isotopic analyses may be relatively expensive but this may be offset by the fact that careful studies of representative aquifers may allow the overall sampling frequency to be reduced for larger areas. Isotope tracer methods are useful indicators

Table 7.1 Principal indicators used in the monitoring of natural (baseline) properties of groundwater.

Parameter	Role and characteristics for baseline monitoring
pH	Indication of buffer capacity of the system. Should be invariant in most natural systems. Key changes are possible near and below pH 5.5. Sudden changes may be a cause for concern
SEC	Main parameter for monitoring and proxy for TDS and salinity. Careful temperature calibration needed. To detect small perturbations
Eh	Main tool for redox characterisation. Rapid changes indicate change from aerobic to anaerobic status
DO	Indicating natural changes in consumption of oxygen in the natural system due to water-rock interaction
TOC	Low baseline concentrations in most natural groundwaters
Cl	Inert and mobile tracer of salinity and, with SEC, for detecting first indications of perturbations
SO_4	Natural concentrations vary due to oxidation of sulphides and gypsum dissolution
HCO_3	Indication of inputs from soils and land use and, in deeper groundwater, following cation exchange
NH_4	Very low natural baseline expected except in reducing groundwaters
Na	See Cl and also HCO_3 above
K	Characterisation of the natural system due to water-rock interaction (silicates)
Al	Low in neutral groundwater, but monitoring important in waters with pH ≤ 6.0
As	Potential for elevated concentrations in some aerobic and also in reducing sections of clastic sediments
Cr	Potential for elevated baseline concentrations in some aerobic groundwaters
F	Naturally high concentrations in deep, older groundwaters (mostly carbonate aquifers)

of past and present recharge conditions and especially water age. They can help to answer questions relating to the age structure and the mixing components of a water body and with the validation and calibration of flow and transport models. The principal roles of isotopic tracers in groundwater baseline-monitoring are considered further in Loosli and Purtschert (2002). Further background to the application of isotopic tracers in baseline studies can also be found in Chapter 5 in relation to several reference aquifers.

7.3.4   *Selection of monitoring sites*

The installation of purpose-drilled monitoring wells to specified depths, with known screened intervals, offers the best chance of obtaining samples that are reasonably representative of conditions in the aquifer, although purpose designed networks may not always be feasible. When selecting the borehole drilling methods, it must be clear that water samples from boreholes must be free of drilling fluids.

When using already available boreholes, it is important that they have good supporting geological information and construction details. Short-screen lengths or partial penetrating in the aquifer are preferable. However, it must be taken into account that short-screened boreholes may have a small yield making sampling problematic when permeability is low.

The use of large diameter and/or long-screened wells for monitoring purposes should be avoided, and mild steel cased boreholes should be used with caution. Also, the use of existing water supply wells for monitoring purposes may lead to the collection of water samples that may be a mixture of water from different aquifer layers, and sample quality may be variable with time and pumping rates.

A knowledge of the vertical distribution of groundwater quality is very important to provide a three-dimensional understanding of the flow system (Buckley 2001); construction of the borehole itself is very likely to have disturbed the local groundwater flow regime. This is especially likely for boreholes open over much of their length and situated in recharge or discharge areas giving rise to significant vertical components of groundwater flow. A borehole may also penetrate two or more aquifers separated by impermeable strata. Different hydraulic heads may produce upward or downward components of flow, and the borehole itself may induce cross-contamination by permitting flow from one aquifer to the other. Such boreholes are generally very misleading for baseline monitoring purposes, and representative groundwater samples are unlikely to be obtained, whatever sampling method is used. Vertical sampling of the aquifer under unconfined conditions may identify a gradation in water quality, with modern, often polluted waters overlying older pristine groundwater. Understanding this stratification through logging and monitoring is important for the overall protection of the groundwater resource.

7.3.5   *Groundwater sampling considerations*

Representative water samples need to be stored and transported to the laboratory for analysis with minimal disturbance. The procedures used to collect, store and analyse groundwater-quality samples, and to evaluate analytical results should ensure that the data are of the type and quality necessary to meet the objectives of the monitoring programme. Groundwater sampling campaigns should follow international protocols but optimised for local conditions.

It is good practice to use a multiport flow-through cell connected in-line to obtain reliable geochemical field data without risk of aeration. Measurements of the principal field parameters – pH, temperature (T), specific electrical conductance (SEC), reduction-oxidation potential (Eh) and dissolved oxygen (DO) should be made from the discharge during pumping conditions and recorded once a stable reading is achieved. Redox readings are measured by platinum electrode and, are reported relative to the standard hydrogen electrode (SHE); however under aerobic conditions Eh measurements will be meaningless since the Platinum electrode (Pt electrode) behaves as a pH electrode (Whitfield 1974).

Samples for subsequent major, minor, trace element analysis should be filtered in the field through 0.45 μm membrane filters into two separate acid-washed high-density polyethylene bottles. Sample preservation for both inorganic and isotopic analysis should follow protocols advised by the laboratory where the samples are analysed. It is usual to stabilise trace metals to approximately pH 1.5 by adding high-purity HNO_3 or HCl (1:100), depending on the analytical procedure being used.

7.4   Baseline monitoring and policy considerations

Demonstrating the purity and wholesome properties of groundwater is of interest to everyone, especially the general public who are often sceptical of the quality of treated waters. Water which can be demonstrated to be free of contaminants requires recognition and special protection. Thus baseline monitoring is aimed strategically at conservation of pure water bodies and not just at the needs of compliance.

The European Parliament and Council Directive established a framework for Community action (in 2001) in the field of water policy: the WFD. This legislation includes the Groundwater Directive 80/68/EEC (EU 1980a) and The Drinking Water Directive 98/83/EC (EU 1980b), among others. Within the WFD all Member States are required to assign groundwater bodies to river basin districts and establish competent authorities, using existing structures or creating new ones, and establish administrative arrangements to ensure that the Directive is implemented effectively within each river basin. These competent authorities are then required to make institutional arrangements to enable the planning, monitoring and enforcing of the requirements of the Directive.

The guidelines of the Directive suggest that an initial surveillance monitoring programme supplement and validate the impact assessment procedure and provide information for use in the assessment of long-term trends both as a result of changes in natural conditions and through anthropogenic activity. The results of this programme then lead to operational monitoring programme in order to establish the chemical status of all groundwater bodies determined as being at risk and to establish the presence of long-term anthropogenic induced upward trend in the concentration of any pollutant.

The WFD requires the classification of all European water according to quality status. One of the basic elements for that is the definition of water types and, limited to surface waters, the definition of reference conditions as a basis for the assignment of the ecological quality class. Therefore, no reference conditions are requested for groundwater, being the assignment of the quality classes classified as 'good or not good'. Baseline quality monitoring should be assigned as the basis to establish reference conditions and leading to a more reliable classification of the groundwater-quality status.

7.5 Current status of groundwater-quality monitoring in Europe

Preservation of good quality groundwater resources is one of the main environmental concerns in Europe, in part because of the particularly high population densities. The relative scarcity of natural, pure freshwater, associated with the risk of its gradual contamination, makes groundwater one of the most important resources for present and future generations, and requires an increased effort to protect this unique resource from further degradation. The assessment of groundwater quality may be achieved through adequate groundwater monitoring, which is crucial for effective groundwater management. European countries follow different approaches towards groundwater-quality monitoring and there is still a long way towards a common implementation strategy.

The groundwater monitoring programmes carried out in the European countries considered in this paper can be divided into three categories:

1 Surveillance monitoring
   (a) to supplement and validate the impact assessment procedure
   (b) to provide information for the assessment of long-term trends due to changes in natural conditions

   (c) to provide information on anthropogenic activity and for improving monitoring networks;
2 operational monitoring
   (a) to be undertaken within two surveillance monitoring periods
   (b) to establish the chemical status of bodies at risk
   (c) to establish the existence of any upward long-term trend in the pollutant concentration;
3 investigative monitoring
   (a) to be applied when operational monitoring has not yet been established and water bodies are at risk
   (b) to be used to verify the impacts dimension and to study unknown reasons of critical chemical status.

The objectives vary with the monitoring type followed but could be summarised as follows:

1 To characterise groundwater quality status;
2 to identify changes and trends in groundwater quality over time;
3 to identify existing groundwater problems with recognition of potential contamination sources;
4 to determine whether the goals of implemented remediation programmes are being met.

In most of the 13 countries in this investigation, the drinking water quality standards and related legislation are still used to manage groundwater-quality. Some countries are still setting up their national groundwater-quality monitoring networks following WFD recommendations (e.g. Portugal, Belgium, France, Spain) but others have already defined or are currently defining baseline values (e.g. Malta, Czech Republic, Poland, Estonia, UK) as a basis for the interpretation of the observed spatial and temporal variations.

The synthesis of different national monitoring groundwater programmes is based on national reports and may rapidly change. The objective is not to show in detail what is done in each country – large variations are even possible within a country – but to show the diversity of conditions that exist prior to the implementation of the EU WFD.

7.5.1 Belgium

In Flanders a regional groundwater-quality monitoring network comprising all aquifers has been established and has recently been extended. But the majority of the monitoring wells are still only being used for piezometric measurements. However, this may change in the near future. Besides, a dense shallow groundwater-quality monitoring network has been established for monitoring nitrate pollution; but complete chemical analyses are being performed on these wells.

Local and private groundwater monitoring wells are used to monitor specific potential contamination sites, with the measurements mostly restricted to parameters related to contamination.

The water supply companies often have efficient monitoring networks in the protection areas around their pumping wells. These networks have limited geographical extent and comprise specifically drilled monitoring wells, production wells and abandoned production wells.

7.5.2 Bulgaria

The Ministry of Environment and Water (MoEW) of Bulgaria is responsible for the setting up and implementation of groundwater monitoring networks. The country is covered by 35 observatory centres that supervise over 2000 monitoring stations.

Systematic hydrological measurements are available since 1920, while hydrogeological observations started later, in 1958. More than 1200 voluntary observers contribute to the maintenance of these networks while full-time personnel work only in some of the most important stations. The groundwater monitoring network includes 508 observation points: 106 springs, 340 deep and shallow wells and 62 pumping stations, used for measuring spring discharge, groundwater table and groundwater chemical status, respectively.

The National Groundwater Quality Monitoring Network managed by National Institute of Meteorology and Hydrology (NIMH) and by the Executive Environmental Agency (EEA) at the MoEW consists of about 400 observation points (120 springs and 280 boreholes). The boreholes belong to pumping stations and in most cases are sampled twice or four times a year. A limited number of chemical species are monitored:

- Field data: pH, temperature and specific electrical conductivity (SEC);
- major and trace elements analysed in the laboratories: alkalinity as HCO_3, NO_3, NO_2, NH_4, PO_4, Ca, Mg, Na and K, SO_4, Cl, Mn, Fe, hardness, COD and dissolved oxygen.

Long series of data exist for the major elements while trace element data are more limited.

7.5.3 Czech Republic

The Czech Republic currently has one of the most complete groundwater monitoring programmes in Europe. The programme started in 1984 with the monitoring of springs and shallow boreholes and was extended in 1991 by the monitoring of deep boreholes. A total of 462 sites are monitored including 137 springs, 147 shallow boreholes and 178 deep boreholes. Groundwater samples are collected twice a year in spring and autumn.

The monitoring programme includes the regular analysis of:

1 Physico-chemical parameters (twice per year): turbidity, colour, sediment, odour, SEC, acid neutralisation capacity, pH, DO, oxidation reduction potential (ORP), dissolved organic carbon (DOC), total hardness, SiO_2, COD, CO_2, NO_2, NO_3, NH_4, PO_4, SO_4, HCO_3, Cl, F, Mn, Fe, Mg, Ca, K, Na, Li, cyanide, phenols and humic substances;

2 heavy metal and metalloid analysis (twice per year): Al, Cr, Zn, Pb, Cu, Ba, Sr, Cd, Ni, As, Mo, Be, B, Co, Se, V, Sb and Hg;

3 organic compounds (twice per year): polycyclic aromatic hydrocarbons (PAH), volatile organic compounds, chlorinated hydrocarbons, pesticides and other dangerous substances;

4 radioactivity (annually).

All the data are stored in a searchable database and client applications include data loading, data check, data retrieval and data presentation. Technical documentation on monitoring sites completes the database. The national agency responsible for the groundwater monitoring programme is the Czech Hydrometeorological Institute.

7.5.4 Denmark

Drinking water in Denmark is obtained almost exclusively from groundwater. On a national scale, the exploitable resource is greater than the present abstraction but climate change, increasing contamination and changes in land use may adversely affect the future groundwater resources. Some areas close to the greater cities are overexploited. Groundwater-quantity monitoring is carried out in a national network of approximately 100 wells (GEUS 2005). Additionally, there are a number of regional monitoring wells and the water supply companies monitor piezometric heads in their abstraction wells and around main well fields. Currently, an important hydrological modelling project is carried out in cooperation between GEUS and the regional water authorities.

Groundwater-quality monitoring in Denmark is based on detailed groundwater chemical analysis in 73 monitoring areas, 5 agriculture catchments and on the results of the regular groundwater-quality monitoring by the water supply companies. Overall, this provides the available information on groundwater chemistry and pollution at the national scale.

The groundwater monitoring is established within small catchment areas ($5-50\,km^2$) with one abstraction well (volume monitoring) to represent flowing groundwater and up to 22 monitoring wells (line monitoring in the main aquifer and point monitoring in secondary aquifers).

Sampling in these areas involves measurement of the following parameters:

1 Sixteen main groundwater components (DO, NO_2, NO_3, NH_4, Ca, Na, Mg, K, HCO_3, Cl, SO_4, Fe, Mn, Non-volatile organic carbon (NVOC), total P and aggressive carbon dioxide). F, Sr, total nitrogen, orthophosphate/phosphorous, hydrogen sulphide and methane have also been part of the monitoring programme until 2006, but will be left out from 2007 onwards due to financial restrictions and because it was considered they provided limited information;

2 Eight trace elements (Al, As, Cd, B, Ni, Pb, Cu and Zn). Until 2006, also Sb, Ba, Co, Cr, I and Se were monitored. Earlier also Br, CN, Li, Hg, Tl, Sn, Ag and V were part of the monitoring programme;

3 Eighteen organic micropollutants (aromatic hydrocarbons and phenolic compounds);
4 Thirty-five pesticides and metabolites have been regularly monitored in the national monitoring programme.

7.5.5 *Estonia*

The Estonian national groundwater monitoring programme began in the mid-1950s and includes observation records of over 20 yr for most aquifers. It is divided into three monitoring subprogrammes with different objectives, covering all the country and its aquifer systems:

1 Groundwater basic monitoring programme;
2 groundwater monitoring of agricultural areas;
3 groundwater monitoring of industrial areas.

The groundwater basic monitoring programme includes seven monitoring districts each of which has different hydrogeological conditions and human impacts (volume of abstraction, type of industries, agriculture practices). It includes 80 investigation sites and 340 monitoring wells for the whole country and different aquifer systems. The monitoring districts have been divided, according to the dominant groundwater stress or vulnerability, into three groups:

1 Aquifers with natural conditions;
2 aquifers with intensive abstraction;
3 aquifers affected by impact of the dewatering of mines and quarries or by industrial activities.

The groundwater basic monitoring programme includes:

1 The measurement of the groundwater level in 306 wells with a frequency of 3 or 5 times a month in shallow aquifers, and once a month in deeper aquifers;

2 the groundwater sampling for chemical analysis from 22 wells once a year (for deeper aquifers, with a slower groundwater movement and stable chemical compositions) or twice a year (for shallow aquifers with high rates of recharge, with the samples being taken in spring with high water levels and in summer, after the dry season);

3 the chemical analysis includes T, Eh, SEC, pH, Na, K, NH_4, NO_3, NO_2, Ca, Mg, Fe, Cl, SO_4, CO_3, HCO_3, total hardness, SiO_2, CO_2, COD. In some cases, a reduced number of analyses are carried out, usually including specific parameters, such as Cl, NH_4, NO_3 or SO_4.

7.5.6 *France*

The groundwater monitoring history in France is relatively complex and the situation is evolving fast, following the recent 'Loi sur l'eau' (2006) and the WFD. Many institutions are still nowadays involved in monitoring of groundwater quality, including public institutions associated with different Governmental Ministries (mainly Health, Industry and Environment), private companies and independent syndicates or associations. Until 1999, and with the exception of the Departmental Divisions of Social and Sanitary Affairs (DDASS), there were no precise rules for defining groundwater-quality monitoring practices, nor any protocol on the parameters to be measured or the monitoring frequency.

The DDASS are responsible for monitoring the groundwater quality for public water supply and the analytical results are included in a database called SISE-EAUX. However, this database is difficult to use, due to the great diversity of sampling points (springs, boreholes, taps, pipes, etc.) and to the different levels/types of hydrochemical investigations.

The Regional Environmental Authorities (DIREN), the Water Authorities (Agences de l'Eau), private companies and the territorial communities associated with national technical authorities are all involved in groundwater monitoring. The French Geological Survey (BRGM) and the universities also carry out some monitoring during specific studies and for limited periods of time. However, all of them follow their own monitoring methodology, which may also change with time and project objectives.

The national groundwater monitoring programme (ONQES, Observatoire National de la Qualité des Eaux Souterraines) has recently tried to combine all these networks at regional, local or river basin scale covering the whole country [Institut Français de l'Environnement (IFEN) 1994].

In 1992, a national network including the Health and Environment Ministries, the French Water Agency, the Fishing Council, IFEN, French Research Institute for Exploitation of the Sea (IFREMER), Électricité de France (EDF), Météo France and Bureau de Recherches Géologiques et Minières (BRGM) was created. One of the objectives was to gather homogeneous and assessable data and to create databases at the river basin scale. The sampling sites were selected to be around drinking water supply boreholes in every kind of aquifer. The sampling frequency for basic monitoring programmes varies between one and four times a year. The parameters observed are listed in an ordinance of the French Health Ministry that defines the detection limits, analytical reference methods and the percentage of precision. More than 200 institutions are involved in sampling and analysing the groundwater

quality data. These activities are not standardised and there is also a lack of national quality control and assurance procedures. There are internal obligations for the laboratories to use comparable standardised regulation for precision and accuracy.

The Water Department of the Ministry of Environment coordinates the programmes for groundwater monitoring activities. The Ministry is assisted by BRGM, which manages the national groundwater database (ADES), and the Ministry of Social Affairs and Integration, which is responsible for sampling and analysing procedures. The monitoring networks show that the principal groundwater quality problems in France are related to high concentrations of arsenic, selenium, nitrate and pesticides in groundwater. However, at present, a new evaluation protocol for groundwater quality (SEQ) is in preparation by Water Authorities (Agences de l'Eau).

7.5.7 *Groundwater monitoring in Germany*

In the Federal Republic of Germany, the monitoring of groundwater quality is the responsibility of the Federal states (Länder). To this end, in recent decades, the Länder have systematically developed networks of groundwater monitoring sites. The aims of groundwater monitoring are:

● to ensure the timely detection of adverse changes in quality;

● to develop targeted remediation and minimisation strategies depending on the causes of pollution; and

● to assess the effectiveness of such protective measures.

In addition to the various measurement networks devised for the requirements of specific Federal Länder, **two national networks** have now also been created. As a rule, these draw on existing monitoring sites in the Länder networks.

Both networks supply the data basis for reports by the Federal Republic of Germany to the European Union and the European Environment Agency (EEA). This monitoring network supplies the essential data for Germany's reporting to the EEA in Copenhagen. The Federal Länder have asked the Federal Environmental Agency to prepare these reports on their behalf and submit them to the EEA. The monitoring network was compiled in collaboration with the Federal Länder, based on the following criteria:

• it should be a representative network providing an overview of the quality of groundwater throughout the whole of Germany;
• the Federal Länder have stipulated that it should be comprised of approximately 800 monitoring sites;
• these monitoring sites should be evenly distributed throughout the entire territory of the Federal Republic of Germany; and
• should be located primarily in the uppermost main groundwater aquifers.

Each year, the Federal Länder supply the Federal Environmental Agency with the monitoring results from this network. The Federal Environmental Agency records and analyses this data. At the end of each year, the Federal Environmental Agency compiles the data from the Länder in accordance with the requirements of the EEA and submits it to Copenhagen.

The federal states (Länder) are requested to provide data on the following parameters and substances; in general sampling should take twice a year but at least once a year: temperature, pH; electrical conductivity, O_2; NH_4; NO_2; NO_3; $O-PO_4$; Cl; SO_4; B; DOC, K;

Na; Ca; Mg, heavy metals/metals/metalloids, Al; As; Pb; Cd; Cr; Fe; Cu; Mn; Ni; Zn, aliphatic halogenated hydrocarbons and pesticides.

In addition, the federal states provide data to characterise the site and its catchments area as altitude of site and filter position, type of sampling site (well, spring, etc.), land use, hydrogeology (stratigraphy, petrography) and type of aquifer (porous rock, fissured rock, karst).

A nitrate monitoring network was devised by the Federal Länder with the sole purpose of fulfilling the specific monitoring requirements of the EC Nitrate Directive. Under the 1991 Nitrate Directive, Member States are obliged to carry out action programmes to minimise water pollution caused by nitrate from agricultural sources. The monitoring data is intended to demonstrate how the action programmes have affected groundwater quality. Reports must be prepared every 4 years and submitted to the European Commission. The following criteria apply to the selection of monitoring sites:

• The sites should be in the groundwater aquifer close to the surface. Analysis focuses primarily on the uppermost aquifer.
• The monitoring sites should indicate significantly elevated nitrate levels (greater than 50 mg L^{-1}, but at least greater than 25 mg L^{-1} NO_3).
• The elevated nitrate levels must be clearly traceable to agricultural sources.
• The selected monitoring sites should be representative of the largest possible catchment area, that is, they should demonstrate the impacts of diffuse substance inputs.

The monitoring network comprises around 180 sites from which samples are generally taken two to four times each year. The nitrate monitoring network selectively records groundwater pollution in contaminated areas (worst case scenario), and unlike the EEA monitoring network, is not representative of groundwater pollution in Germany.

7.5.8 *Malta*

The monitoring of groundwater quality in Malta dates back to 1944 (also with some earlier data). The Maltese Regulatory Authority (MRA), as the Regulator, is responsible for coordinating groundwater monitoring while the Water Services Corporation is the implementing agency. The principal aquifer systems monitored in Malta are all in fractured carbonate rock.

Over 95% of Malta's aquifers are monitored for groundwater quantity and quality. The network density is variable depending on the number of wells in operation. Monitoring points include boreholes specifically drilled for gauging purposes and decommissioned boreholes that are no longer in use for production purposes. Most of the monitoring boreholes are located in the mean sea-level aquifers (20 in Malta and 6 in Gozo) because of their strategic importance in terms of water resources. The perched aquifers are only monitored for quality and quantity at the point of discharge.

Groundwater quality is usually evaluated monthly and the parameters measured include piezometric level, pH, SEC, Cl and NO_3. A new network is being proposed and takes into account new monitoring requirements set out by the WFD. It will have a minimum network density of one monitoring station per 16 km^2 of aquifer and a minimum of three monitoring points per aquifer. Monitoring stations will be fixed and monitoring will not depend on whether a source is connected to public supply or not, ensuring a continuous track of data.

Baseline values are currently being defined but EU maximum admissable concentrations (MAC) values have been adopted as standard values. When groundwater does not comply with those values, the borehole is decommissioned from the public supply network or the water treated further before being used for drinking purposes. The main identified groundwater quality problems are related to high nitrate concentrations and localised salinisation problems.

7.5.9 *Poland*

The National Groundwater Monitoring in Poland is integrated in the Environmental Protection Programmes. The legal basis for the national environmental monitoring system was established in 1991, after the Polish Parliament approved the State Inspector of Environmental Protection Act.

Presently in Poland there are three different groundwater monitoring networks at national scale, as well as a number of regional and local monitoring networks. The national scale groundwater monitoring networks include:

1 monitoring of groundwater quantitative status;

2 groundwater level monitoring network;

3 monitoring of groundwater chemical status;

4 groundwater-quality monitoring.

The groundwater-quality monitoring is carried out by the Polish Geological Institute (PGI) and works within the framework of the environment quality monitoring programme of the State Environmental Monitoring System.

The PGI set up the Stationary Groundwater Observation (SGO) network to monitor the primary and secondary usable aquifers within the entire country, excluding mineral, therapeutic and thermal waters. The aim of the programme is to determine the fresh-water dynamics, to protect groundwater resources from excessive exploitation, to prevent groundwater quality degradation and also to promote the public access to the results. The SGO network has about 600 monitoring sites, with a number of sites per aquifer system proportional to the estimated groundwater resources. For each observation point there is documentation concerning the geology, hydrogeological parameters and data regarding the environment (location, management method, land use, etc.).

The monitoring is made regularly, observing the groundwater, assessing and interpreting the groundwater-quality changes. The aim of groundwater monitoring is to support the actions leading to limitation of any negative influence of human impact on the groundwater. The groundwater-quality monitoring network has over 700 sites (exploitation wells, observation wells, dug wells and springs), monitoring various hydrological units. The proportion of sites monitoring shallow unconfined groundwater and deep groundwater are 54.6% and 45.4%, respectively. The sampling of the monitoring network is carried out by the PGI once a year between July and September.

The Central Chemical Laboratory of the PGI in Warsaw, holding a quality certificate, carries out the measurement of the groundwater parameters. These include the following physico-chemical indicators: acidity, alkalinity, Al, NH_4, As, Ba, CO_3, HCO_3, B, Br, Cd, Ca, carbonate hardness, Cl, Cr, Cu, DOC, SEC, F, hydrocarbons, Fe, Pb, Li, Mg, Mn, Mo, Ni, NO_3, NO_2, pH, PO_4, K, SiO_2, Sr, SO_4, Ti, TDS, total hardness, V and Zn.

An annual report from the groundwater monitoring studies is presented to the Head Inspector of Environmental Protection. Each report contains the review of: (1) groundwater quality (as well as changes relative to previous years); (2) water quality in terms of quality indicators (physical and chemical parameters); (3) water quality in terms of hydrogeological stages; (4) water quality in areas of different land use and (5) water quality as a function of depth of occurrence in the aquifer.

Results of the processed measurements along with short information about the monitoring system are periodically published in the series of the Environmental Monitoring.

7.5.10   Portugal

Groundwater resources support about 60–70% of the water supply in Portugal. However, some of the aquifers are highly vulnerable to groundwater contamination and require good groundwater-quality monitoring programmes. The Water Institute [Instituto da Água (INAG)] coordinates all activities related to the National Surveillance Groundwater Quality Monitoring Program of raw water and runs the national database. The Commissions of Coordination and Regional Development (CCDR) are responsible for carrying out the sampling campaigns and collaborate with INAG for database management and final reporting.

The existing National Surveillance Groundwater Quality Monitoring Program covers all the main aquifer systems in Portugal with a total of 657 monitoring sites for the whole country. From these sites, 145 are located in the sedimentary basin of Tagus and Sado rivers, 152 in the western sedimentary border, 99 in the southern sedimentary border and the remaining 261 are located inland in areas of undifferentiated crystalline rocks, with a minimum number of two monitoring sites per aquifer system.

The density and spatial distribution of the network was defined based on the type of aquifer media (porous, karstic or crystalline), flow direction, sampling points available (springs and boreholes preferred to wells) and on the aquifer vulnerability to groundwater contamination. For a better control of the water supplied to the end users and also for economical and practical reasons, public water supply boreholes were included wherever possible as monitoring sites.

The monitored parameters include

• physical parameters: T (measured every 6 months);

• chemical parameters: pH, SEC, DO, Cl, SO_4, PO_4, NH_3, NO_3, NO_2, TOC Ca, Mg, Na, K, HCO_3, Al, Mn, total Fe, Cu, Cd, Hg, Ba, B, As, Pb, CN, Cr, F, Se, Be, Co, Ni, Mo, V, U, trichloroethylene, tetrachloroethylene, toluene, total pesticides and individual pesticides (such as atrazine; simazine; 2,4,5-T; 2–4-D) and total hydrocarbons (measured every 6 months);

• biological parameters: total and fecal coliforms, fecal coliforms, fecal streptococus (measured every 6 months).

The water temperature is the only parameter measured in the field and the rest are analysed in a certified laboratory. All the data are available on line at the web site www.snirh.inag.pt/.

7.5.11 *Slovakia*

In Slovak Republic (49,000 km²; 5.4 million inhabitants) groundwater is used for drinking water supply for about 85% of the population. According to the National Water Balance Report, Groundwater, in 2003 45% of the Slovak population was dependent on karst groundwater resources from the mountainous regions – up to now showing good qualitative status, but also with high vulnerability to potential contamination. The overall groundwater-quality monitoring programme is under the responsibility of the Slovak Hydrometeorological Institute (SHMI; http://www.shmu.sk/sk/) which coordinates all activities related to the National Groundwater Quality Monitoring Network and runs the national database. Water users – mostly large waterworks – are running their own monitoring programme according to legal requirements for drinking water suppliers, and the results of this monitoring are collected and reported by the Slovak Water Research Institute (www.vuvh.sk).

Slovak National Groundwater-Quality Monitoring Network with a total of 376 monitoring sites is designed to cover the main aquifer systems. Such a groundwater quality monitoring network was designed and performed with slight changes since 1982. The density and spatial distribution of the monitoring network was selected according to the type of aquifer (porous, karstic, crystalline, flysh) and resulting basic vulnerability to potential contamination.

One hundred and twenty-two monitoring sites are located in Quaternary alluvial fans of the main rivers. They are sampled twice a year, mostly in spring (high-water) and autumn (low-water stage). Extra monitoring with 84 sites concentrates on the Zitny Ostrov Quaternary sediments along the Danube river with high thickness and water-supply importance. Sampling of these wells is performed two or four times a year. In the mountainous areas, 38 karstic springs are monitored four times a year, while another 24 springs in other lithologies (mostly from crystalline – granitic rocks – or flysh type rocks) are monitored only once per year. From this basic groundwater-quality monitoring network, 81 sampling points are already exploited for public water supply (35 wells and 46 springs). The monitored parameters include:

• in situ measurements: groundwater level, oxygen content/saturation, pH, specific electric conductivity, groundwater temperature, air temperature, acid neutralisation capacity, base neutralisation capacity, colour, odour, sediment content;
• basic chemical substances: Na, K, Ca, Mg, Mn, Fe-total, Fe^{2+}, NH_4^+, NO_3^-, NO_2^-, Cl^-, SO_4^{2-}, PO_4^{3-}, SiO_2, HCO_3^-, COD-Mn, H_2S;
• trace elements: As, Al, Cd, Cu, Pb, Hg, Zn, Cr, Ni;
• common organic compounds: total organic carbon – TOC;

A supplementary group of determinants (not regularly measured for each site) include: basic physio-chemical substances (H_2S, CN^-), common organic compounds, aromatic hydrocarbons, chlorinated solvents and polyaromatic hydrocarbons.

Apart from the in situ measurements samples are analysed in a certified laboratory. The data are not currently available on-line and at the web site www.shmu.sk the public can find only maps and general reports for individual years.

7.5.12 Spain

Prior to the 1985 renovated Water Act groundwater was mostly a private domain and the different River Basin Water Authorities (Confederaciones Hidrográficas, created in 1926) were not involved. The Geological Survey of Spain (IGME) established a national network starting in the 1960s that by the late 1990s comprised 1150 groundwater-quality sampling points in porous formations, 400 in karst media and almost 1400 in other diverse aquifers. Generally the data set contains 16 parameters for water-quality and only chloride and specific electric conductivity for a network of points in coastal aquifers selected to monitor saline intrusion. Most sampling points are wells and springs. Sampling frequency is twice a year (spring and autumn). All data are organised in a database called AGUAS, that reunites 13 interconnected regional databases (www.igme.es). It is easily accessible to the public through the SIAS (Groundwater Information System). Monitoring in many areas was discontinued in the late 1990s.

The 1985 Water Act declared all water – including groundwater – a public domain and the responsibility for monitoring and management was given to the different Water Authorities (depending on the General Directorate for Water of the Ministry of the Environment) in the case of river basins extending over more than one region – Spain is divided into Autonomous Regions with competence on the territory – and on Regional Water Authorities when all the river basin is inside the region. Often the Water Authorities have reacted slowly. In some cases they have established their networks and databases in the whole territory (Ebre, Xùquer, Internal Basins of Catalonia) or in some relevant aquifers (the case of Doñana), but in other cases they have been slow to replace the IGME network or to pass to it the funds to continue the monitoring. Some gaps in the time series data have resulted. Attention has been given to groundwater level – even with automatic data acquisition equipment in some cases – but groundwater-quality monitoring lags well behind and in many aquifers data avaliable are limited to those derived from results from studies of the IGME and University teams, or occasional surveys contracted by the Water Authorities. Data acquisition is mainly related to water planning and more recently to respond to the requirements of the European WFD which has proved a significant driver for establishing and extending groundwater-quality monitoring. Access to the water quality data for the public is variable from Water Authority to Water Authority and still there is no central data centre. Current progress is noticeable.

7.5.13 Switzerland

More than 80% of Switzerland's drinking and process water is sourced from groundwater. Quantitative and qualitative monitoring of groundwater is therefore required in order to ensure the long-term preservation of drinking water resources (FOEN 2006). The National Ground Monitoring Network (NAQUA) was implemented in 1999 and is nowadays used to assess the impact of certain substances used in human activities on the evolution of groundwater quality. NAQUA comprises two monitoring networks:

1 National Monitoring Network of Groundwater Quality Trend (NAQUATrend), which is a long time monitoring network with 50 representative sites across Switzerland. The sampling frequency is dependent on the hydrogeological and geochemical variability of each well;

2 National Monitoring Network of Groundwater Quality (NAQUASpez), which is a network with 508 sampling sites (260 springs, 243 groundwater wells, 5 piezometers) operated by the Swiss Federal Authorities in collaboration with the agencies responsible for groundwater protection in each canton. It is designed to permit targeted studies of potential groundwater pollutants.

The monitoring sites are selected and distributed for each specific sampling/testing programme according to the following criteria (1) type: groundwater or spring water; (2) geographical location and elevation: Alps, Midlands, Jura, Border Areas; (3) land use in catchment area: all kinds of agriculture (vegetables, pastoral farming, vine, forest), housing estates, traffic routes; (4) geology/hydrogeology: gravel or hard rock; (5) impacts: pristine or anthropogenic influenced; (6) casing: metal or synthetic material and (7) the purpose for which groundwater is used. But for each programme, a certain number of monitoring stations are included that are unlikely to show any impact of human activities. The analyses from these sites provide baseline values.

The sampling and analytical methods have to meet specific requirements. Sampling involves direct measurement of the following parameters in the field: spring discharges/groundwater levels, SEC, groundwater temperature and dissolved oxygen/oxygen saturation levels. In a preliminary phase other natural and anthropogenic components such as tritium, pesticides, zinc, benzene, nitrate, carbonate – CO_2 equilibrium, CFCs, fluorine, dissolved inorganic carbon (DOC), pH and temperature are also measured, with the purpose of defining the present situation. In a second phase, long-term trends are monitored for specific components.

Sampling frequency is adapted to the specific characteristics of each monitoring station and takes into account the likelihood of anthropogenic influences, the hydrogeological characteristics of the site and the variability of hydrochemical parameters. All physical and chemical data is recorded in a GIS-supported database, reviewed and assessed.

7.5.14 United Kingdom

In England and Wales, the Environment Agency (EA) is the Government appointed organisation responsible for monitoring and managing groundwater according to the relevant European legislation. In Scotland, the Scottish Environmental Protection Agency (SEPA) is responsible; Northern Ireland also has devolved responsibility. Groundwater use and abstraction occurs to some extent in all parts of the country, and so cannot be neglected anywhere, but is most extensive in southern and eastern England. The water companies that supply groundwater also conduct some monitoring for their own purposes, although often this is focused on supply monitoring rather than raw resource monitoring.

In England and Wales the monitoring programme is currently operated by the eight regions controlled by the Environment Agency. The density and frequency of monitoring and the parameter suites vary from region to region. The most extensive programmes, covering major aquifers with high population densities, measure suites of up to 90–100 chemical parameters, up to four times per year. In other areas monitoring is much sparser and can be as little as once a year for a smaller suite (20–30 parameters) and for a network of lower sampling-point density.

Since the introduction of the WFD the Environment Agency, in conjunction with the British Geological Survey (BGS), has been actively reviewing its monitoring practice nationally and has begun to implement a revised national strategy designed to ensure a consistent approach to monitoring across all regions, sufficient to ensure compliance with the requirements of the WFD. A national programme of baseline surveys of all major and some minor aquifers has also been conducted by BGS on behalf of the Agency.

When fully implemented it is likely that the completed network will comprise approximately 3500 sampling points across England and Wales. The complete set of suites will include over 200 aqueous species or physico-chemical parameters, but not all will be measured everywhere or everytime. Land-use data will be used to guide decisions about which parameters need to be measured and in which areas. The nature and response-time of the aquifer and of the particular species guide the choice of sampling frequency.

7.6 Conclusions

From the above discussion of the monitoring situation in the different European countries it is clear that no standard procedures yet exist in Europe and even within each country there are often differing practices and approaches region by region. Many countries have patchy regional coverage and have lost the continuity in monitoring due to successive reorganisations and changes in responsibilities. Monitoring programmes, especially for contaminants, have also been compounded by changing priorities, responding to new classes of contaminants and financial limitations. It is rare to find long runs of monitored data.

Against this background there is a pressing need for harmonisation across Europe to match the level of consistency found for example in the NAWQA programme in the USA. As monitoring programmes develop as part of the implementation of the Groundwater Directive, there is also the opportunity to pay attention to the baseline quality status and to be able to build in a geochemical approach which will ensure that what constitutes the truly natural component of water quality is recognised so that early warning of human impacts can be recognised.

Groundwater baseline monitoring contributes to an understanding of the regional groundwater quality. It may be used to determine the background levels for studying natural processes and the regional groundwater-quality variations. These values are reference levels for the study of large-scale and long-term anthropogenic impacts. For aquifers with short residence times the emphasis needs to be placed on the infiltrating recharge waters and the changes in its composition due to water–rock interactions and aquifer hydrodynamics.

The development of an understanding of groundwater-quality evolution and ground-water regime will contribute towards an optimal management of groundwater resources. This needs a monitoring network of suitable boreholes and springs to be sampled periodically, a few times a year to every 1 or 2 years.

The main aspects of a baseline monitoring programme are summarised conceptually in Fig. 7.1 showing how monitoring leads to identification of pollution early warning and feeds into policy on water quality management. In summary therefore a groundwater baseline quality monitoring programme should:

• establish the natural baseline quality of groundwater resources, understand the flow system (groundwater recharge, flow paths, timescales, water balances) and identify the principal geochemical processes within the aquifer; use median and upper baseline (97.7 percentile for groundwater bodies);

• monitor for future changes in the groundwater baseline quality over time;

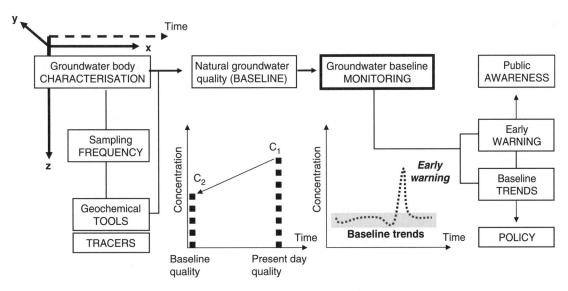

Fig. 7.1 Schematic representation of the groundwater baseline monitoring approach.

- identify past baseline trends in the concentrations of groundwater constituents;
- provide early warning of human impacts and land management practices on groundwater quality especially through mobile constituents;
- provide enough geochemical data to establish cause–effect relationships;
- provide advice on aquifer management and/or remediation actions in order to guarantee the sustainability of groundwater resources.

The minimum requirements of a groundwater baseline quality monitoring network are: (1) detailed knowledge of the local hydrogeology; (2) the selection and/or construction of the sampling sites; (3) the range of variables and tracers to monitor; (4) the verification of the laboratory analytical protocols and detection limits; (5) the choice of the sampling protocol and preservation methods; (6) definition of sampling frequency; (7) the decision of which type of data analysis and interpretative approaches should be used so that early warning is provided in time for decision making and sustainable aquifer management.

References

Boulding, R. (1995) *Practical Handbook of Soil, Vadose Zone and Ground-Water Contamination: Assessment, Prevention and Remediation*. CRC Press Inc., Boca Raton, Florida, 948 pp.

Broers, H.P. and Buijs, E.A. (1996) Trends in the grondwaterkwaliteit van Drenthe; analyse van 12 jaar meetgegevens van landlijke meetpunten in de providence Drenthe. TNO Groundwater en Geo-Energie, rapport GGR 96–80 (A).

Buckley, D. (2001) Application of geophysical borehole logging techniques to examine coastal aquifer palaeohydrogeology. In: Edmunds, W.M. and Milne, C.J. (eds) *Palaeowaters of Coastal Europe; Evolution of Groundwater since the Late Pleistocene*. Geological Society, London, Special Publication 189, pp. 251–70.

Custodio, E. (2006) Groundwater protection and contamination. In: Llamas, M.R., Ragone, S. and de la Hera, A. (eds) *Groundwater Sustainability*. National Groundwater Association, Westenville, Ohio.

Edmunds, W.M. and Gaye, C.B. (1994) Estimating the spatial variability of groundwater recharge in the Sahel using chloride. *Journal of Hydrology* 156, 47–59.

Edmunds, W.M. and Smedley, P.L. (2000) Residence time indicators in groundwaters: The East Midlands Triassic Sandstone aquifer. *Applied Geochemistry* 15, 737–52.

EU (European Union). (1980a) The Dangerous Substances Directive (DSD), Council Directive 80/68/EEC of 17 December 1979 on the protection of groundwater against pollution caused by certain dangerous substances.

EU (European Union). (1980b). The Drinking Water Directive (DWD), Council Directive 98/83/EC of 3 November 1998, concerning the quality of water intended for human consumption.

EU (European Union). (2000) Directive 2000/60/EC of the European Parliament and of the Council, 72 pp.

EU (European Union) (2006) Directive 2006/118/EC of the European Parliament and of the Council, 13 pp.

Gaye, C.B. (2001) Isotope techniques for monitoring groundwater salinisation. *Hydrogeology Journal* 9, 217–18.

GEUS (Geological Survey of Denmark and Greenland) (2005) Groundwater. In: NOVANA – National Monitoring and Assessment Programme for the Aquatic and Terrestrial Environment. Programme Description – Part 2, pp. 49–67. NERI Technical Report No. 537, 2005. National Environmental Research Institute. www.dmu.dk/NR/rdonlyres/5EC5E735-724E-4769-A560-D53613C89536/0/FR537

Gooddy, D.C., Clay, J.W. and Bottrell, S.H. (2002) Redox-driven changes in porewater chemistry in the unsaturated zone of the chalk aquifer beneath unlined cattle slurry lagoons. *Applied Geochemistry* 17, 903–21.

FOEN. (2006) National Groundwater Quality Monitoring Network (NAQUA). www.bafu. admin.ch/

Loi sur l'eau (2006). Loi n°2006–1772 sur l'eau et les milieux aquatiques promulguée le 30 décembre 2006 (J.O. du 31/12/2006). Ministre de l'Ecologie et du Développement Durable, France.

Loosli, H.H. and Purtschert, R. (2002) How can sustainability in groundwater use be determined? Goldschmidt Conference Abstracts, Davos, 66, A468.

Moss, P.D. and Edmunds, W.M. (1992) Processes controlling acid attenuation in the unsaturated zone of a Triassic Sandstone aquifer (UK), in the absence of carbonate minerals. *Applied Geochemistry* 7, 573–83.

NAWQA. (2004) Water Quality in the Nation's Streams and Aquifers – Overview of Selected Findings, 1991–2001. United States Geological Survey. USGS Circular 1265.

Ohse, W., Matthess, G., Pekdeger, A. et al. (1984) Interaction water-silicate minerals in the unsaturated zone controlled by thermodynamic equilibria. In: Eriksson, E. (ed.). *Hydrochemical Balances of Freshwater Systems*, Proc.

Symposium Uppsala Sept. 1984. *IAHS Publication* 150, pp. 31–40.

Rouhani, S. and Hall, T.J. (1998) Geostatistical schemes for groundwater sampling. *Journal of Hydrology* 81, 85–102.

Parker, J.M., Young, C.P. and Chilton P.J. (1991) Rural and agricultural pollution of groundwater. In: Downing, R.A. and Wilkinson W.B.(eds) *Applied Groundwater Hydrology*. Clarendon Press, Oxford, pp.149–63.

Stuyfzand, P.J. (1984) Groundwater quality evolution in the upper aquifer of the coastal dune area of the western Netherlands. In: Eriksson, E. (ed.) *Hydrochemical Balances of Freshwater Systems*. Proc. Symposium Uppsala Sept. IAHS Publication 150, pp. 87–98.

Whitfield, M. (1974) Thermodynamic limitations on the use of the platinum electrode in E_H measurements. *Limnology and Oceanology* 19, 857–65.

Williams, G.M., Young, C.P. and Robinson H.D. (1992) Landfill disposal of wastes. In: Downing, R.A. and Wilkinson, W.B. (eds) *Applied Groundwater Hydrology*. Oxford Science Publications, Oxford University Press, 340 pp.

Vrba, J. (1989) Protection and pollution of ground water resources. Department of Technical Co-Operation for Development, United Nations, Interregional Seminar on Water Quality Management in Developing Countries, ISWQM/SEM/4, 59 pp.

8 Natural Groundwater Quality: Policy Considerations and European Opinion

E. CUSTODIO, P. NIETO AND M. MANZANO

The effective application of the EU Water Framework Directive (WFD) to groundwater in its territory asks for a clear and scientific understanding of groundwater background or baseline quality, in order to be compared to a given current situation. This was at the heart of the research project 'BaSeLiNe' which was conducted by scientific institutions from a consortium of fifteen European countries. Attitudes to natural baseline quality were examined using a questionnaire, which was distributed to experts on groundwater issues (groundwater managers, end-users and regulators, and stakeholders) selected from each participating country. Results obtained from the 69 questionnaires are presented together with the ensuing discussions. Most answers refer to municipal water supply issues in relatively humid areas, but not exclusively. Environmental and agricultural issues are under-represented, but not absent. Groundwater is, in many cases, the main or the only available water source for human survival and activities, and is generally of good quality, although a progressive deterioration, due to an increasing content of some components is becoming noticeable. Groundwater protection is considered a long-term key issue that needs regulation, management and planning. This implies the joint participation of public institutions, private organisations and stakeholders, including institutions responsible for the environment. Groundwater protection should be subject to greater public information and involvement. However, at policy-making levels little awareness of the characteristics and use of groundwater and its protection exists. Groundwater-quality protection is envisaged from improved, less stressful land use and better farming practices. The costs involved for implementing protection measures must be supported directly or indirectly by water users, but there is still a reluctance to pay for quality improvements and subsidies are still sought. In many European countries there is some positive participation and co-responsibility of river basin districts, groundwater users' associations and stakeholders in groundwater management, although documented experience is scarce. Improved governance and institutional capacity is now needed to attain sustainable management involving all sectors.

8.1 Introduction

The WFD (2000), enacted in year 2000, is an important and ground-breaking legislation aimed at preserving and improving the water environment as a necessary step to sustainability. Its contents are neither easily perceived nor understood by citizens, and even hydrogeologists react negatively to some of its concepts. Implementation will be a slow process, during which time many aspects will need to be redefined and incorporated. A scientific approach is needed together with a suitable means of developing its concepts through parallel European research projects. The BaSeLiNe project conducted with a group of European countries (1999) is one of these.

The BaSeLiNe project has considered among its objectives an approach to policy and end-users, and is comprised of scientists, policy-makers, regulators, water industry managers and stakeholders to reach a consensus and state of the art. The aim was to find a common approach for the needs of the Community, with regard to groundwater baseline quality and the ways in which scientific and technical studies can address issues of protecting pristine waters and minimising their contamination. This involves defining strategies to use and disseminate data, and also to provide advice on data required to meet problems on local and regional scales. It is necessary to take account of the environment's good status, as defined and required in the European WFD (2000), and currently incorporated into Member States' national laws. Groundwater, a major freshwater resource that has been developed relatively recently, poses new and specific problems due to its intrinsic characteristics (Custodio 2005) and there is general lack of experience in addressing the special nature of groundwater at a managerial and policy-making level. This is the reason for the preparation of a Groundwater Daughter Directive, whose drafting has been slow and difficult but brought to completion in early 2007.

To make positive steps and work in the right direction, scientists/experts and stakeholders should work jointly, integrating their own particular domains (scientific research and policy decisions) in a complimentary manner. To accomplish this, within the BaSeLiNe project, a strategic advisory group was set up, comprised of experienced end-users and policy-makers from each of the participating countries, although the water supply aspects dominated in this group since they were the most organised assemblage in the water sector. However agriculture consumes up to 90% of the often-scarce freshwater resources in the Mediterranean regions and the farmer's voice is less organised and more difficult to integrate into advisory teams. In fact, farmers mostly organise to defend sectorial strategies rather than technical issues, which involve direct and covert subsidies and a poorly known economic framework (Hernández-Mora and Llamas 2000; AIH-GE 2003). Groundwater is also important and a key issue over much of Europe in sustaining river baseflow and wetlands, both in humid and dry climates, but there is a lack of specialised water representatives. Then, there is a bias that was known to the authors from the beginning that affects more dramatically water quantity aspects than the water-quality aspects considered in this paper, and emphasises the more humid Central and Northern Europe issues over the drier and more stressed Mediterranean region.

8.2 Methods adopted to involve end-users

In order to involve end-users and policy-makers in the project a number of consecutive steps were adopted: (1) personal approaches and contacts during meetings held at a local and national scale within each country involved; (2) at a European scale, one meeting gathering one or two representatives from each country group appointed by the respective national group; (3) bringing together national policy representatives from across Europe at a workshop held in Funchal, Madeira, Portugal, in October 2002.

In order to obtain comparable information and to draft a document for discussion, the Spanish team, who was responsible for this activity, prepared a questionnaire, which was discussed and finally agreed upon by the national teams. The objectives were to know what the selected stakeholders understood as groundwater quality, the baseline concept and their current and future application within the WFD. This questionnaire was sent to about 150 people selected from the BaSeLiNe consortium countries. The distribution within each country was made by the respective national team, who contacted the people and sometimes helped them fill in the forms. A response of 69 out of 150 questionnaires was achieved (45%), reflecting the partners' interest in this matter, the choice of targets and follow-up from the BaSeLiNe team. Most of the people who replied were from water supply companies, regulatory bodies and environment agencies/organisations with regional responsibilities. This introduces the bias already mentioned, since other users (agriculture, in-house use, recreational and the environment) are under-represented, but do reflect the two main interests – supply and regulation, especially in the more humid Central and Northern Europe.

8.3   The questionnaire

8.3.1   *Preparation*

Since the time and economic resources required were not available, carrying out a formal inquiry analysis approach was discarded, and the sample was forced to be relatively small and conditioned by the scarce number of representative water institutions in a given territory, as well as their bias towards town supply and regulatory functions. Consequently, the questionnaire was drafted informally.

8.3.2   *The sample*

The respondents to the questionnaires were from a background including water suppliers, policy-makers, regulatory agencies, water associations' representatives, hydrogeologists, groundwater experts and water engineers. There was some bias in the answers due to social and political constraints in some countries and sectors, and some respondents were not used to such enquiries. Some sectors such as agriculture and environmental protection were not adequately represented and they were addressed through the answers of regulatory bodies. No responses were obtained from those organisations more directly concerned such as conservation and wildlife interests. Stakeholders were poorly represented, but for a few groundwater users' associations mostly interested in water quantity aspects. The answers are a variable mix of the corporate point of view with respondent's own thinking.

8.3.3   *Contents*

The questionnaire was directed both towards the specific project objectives and the wider application of the WFD to groundwater. The field of answers was restricted in order to get comparable responses. For instance, this meant avoiding some bias relating to the water resources environment of Spain, the country responsible for this activity. Spain is influenced, to some extent, by problems in semi-arid areas where intensive aquifer development for irrigation is a prevailing issue (Custodio 2002; Llamas and Custodio 2003) when compared with other European areas where crop irrigation is less important and pollution is often a more serious threat due to thin unsaturated areas on top of shallow aquifers. As well as the larger more widespread deep European aquifers, smaller aquifers often used for municipal supply were taken into consideration. It was agreed that technical, economic, social and administrative aspects would be considered as equally important issues to address current problems of development, intensive exploitation and sustainability (Burke and Moench 2000; Custodio et al. 2005; Rogers et al. 2005).

Monitoring in relation to groundwater baseline quality characterisation has been the subject of a separate activity within the project (see Chapter 7), but it was also decided, in conjunction with the Portuguese team who were responsible to use the questionnaire to address monitoring aspects, which are closely related to end-users.

The questionnaire contained five chapters with four to seven sections, each section with three to five questions (see Table 8.1). Both oriented and non-oriented answers were possible. Oriented answers consisted of a closed list of four to six possible choices; some were provocative, while others were an indefinite alternative to indicate some doubt about the answer. In some cases, choices were restricted, but in others more than one alternative could be selected. Non-oriented answers consisted in encouraging free-format comments to produce nuances,

but this was seldom used by respondents. The final questionnaire had in all 30 pages with 125 questions (BaSeLiNe 1999).

The first chapter of the questionnaire deals with introductory aspects, mostly related to characteristics of the respondent. The next three chapters address groundwater policy, groundwater quality and baseline quality understanding, as well as the regulations and norms affecting the development and use of groundwater resources. The last chapter is devoted to groundwater baseline quality monitoring.

8.3.4 *The answers*

The respondents selected to fill out the questionnaire and to participate in the meetings shared many characteristics, but they also presented large differences. Table 8.2 shows their distribution by country and activity. All respondents were deeply involved in groundwater issues and played a relevant role in the water sector at their respective scale. They were public, private or mixed organisations, and their scope was municipal, regional or national. People filling out the questionnaires included hydrogeologists, engineers, chemists, lawyers and managers.

Countries involved in this consultation have diverse conditions – in size, climate, population, in the importance of agriculture, animal husbandry and irrigation, human activities and/or economic characteristics – creating aquifer management conditions unique for each. Sometimes there are strong differences, for instance, between a small-sized coastal aquifer in semi-arid Southern Europe and a large continental aquifer in a temperate industrialised area. However, some common features can be found. This meant facilitating the real application of the

Table 8.1 Sections of the Groundwater Baseline Management Questionnaire.

1	**Introductory aspects**	3.3	Elasticity of demand to groundwater quality [3]
1.1	Relevance of groundwater quality for human supply in your organisation/company [4]*	3.4	Origin of funds [3]
1.2	Main groundwater quality problems [3]		
1.3	Assessment of public attitude [4]	**4**	**Administrative and policy issues**
1.4	Influence of groundwater quality standards/directives [4]	4.1	Groundwater quality regulations and laws [4]
		4.2	Main legal/administrative problems [4]
2	**Groundwater quality management issues relative to baseline**	4.3	Social appraisal of groundwater quality [4]
		4.4	Solution to conflicts [4]
2.1	Groundwater abstraction works [well construction, operation and maintenance] [4]	4.5	Role of users' associations/ citizen groups [5]
2.2	Aquifer protection and land use [4]	4.6	Role of the water/environmental authority [4]
2.3	Adequate knowledge/understanding of the aquifer system [4]	4.7	What is expected needed from the EU Water Framework Directive ? [5]
2.4	Attitude towards original groundwater quality problems [4]		
2.5	Environmental issues and public attitude [4]	**5**	**Groundwater quality monitoring**
2.6	Groundwater baseline quality sustainability [short- vs. long-term issues] [4]	5.1	Groundwater monitoring [5]
		5.2	Groundwater quality monitoring [4]
		5.3	Well monitoring for groundwater quality [4]
3	**Economic issues**		
3.1	Cost of groundwater quality [4]	5.4	Study of aquifers, specially coastal ones [4]
3.2	Pollution and opportunity cost [4]		

*Bracketed figures indicate the number of questions included in the item.

WFD within a feasible timeframe and with a reasonable effort.

8.3.5 Procedure to recover information

A summary of the procedure is given in Fig. 8.1. First, collected data were handled and treated at country level. A table was prepared with all the original results listed straight from the questionnaires, plus some elementary rates calculated using these data to single out particularities from the whole. The first page of the results for Spain (Nieto et al. 2003), is shown in Table 8.3. These results were presented and discussed with the national partners involved at national meetings. The first primary conclusions were taken into account to prepare the corresponding 'national summary' draft by each country team.

Later, the same calculations were applied with the same purpose to the complete dataset from all other countries and a sample of these summary results is illustrated in Table 8.4. Extreme scores may easily single out some relevant results, some of which are gathered in Table 8.5.

The study of all this information gave rise to a partial report which was thoroughly

Table 8.2 Respondents to the questionnaire by countries and the main characteristics of the entities.

Country	Status (no.)			Activity (no.)					Scope			Number of questionnaires
	Public	Private	Mixed	Supply	Regul./Manag.	Consultant	Research	Other	Nat./Intern.	Regional	Local	
Belgium	7	0	0	6	1	0	0	0	0	6	1	7
Bulgaria	1	0	1	1	1	0	0	0	1	1	0	2
Czechia	2	0	1	2	1	0	0	0	1	1	1	3
Denmark	5	0	0	0	3	1	1	0	2	2	1	5
France	3	0	0	0	2	0	0	1	0	3	0	3
Poland	10	0	0	5	5	0	0	0	2	5	3	10
Portugal	6	1	1	3	4	0	0	1	2	4	2	8
Spain	10	3	3	3	12	0	1	0	1	13	2	16
Switzerland	5	0	0	3	2	0	0	0	1	2	2	5
UK	4	6	0	4	4	1	1	0	2	8	0	10

Two of the involved countries were not able to supply answers.

no. = represent number of questionnaires.

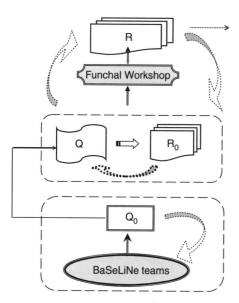

Fig. 8.1 Schematic representation of the methodology applied, bottom up. Arrows indicate the feedback processing. The questionnaire produced by the Spanish team and agreed by the BaSeLiNe teams goes to the selected persons. The answers (Q_0) go to the study stage (in the middle) where the combination of these questionnaires (Q) and the national reports are synthesised in a draft report (R_0). The final result was the report (R) after interaction and discussion in the Funchal Workshop. The overall results are presented in this chapter.

discussed by the participants (39 people) of the different countries that attended the discussion during the project meeting held in 2002 in Funchal, Madeira (Portugal). Valuable suggestions emerged, which gave rise to the final report summarised below.

The comments and suggestions forwarded and received complemented and improved the interpretation and allowed upgradings and nuances to the draft reports. This method required consecutive feedbacks and a deep commitment from all partners throughout the project.

8.3.6 *Interpretation*

The particular characteristics of each respondent were ignored and not taken explicitly into account when analysing the results. Arithmetic averages were calculated by summing all answers to each question to give the value T but no detailed statistical analysis was conducted.

Since in many cases several of the suggested options have been selected simultaneously for each question, T represents how many times each one of these options was selected. Two simple arithmetic ratios are calculated with T to obtain an average result for each option:

1 The **particular** or **internal** average: T over the total amount of **options selected** in this question, that is, summing up all individual Ts. The addition of all these particular averages gives 100 (%);

2 The **general** or **external** average: the same number T over the total number of questionnaires; the addition of these ratios for each question may give a number exceeding 100%. This ratio gives a more general idea about the option preferred for each question.

The greater the number of questionnaires containing no answer for a given question, and/or the number of different options selected at a time on many questionnaires (the most frequent case) the greater the difference between the two ratios. The analysis performed focused only on the general or external ratio because it was considered to be more interesting for project goals. In certain cases the internal ratio was also taken into account if it helped in the interpretation. Examples are shown in Tables 8.3 and 8.4.

Quantitative and qualitative results were derived both from what was indicated and what was not indicated throughout the

Table 8.3 Summary results from the Spanish questionnaires: first page (after Nieto et al. 2003).

Questionnaire of groundwater baseline management: SPAIN

Number of questionnaires { sent: S=30; answered: A=16 } A/S = 53%

NA = Sections with no answer
T = Total of answers

		Set of answers: 1 to 16																T	NA	T/ΣT		T/A	
1 Introductory aspects																							
1.1 Relevance of groundwater quality for human supply in your organisation/company																							
a The majority of water quality problems you face are:	long-term			1		1	1	1		1				1	1		1	8		8/23	35%	8/16	50%
	medium-term	1			1		1			1	1	1	1	1		1		9		9/23	39%	9/16	56%
	short-term		1				1				1				1			4		4/23	17%	4/16	25%
	just the present										1				1			2		2/23	9%	20/16	13%
	no problems																	0		0/23	0%	0/16	0%
	other																	0		0/23	0%	0/16	0%
	No. of answers in this section	1	1	1	1	1	3	1	0	2	3	1	1	2	3	1	1	23	2				
b Is groundwater quality related to operation / management options?	yes		1	1	1		1	1					1	1	1			9		9/16	56%	9/16	56%
	sometimes					1	1			1	1	1				1		6		6/16	38%	6/16	38%
	no	1																1		1/16	6%	1/16	6%
	other																	0		0/16	0%	0/16	0%
	No. of answers in this section	1	1	1	1	1	2	1	0	1	1	1	1	1	1	1	1	16	2				
c Groundwater abstraction/ use is:	continuous	1		1	1	1	1	1		1	1		1	1	1			11		11/20	55%	11/16	69%
	seasonal										1		1	1	1	1	1	6		6/20	30%	6/16	38%
	back up											1						1		1/20	5%	1/16	6%
	other		1										1					2		2/20	10%	2/16	13%
	No. of answers in this section	1	1	1	1	1	1	1	0	1	2	1	3	2	2	1	1	20	2				
d Groundwater use:	urban purposes	1	1	1	1	1	1	1		1	1				1	1		11		11/49	22%	11/16	69%
	drinking water						1	1			1			1	1		1	6		6/49	12%	6/16	38%
	industrial supply			1	1	1	1	1		1	1				1	1	1	10		10/49	20%	10/16	63%
	cooling						1				1				1	1		4		4/49	8%	4/16	25%
	irrigation			1	1	1	1			1	1			1	1	1	1	10		10/49	20%	10/16	63%
	animal use						1				1							2		2/49	4%	2/16	13%
	other			1		1	1			1				1		1		6		6/49	12%	6/16	38%
	No. of answers in this section	1	1	4	3	4	7	3	0	4	6	0	0	3	4	6	3	49	4				
1.2 Main groundwater quality problems																							
a Do you have groundwater quality problems?	no relevant problems		1						1			1						3		3/19	16%	3/16	19%
	in the past				1										1			2		2/19	11%	2/16	13%
	they are appearing				1	1	1			1	1			1	1		1	8		8/19	42%	8/16	50%
	not known																	0		0/19	0%	0/16	0%
	others	1		1			1	1					1			1		6		6/19	32%	6/16	38%
	No. of answers in this section	1	1	1	2	1	2	1	1	1	1	1	1	1	2	1	1	19	1				

Table 8.4 Example showing the first summary results obtained for all countries (after Nieto et al. 2005).

Questionnaire on groundwater baseline management: ALL COUNTRIES
Number of questionnaires answered: A = 68

T = Total of answers

| | COUNTRY | | | | | | | | | | T | Main values obtained | |
---	E	P	B	DK	F	GB	PL	CH	BU	CZ		T/ΣT	T/A
1 Introductory aspects													
1.1 Relevance of groundwater quality for human supply in your organisation / company													
a The majority of water quality problems you face are:													
long-term	8	5	5	5	2	8	5	3	0	2	43	43/97 44%	43/68 63%
medium-term	9	1	2	3	1	3	3	2	0	1	25	25/97 26%	25/68 37%
short-term	4	2	2	0	1	3	2	4	0	0	18	18/97 19%	18/68 26%
just the present	2	0	0	0	0	0	0	0	1	0	6	6/97 6%	6/68 9%
no problems	0	0	1	0	0	0	0	1	0	0	2	2/97 2%	2/68 3%
other	0	1	0	0	0	1	1	0	0	0	3	3/97 3%	3/68 4%
No. of answers in this section	23	9	10	8	4	15	14	10	1	3	97		
b Is groundwater quality related to operation / management options?													
yes	9	1	6	0	2	4	3	1	1	3	30	30/61 49%	30/68 44%
sometimes	6	3	1	4	0	3	3	2	0	0	22	22/61 36%	22/68 32%
no	1	1	0	1	1	0	2	2	0	0	8	8/61 13%	8/68 12%
other	0	1	0	0	0	0	0	0	0	0	1	1/61 2%	1/68 1%
No. of answers in this section	16	6	7	5	3	7	8	5	1	3	61		
c Groundwater abstraction / use is:													
continuous	11	2	5	5	0	5	8	5	1	3	45	45/62 73%	45/68 66%
seasonal	6	0	1	0	0	0	1	1	0	0	9	9/62 15%	9/68 13%
back up	1	2	0	0	0	0	1	0	0	0	4	4/62 6%	4/68 6%
other	2	0	0	0	0	2	0	0	0	0	4	4/62 6%	4/68 6%
No. of answers in this section	20	4	6	5	0	7	10	6	1	3	62		
d Groundwater use:													
urban purposes	11	1	0	1	0	0	3	3	1	1	21	21/102 21%	21/68 31%
drinking water	6	2	5	2	0	0	3	3	1	2	24	24/102 24%	24/68 35%
industrial supply	10	1	1	2	0	0	4	3	1	1	23	23/102 23%	23/68 34%
cooling	4	0	0	0	0	0	1	2	0	0	8	8/102 8%	8/68 12%
irrigation	10	0	0	1	0	0	0	2	0	0	13	13/102 13%	13/68 19%
animal use	2	0	1	0	0	0	0	1	0	0	3	3/102 3%	3/68 4%
other	6	0	0	1	0	0	0	2	0	0	10	10/102 10%	10/68 15%
No. of answers in this section	49	4	7	8	0	0	11	16	3	4	102		

Table 8.5 Questions with the highest scores for all countries (after Nieto et al. 2005).

Number of questionnaires answered: 68   T = Total of answers
                                               Σ T = Total of answers in the section

Question	\multicolumn{10}{c}{COUNTRY}										T	\multicolumn{2}{c}{Main values obtained}		
	E	P	B	DK	F	GB	PL	CH	BU	CZ	T	T/ΣT		T/A	
1.1.c	11	2	5	5	0	5	8	5	1	3	45	45/62	73%	45/68	66%
1.3.c	15	8	5	5	3	6	7	4	1	3	57	57/69	83%	57/68	84%
2.3.d	13	5	4	2	2	7	1	3	1	2	40	40/57	70%	40/68	59%
2.6.a	13	8	6	4	2	8	6	5	0	2	54	54/64	84%	54/68	79%
2.6.c	10	7	4	4	1	4	5	5	1	2	43	43/63	68%	43/68	63%
2.6.d	11	7	4	4	2	8	10	2	1	2	51	51/62	82%	51/68	75%
4.3.c	15	7	4	4	2	8	5	3	2	1	51	51/65	78%	51/68	75%
5.1.a	9	7	5	3	1	7	5	5	2	1	45	45/64	70%	45/68	66%
5.2.b	10	5	6	5	3	7	4	4	0	2	46	46/68	68%	46/68	68%
5.2.c	16	8	6	5	2	8	7	4	2	3	61	61/64	95%	61/68	90%

Question	Content	Selection
1.1.c	Groundwater abstraction use is:	Continuous
1.3.c	What do you think about providing information on groundwater quality to users?	Beneficial
2.3.d	Are some of your technical staff capable of dealing themselves with groundwater quality problems that may appear?	Yes
2.6.a	Does long-term groundwater quality sustainability concern you?	It is a key Issue
2.6.c	Must groundwater baseline quality sustainability be the subject of strict planning and regulation?	Yes
2.6.d	Are you in favor of keeping and protecting good quality palaeo-water primarily for drinking purposes?	Yes
4.3.c	Is groundwater quality a serious matter?	Very serious
5.1.a	Do you operate a groundwater monitoring network?	Yes
5.2.b	Early warning of groundwater quality changes is:	Essential
5.2.c	Investment in monitoring for early warning of water quality changes is:	Important

consecutive questions. Relevant particularities may be picked up straight from the calculations, such as items with the highest percentage values of their average (Table 8.5). This is considered a sort of 'wide agreement' from the participants on these particular issues, be it a definite aspect or the general lack of interest in it. A similar outcome can be obtained from answers with the lowest percentages, although in this case the meaning is not as direct as before and must be considered with care.

8.4   Relevant results

The results are presented in terms of scores and then as individual comments. Something that was not explicitly included in the questionnaire, but did appear recurrently in later meetings, is the importance of small and shallow aquifers for human supply. Overall the European study had mostly focused on the natural reference quality of large and deep aquifers. While large deep aquifers may still contain palaeowater (>10,000-years old), or at least pre-industrial water, in small aquifers groundwater is essentially young (a few to some tens of years) and reflects recent distributed recharge, or surface water recharge, including irrigation return flows and leakages. The baseline concept becomes more difficult to establish for aquifers with a rapid turnover, although in shallow aquifers baseline water may still be retained in some areas from pre-industrial times, under close-to-natural conditions or may originate from recent recharge in areas with low human pressure or from negligible impacts from polluted surface water.

As stated earlier it is important to take into account that results are derived mostly from organisations, but they rarely represent the groundwater final user and other stakeholders, since it was difficult to have access to them and obtain their collective opinion within the scope and possible actions of the project. This also applies for irrigated agriculture using groundwater and the environment depending on groundwater, such as river base flow, springs and wetlands. The qualitative comments may be summarised as follows.

8.4.1 On groundwater issues

Groundwater is used as a continuous water supply source for all types of human demands. Sometimes it is the only water resource available and generally, although not always, it is of good quality. The protection of wells and springs used for supply is a paramount regulatory and managerial objective to address the diverse problems faced by human impacts.

In many cases there is a slow but continuous deterioration of groundwater quality and hardly any improvement is observed. This may be real or just a consequence of paying more attention to quality or a better and more accurate monitoring, accompanied by the improvement of analytical capabilities over recent decades. Diffuse contamination by agrochemicals, and also from evaporative concentration of applied water in southern areas of Europe, is probably the most common cause of negative quality trends.

Occasional high concentrations of some natural components in groundwater, such as hardness, may give preference to surface water for human supply, despite its greater vulnerability to pollution and its susceptibility to rapid degradation. This is true for large utilities operating relatively large plant with trained personnel, but not for small facilities that prefer a reliable constant source (groundwater) instead of a good but variable and vulnerable quality source, such as surface water.

It is apparent that improved information concerning groundwater and its natural baseline quality is important and is acknowledged by users. Nevertheless this information regarding origin, quality and other intrinsic properties should be adequately presented and prepared in order to make easier and less biased their interpretation.

When limits/guidelines are exceeded, organisations and those most concerned refer mostly to financial considerations and technical solutions to redress the damaged situation, rather than the origin of the problem. It was recognised that some problems could be prevented if defensive monitoring were employed within groundwater catchment areas.

The European Union Directives are generally welcomed and are considered as both adequate and essential to ensure stabilisation and improvements in aquifers. However it is indicated that they would be better accepted if they address more real-world problems and provide practical solutions to local situations. This means that the dominantly environmental objectives of the Directives are not always well understood by some citizens, who expected a solution to their local problems.

Professional hydrogeologists are mostly employed in private consultancy and in some regulatory agencies rather than in water companies, groundwater development organisations and environmental institutions. However, many well-qualified graduates in hydrogeology and groundwater hydrology are looking for the opportunity to apply their knowledge. At the same time organisations with managerial and policy responsibility on groundwater are clearly understaffed and prefer contracting outside studies and services. This means that they do not hold their own expertise on important issues relating to groundwater. Moreover they tend to favour large water schemes – even if they are more expensive and less effective – instead of simple and less dramatic solutions, which may be more numerous and need more administrative work to be implemented.

8.4.2 On groundwater management issues related to baseline quality

The sustainability of groundwater quality is considered a key issue and requires long-term commitment. Both public and private organisations declare that they are concerned by good quality groundwater being negatively impacted and by the fact that good quality groundwater reserves, and even palaeowaters, are too often used for purposes that do not require water of high quality. This issue should be the subject of public information and also improved studies, as well as planning and regulation, clearly advocating for the protection and preservation of high-quality water – including palaeowater – for drinking purposes. This means that abstraction of groundwater for different purposes should be managed, and this includes adequate water pricing.

Existing knowledge on hydrogeological and technical advantages and problems related to aquifer management seems adequate, but policy-makers, managers and large-scale water producers may not be fully aware of the consequences of their actions. This explains their preference for surface waters. A corporate lobbying for more widespread attention to groundwater resources is needed, and this should be supported by professional knowledge and through expert hydrogeologists.

Threats to groundwater quality and quantity are well known to professionals and, therefore, there is a clear awareness of the risks. This explains that most water supply companies have or are asking for well-head protection zones. But, in spite of existing regulations, actual accomplishments may be poor; also regulations do not have the full support from water authorities. Aquifer protection can and has to be improved by a combination of factors that include changes in land use as the main solution. This includes better farming practices, with improved – or restricted – use of agrochemicals and more efficient irrigation practices in areas with scarce water resources.

There is a need for water industry personnel to better understand how groundwater quantity and quality occur, behave and evolve. However, employed professionals are considered as well qualified on groundwater issues as on related domains. This is less clear for technicians. No compulsory hydrogeological training scheme exists for technicians, who may work in small drilling companies, or in monitoring or surveying. Some environmental agencies are currently developing guidance in borehole design, construction and abandonment of wells, but low cost criteria are always a very strong factor, at the expense of water quality.

8.4.3 On economic issues

Aquifer pollution is considered a very important economic issue, which may involve high costs and losses, but opinions differ on possible ways to finance remediation programmes, especially for diffuse contamination cases.

Most drinking water companies are in favour of spending more on guaranteeing a good enough quality groundwater supply, provided this is recognised by the regulator and that costs are recoverable from end-users. However, the public would expect a lot from any increase in water prices, which, in many cases, are already considered high enough. Water companies doubt that the public is willing to pay more for a better water quality.

Local/regional regulators should probably be more proactive in aquifer protection, in general, and in the protection of well catchment areas, in particular, to prevent the loss of supplies. Water-quality regulations are

stringent but the actual application often depends on real circumstances that point differently, or lack support, or suffer from insufficient or inadequate institutional framework. It is not rare that other institutions make regulations for diverse purposes that go directly or indirectly against groundwater-quality protection. Thus, there is a need for integrated land and water management.

It is suggested that preserving good groundwater quality could be subsidised partly by users and partly through general taxation. This might be put into practice by the appropriate water agency. However, there exist large differences among the diverse European countries and even regions, and this might need further study to find common available approaches.

8.4.4 On administration and policy issues

Although current regulations already cover the main groundwater-quality issues, little is known on how effective they are and how extensive is their implementation. Improved current regulations need a greater weight to be given to specific conditions, introducing technical progress in monitoring and making more explicit the link with health problems and the social benefit obtained.

It is expected that the European WFD, and the Groundwater (Daughter) Directive, will recommend improvements on protection and management of aquifers. They should highlight the areas in which rules are currently lacking, especially where toxicological data are sparse and where responsibilities are not clear, forcing authorities to assume their responsibility according to the subsidiarity principle, one of the cornerstones of European legislation. Following this principle action should be taken at the level closer to the citizen.

Some water authorities/regulators already have a strong influence on groundwater management. These influences are perceived as helpful by public associations, as domineering by private companies, and even as excessive by some irrigation groups. All agree however that water authorities/regulators do have an important role in promoting groundwater sustainability. Their current activity is often seen as primarily reactive and disrupted by insufficient human, financial and managerial resources. It is felt that water-quality problems are increasing and need intermediate-term action, while staff are mostly forced to deal with pressing daily problems and emerging issues. At the same time, increasing bureaucratic complexity tends to make decision-making a slow process, sometimes separated from real-world problems. This situation is common in regulatory agencies, which have a complex structure, but are rather inefficient in carrying out their duties and some have poorly trained and motivated personnel.

Regulations may assign diverse groundwater resources to different uses according to its baseline quality, but it is not clear whether society understands the issues behind adopting a multiple approach to water quality and how this might be achieved. A necessary step is improved public awareness of water environmental issues, including dialogue and involvement among the public served, the supply companies and the institutions representing the environment.

It is expected that the European WFD will enforce existing and new regulations leading to a better assessment of groundwater sources and their quality, maybe through the creation of a European Water Authority.

The creation of river basin districts, with effective surface/groundwater integration, is also considered to be another important and useful approach. A real, transparent and holistic view of costs and their application – who pays – is needed. The water price should reflect the true cost, and the EU should be more involved in this. However, water companies from countries recently joining the EU are asking for substantial financial aid from the EU and the same happens with farmers. It is pointed out that the public should get a higher representation in the various institutions and in the decision-making process.

Improving research – with clear objectives – are pointed out as important, together with some other technical proposals, such as regulating laboratory capabilities.

8.4.5 *On groundwater-quality monitoring*

Monitoring is an essential element in groundwater-quality assessment programmes. Several different organisations are commonly involved in monitoring at national, regional and local levels. The multiplicity of aims and activities lead to several approaches, densities and protocols for network design and development, even for the same aquifer. This complex situation leads to a certain confusion related to what should be monitored and how frequently, which is an issue that needs to be addressed within a national and/or European strategy.

Professional advice is needed on what should be monitored, because this may differ from aquifer to aquifer and from basin to basin, depending on the local situation within each country/region. To understand present and to foresee future changes in groundwater quality, the practice of monitoring with limits and thresholds adjusted to regulated values have to be substituted by practices, which are wider in scope than those currently in place. Modern analytical methods allow very low detection limits to be reached, providing an early warning of impacts on quality. It is no longer satisfactory to work to threshold values, the 'less than' values set by drinking water standards. It is possible now to monitor subtle changes in water quality that give early warning of impacts on the aquifer. A series of indicator elements/species need to be agreed on and prioritised to include major ions, key trace elements, microbiological components and some trace organics and (radio)isotopes, to be monitored across the surveillance network with the same frequency. Particular attention should be given to specific techniques to determine groundwater residence times. This should be suitable to identify both, changes in baseline quality, as well as existing or coming pollution impacts.

Adequate operational procedures and the maintenance of high-quality monitoring programmes, including early warning of groundwater changes, will require significant financial resources. Their cost effectiveness depends more on the actual importance of the aquifer's use than on the aquifer's size.

Even though a wide range of tools and sampling devices are commonly applied, usually the simplest are employed, and most are used to obtain water samples for common laboratory analyses. More elaborate devices are used only for special and research studies.

Simple technologies are often applied for the construction of wells with little regard to preservation of water quality and durability. Also norms for well and borehole abandonment are often poor or non-existent. It is important to recognise that actual practices depend greatly on the particular conditions of each borehole and aquifer, requiring a specific project.

8.5   Conclusions

The results of the surveys show that there is a general agreement on the importance of having and maintaining good quality groundwater, especially for human consumption. However this is often not reflected, in practice, through effective groundwater management, especially for its protection. This inconsistency is derived from the rather poor common knowledge of aquifer properties, their functioning and behaviour. Experts are often not involved at the decision-making level. This results in poor performance of water authorities and regulatory agencies, lacking the human and economic resources and whose hydrogeological expertise is limited. Moreover the public often accepts, generally under strong advertising pressure, that the solution to their present and future problems of water quality is to use bottled water for drinking purposes. The reaction of groundwater to external influences (changing water levels and rates of quality impacts and remediation) is slow. This is a further factor that makes aquifer management difficult for untrained people who are working under short-term goals. Decision-making pays more attention to 'urgent' daily problems and on-the-spot pressure than to aspects that develop slowly, even when they are correctly explained and presented by specialists. Hydrogeological studies warn against the pervasive risks of groundwater-quality degradation, and provide well-documented case studies. But there are exceptions, either at local or national levels that show how to deal with current situations and risks as well as how to face future problems.

A long-term objective for aquifer management is a desirable social goal but this contrasts with the short- and medium-term interests of most traditional water organisations, government teams and political parties. Institutional aspects represent a main weakness of aquifer management, and the effective participation of water users is seen as a requirement of improving management and governance. The idea of participation is received positively by respondents but not all of them are sure of the benefits. Public information is also seen as a possible contribution to correct groundwater management.

Aquifer monitoring is considered a key tool for groundwater protection. Technical improvements are urgently needed, paying attention to automatic equipment, making it more applicable and economically friendly for water companies and institutions, in a similar way as the advances that took place some years ago for surface water monitoring. Equipment design has to be purpose-made, however, taking into account the very different circumstances of groundwater with respect to surface water. The EU has a role in integration and standardisation. On technical grounds this means the need to define the minimum requirements for aquifer monitoring by means of adequate guidelines. Financially this entails devising some support to foster prompt application.

As a result of monitoring, citizens should receive information explaining the state of the groundwater and how it evolves over time, by means of objective approaches that avoid the bias that may be derived from commercial, corporate, political or other interests. The importance of recognising water

with high natural baseline quality is part of this process and this will lead to improved strategies for groundwater protection. Better and sound information from water authorities and water companies will probably, and hopefully, lead to increasing users' involvement in groundwater management affairs.

Future action should improve the balance of stakeholder representation by a better representation of users other than water supply, such as farmers and environmentalists, and the groundwater users' organisations.

References

AIH-GE (2003) Presente y futuro del agua subterránea en España y la Directiva Marco Europea [Present and future of groundwater in Spain and the European Framework Directive]. International Association of Hydrogeologists – Spanish Chapter/IGME. Madrid, pp. 1–540.

BaSeLiNe (1999) Natural baseline quality in European aquifers, a basis for aquifer management. Final Report Project EVK-1-1999-0006.

Burke, J.J. and Moench, M.H. (2000) *Groundwater and Society: Resources, Tensions and Opportunities.* United Nations, New York, pp. 1–170.

Custodio, E. (2002) Aquifer overexploitation, what does it mean? *Hydrogeology Journal* 10, 254–77.

Custodio, E., Kretsinger, V. and Llamas, M.R. (2005) Intensive development of groundwater: Concept, facts and suggestions. *Water Policy* 7, 151–62.

Hernández-Mora, N. and Llamas, M.R. (2000) *La economía del agua subterránea y su gestión colectiva [Groundwater economy and their collective management].* Ediciones Mundi Prensa, Madrid, pp. 1–549.

Llamas, M. R. and Custodio, E. (eds) (2003) *Intensive Use of Groundwater: Challenges and Opportunities.* Balkema, Lisse, NL, pp. 1–471.

Nieto, P., Custodio, E. and Manzano, M. (2003) La calidad natural de referencia del agua subterránea: percepción en España [Natural BaSeLiNe groundwater-quality: perception in Spain]. In: Presente y Futuro del Agua Subterránea en España y la Directiva Marco Europea. *AIH-GE/IGME*, Madrid, pp. 337–43.

Nieto, P., Custodio, E. and Manzano, M. (2005) Baseline groundwater quality: A European approach. *Environmental Science and Policy* 8, 399–409.

Rogers, P.P., Llamas, M.R. and Martínez-Cortina, L. (2005) *Water crisis: Myth or reality?* Fundación Marcelino Botín, Francis & Taylor/Balkema, London, pp. 1–333.

WFD (2000) Water Framework Directive. 2000. Directive 2000/60/CE of the European Parliament. (ECOJ 22 December 2000).

9 The Chalk Aquifer of Dorset, UK

W. M. EDMUNDS AND P. SHAND

The Dorset Chalk contains some of the highest quality chalk groundwater in the UK. However, much of the area is given over to farming, the long-term impacts of which may impact on the natural groundwater quality. The main observed effect of agricultural activity is the increase in nitrate over several decades from a baseline value of around 1 mg L^{-1} in the early twentieth century to concentrations up to the maximum acceptable concentrations of 11.3 mg L^{-1} NO_3–N. In addition, a rise in potassium from agricultural sources above the natural background is demonstrated. Apart from these examples of diffuse pollution, there is no evidence for widespread contamination from urban and rural wastes or other point sources of pollution.

The natural baseline is expressed as a range of values which are controlled by rainfall inputs and by the geology of the area. There are only subtle changes in this baseline in the main body of the groundwater in the unconfined aquifer, for example, the Cl concentrations which vary with distance from the coast. The redox change to a reducing environment as the water moves under the confining Palaeogene strata produces increases in concentrations of Fe and Mn. This environment may also lead to natural denitrification with possible net benefits for abstraction zones. Details of the natural variations and the overall statistical data for

91 considered constituents in the groundwaters of the area are presented.

It is concluded that the properties of groundwater in the Dorset Chalk are overwhelmingly determined by natural reactions between rainwater reacting with the bedrock, especially in the topmost few metres. It is generally of high quality but is being progressively impacted by agrochemicals in unconfined areas, the effects of which are evident even in some of the deeper groundwaters. Waters of pristine (pre-industrial) quality are rare in the basin and restricted essentially to the confined aquifer.

9.1 Introduction

The Chalk aquifer is the most important single aquifer in Europe and of special significance in the UK and France where it forms basins underlying the cities of London and Paris (Edmunds et al. 1992). This paper focuses on one of the least impacted areas of Chalk in the UK as a type area of the aquifer. The Dorset Chalk area lies at the western end of the Wessex Basin and is centred on the town of Dorchester (Fig. 9.1). The basin is geologically a synclinal structure, which has its northern limb forming the high ground overlooking the Vale of Pewsey and its southern limb forming the high ground of the Purbeck monocline overlooking the

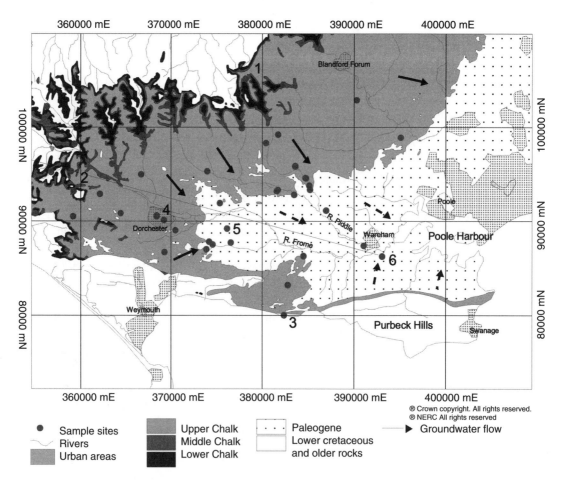

Fig. 9.1 The Dorset Chalk of the Wessex Basin showing groundwater flow towards Poole Harbour beneath confined Palaeogene strata. Main sample sites (this study) are shown together with representative sites (1–6), which are shown in Table 9.4.

coast near Lulworth. The river catchments of the Frome and the Piddle drain the centre of the syncline structure. The area comprises typical Chalk downland, common to much of southern England. In the east, towards Poole Harbour, the Chalk is overlain by younger Palaeogene sediments (sands, gravels and clays), which provide the contrasting heathland scenery.

The area of Chalk downland forms the core area for evaluating the regional baseline groundwater quality. Selected data for groundwater (springs and boreholes) from this region are used to provide areal coverage and to investigate trends in quality over recent decades. The geochemical properties of the Chalk are relatively uniform; results from the Dorset Chalk may be used for comparison with other regions of the Chalk of southern Britain.

The Chalk aquifer is a major aquifer, containing large volumes of high quality potable water. It also provides an essential source of baseflow to maintain river quantity and quality. Its principal economic uses are its highly important role in providing high quality drinking water for both public and private water supplies, agricultural use, such as watercress and fish farming, and for industry. Specialised local users also include hospitals, breweries and food processing. The UK Environment Agency is responsible for the licensing of abstractions to supply these needs, with special regard for environmental concerns such as maintenance of wetlands.

Current issues in the area relate mainly to the over-abstraction of groundwater that may lead to unacceptably low flow in rivers and the drying out of wetlands. The Dorset Chalk is believed to contain some of the highest quality Chalk groundwater in the country. However much of the area is given over to farming and the impacts on groundwater quality of long-term farming, especially since the Second World War, are not fully known, although concerns over rising nitrate concentrations are being addressed. As well as diffuse pollution, there is the potential for point source pollution from urban and rural wastes, including landfills, industry and farm sources.

9.2 Geology and hydrogeology

The Chalk of Dorset, notably South Dorset, has undergone extensive deformation as part of the Alpine earth movements, and the structure is dominated by an east–west, faulted monocline. The zone of associated faults and folds, orientated E–W and NNW–SSE, is known as the Purbeck disturbance. The faults associated with the zone of disturbance have an important control on the groundwater flow regime of the area. Flow and subsequent preferential solution along the NNW–SSE orientated discontinuities has produced zones of enhanced hydraulic conductivity.

The Upper Chalk (up to 260 m) forms the main outcrop over most of the area. Thin outcrops of Middle and Lower Chalk (up to 41 and 57 m respectively) are also present in the upper valleys and along scarp edges. To the east, the Chalk is progressively confined by Palaeogene sand and clay deposits which may reach a thickness of more than 100 m in the east of the area. The thickness of these confining strata limit the extent that the confined Chalk is economically developed for potable supply.

Until around 8,000 years ago, sea levels were much lower than today and the coastline of Dorset lay at least 50 km to the south of the present-day coastline (Edmunds et al. 2001). This situation which had lasted for over 80,000 years, was the result of a smaller ocean volume due to the formation of extensive ice caps. The lowered sea levels would have resulted in the development of new base levels for surface and groundwater flow (Velegrakis et al. 1999). The land area was affected by a periglacial climate for much of the Pleistocene. These cold conditions, with permafrost, accelerated mechanical weathering of the chalk and enhanced the near-surface permeability. Solution features (dolines) abound near the junction between the Chalk and Palaeogene strata. These features are believed to have been exposed following erosion of the feather-edge of Palaeogene sediments. They resulted in locally enhanced aquifer transmissivities.

The hydrogeology of the aquifer can be considered in three areas; the southern

Chalk outcrop of the Purbeck Hills and Lulworth, the northern Chalk outcrop and the valleys of the Frome and Piddle and finally the confined Chalk in the Wareham and Poole Harbour areas (Fig. 9.1). The Chalk is underlain by the Upper Greensand, which is in hydraulic continuity with the overlying Chalk. Transmissivity of the Chalk ranges from 500 to 1,000 m^2 d^{-1} and storage coefficients are between 5×10^{-4} and 3.5×10^{-2}. Pumping tests indicate typical transmissivity values to be much lower at interfluve locations, less than 50 m^2 d^{-1} (Allen et al. 1997).

The confining Palaeogene beds comprise the Poole Formation, which consists of an alternating sequence of sands and clays; groundwater within the Poole Formation forms an important minor aquifer. Underlying the Poole Formation is a variable thickness of London Clay. The London Clay has a sandy unit, the West Park Farm Member, at its base in contact with the Chalk. It is likely that this unit is therefore in hydraulic connection with the underlying Chalk and that some leakage could occur between the Chalk and the Palaeogene deposits, although there is little geochemical evidence to suggest this. In the vicinity of Empool, where the Chalk is confined by the Palaeogene deposits, values of transmissivity range from 2,000 to 15,000 m^2 d^{-1}, as a consequence of solution enhancement.

In southern Dorset, the highly folded chalk typically has a lower porosity and intergranular permeability than the less-deformed chalk of the northern outcrop. The tectonically hardened chalk of south Dorset, has generally lower transmissivity and storativity, by which it is inferred that fractures have been modified by chemical or mechanical diagenetic processes (Allen et al. 1997).

In contrast, on a localised scale, faulting appears to have increased transmissivity along dry river valleys, for example, at Lulworth.

Groundwater flow divides are discordant with the surface water drainage pattern in the southern part of the aquifer. In the vicinity of Lulworth, this discordance is the product of significant groundwater discharges at the coast occurring in the form of springs located at Lulworth Cove and Arish Mell.

Fluid logging in the confined Chalk (Buckley 1996) indicates that the main horizons of groundwater flow within the Chalk are in the uppermost 50 m. Fluid temperature logs have been used to demonstrate the depth of groundwater circulation. Most of the circulation is restricted to the Upper Chalk (notably the part classified as Portsdown Chalk under the current nomenclature) although minor temperature changes may be detected at greater depths, up to −300 m OD (Ordnance Datum). The groundwater at these greater depths is relatively fresh due to flow in the late Pleistocene (Edmunds et al. 2001) although the transmissivities are low.

9.3   Geochemistry of the Chalk

Chalk is a marine sediment which accumulated slowly on the sea bed, principally comprising the fossilised coccoliths of planktonic foraminifera. Some silica skeletons also contributed and these have undergone diagenesis to form flint bands in the Upper and Middle Chalk. The microscopic texture of chalk from the region (Fig. 9.2) shows that the microfossil remains are well preserved. The high intergranular porosity (between 20% and 45%) is also apparent,

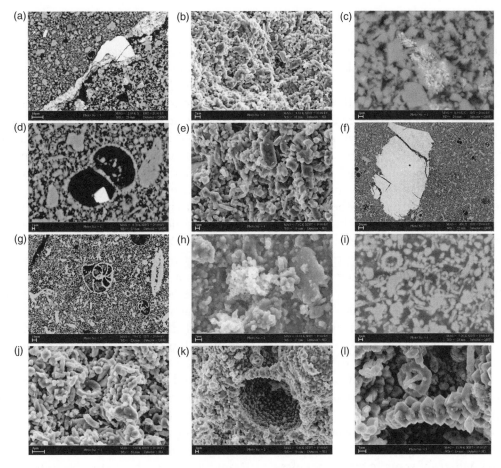

Fig. 9.2 SEM micrographs of the Chalk and Chalk surfaces from Lulworth and Shapwick boreholes: (a) Lulworth (93.8 m) Magnetite and pyrite vein filling with large crystal of sphalerite (ZnS). Open and recrystallised Chalk fabric, slightly different on either side of fracture; (b) Lulworth (47.9 m) Sugary fabric with recrystallising Chalk debris – high porosity retained; (c) Intergrowths of pyrite and recrystallised pyrite Lulworth (93.8 m); (d) Lulworth (47.9 m) Foraminifera with late crystal of magnetite growing inside recrystallised skeleton; (e) Lulworth (93.8 m) Newly crystallised calcite rhombohedra on a matrix of microfossil fragments (some coccoliths still recognisable); (f) Lulworth (47.9 m) Overall texture of the Chalk at Lulworth with a phosphatic fragment of fish scale (g) Shapwick. Texture of Chalk showing high porosity and preserved chambered formainifera; (h) Lulworth (47.9 m) Intergrowth of clays (smecite) and gypsum on surface of a fracture; (i) Blackscattered electron image of Chalk texture – mainly open structure with only slight recrystallisation (Shapwick); (j) Shapwick High magnification view of the Chalk matrix at Shapwick; (k) Lulworth (93.8 m) Recrystallisation of foraminifera showing the overgrowth of calcites on skeleton. However, some coccoliths are still well preserved in this tectonised zone of the Chalk; (l) Detail of slide (k).

with the diameter of the pores around 1 μm in size. This provides a large surface area for chemical reactions and this intimate interaction between water and rock largely controls the baseline chemistry.

The Upper and Middle Chalk microfossil debris consist of relatively pure calcite (calcium carbonate). However small amounts of other elements (Mg, Sr, Mn, Fe) are present in the calcite structure, which helped to stabilise the carbonate in the marine environment. These impurities in the carbonate lattice play an important role in the evolution of the groundwater chemistry, since they are slowly released to the groundwater as the chalk recrystallises under freshwater conditions.

The Upper and Middle Chalk also contain 1–5% non-carbonate fraction, rising to 5–12% in the Lower Chalk. This fraction consists of quartz, clay minerals (montmorillonite, white mica and in the Lower Chalk kaolinite), as well as phosphate minerals (francolite, fluorapatite).

Groundwater in the Chalk occurs both in the pore spaces of the matrix and also in fractures, where most of the flow takes place. Water in the pore spaces remains relatively immobile on account of the low intergranular permeablilty, several orders of magnitude less than the permeability of the fractured rock. Water moving along the fractures brings in new (reactive) water, which dissolves fresh chalk, but may also precipitate authigenic minerals on the fracture surfaces. There is slow diffusive exchange between the porewater in the matrix pore space of the Chalk and mobile water in the fractures, which has a strong influence on the resulting groundwater chemistry with time as water moves through the aquifer.

9.4 Rainfall chemistry

Rainfall chemistry can be regarded as the primary input and starting point for explaining the baseline quality. For some elements it may well be the major source of solutes with very little being added during infiltration and groundwater flow. For this area no climatic stations, which routinely measure rainfall chemistry, exist and therefore a station in Sussex (Barcombe Mills) is used. This is in a similar geographic position relative to the coastline and with a similar rainfall amount to Dorset.

The analyses for this station are given in Table 9.1 for the major elements. In addition, these have been multiplied by a factor of 3, which roughly accounts for the likely concentration due to evapotranspiration under the prevailing climatic conditions. These values may be used as a guide for comparison with groundwaters below. It is important to note that Cl is inert and groundwater concentrations of Cl are largely rainfall derived. The Mg concentrations in rainfall exceed Ca and reflect the higher ratios in sea water aerosols. Ammonium is the main N source being mainly derived from agricultural sources. The background concentrations (as N) at the present day, after allowing for evapotranspiration, may be as high as 4 mg L^{-1}.

9.5 Data for Dorset Chalk groundwaters

Extensive searches failed to locate early analyses carried out on public or private water supplies except during the past two decades of the twentieth century. Unpublished data on interstitial water from core material from scientific drilling at Lulworth is used to

Table 9.1 Rainfall chemistry for Barcombe Mills. Average of 3 years 1996–1998.

	Units	Rainfall	Rainfall × 3
pH	pH scale	4.9	
Na	mg L^{-1}	7.44	22.3
K	mg L^{-1}	0.24	0.73
Ca	mg L^{-1}	1.84	5.5
Mg	mg L^{-1}	4.21	12.7
CL	mg L^{-1}	5.61	16.8
SO$_4$	mg L^{-1}	1.07	3.2
NO$_3$–N	mg L^{-1}	0.38	1.15
NH$_4$–N	mg L^{-1}	1.34	4.01
Total N	mg L^{-1}	1.13	3.85
Rainfall	mm	678	

determine input histories. Data are available from the UK Environment Agency's network for strategic monitoring of groundwater quality. However in the absence of a nationally consistent programme, the data are incomplete and generally only comprise partial time series. Data used here are from 26 selected sites (one analysis from the mid-1990s) in the Environment Agency's monitoring network. These have been used (1) for statistical purposes to supplement new sample data and (2) to determine trends at the decadal scale that aid interpretation of trends away from baseline conditions.

A sampling programme for 30 sites was undertaken to establish baseline conditions for the present study during late autumn 2000 from representative sites in the Dorset area shown in Fig. 9.1. Samples have been analysed for a full range of inorganic species and, in addition, field measurements (Eh, DO, pH, temperature and SEC) are used in the interpretation. All samples were filtered in the field and acidified with nitric acid (1%) to stabilise trace elements.

9.6 Principal hydrochemical characteristics of groundwater in Dorset and baseline definition

9.6.1 Summary statistics

The summary statistics for Dorset groundwaters are given in Table 9.2 with maxima, minima and median values calculated from all data (using half detection limit values where appropriate). A total of 59 samples were available from both British Geological Survey and Environment Agency datasets and a further 30 analyses using the new sampling sites. Many of the data are discussed below but for some elements there is no further discussion, especially for trace elements below detection limits.

9.6.2 Indicators of anthropogenic pollution

Three indicator organics (atrazine, chloroform and trichloroethylene) were used to indicate the presence of any unequivocal

Table 9.2 Field parameters, isotope data and range of major and minor element concentrations in groundwaters of the Dorset Chalk.

Parameter	Units	Min.	Max.	Median	97.7%ile	n
T	°C	9	17.2	11.4	16.4	58
pH		6.94	7.58	7.21	7.52	32
Eh	mV	199	552	517	548	13
DO	mg L^{-1}	<0.1	11.01	7.61	10.5	56
SEC	µS cm^{-1}	343	1177	586	857	31
δ^2H	‰	−45	−36	−39	−36	16
δ^{18}O	‰	−6.8	−5.82	−6.36	−5.8	16
δ^{13}C	‰	−15.7	−5.6	−14.4	−7.8	16
Ca	mg L^{-1}	50	125	105	120	34
Mg	mg L^{-1}	1.7	19.4	2.5	18.1	34
Na	mg L^{-1}	6	155	10.6	85.2	34
K	mg L^{-1}	0.9	7	2.3	6.6	8
Cl	mg L^{-1}	14	223	21	83.3	59
SO$_4$	mg L^{-1}	2.5	43	13	32.8	35
HCO$_3$	mg L^{-1}	107	324	269	313	30
NO$_3$–N	mg L^{-1}	0.05	12.00	6.20	11.2	34
NO$_2$–N	mg L^{-1}	<0.001	0.0569	<0.001	0.043	58
NH$_4$–N	mg L^{-1}	<0.003	0.310	0.011	0.131	58
P	mg L^{-1}	10	163	44	118	30
TOC	mg L^{-1}	0.49	5.78	1.18	5.3	30
DOC	mg L^{-1}	0.25	4.16	0.75	3.8	56
F	mg L^{-1}	0.048	4.3	0.08	1.7	31
Br	mg L^{-1}	0.057	0.289	0.084	0.222	31
I	µg L^{-1}	2	20.1	3	18.7	31
Si	mg L^{-1}	3.3	12.2	4.7	11.6	56

modern contamination. Atrazine was detected at two sites (both springs) at approximately twice the detection limit. Chloroform was detected in three springs (different to the above) at up to eight times the detection limit and in one borehole at twice the detection limit. Trichloroethylene was detected at one site (the same site as for chloroform) at just above the detection limit. All other sites contained no traces of the three artificial chemicals used as indicators. Nitrate

median concentrations are high (median of 6.20 mg L^{-1}) and indicate more widespread human influence from agriculture, which is discussed further below.

9.6.3 Natural concentrations and baseline distributions

Using the summary statistics, knowledge of the first order extent of modern contamination, and the cumulative frequency

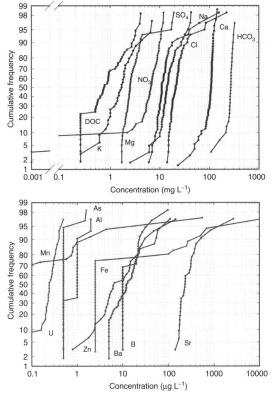

Fig. 9.3 Cumulative frequency (log-probability) plots for selected major and trace elements in Dorset Chalk groundwaters.

(log-probability) plots (Fig. 9.3), it is possible to define baseline characteristics from a purely statistical perspective. The cumulative frequency plots show the range and distribution of concentrations for selected elements/species. The median (50%) and 97.7 percentile are used as reference values and these are summarised for all elements in Tables 9.2 and 9.3. These individual values are useful for comparison between areas but it should be stressed that the baseline is represented by a range of compositions in any given aquifer. For the vast majority of elements, the distribution represents geological

and geochemical controls on the groundwater compositions, discussed below. The slope, curvature and overall shape of the lines is indicative of different categories of control (see Chapter 1).

The statistical summaries (Tables 9.2 and 9.3) include many of the common elements, including many of those ranked as hazardous by the EU Drinking Water (DW) Standards. Individual concentrations of NO_3, F, Fe, and Mn exceed the DW standards for groundwaters in some waters, but for most elements the observed values are consistently less than DW limits. This dataset comprises natural (raw) waters and includes some sources not used for drinking purposes and these may contain high values of certain elements. Several elements in this summary are toxic (e.g. Tl, Be) and fortunately their natural abundance and mobility is very low. The present definition of low baseline concentrations can be used to assess contamination incidents by uncommon elements that may occur in the future. Other elements, notably F, Se and I are essential elements, although their deficiency or excess above certain concentration ranges may lead to health problems.

Nitrate distribution indicates that around 80% of the groundwater samples contain concentrations above 5 mg L^{-1} NO_3–N. This indicates an extensive and pervasive penetration of agrochemicals and/or organic fertilisers into the chalk aquifer above baseline values and this issue is discussed further below.

It is difficult to define the extent to which the major ions other than nitrate may have been influenced by human activity. Potassium is one other element that may have a significant anthropogenic signal from agrochemicals or organic fertilisers. A plot of K versus Na (Fig. 9.4) shows that the

Table 9.3 Trace element concentrations in groundwaters of the Dorset Chalk.

Parameter	Units	Min.	Max.	Median	97.7%ile	n
Ag	µg L^{-1}	<0.05	0.16	<0.05	0.07	30
Al	µg L^{-1}	<1.0	2.0	1.0	2.0	30
As	µg L^{-1}	<1	1.60	<1	1.24	56
Au	µg L^{-1}	<0.05	<0.05	<0.05	<0.05	30
B	µg L^{-1}	<20	<21	<22	74.8	31
Ba	µg L^{-1}	5.00	99.0	12.0	40.5	57
Be	µg L^{-1}	<0.05	0.25	0.03	0.10	30
Bi	µg L^{-1}	<0.05	0.07	0.03	0.06	30
Cd	µg L^{-1}	<0.05	0.06	<0.05	<0.05	30
Ce	µg L^{-1}	<0.01	0.010	<0.01	0.010	30
Co	µg L^{-1}	<0.01	14.5	0.02	8.08	30
Cr	µg L^{-1}	<0.5	<0.5	<0.5	<0.5	30
Cs	µg L^{-1}	<0.01	0.020	<0.01	0.013	30
Cu	µg L^{-1}	0.2	13.7	2.20	11.2	30
Dy	µg L^{-1}	<0.01	<0.01	<0.01	<0.01	30
Er	µg L^{-1}	<0.01	<0.01	<0.01	<0.01	30
Eu	µg L^{-1}	<0.01	<0.01	<0.01	<0.01	30
Fe	µg L^{-1}	<5	20007	2.50	4901	35
Ga	µg L^{-1}	<0.05	<0.05	<0.05	<0.05	30
Gd	µg L^{-1}	<0.01	<0.01	<0.01	<0.01	30
Ge	µg L^{-1}	<0.05	0.08	<0.05	0.08	30
Hf	µg L^{-1}	<0.02	<0.02	<0.02	<0.02	30
Hg	µg L^{-1}	<0.1	<0.1	<0.1	<0.1	56
Ho	µg L^{-1}	<0.01	<0.01	<0.01	<0.01	30
In	µg L^{-1}	<0.01	<0.01	<0.01	<0.01	30
Ir	µg L^{-1}	<0.05	<0.05	<0.05	<0.05	30
La	µg L^{-1}	<0.01	0.02	<0.01	0.01	30
Li	µg L^{-1}	0.20	20.00	0.85	14.0	31
Lu	µg L^{-1}	<0.01	<0.01	<0.01	<0.01	30
Mn	µg L^{-1}	<0.04	555	<0.04	150	33
Mo	µg L^{-1}	<0.1	2.50	<0.1	1.17	30
Nb	µg L^{-1}	<0.01	0.01	<0.01	0.01	30
Nd	µg L^{-1}	<0.01	0.02	<0.01	0.01	30
Ni	µg L^{-1}	<0.2	20.3	0.35	16.2	30
Os	µg L^{-1}	<0.05	<0.05	<0.05	<0.05	30
Pb	µg L^{-1}	<2.0	<2.0	<2.0	<2.0	30
Pd	µg L^{-1}	<0.2	<0.2	<0.2	<0.2	30

Table 9.3 Continued

Parameter	Units	Min.	Max.	Median	97.7% ile	n
Pr	µg L^{-1}	<0.01	<0.01	<0.01	<0.01	30
Pt	µg L^{-1}	<0.01	<0.01	<0.01	<0.01	30
Rb	µg L^{-1}	0.71	1.96	1.31	1.84	30
Re	µg L^{-1}	<0.01	0.01	<0.01	0.01	30
Rh	µg L^{-1}	<0.01	<0.01	<0.01	<0.01	30
Ru	µg L^{-1}	<0.05	<0.05	<0.05	<0.05	30
Sb	µg L^{-1}	<0.05	0.98	<0.05	0.47	30
Sc	µg L^{-1}	1.19	2.99	1.58	2.50	30
Se	µg L^{-1}	0.5	2.50	0.65	1.87	56
Sm	µg L^{-1}	<0.05	<0.05	<0.05	<0.05	30
Sn	µg L^{-1}	<0.05	0.21	0.07	0.18	30
Sr	µg L^{-1}	142	2680	233	1407	31
Ta	µg L^{-1}	<0.05	<0.05	<0.05	<0.05	30
Tb	µg L^{-1}	<0.01	<0.01	<0.01	<0.01	30
Te	µg L^{-1}	<0.05	0.10	<0.05	<0.05	30
Th	µg L^{-1}	<0.05	<0.05	<0.05	<0.05	30
Ti	µg L^{-1}	<10	<10	<10	<10	30
Tl	µg L^{-1}	<0.01	0.04	0.01	0.03	30
Tm	µg L^{-1}	<0.02	<0.02	<0.02	<0.02	30
U	µg L^{-1}	<0.05	0.50	0.24	0.43	30
V	µg L^{-1}	<1	<1	<1	<1	30
W	µg L^{-1}	<0.1	<0.1	<0.1	<0.1	30
Y	µg L^{-1}	<0.01	0.06	<0.01	0.05	30
Yb	µg L^{-1}	<0.01	0.01	<0.01	<0.01	30
Zn	µg L^{-1}	0.8	145	6.40	80.5	30
Zr	µg L^{-1}	<0.5	<0.5	<0.5	<0.5	30

ratio found in pristine porewaters from the Lulworth borehole has a lower K/Na ratio (0.067) than the upper portion of the borehole (0.10), which contains significant modern nitrate. A further significant enrichment of potassium in the pumped groundwaters is also found (0.12). This suggests that, like nitrate, K enrichment is widespread, from a similar source of diffuse pollution. The cumulative frequency curves for both K and NO_3 (and possibly for DOC) are steeper than for Na and Cl. The latter are controlled by the rainfall inputs and the steeper slopes for K and NO_3 are indicative of control by pollution.

9.6.4 Regional hydrochemical properties of the groundwaters

The trilinear plot (Fig. 9.5) and cumulative frequency plots (Fig. 9.3) show that Ca and HCO_3 are the major constituents of the

Dorset Chalk groundwaters. An upper limit is apparent for the concentrations of these two ions imposed by the maximum solubility of CaCO$_3$. Na and Cl are next in abundance, the behaviour of both these elements being similar, and both derived principally from rainwater. Some 10% of the groundwaters contain slightly greater salinity than derived from rainfall, being derived from

natural sources as discussed later. Most other ions (Fig. 9.3) show lognormal distributions. The major exceptions are nitrate and Fe and Mn, which exceed 100 and 1 µg L^{-1} respectively in the deeper confined groundwaters.

9.7 Downgradient evolution in water quality

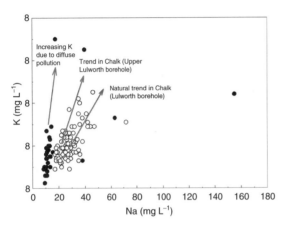

Fig. 9.4 Plot of K versus Na showing the trends in natural waters (Lulworth) and the likely trends as a result of diffuse pollution.

The natural baseline characteristics of the Dorset groundwaters can be further explained with reference to the trilinear diagram (Fig. 9.5) and with reference to the cross section through the area, from Powerstock in the west to Wareham in the east (Figs. 9.6 and 9.7). The trilinear diagram plots the groundwater data in two ways, first as a simple plot of the Dorset data given above and secondly as in relation to other data from Dorset and Berkshire (for reference) are plotted. These data are taken from additional sources including Edmunds et al. (1987). Almost all groundwaters are dominated by calcium, this being clear evidence that the

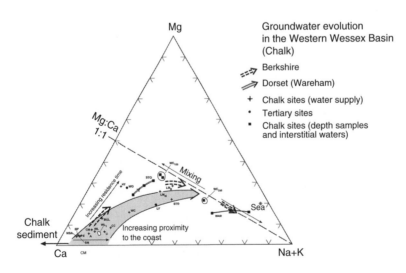

Fig. 9.5 Trilinear cation plot showing the relative proportions of major cations in Chalk groundwaters.

main control is the initial (congruent) solution of the chalk as water enters the ground; these groundwaters are only slightly enriched in Mg compared with the chalk itself. Some slight initial elevated Mg in dilute groundwaters compared to Chalk may be due to the influence of the rainfall (Table 9.1). As the groundwaters evolve along flow lines and with depth (see also Fig. 9.6), an increase in Mg is seen due to incongruent solution of the chalk. The increase in Mg and Mg/Ca, is indicative of increasing residence time as observed in Berkshire (Edmunds et al. 1987) following continuous reaction between groundwater and the rock. On this basis, the groundwater beneath Wareham is the oldest groundwater in the area.

The downgradient trends show that groundwater chemistry evolution within the groundwater body, formed of a single chalk lithology, is a function of spatial distribution and also depth and time, as well the changes in lithological facies. The lines of section (Figs. 9.6 and 9.7) commence in unconfined Chalk but show the evolution towards the deeper confined aquifer. Most boreholes are in the range of 30–100 m in depth and some springs are also included. On the diagram the springs (s) are indicated, as are the two deepest boreholes at Dorchester Hospital (DH) and Eldridge Pope Brewery (EP). The latter might be expected to be the most 'evolved' of the unconfined set of groundwaters.

Chloride is inert and is likely to have its source in rainfall, ancient formation waters (old sea water for example) or from contaminant sources. There is an increase in this area towards the coast from 16 to about 40 mg L^{-1} Cl, consistent with evaporation of the rainwater. Some of the groundwater in the confined aquifer also has higher Cl,

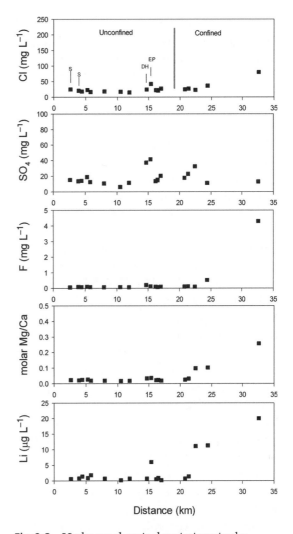

Fig. 9.6 Hydrogeochemical variations in the Chalk groundwaters along a line of flow from Powerstock to Wareham. Line of section in Fig. 9.1

indicating that the older water contains traces of original salinity derived probably from the chalk formation waters (Edmunds et al. 1987). The concentrations of sulphate (median 14 mg L^{-1}) are in excess of those from rainfall (3.2 mg L^{-1}) and represent a geological or pollution source. It is apparent that the two deepest groundwaters at Dorchester also have high SO$_4$ and so a geological origin is suggested. The SEM sections (image j, Fig. 9.2) indicate that gypsum, derived from the oxidation of pyrite, may be present in the aquifer at intermediate depths. Dissolved oxygen, nitrate and ammonium illustrate the redox (oxidation–reduction) characteristics of the groundwaters (Fig. 9.7). Dissolved oxygen concentrations are uniformly high in the unconfined groundwaters (overall median value of 7.6 mg L^{-1}). These values represent a loss of only 30–40% of oxygen from atmospheric values. There is a distinct change in the oxygen concentrations near the unconfined/confined contact. To the east of this point, termed a 'redox boundary', the groundwaters become reducing as they are confined. Once the oxygen has been consumed, the nitrate is then reduced very quickly. The nitrate concentrations are uniformly high across the unconfined aquifer, even at two deeper sites near Dorchester. This suggests that contamination from agricultural sources has penetrated to depth. It is of interest that the lowest NO$_3$ concentration is from the upland area of Bulbarrow Down where the catchment is mainly unfertilised grassland. In the reducing aquifer, an increase in NH$_4$ is observed; this is a natural phenomenon and unrelated to groundwater pollution, with ammonium released from clay minerals. Ferrous iron then increases in concentration in the reducing environment.

As well as the Mg/Ca ratio, fluoride and lithium concentrations also illustrate the hydrogeochemical evolution across the aquifer. Low fluoride concentrations at or below 100 µg L^{-1} characterise the unconfined groundwaters and reflect rainfall input concentrations. Slightly higher Mg/Ca and also F from one site are found for the deeper

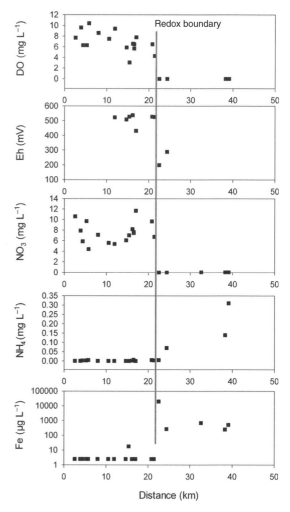

Fig. 9.7 Redox characteristics of the Chalk groundwaters (DO, Eh, NO$_3$, NH$_4$, Fe) along a line of flow from Powerstock to Wareham. Line of Section in Fig. 9.1.

Dorchester boreholes, further supporting the idea that the deeper waters are more evolved and therefore considerably older. A significant increase in Mg/Ca, F, Li and F can be detected along the flow path, this being a sign that slightly older water is being entrained as the groundwater evolves. This effect is then shown very clearly in the deeper confined (although still fresh) ground-waters beneath Wareham.

9.8 Quality changes with depth and time

9.8.1 Depth profiles

The samples obtained from pumping boreholes and springs are typically mixtures of water from different flowpaths and depths and represent the average chemical composition of the groundwater in large volumes of the aquifer. It is possible to obtain improved resolution of stratification in chemistry or age from samples taken either from interstitial waters (waters contained in the rock pores) or from depth samples taken from the borehole water column. Interstitial waters are obtained by high-speed centrifugation (Edmunds and Bath 1979) and allow continuous detailed chemical profiles to be built up. Examples are available from Dorset of interstitial water profiles taken from both the unsaturated and saturated zones as well as borehole depth profiles of conductivity and temperature, which have provided the basis for depth sampling for chemistry.

9.8.2 Unsaturated zone profiles

Interstitial water profiles are available for a borehole drilled in 1977 and 1979 beneath rough grassland at Gussage (SY 9938 1094) at a site previously cored in 1970 to provide an indication of the rate of movement of water through the unsaturated zone (Geake and Foster 1989). The radioactive isotope tritium, which was produced during thermo-nuclear (hydrogen bomb) testing in the 1960s, acts as a tracer for the water molecule, with a maximum concentration

occurring in 1963. These profiles illustrate at what rate water has moved towards the water table. The preservation of the tritium peak, although attenuated by radioactive decay over time, indicates that the bulk of the water moves downwards at around $1 \, \text{m yr}^{-1}$ through the pore spaces and microfractures, although undergoing some dispersion. Nevertheless, it was shown that at some locations, where larger fractures occur, that some bypass flow takes place allowing a fraction of the water to move more rapidly.

9.8.3 Saturated zone profile (unconfined aquifer)

Interstitial waters from a research borehole at West Lulworth, drilled in the late 1970s to a depth of $-170 \, \text{m}$ OD provides a very detailed profile of the changes in the water quality with depth (Fig. 9.8). Temperature and electrical conductivity (SEC) logs subsequently carried out in 1997 provide an additional control on the groundwater flow and chemistry at this site. It can be seen that the base of the present-day flow system is at approximately $-65 \, \text{m}$ OD as defined by the temperature profiles. Below this depth, the temperature increases linearly in line with the geothermal gradient; above this depth the water temperature is disturbed by flow. A slight decrease in fluid conductivity is recorded at below $-65 \, \text{m}$ OD indicating that the Chalk contains freshwater to its total depth. This is in line with deeper penetration of freshwater found elsewhere near the coastline of southern England (see below).

Higher nitrate concentrations between 5 and 15 mg L^{-1} NO_3–N are found above $-65 \, \text{m}$ confirming the penetration depth of modern groundwater. Below this depth nitrate

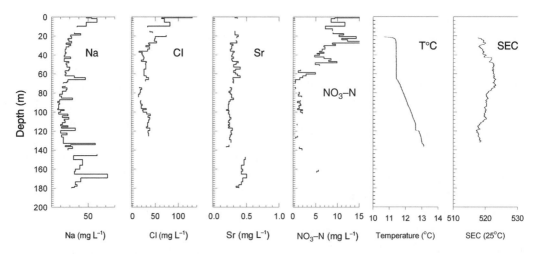

Fig. 9.8 Interstitial water chemistry, temperature and SEC profiles from the research borehole at West Lulworth. Site 3 Figure 9.1.

concentrations are still detectable and are in the range 0–2 mg L^{-1}. No radiocarbon data are available for this water but stable isotope results indicate a modern signature rather than palaeowater (unpublished data). Some suggestion of more mature water is indicated by higher Sr and Na in the pore water profile at a depth below –140 m. It is probable that groundwater below –65 m OD to the total depth (–170 m) represents water which was moving offshore during the lower sea levels some 8,000–10,000 years before present (BP), comparable to that found in the South Downs of the UK (Edmunds et al. 2001). This water remains separate from the main groundwater circulation of the present day. This water represents a good reference for the baseline composition for the Chalk of the area.

9.9 The age and quality of water in the confined Dorset Chalk

Several indicators are available to help understand the residence time of water in the Chalk such as the seasonality of springs, records of tritium, the presence of contaminants introduced by man since the industrial era, as well as radiocarbon dating for the older groundwaters; the ratios and absolute concentrations of major or trace elements also gives an indication of relative age. The initial age distribution in the Chalk groundwaters has inevitably been disturbed by pumping with water moving preferentially towards abstraction areas, distorting the age structure.

Tritium measured in the unsaturated zone suggests that groundwater moves towards the water table at approximately 1 m yr^{-1} and that water at the immediate water table will therefore have an age proportional to the depth of the unsaturated zone – typically measured in decades. In areas of the Chalk with strong fracture systems, for example along valleys, preferential, more rapid flow may occur. The presence of high nitrate concentrations in most of the Chalk outcrop areas (and rising concentrations in springs and boreholes) supports these observations and the influence of agricultural practices over the past 40–50 years, involving high rates of agrochemical application.

Confined groundwater in the Chalk aquifer beneath Poole Harbour is relatively fresh (<1000 µS cm⁻¹ down to a depth of at least –240 m OD in the Stoborough borehole). Fluid temperature logging in the basin suggests that present-day groundwater circulation in the deeper Chalk aquifer exists down to a depth of –170 to –180 m OD (Edmunds et al. 2001). This is considerably deeper than elsewhere along the South Coast of the UK and is believed to be controlled by the local geological structure. Interstitial waters from terminal cores taken from Stoborough and

Wareham boreholes indicate that freshwater occurs to depths of –250 m OD and with only minor residual salinity derived from saline formation water. The original saline formation water was probably expelled from the sediment during the Alpine folding, which affected this part of Britain 20–40 million years ago. However, detectable radiocarbon of 1–2 pmc (per cent modern carbon) in these samples implies that the bulk of the freshwater is late Pleistocene in age, in excess of 10,000 years BP.

Representative analyses of the groundwater from beneath Wareham are shown in Table 9.4 (Stoborough borehole). The evolved geochemical characteristics of the water

Table 9.4 Representative analyses of groundwater from Dorset. The locations are shown in Fig. 9.1 (Woodsford Farm and Stoborough are in the confined aquifer).

Locality	Units	Bulbarrow Farm Farm borehole	Winterbourne Abbas Public supply	Lulworth Spring Spring	Eagle Lodge BH2a Public supply	Woodsford Farm Farm borehole	Stoborough Deep borehole
T	°C		11.6	11.4		11.4	15
pH		7.38	7.23	7.42	7.13	7.02	7.5
DO	mg L⁻¹	5.3	9.6	6.5	5.7	0	0
SEC	µS cm⁻¹	492	558	584	586	343	683
Ca	mg L⁻¹	87.8	110	94	114	23	60
Mg	mg L⁻¹	1.73	2.28	3.82	2.57	2.24	17.2
Na	mg L⁻¹	8.4	12.1	23	13.3	9.3	63
K	mg L⁻¹	0.9	1.8	1.9	2.3	0.25	3.3
Cl	mg L⁻¹	15.2	20.3	36.6	22.2	22.7	85
SO₄	mg L⁻¹	12.3	13.5	13.3	15.3	32.4	8
HCO₃	mg L⁻¹	214	249		271	107	277
NO₃–N	mg L⁻¹	3.31	7.9	6.31	7.64	<0.1	<0.1
NH₄	mg L⁻¹	0.0047	<0.003	0.0045	0.0072	0.0063	0.14
DOC	mg L⁻¹	0.5	0.91	1.44	0.94	0.87	1.95
F	mg L⁻¹	0.064	0.08	0.087	0.07	0.09	
Si	mg L⁻¹	6.8	4.22	4.38	4.69	3.3	
Ba	µg L⁻¹	5.5	7.4	8.2	11.9	11.7	
Fe	mg L⁻¹	<0.005	<0.005	<0.005	<0.005	20	0.25
Li	µg L⁻¹	3.0	0.8	0.7	0.8	11.1	
Mn	µg L⁻¹	1.94	<0.04	<0.04	<0.04	555	<0.04

(high F, Sr) confirm that despite the freshness, the water is of considerable age.

9.10   Trends in water quality parameters

Long-term datasets are not widely available from the region in general, and it is feared that much of the time series data from public supply wells prior to the 1980s has not survived. However, some records have been identified from the Sutton Poyntz spring (Limbrick 2003) dating back to the late nineteenth century as well as from the breweries in Dorchester and Blandford Forum.

A 17-year record for a limited number of parameters is available for a shallow borehole at West Houghton fish farm (Fig. 9.9). The Cl record shows no significant increase over the measurement period. The initial data seem rather noisy and this is attributed to analytical problems. The mean value for this inland site (16 mg L^{-1} Cl) is about 5 mg L^{-1} lower than for sites nearer the coast, reflecting the decrease in maritime influence. Nitrate (as NO_3–N) increases from 2.5 to 5 mg L^{-1} over the measurement period, with evidence also of seasonal changes. For this site there is also a record of dissolved oxygen which shows some oscillation with time. This has no apparent explanation, nevertheless strongly aerobic conditions are maintained over the whole period.

Miscellaneous records are available in BGS files dating back to the immediate post-war period. Of relevance is analysis of nitrate (1.5 mg L^{-1} NO_3–N) at Alton Pancras new borehole (1946), which just pre-dates the introduction of intensive use of fertilisers and the ploughing up of Chalk downland.

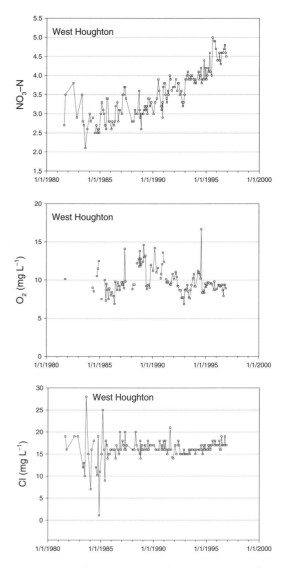

Fig. 9.9   Trends in chemistry from 1981–1997 for the West Houghton borehole.

Further data are available back to the early twentieth century (Whitaker and Edwards 1926) and a summary of these is given in Table 9.5. It can be seen that pre-First World War values ranged from 0.73 to 1.3 mg L^{-1} NO_3–N and little difference is found between

Table 9.5 Summary information from archive data taken from BGS records.

Site	Date	Nitrate NO$_3$–N (mg L^{-1})	Cl (mg L^{-1})
Corfe Mullen well	1908	1.0	30
Durweston	1911	1.3	17
Upwey	1910	0.88	23
Sutton Poyntz	1913	0.73	19
Alton Pancras	1946	1.5	–

these values and the 1946 data, suggesting little change in land use between the two World Wars.

9.11   Summary of the baseline quality

Dorset is a type location for demonstrating groundwater quality evolution in the UK and European Chalk. It is exposed to the influence of the prevailing westerly winds which influence rainfall and input water quality, and it lies in an area free of any major urban or industrial influences. It is, however, an area which contains a high proportion of arable farmland. A high proportion of water supply is derived from groundwater, although compared to other areas of the UK the demands on the aquifer are not stressed (i.e. abstraction does not exceed recharge).

The main aspects of baseline quality can be summarised with reference to the cross section Fig. 9.10. Rainfall chemistry influences are apparent in generating the chemical background of inert constituents such as Cl; this influence is apparent in the groundwater as a gradual reduction in Cl concentrations away from the coastline. The baseline quality for several other elements within the unconfined aquifer such as F is also likely to be influenced by rainfall after allowing for evapotranspiration. The main

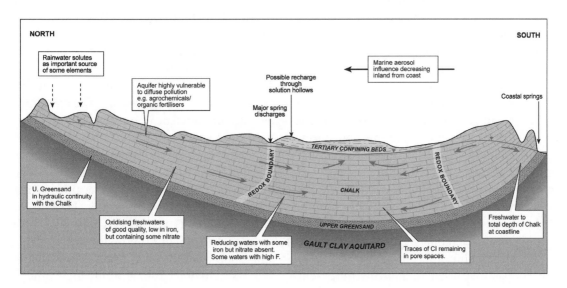

Fig. 9.10 Conceptual diagram of the Chalk aquifer of Dorset highlighting the main geochemical processes controlling water quality.

control on the baseline quality however is the geochemistry of the chalk sediment. Water interacting with the chalk produces hard water where Ca predominates over all other cations, HCO_3 being the main anion. Variations in the natural chemistry also take place with increasing residence time and especially due to redox changes as groundwater moves beneath the Palaeogene cover. High natural concentrations of Fe, Mn and F may be found in the deeper freshwaters, not currently used for water supply purposes. In general, the residence time of groundwaters in the Dorset Chalk can be measured in decades, although it is of the order of several months for flows to springs. Fresh groundwater occurring at depth in the Chalk beneath the Palaeogene strata, extending beneath Poole Harbour, has a residence time of the order of 20,000 years.

The natural baseline is represented by a range of concentrations for any given element, these being controlled by climatic and geological factors. Human impacts are mainly visible in the presence of indicator contaminants such as nitrate and artificial organic chemicals and it is very difficult to discern any effects of human activity in other solutes. As far as nitrate is concerned, evidence from pore waters at depth (Lulworth borehole) and from the inspection of old records, indicates that baseline concentrations were below 2 mg L^{-1} NO_3–N and most probably around 1 mg L^{-1}, before the onset of intensive farming in the latter half of the twentieth century. Nitrate is highly mobile under oxidising conditions, and its widespread presence at concentrations well above the baseline value, increasing progressively over the past two decades, indicates the extent to which the aquifer is now influenced by farming practices. Trace organic compounds are only detected locally in spring samples and are likely to be less mobile and less stable than nitrate, having undergone degradation and/or adsorption. The effect on the overall chemical quality is more difficult to demonstrate, although the increase in potassium concentrations (K/Na ratios) may be shown to relate to agriculture rather than to natural geochemistry. Some increase in sulphate may also be due to human impacts although in Dorset no definitive evidence is available to support this.

The summary tables and the plots illustrated show ranges and individual concentrations for the wide range of chemical elements and constituents considered in the study. All the common elements including most of those ranked as potentially hazardous by water quality (EC) guidelines are included in the tables. In addition, data are summarised for a wide range of potential contaminants, which may be of natural or introduced origin. The median value may be used as a reference value for purposes of comparison, but it is emphasised that the baseline range may extend across the 2.3–97.7 percentile range, over an order of magnitude for some elements, as a result of purely natural geochemical processes. It is found that individual values for NO_3, F, Fe and Mn may exceed the maximum permissible concentrations for drinking water in some waters. Of these F, Fe and Mn are typically of natural origin. The summary range of baseline concentrations is of strategic importance providing values against which potential pollution problems may be assessed. Several elements in this category are highly toxic (e.g. Tl, Be) but their natural abundance and mobility in the Chalk aquifer is very low.

In conclusion, the groundwater quality in Dorset is overwhelmingly controlled by natural reactions between rainwater and the

Chalk, following which further geochemical evolution takes place along flow pathways. It is also clear, however, that the influence of diffuse pollution from agriculture is widespread, especially in the build up of nitrate above a low baseline of around 1 mg L^{-1}. At present, this poses no immediate threat to public supplies although a reversal of the current upward trends must be sought. This may take place over coming decades with better source control and improved legislation, but also within the confined region sufficient attenuation may be provided by the natural reducing properties of the aquifer.

References

Allen, D.J., Brewerton, L.J., Coleby, L.M. et al. (1997) The physical properties of major aquifers in England and Wales. British Geological Survey Technical Report WD/97/34.

Buckley, D.K. (1996) A review and interpretation of geophysical logging performed for the Wareham Groundwater Project. BGS Technical Report WD/96/45C. British Geological Survey.

Edmunds W.M. and Bath A.H. (1976) Centrifuge extraction and chemical analysis of interstitial waters. *Environmental Science and Technology* 10, 467–72.

Edmunds, W.M., Cook, J.M., Darling, W.G. et al. (1987). Baseline geochemical conditions in the Chalk aquifer, Berkshire, UK: A basis for groundwater quality management. *Applied Geochemistry* 2, 251–74.

Edmunds, W.M., Darling, W.G., Kinniburgh, D.G. et al. (1992) Chalk groundwater in England and France: Hydrogeochemistry and water quality. Research Report SD/92/2. British Geological Survey, Keyworth.

Edmunds, W.M., Buckley, D.K., Darling, W.G. et al. (2001) Palaeowaters in the aquifers of the coastal regions of southern and eastern England. In: Edmunds, W.M. and Milne, C.J. (eds) *Palaeowaters in Coastal Europe: Evolution of Groundwater since the Late Pleistocene.* Geological Society, London, Special Publication, 189, pp. 71–92.

Geake, A.K. and Foster, S.S.D. (1989) Sequential isotope and solute profiling in the unsaturated zone of the British Chalk. *Hydrological Sciences Journal* 34, 79–95.

Limbrick, K.J. (2003) Baseline nitrate concentration in groundwater of the Chalk in south Dorset. *Science of the Total Environment* 314–16, 89–98.

Velegrakis, A.F., Dix, J.K. and Collins, M.B. (1999) Late Quaternary evolution of the upper reaches of the Solent river, southern England, based upon marine geophysical evidence. *Journal of the Geological Society* 156, 73–87.

Whitaker, W. and Edwards, W.N. (1926) *Wells and Springs of Dorset.* HMSO, London.

10 Groundwater Baseline Composition and Geochemical Controls in the Doñana Aquifer System, SW Spain

M. MANZANO, E. CUSTODIO, M. IGLESIAS AND
E. LOZANO

The geochemical controls of groundwater baseline chemistry of the Doñana aquifer system (DAS – SW Spain) have been characterised using major and minor inorganic components and environmental isotopes. In unconfined areas, groundwater baseline is controlled mainly by rainwater composition, which is of Na–Cl type, equilibrium with silica and dissolution of soil CO_2, Na– and K– feldspars, and $CaCO_3$ where present. The resulting mineralisation ranges from very low to moderate. Groundwater baseline changes from the unconfined areas to the confined sector mostly by mixing with old marine water, Na/Ca–(Mg) cation exchange, sulphate reduction and calcite dissolution/precipitation. The resulting salinity ranges from 1 mS cm^{-1} up to 80 mS cm^{-1}.

In the shallower layers (<40 m) of the unconfined areas, groundwater baseline has been modified by different human activities, as shown by the presence of agrochemicals (nutrients, metals, pesticides) and of industry-derived airborne pollutants (mainly metals coming from the nearby Huelva industrial site and from open-pit mining sulphide

exploitation). This is supported by tritium based ages, which show residence times greater than 40 years for flow lines deeper than 35–40 m.

Groundwater composition in the confined areas is mostly naturally derived (baseline), as supported by ^{14}C ages ranging from 1 to more than 15 kyr. Agriculture-derived pollutants are present in groundwater below irrigation areas exploiting groundwater confined layers close to the northern boundary of the aquifer confined under the marshes, as well as the unconfined aquifer.

10.1 Introduction

The DAS is situated close to the southwestern Atlantic coast of Spain, between the Guadalquivir River and the Portuguese border, and covers an area of about 2700 km^2. The area belongs to the provinces of Huelva, Sevilla and Cádiz. The main towns are Sevilla (1.6 × 10^6 inhabitants), some 100 km to the NE; Cádiz (0.2 × 10^6 inhabitants), some 60 km to the SE; and Huelva (0.05 × 10^6 inhabitants), situated some 70 km to the West. However, most of the Doñana core area is uninhabited, except for people in charge of the environmental protection, some researchers, the occasional developers of some natural resources, and visitors.

Doñana is a Ramsar Convention protected area for wildlife and waterfowl. It comprises one of the largest valuable environmental zones in Spain: some 1100 km² are protected under the legal statutes of a National Park (human activities are severely restricted) and a Natural Park (some controlled farming and light tourism activities are allowed, such as pine cone collection, charcoal production and walking). (Fig. 10.1).

Outside the protected zones, large areas are used for irrigation, established late in the 1970s when a groundwater-based irrigation farming development plan started. It was sponsored by the United Nations Food and Agriculture Organization and the Spanish Government. Tourism is also an important economic activity. It is mainly beach-based and concentrated both seasonally (spring and summer) and spatially (Matalascañas and Mazagón coastal resorts with a capacity of around 300,000 people). A religious pilgrimage concentrates around 1 million people during one week at El Rocío village every year, close to the marshes and other fragile areas of the National Park. This causes a serious impact on the environment, mainly from the great number of vehicles and waste dumping.

The climate is Mediterranean sub-humid with Atlantic influence characterised by dry

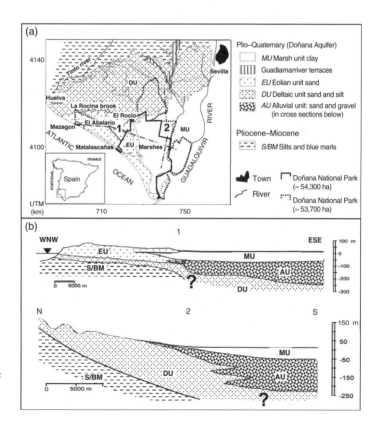

Fig. 10.1 Location, geology and geometry of the Doñana aquifer system. The aquifer limits coincide with those of the Plio-Quaternary sediments.

summers and wet winters. Mean rainfall, concentrated between October and March, is 500–600 mm, but has a very high inter-annual variability, from 1100 mm to 250 mm. Mean yearly temperature is around 17°C near the coast and 18°C in the centre of the area. Values above 35°C are often reached during July–August, while values under 0°C are seldom attained during December–January. Temperature shows very small inter-annual variation. The mean number of annual sun hours is close to 3000.

Two large rivers, the Tinto River to the NW, and the Guadalquivir River to the E, surround the aquifer area, but they do not contribute water to it. Surface water inside the territory is limited to a couple of small groundwater fed permanent rivers and some seasonal brooks, and to some permanent and many small temporal lagoons in small depressions in which the water-table crops out. Groundwater is the only permanent water source both for human and environmental uses.

Intensive groundwater exploitation since early in the 1980s, mostly for irrigation and to supply touristic areas, but also locally for some environmental uses, led to a groundwater level drawdown and to the development of conspicuous localised piezometric depression cones. The resulting water–table drawdown has reduced the aquifer's natural discharge through springs, streams and phreatic evapotranspiration, thus inducing slow vegetation changes and modifying the hydrological behaviour of wetlands (Coleto 2004; Manzano et al. 2005a).

Groundwater chemistry in most of the aquifer has been studied previously in several projects focusing on the aquifer behaviour characterisation. This information has been integrated with new studies carried out

in the framework of the EU project BaSeLiNe (2003), which focused on the natural (baseline) groundwater composition and its origin.

10.2 Geological and hydrogeological setting

The DAS comprises Plio–Quaternary, mostly unconsolidated, fine sediments deposited in fluvio-marine (alluvial, deltaic, aeolian, estuarine and marshy) environments, which overlie Pliocene and Miocene silts and marls. The Miocene to Quaternary sediments form a single multi-layered series between a few metres thick inland to more than 200 m thick at the coastline.

The aquifer system has a total surface area of around 2700 km^2, some 1800 km^2 of which comprises clay-rich marshland. The rest is mostly a sandy area, especially the coastal fringe, (some 70 km in length and a few to 30 km wide), going from the Guadalquivir river mouth to the Tinto river estuary, near the town of Huelva (Fig. 10.1).

The aquifer geometry and structure are shown in the cross sections of Fig. 10.1:
● From N to S: the lithology and thickness of the Quaternary units are variable. It consists of a thick Deltaic Unit of silts and sands (DU in Fig. 10.1), which is replaced near the northern border of the marshes by gravelly layers of the Alluvial Unit (AU in Fig. 10.1). This unit thickens to the south, has interlayers of clays and reaches the centre of the marshes. Further to the S, the gravels change gradually into sands, silts and clays, overlapping similar Miocene and Pliocene sediments (marls, silts and sandy silts), and giving way to a thick (50 to >300 m) sequence of very fine sediments (Salvany and Custodio 1995). The alluvial sediments are covered by

a sequence (50–70 m thick) of estuarine and marshy clays (Marsh Unit, MU in Fig. 10.1), nowadays separated from the sea by a recent littoral sand spit.

● From NW to SE: on top of the Miocene and Pliocene marine marls and silts there is a thin (<20 m) deltaic layer with gravels, which is covered by a thick (c.80 m) sand unit of Pleistocene to Recent age. This unit comprises fluvio-marine sediments overlain by alluvial and then aeolian sands. The aeolian layers have several interlayered clay bodies of lacustrine origin, sometimes with peat. Inspite of its diverse origin, the whole sequence is named Aeolian Unit (EU in Fig. 10.1).

At the regional scale the surface of the aquifer system shows two lithologic domains: a clayey one in the marsh area, and a sandy one to the N and W of the marshes. The sandy domain roughly behaves as a water-table aquifer which in some places has two layers, a coarse one at depth and an overlying fine sand unit. Under the clayey domain a large volume of confined aquifer develops. Recharge to the aquifer occurs by rain infiltration in the sandy areas to the N and W of the marshes. There is also recharge through excess irrigation water in the agricultural areas, which however are irrigated with local groundwater and thus do not imply additional water.

Also at a regional scale, groundwater flows mostly to the S and SE – that is, to the area confined under the marshes–, to a main brook called La Rocina, and to the ocean. Under natural conditions, discharge takes place as seepage to the ocean, to the streams and to the many small phreatic wetlands situated on top of the aeolian sands, as phreatic evapotranspiration, as upward flows near the boundary (ecotone) between the sands and the clays, and perhaps through the Quaternary

clays in the confined area, although this is probably a minor term of the water balance. The SE sector of the confined aquifer (Alluvial Unit and Marsh Unit) contains almost immobile, old connate marine water which has not been flushed out on account of the low hydraulic head prevailing since the late Holocene sea level stabilisation some 6 ka BP (Manzano et al. 2001).

However, intensive groundwater abstraction has partially depleted natural discharge during the last three decades. Agricultural wells are concentrated near and along the contact between the sands and the clays, so that nowadays a large proportion of recharge water is pumped out from the unconfined area, largely in a zone close to the marshes (Custodio and Palancar 1995; Trick and Custodio 2004). Intensive and localised pumping has decreased phreatic evapotranspiration and natural seepage to springs and wetlands (Manzano et al. 2005a). Also, some local flow reversals between the Alluvial Unit confined under the clays and the overlying Quaternary clays, have been recognised to the NE of the marshes. They are contributing to the salinisation of formerly fresh groundwater irrigation wells.

The aquifer permeability ranges from moderate to low, with the most permeable materials (alluvial and aeolian sands with some thin interlayered gravel layers) cropping out to the west and north of the system.

10.3 Materials and methods used

This groundwater baseline study focuses on the western sector between La Rocina brook, the Atlantic Ocean and the centre of the marshes (see Fig. 10.1). Although less intensive, the study was also extended to a regional groundwater flow path from the northern unconfined sector to the centre of the marshes.

Both historical and newly obtained chemical data were used to characterise the baseline composition of groundwater. Previous information on major components and environmental isotopes is abundant (Baonza et al. 1982; Poncela et al. 1992; Iglesias 1999; Delgado et al. 2001; Lozano et al. 2001, 2005; Manzano et al. 2001; Lozano 2004). For the purposes of the present study, about one hundred new samples were collected within the BaSeLiNe project during two campaigns (November 2000 and July 2001). They were analysed for major and some minor components, trace elements, and ^{3}H, δ^{18}O, δ^2H, δ^{13}C and ^{14}C. Most of the samples are from single-screen boreholes, a few are from multi-screened boreholes or open wells, and a few samples are lagoon waters.

The samples were collected in double-cap polyethylene bottles. Samples for trace elements were filtered with fibreglass 0.45 μm Millipore filters, acidified to pH less than 2 and kept refrigerated at around 4°C until analysis. The analyses were performed at the British Geological Survey (BGS, Wallingford, UK) and also at the Geological Survey of Spain (IGME, Tres Cantos laboratory in Madrid, Spain), using ICP-AES in both laboratories. Ten duplicates were used to check the coherence between both laboratories. In addition, 24 samples (fresh and brackish waters) were analysed for Br and Cl in the IGME by slow flow ion chromatography to characterise the origin of water salinity.

Tritium was analysed at the CEDEX (Centro de Experimentación de Obras Públicas) laboratory in Madrid (Spain); some analyses of δ^{18}O and δ^2H were performed at the CSIC (Consejo Superior de Investigaciones Científicas) laboratory in Granada (Spain) and the rest were analysed at the CEDEX; δ^{13}C and ^{14}C were analysed at the University of Bern (Switzerland).

Time series data for Ca, Mg, Na, K, SO$_4$, HCO$_3$, pH, EC and groundwater levels from the water quality monitoring networks of the IGME and the IARA (Instituto Andaluz de Reforma Agraria, Andalusian Government) have also been used for the study of temporal and spatial chemical trends. The period covered is from 1973 to 2000. Measurement frequencies are variable: the decades between 1970 and 1980 are poorly represented, with few analyses, while the best-covered period (from 1990 to 2000) has monthly or even fortnightly measurements. Some 40 wells and boreholes have been studied, their datasets ranging from 100 to 400 measurements. These data are not shown in this paper, only the main conclusions from the studies.

Simple statistical treatment, classical hydrogeological graphs and hydrogeological cross sections have been used to identify and illustrate chemical water types, their spatial distribution and the existence of time/space chemical trends. The code SPSS8.0 (SPSS Inc. 1998) has been used for the statistical study of the most significant chemical variables (average, median, mode, standard deviation, variance, rank, maximum, minimum, and the 25, 50, 75 and 97.7 percentiles) and for principal component analysis (PCA).

The code PHREEQC-Version 2 (Parkhurst and Appelo 2002) has been used to check the hydrogeochemical conceptual model proposed for groundwater baseline evolution along flow paths.

10.4   Sources of chemical baseline levels in the Doñana aquifer

The primary sources of the chemical baseline levels groundwater chemistry are rainfall inputs and aquifer mineralogy. Both have been characterised in the study. As polluting sources and pollution processes are almost ubiquitous, potential pollution sources are also listed, and the presence of pollution indicators has been studied.

10.4.1   Rainfall and atmospheric fallout chemistry

Rainfall chemistry can be considered the minimum baseline concentrations in groundwater. The data available for the studied area come from three different sources and six different sites, randomly placed at different distances from the coast. Up to now, no attempt has been made to quantify the local landward chemical gradient of rainfall chemistry, although there is a recent study for chloride concentrations in the Iberian Peninsula rainfall (Alcalá and Custodio 2004).

Existing data were integrated and elaborated by Iglesias et al. (1996). Because of the different sampling places and sources, one of the most adequate ways of looking at the significant values is through simple statistical treatment. The 25, 50 and 75 percentile values of major elements, pH and electrical conductivity (EC) are shown in Table 10.1, which illustrates the characteristic compositions of the 34 samples available.

Local rainwater is of Na–Cl type. Most of the ionic ratios clearly show that the main solute source is of marine origin, both as wet and dry deposition. The SO_4/Cl ratio, in equivalents, shows a sulphate excess ($SO_4/Cl = 0.2$–0.9) with respect to seawater ($SO_4/Cl = 0.1$). Using Cl mass balance between rain and phreatic water, only about 30% of S in rainwater can be accounted for from a marine origin. In areas without agricultural influence, the rest (70%) may be partly attributed to atmospheric supply of sulphide particles from the W of the aquifer recharge area (the dominant wind direction). The proposed original source for these particles is either open-pit mining activities in the numerous sulphide mines of the Portuguese–Spanish Pyrite Fringe, which crops out to the N and NW of Doñana, and/or industrial emissions from the Huelva industrial site, situated some 30 km to the NW. Part of the studied sector is under cultivation, so agrochemicals are also a local source of sulphur to groundwater, although atmospheric

Table 10.1 Characteristic composition of 34 rainwater samples taken at different places and periods over the western Doñana recharge area.

Component (mg L^{-1})	Percentiles		
	25%	50%	75%
HCO$_3$	6.1	8.54	16.46
SO$_4$	1.92	3.37	8.17
CL	4.26	7.8	11.35
NO$_3$	0.06	0.12	0.62
Na	3	4.14	6.21
K	0.39	0.78	1.12
Ca	1.6	2.2	4.8
Mg	0.48	0.73	1.57
NH$_4$	0	0.18	0.54
CE µS cm^{-1}	37	50	69
pH	5.8	6.2	6.7

Note: Values incorporate dissolved dry deposition.
Source: After Iglesias et al. (1996).

input of sulphide particles has also to be considered.

10.4.2 *Aquifer mineralogy*

Aquifer mineralogy is dominated by amorphous silica, with some spatial differences due to the presence of K- and Na-feldspars (microcline and albite); illite, chlorite and kaolinite are minor components. Carbonate, mostly $CaCO_3$, is present either as detrital grains or as shell remains, except in the upper sand layers of the western sector, where they have already been leached by rainfall infiltration.

10.4.3 *Pollution sources*

Although less accurately known, urban, agricultural and industrial water contamination sources are abundant in the aquifer recharge areas (Manzano et al. 2005b):
• Farming (agriculture and some animal farming) is widely developed on the aquifer outcrop area causing diffuse soil and both surface and groundwater contamination by nutrients, pesticides and heavy metals. Agrochemically derived SO_4, NO_3, Co, Cu, Br and Zn have been identified in shallow and intermediate depth groundwaters, according to previous and present studies;
• the highly contaminant agro-industry (mostly olive related) is concentrated around some villages on the northern outcrop of the aquifer. They generate a well-known surface water (rivers and streams) pollution problem by mostly difficult to degrade organic matter from olive oil production and also Na coming from olive treatment with NaOH;
• most of the sewer network has been developed in the last 15–20 yr, and some small villages and factories still dump untreated waste water to the local streams and brooks and into infiltration areas;
• petroleum refining and pyrite burning, among other highly polluting activities, have been active since decades ago at the Huelva industrial site, some 30 km to the W of the recharge area.

Some pollution indicators have been studied and reported below. Though being a potential polluting source, up to now, marine intrusion has not been detected in the coastal area.

10.5 Hydrochemical characteristics of groundwater

10.5.1 *Water types*

A first approach to groundwater chemistry at a regional scale shows that water salinity ranges over three orders of magnitude. When plotting all the samples in logarithmic or similar plots (Fig. 10.2) it seems as if most of the waters were either the result of mixing or an evolution between two members:
1 A freshwater member, with EC (electrical conductivity) less than $1\,mS\,cm^{-1}$ and mostly of Na–Cl type (locally Na–Ca–Cl–HCO_3);
2 A saline water member, with EC = 1–$80\,mS$ cm^{-1} and mostly of Na–Ca–Cl–HCO_3 type.

The freshwater member corresponds to the most characteristic groundwater found in the western water-table area down to 40–50 m depth. The saline one is characteristic of the area confined under the marshes and also of some shallow layers in the Aeolian Unit around phreatic lagoons.

Though many samples are really the result of vertical mixing between fresh groundwater, flowing upward through or close to the marshy clays, and saline pore water, many others are mostly the result of local

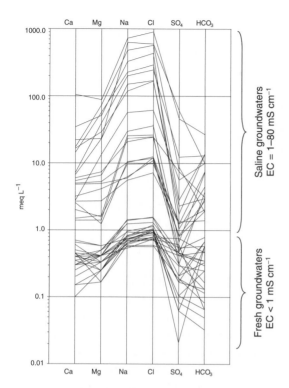

Fig. 10.2 Representative fresh and saline groundwaters from the Doñana area. Most of them have the same Na–Cl chemical facies, though locally some Na–Ca–Cl–HCO₃ waters may exist.

evaporative concentration in shallow lagoons, both directly or by dissolving previously deposited salts. Locally, water-table groundwater to the seaward side of the coastal dunes gets its moderate salinity mostly from marine spray.

Baseline composition ranges of fresh and saline waters were obtained through a basic statistical study. However the characteristic values and ranges are not described here, as the most relevant results arise from the interpretation of their regional evolution and changes with depth, as well as from the

study of the geochemical processes that control the calculated ranges.

10.5.2 Temporal and spatial groundwater chemistry evolution and causes

10.5.2.1 Evolution along main flow paths

Groundwater chemistry evolves along a flow path as shown in Fig. 10.3. Shallow and intermediate depth groundwater in the western water-table area is of Na–Cl type and has low mineralisation (EC less than 0.5 mS cm⁻¹). After previous and contemporary studies this water has been interpreted as the result of slightly evaporated rainwater, dissolving soil CO_2 and Na/K feldspars (Fig. 10.4). At greater depths (>50 m) groundwater mineralisation increases and water becomes of Ca–HCO₃ type as a result of calcite dissolution.

Flowing to the east, as groundwater approaches the marine and estuarine sediments it becomes brackish and saline mostly because of mixing with saline groundwater. A saline groundwater body (>40 mS cm⁻¹) is found in the coastal area, sometimes more concentrated than seawater. Recirculation of evaporated seawater when the marshes were open to the sea is a possible origin (Konikow and Rodríguez-Arévalo 1993).

A broad mixing zone develops from NW to SE under the marshes, but its geometry is not well known due to the difficulties of drilling adequate observation boreholes in the marsh area and of obtaining representative samples in a multi-layer aquifer. A number of baseline changes can however be identified at a local scale including: (1) increasing content of marine related water; (2) equilibrium with calcite; (3) sulphate reduction; (4) cation exchange between Na in solution and Ca/Mg in the solid phase.

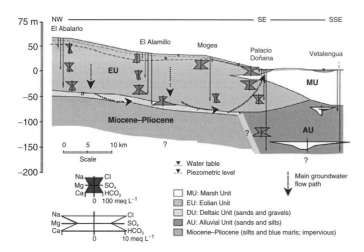

Fig. 10.3 Groundwater chemical evolution W–E along main flow paths. The section goes from El Abalario, in the water-table western area, to the marshes (see Fig. 10.1).

A chromatographic spatial distribution is difficult to observe because of the few available sampling points southward, multi-screened boreholes and pumping wells in the marshes, which are not adequate to get samples following flow paths.

This chemical cross section is representative, in general terms, of groundwater evolution along flow at a wider scale. In the water-table area to the N of the marshes, where the aquifer is thin (<50 m) and consists of silts and siliceous and carbonate silty sands, groundwater is of the Ca–HCO_3 or Na–Ca–HCO_3–Cl type, but as it reaches the confined area groundwater evolves rapidly to a Na–Cl type and becomes brackish and finally salty.

10.5.2.2 Variations in depth and their causes

The evolution of major elements with depth in the western water-table area has already been introduced in Figs. 10.3 and 10.4. The combination of major elements with some minor and trace elements supports the hydrogeochemical model already proposed and adds new information.

Figure 10.5 summarises the statistical parameters for the elements whose concentrations vary significantly with depth. The term Upper Unit refers to the upper and central parts of the Aeolian Unit in cross section 1 of Fig. 10.1, where carbonates are absent. Lower Unit refers to the lower part of the Aeolian Unit plus the sediments of the Deltaic Unit, where carbonates exist. Almost all the components have a wider range of concentrations in the Upper Unit than in the Lower Unit. Very different situations can be observed for different elements when taking into account the median or the 75 percentile: •Ca, HCO_3, Ba, Sr and Mn are clearly more abundant in the Lower Unit. Except for Mn, all of them can be accounted for by the presence and dissolution of biogenic carbonate in this unit, while they are absent in the Upper Unit; •Na, Cl and Mg show very similar concentrations in both units, but they are slightly more abundant in the Upper Unit. This points to recharge water as the single common source. Mg evolution shows the apparent absence of dolomite in the Lower Unit, although Iglesias (1999) mentioned that this mineral is present in the northern part (see Fig. 10.4).

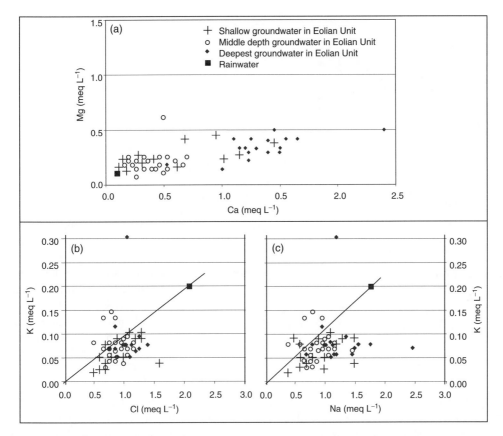

Fig. 10.4 (a) Mg versus Ca content in groundwater of the western water-table area as a function of depth. The aquifer is formed by siliceous sands and carbonate-containing sediments are present only in the deepest layers. The evolution of the Mg/Ca ratio suggests that there is only calcite, and dolomite is not present. (b) K versus Cl content in groundwater of the same area. The line separates the samples where Cl and K have a common origin (rainwater; samples along and below the line) and those including an additional source of K (samples above the line). (c) K versus Na content in the same groundwaters. The figure shows that there is a common source for K and Na in the groundwaters above the line, and this points to feldspar dissolution in the shallow and middle layers of the aquifer. Those processes have been checked by numerical modelling. (Redrawn after Iglesias 1999.)

• SO_4, NO_3, Co, Cu and Zn are clearly more abundant in the Upper Unit. The possible lithological sources (silicates) in both units are the same, so in the Upper Unit there is at least an additional source for these elements, which is considered to be recent, since they are not found yet in the Lower Unit. In the next section, two possible (anthropogenic) sources are introduced to support these observations.

10.5.2.3 Temporal variations and their causes

After the combined study of chemical and groundwater level time series, and their

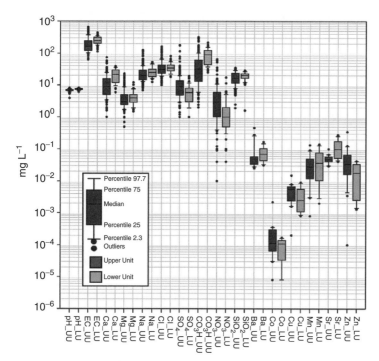

Fig. 10.5 Box plots of those elements with concentrations varying significantly with depth in the western recharge area. The Upper Unit is formed of siliceous aeolian sands without carbonates; the Lower Unit is formed by siliceous littoral sands and gravels with carbonates.

integration into the available knowledge about groundwater development and soil-use changes in different sectors, only a few out of the 40 analysed wells show visible temporal chemical variations. Changes are, however, very subtle, and there is not the same degree of certainty about the different proposed causes of evolution, so complementary work has to be done in some cases.

Two main types of evolution and proposed causes have been observed in two different zones (Fig. 10.6).

a) In the agricultural area to the NE of the marshes (the transition zone from the unconfined sands to the clays) groundwater development is intensive. A slight groundwater salinisation due to pumping-induced displacement of old saline water from the S towards the agricultural wells in the N is observed. Concentrations of Cl, Na, Mg and

SO_4 increase with time, because the saline water inflow fraction increases.

Most of the wells do not show permanent groundwater salinisation, but oscillating displacements of the mixing zone linked to seasonal pumping patterns seems to control groundwater chemistry. This cannot be observed with standard 'concentration versus time' plots, but it shows up when ion concentrations are plotted versus Cl (Fig. 10.6). When Cl increases because of saline water inflow to the well, Na increases in a lower proportion, Ca increases conspicuously and HCO_3 decreases significantly. This points to Na_{liquid}/Ca_{solid} exchange and to $CaCO_3$ precipitation as Ca activity increases. Numerical modelling with PHREEQC has validated the proposed geochemical model.

b) In the area to the NW of El Rocío, agriculture was abandoned early in the 1990s

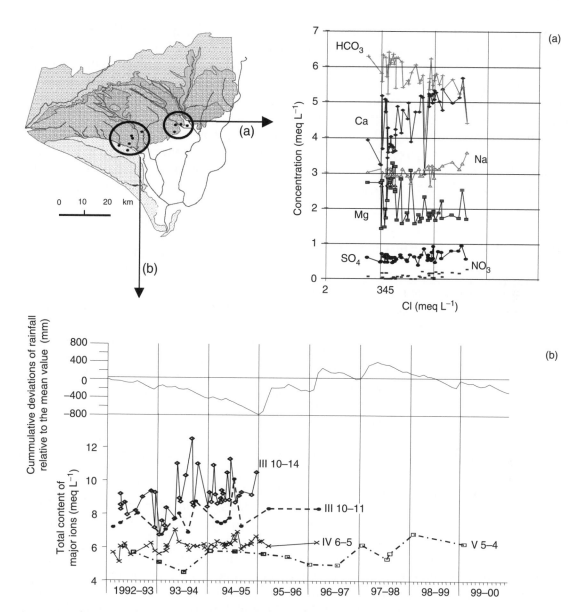

Fig. 10.6 Observed trends of groundwater chemistry temporal evolution: (a) Well in the agricultural area to the NE of the marshes: groundwater salinity increases due to pumping-induced displacement of old saline water from the south towards the agricultural wells in the north. Cl, Na, Mg and SO_4 increase due to growing saline water content; Ca increases due to Na_{liquid}/Ca_{solid} exchange, while HCO_3 decreases because of $CaCO_3$ precipitation as Ca activity increases. (b) Wells in a former agricultural area close to El Rocío. A slight groundwater salinisation that seems to follow a climatic pattern is observed. It is attributed to the input of salts remaining in the shallower part of the unsaturated zone transported by rainwater recharge.

(although re-established in the last few years), the data do not cover this period. A slight groundwater salinisation that seems to follow a climatic pattern is observed. It is primarily attributed to downward displacement by rainwater recharge of the salts remaining in the shallower part of the terrain since the times of active irrigated agriculture. This could be a very common polluting mechanism in the area during the coming years.

10.5.2.4 Pollution indicators and sources

In the upper part of the unconfined areas (less than 40 m depth), groundwater baseline composition has already been impacted by different human activities:

1 *Agricultural pollution.* In the agricultural areas, the presence of NO_3 (up to 65 mg L^{-1} in some shallow samples) denotes that baseline has already been modified by human activity. Besides the rain (with NO_3 concentrations <1 mg L^{-1}), there are two nitrogen sources in the area: agrochemicals applied in agriculture (the principal one, but localised) and livestock farming (secondary but widespread until a few years ago).

In the westernmost, best-studied recharge area (El Abalario), SO_4 and NO_3 ranges and median concentrations are higher in the upper layers of the aquifers than in the lower ones. It is also the case for some trace components such as Zn, Co or Br (see Fig. 10.5). PCA calculations point to a common origin for Zn, Cr, Fe and Ni, which seems to be different to that of Cl, Na, Ca, Ba, Sr, Mn, pH and SiO_2, assumed to be lithological. This suggests that SO_4 and NO_3, Zn, Cr, Fe and nitrogen probably come from agrochemicals, and that they are still in transit to the deep layers of the aquifer.

In the confined areas, ^{14}C groundwater ages range from a few to more than 16 ka (dissolved inorganic carbon – DIC – content not evolves from the recharge areas to the confined aquifer). All the samples available are however admixtures of old and young components, as indicated by the presence of measurable tritium (Fig. 10.7) and NO_3 in most of them (Manzano et al. 2001). A few saline samples from the agricultural zone to the NE of the marshes contained NO_3 (between 0.3 and 100 mg L^{-1}), thus agriculture can account for this contamination. However, many others are from boreholes placed well inside the marshes and far from the farming areas (e.g. borehole 13 in Fig. 10.7). In these boreholes, NO_3 can be explained through (1) the existence of vertical flows along boreholes in the confined areas and (2) the presence of remnants of almost immobile drilling water in the formations surrounding the sampled boreholes.

2 *Atmospheric pollution.* Heavy metals have been found down to 25–30 m depth (Iglesias 1999; BaSeLiNe 2003). In the farming areas, they can also be derived from agrochemicals, but in the protected areas (free of agricultural activity) the increased concentrations of some industrially derived elements in the upper layers of the saturated zone reflect the atmospheric input of pollutants from the nearby Huelva industrial area. This is supported by the calculated tritium ages for the western water-table area, which show residence times (t_r) of a few years in the top 10 m, t_r <35 years for groundwater shallower than 35–40 m and t_r >40 years for groundwater deeper than 40 m. The ^{14}C ages indicate that even the oldest groundwaters in the water-table areas are less than a few centuries old.

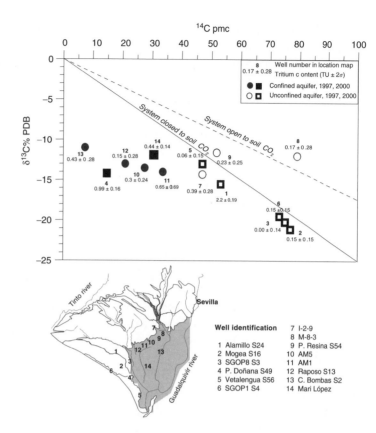

Fig. 10.7 Tritium, ^{13}C and ^{14}C contents in groundwater in the water-table aquifer and in the confined sector under the marshes.

Moreover, the observed SO$_4$ excess in local rainwater relative to marine airborne sources also points to additional sulphur sources in recharge water. Two sources can potentially be contributing: dust and gas emissions from the nearby Huelva industrial site to the W of Doñana, and pyrite-derived dust transported from the many open-pit mining activities in the Spanish–Portuguese Pyrite Fringe, to the N and NW of Doñana.

10.6 The geochemical model of groundwater baseline in the DAS

The studies to characterise groundwater baseline composition and evolution have focused mainly in the western recharge area and, less intensively, in the northern one. The integration of previous studies and the new results have allowed the development of a geochemical model to explain groundwater baseline composition.

The geochemical model of groundwater baseline in the siliceous DAS can be established as a function of the unconfined or confined character of the aquifer as follows:

1 *Groundwater baseline in the unconfined areas of the aquifer.* The following main processes control geochemical evolution:
(a) rainwater composition, which is of Na–Cl type, but with sulphate and other anthropogenically-derived components from the nearby Huelva industrial site;

(b) equilibrium with silica;

(c) dissolution of soil CO_2;

(d) dissolution of Na/K feldspars;

(e) dissolution of $CaCO_3$ (in the deepest layers of the western unconfined area and throughout the northern unconfined area).

The resulting mineralisation ranges from low to moderate (<0.2 to 1 mS cm^{-1}). In the upper part of the unconfined areas (c.<40 m deep), groundwater baseline has already been impacted by different human activities. In the agricultural areas, this is pointed out by the presence of agrochemically derived substances (Br, pesticides, SO_4, NO_3, heavy metals) down to depths of 25–35 m. In the protected areas, where agriculture has never been active, the higher concentrations of some industrially derived elements (mainly metals) in the upper layers of the saturated zone with respect to the deeper ones reflects the atmospheric input of pollutants from the nearby Huelva industrial site. This agrees with the calculated tritium-based ages for the western water-table area, which have residence times less than 35 years for flow paths shallower than 35–40 m, and more than 40 years for deeper flow paths. Radiocarbon-based ages show that even the oldest groundwaters in the water-table areas are younger than a few centuries.

2 *Groundwater baseline evolution from the water-table to the confined areas.* Following the main groundwater flow paths from the unconfined to the confined parts of the aquifer (from N to S and from W to SE), the main chemical change is that groundwater becomes increasingly saline because of mixing with the old marine water trapped both in the confined permeable layers and in the clays. A broad mixing zone develops from NW to SE under the marshes, but its geometry is not well known due to the difficulties

of drilling adequate observation boreholes and of obtaining representative samples in a multi-layered aquifer.

3 *Groundwater baseline in the confined sector of the DAS.* Baseline composition changes mostly due to the following processes:

(a) mixing with modified old marine water;

(b) equilibrium with calcite;

(c) cation exchange [Na/Ca–(Mg)] in moving fresh-saline water fronts;

(d) sulphate reduction (depletion with respect to conservative mixing with sea water);

(e) probably C incorporation from sedimentary organic matter evolution, although the ^{13}C studies did not confirm this.

The resulting salinities range from 1 mS cm^{-1} up to 80 mS cm^{-1}. Although groundwater chemistry in the confined area is mostly naturally derived (baseline), in some agricultural wells exploiting confined layers close to the northern water-table area, groundwater baseline is already being modified by agricultural pollutants.

10.7 Conclusions

The DAS is a thick, detrital and mostly siliceous young system situated in an area that shares intensive groundwater use for agriculture and human supply with strictly protected natural habitats that depend on groundwater. Using the methodology proposed in Chapter 1 of this book, the natural water quality of the Doñana aquifer has been studied. This methodology is based on a geochemical approach, so that through the integrated use of statistics, chemical trends, time tracers and geochemical modelling it has been possible to establish the natural groundwater quality variations and the responsible processes, as well as to identify the areas and potential sources of the pollution processes observed.

The groundwater quality of the DAS and its natural or altered nature may be summarised as follows:

- The unconfined part of the aquifer contains low mineralised freshwater that is of the Na–Cl type, with low hardness, and is slightly acidic in the upper part of the western area. It becomes harder and more alkaline where the terrain still contains remnants of calcite shells, and it is also slightly more saline due to decreasing recharge rate. Anthropogenic effects from airborne pollutants and agricultural activities are limited to the upper part of the aquifer.
- The confined part of the aquifer mostly contains brackish to connate saline water, emplaced in the past. There is a broad mixing zone with the freshwater of the unconfined aquifer that is currently altered to some extent by the effect of groundwater development.

Thus, the early protection of a large part of the territory since the 1980s has contributed to the preservation of groundwater baseline conditions, which still prevail in parts of the aquifer. However, and beyond the local disturbance of groundwater baseline by agricultural and/or industrial pollution, the natural baseline is currently also being modified in the shallower layers of the protected area due to atmospheric input of pollutants, incorporated into rainwater recharge to the aquifers.

References

Alcalá, F.J. and Custodio, E. (2004) La deposición atmosférica de cloruro al terreno en España. Boletin Geologico y Minero 115, 319–29.

BaSeLiNe (2003) Natural BaSeLiNe quality in European aquifers. A basis for aquifer management. Final report to EU Project EVK1-CT1999–0006.

Baonza, E., Plata, A.A. and Silgado, A. (1982) *Hidrología isotópica de las aguas subterráneas del Parque Nacional de Doñana y zona de influencia*. Cuadernos de Investigaciones, C7, CEDEX, Madrid, pp. 1–139.

Coleto, M.C. (2004) Funciones hidrológicas y biogeoquímicas de las formaciones palustres bipogénicas de los mantos eólicos de El Abalario-Doñana (Huelva). Doctoral Thesis. Autonomous University of Madrid. Canto Blanco, Madrid.

Custodio, E. and Palancar, M. (1995) Las aguas subterráneas en Doñana. *Revista de Obras Públicas (Madrid)* 142 (3340), 31–53.

Delgado, F., Lozano, E., Manzano, M. et al. (2001) Interacción entre las lagunas freáticas de Doñana y el acuífero según los iones mayoritarios. In: Medina, A., Carrera, J. and Vives L. (eds) *Las Caras del Agua Subterránea*. Serie Hidrogeología y Aguas Subterráneas 1–2001. IGME, Madrid, pp. 111–17.

Iglesias, M. (1999) Caracterización hidrogeoquímica del flujo del agua subterránea en El Abalario, Doñana, Huelva. Doctoral Thesis. Civil Engineering. School, Technical. University of Catalonia, Barcelona, Spain.

Iglesias, M., Custodio, E., Giráldez, J.V. et al. (1996) Caracterización química de la lluvia y estimación de la recarga en el area de El Abalario, Doñana, Huelva. IV Simposio sobre el Agua en Andalucía. Almería. II: pp. 99–121.

Konikow, L.F. and Rodríguez-Arévalo, J. (1993) Advection and diffusion in a variable-salinity confining layer. *Water Resources Research* 29, 2747–61.

Lozano, E. (2004) Las aguas subterráneas en los cotos de Doñana y su influencia en las lagunas. Doctoral Thesis. Technical University of Catalonia, Barcelona, Spain.

Lozano, E., Delgado, F., Manzano, M. et al. (2001) Interacción entre las lagunas freáticas de Doñana y el acuífero según los isótopos ambientales. In: Medina, A., Carrera, J. and Vives L. (eds) *Las Caras del Agua Subterránea*.

Serie Hidrogeología y Aguas Subterráneas 1–2001. IGME, Madrid, pp. 379–85.

Lozano, E., Delgado, F., Manzano, M. et al. (2005) Hydrochemical characterisation of ground and surface waters in 'the Cotos' area, Doñana National Park, southwestern Spain. In: Bocanegra, E.M., Hernández, M.A. and Usunoff, E. (eds) *Groundwater and Human Development.* International Association of Hydrogeologists, Selected Papers on Hydrogeology, 6, Balkema, Leiden, pp. 217–31.

Manzano, M., Custodio E., Loosli, H.H. et al. (2001) Palaeowater in coastal aquifers of Spain. In: Edmunds, W.M. and Milne, C.J. (eds) *Palaeowaters in Coastal Europe: Evolution of Groundwater since the Late Pleistocene.* Special Publication, 189, Geological Society, London, pp. 107–38.

Manzano, M., Custodio, E., Mediavilla, C. et al. (2005a) Effects of localised intensive aquifer exploitation on the Doñana wetlands (SW Spain). Groundwater Intensive Use. International Association of Hydrogeologists, IAH Selected Papers, 7, Balkema, Leiden, pp. 295–306.

Manzano, M., Custodio, E., Colomines, M. et al. (2005b) El fondo hidroquímico natural del acuífero de Doñana. V Iberian Congress of Geochemistry/IX Spanish Congress of Geochemistry. In: DVD. Diputación Provincial de Soria Electronic Library, no.6.

Parkhurst, D.L. and Appelo C.A.J. (2002) PHREEQC (Version 2) – A Computer Program for Speciation, Batch-Reaction, One-Dimensional Transport, and Inverse Geochemical Calculations. In: www.brr.cr.usgs.gov.

Poncela, R., Manzano, M. and Custodio, E. (1992) Medidas anómalas de tritio en el área de Doñana. *Hidrogeología y Recursos Hidráulicos* XVII, Madrid, pp. 351–65.

Salvany, J.M. and Custodio, E. (1995) Características litológicas de los depósitos pliocuaternarios del Bajo Guadalquivir en el área de Doñana: implicaciones hidrogeológicas. *Revista Sociedad Geologica de España* 8, 21–31.

SPSS Inc. (1988) SPSS Base 8.0: *Users Guide.* SPSS, Chicago, IL.

Trick, T. and Custodio, E. (2004) Hydrodynamic characteristics of the western Doñana Region (area of El Abalario), Huelva, Spain. *Hydrogeology Journal* 12, 321–35.

11 The Aveiro Quaternary and Cretaceous Aquifers, Portugal

M. T. CONDESSO DE MELO AND
M. A. MARQUES DA SILVA

The Aveiro Quaternary and Cretaceous aquifer systems (northwest coast of Portugal) have been the subject of detailed hydrogeochemical studies during the present investigation. Geochemical data were collected to study the relations between the groundwater chemistry, aquifer mineralogy and patterns of regional flow. The approach followed included the identification of major groundwater geochemical patterns in the aquifer, study of the aquifer natural stratification, determination of the aquifer natural background levels, and definition of the controlling processes responsible for the downgradient changes in water chemistry, focusing on mixing processes and water–rock interactions. This was accomplished by studying and comparing the chemical composition of rain and groundwater, through determination of ionic and molar and calculation of saturation indices relative to some common minerals. Isotopes were used as constraints for baseline definition helping to justify spatial trends, distinguishing the natural baseline from human impact. The results may be used for future groundwater-quality management in the aquifer systems, which are regionally very important for sustainable development.

11.1 Introduction

The present research focuses on the Aveiro Quaternary and Cretaceous aquifer systems, two multilayer coastal aquifers that cover over 1300 km^2 in the northwest part of the Portuguese mainland and the adjacent continental shelf. The two aquifers are an integral part of the same Meso-Cenozoic infilling stratigraphic sequence of the Vouga river sedimentary basin but present distinctive hydrogeological and geochemical conditions.

The hydrogeochemistry of groundwaters in both aquifers was investigated in detail in order to establish their natural or baseline chemistry. Geochemical data were collected to study the relationships between groundwater chemistry, aquifer mineralogy and present and past patterns of regional flow.

The approach adopted included the identification of major groundwater geochemical characteristics, the study of the natural stratification of groundwater quality and the definition of the controlling processes responsible for the downgradient changes in water chemistry, focusing on mixing processes and water–rock interactions. This was accomplished by studying and comparing the chemical and isotopic composition of rain and groundwater, through determination of chemical and isotopic ratios and residence times, calculation of saturation indices relative to the most common minerals and by quantifying the mass balance of solutes along the main flow paths.

This study constitutes the first detailed hydrogeochemical study in the Aveiro Quaternary aquifer at regional scale. Previous hydrogeological studies on the aquifer were carried out only locally (Ferreira 1995; Gonçalves and Vozone 1999) or were related to local contamination/rehabilitation problems (Leitão 1996). Peixinho de Cristo (1985, 1992) wrote the only two available synthesis papers on the hydrogeology of the Quaternary formations. In comparison, many studies over the last 20 years have produced valuable information for the understanding of the Aveiro Cretaceous aquifer. The first references to the hydrogeology of the Cretaceous formations can be found in Zbyszewski (1963), Zbyszewski et al. (1972), Teixeira and Zbyszewski (1976), Barbosa (1981), Saraiva et al. (1983) and Lauverjat et al. (1983). Previous hydrochemical investigations on the Aveiro Cretaceous aquifer were based on major elements and isotopic analysis (Peixinho de Cristo 1985; Marques da Silva 1990; Carreira et al. 1996; Oliveira, 1997; Carreira Paquete 1998; Condesso de Melo et al. 2001, Condesso de Melo 2002) and always revealed a consistent hydrogeochemistry for the aquifer and a remarkable constancy in the groundwater quality. Noble gas analysis and past recharge temperatures were reported for the aquifer by Carreira Paquete (1998).

11.2   Regional setting

The area of northwest Portugal (western Iberian Peninsula), covered by the present research, extends between 40°18′N and 40°57′N latitude, and is bounded to the west by the Atlantic Ocean and to the east by the 8°26′W longitude meridian (Fig. 11.1). It forms a gently sloping coastal plain of about 1300 km² on the Portuguese mainland and offshore beneath the Atlantic Ocean.

The study region corresponds to the lower end of Vouga River Basin (VRB) and is known for the 'Ria de Aveiro', a shallow coastal lagoon with both marine and estuarine waters,

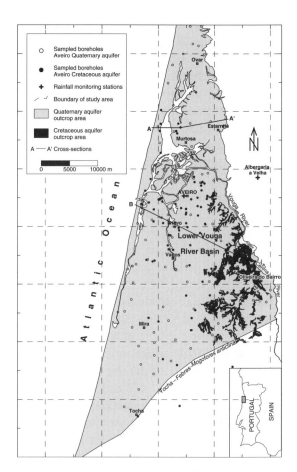

Fig. 11.1 Map of the study area showing rainfall and groundwater sampling sites and the location of the cross sections used for hydrogeological description.

separated from the adjacent sea by an elongated barrier of sand. Today, the lagoon system and adjacent channels encompass an area of approximately 527 km², with extensive mudflats, sandbanks and salt marshes developed in its inter-tidal zone. This diverse area associated with the Aveiro lagoon is regarded as one of the most important wetland areas in Portugal. It provides wintering retreats for more than 20,000 aquatic birds and has been recently classified as Special Protection Area (SPA) under the EC Directive on the Conservation of Wild Birds.

Several rivers flow into the Aveiro lagoon but the river Vouga is the longest (148 km), forming the main tributary within the regional hydrological system. It has a moderate-sized catchment area (3635 km²) with a dominant east–west orientation. The flat, low-lying alluvial plain of the river Vouga completely dominates the local topography and surface geology.

The study region has traditionally been a mainly rural area, but in the last decade widespread industrial development has occurred, leading to rapid social and economic transformations that place a continually increasing demand on water resources in the basin.

Historically, the local populations and industries pumped groundwater from the shallow unconsolidated aquifer formations (Aveiro Quaternary aquifer) but the growing demand for water resources and some contamination problems, led progressively to an increased reliance on groundwater resources from the deeper confined formations (Aveiro Cretaceous aquifer).

Since 1996, integrated surface/groundwater management is being applied to drinking, industrial and agricultural water needs in the VRB. Nevertheless, over-abstraction and point source plus diffuse contamination are increasing pressures on both the Aveiro Quaternary and Cretaceous aquifers and contributing to a general decline in groundwater quality in the region.

11.3 Geological and hydrogeological background

The lower Vouga basin represents the northernmost part of the Lusitanian basin, one of the Atlantic margin rift-basins formed along the western border of the Iberian Peninsula in response to Mesozoic extension and subsequent opening of the North Atlantic Ocean. During the Mesozoic, the depositional conditions varied within the Lusitanian basin. In the southern and western areas of the basin, there was a gradual opening to marine influence, while in the north and east, there occurred a predominant detrital infilling of continental origin.

The bedrock in the region consists of schists of Proterozoic age in most of the studied region, with the exception of the north part, where Proterozoic gneisses, migmatites and Palaeozoic granitoids constitute the actual bedrock. The Precambrian–early Palaeozoic bedrock units are overlain by the Meso-Cenozoic sedimentary cover in most of the study area, being exposed along a NNW–SSE axis close to the border limit with the Iberian Meseta.

The stratigraphic sequence of the basin is discontinuous and ranges in age from late Triassic to Holocene. Quaternary deposits comprise mainly recent alluvial sediments and sand dunes (primarily fine sands) of Holocene age and old beach deposits and fluvial terraces (mainly clays and clayey sands in the upper part and coarse sands and pebbles at the base) of Pleistocene age.

The Quaternary alluvium and terrace deposits unconformably overlie consolidated geological units of Cretaceous age that were deposited in fluvial, deltaic or shallow marine environments under predominantly transitional or continental depositional conditions.

From the hydrogeological point of view, both the Quaternary and Cretaceous units form important multilayer aquifer systems that yield substantial volumes of water. A detailed description of the main hydrogeological units is presented in the following sections and summarised in Table 11.1.

11.3.1 Quaternary hydrogeological units

From the hydrogeological point of view, the Quaternary deposits form a three-layer aquifer system that occurs regionally as a surficial layer or as layers of variable thickness. Three separate hydrogeological units may be defined in the Aveiro Quaternary aquifer (Fig. 11.2).

11.3.1.1 Surficial phreatic aquifer

The surficial phreatic aquifer is composed of modern unconsolidated alluvium, beach and aeolian sands and sand dune deposits, of Holocene age, continually being eroded, transported and deposited. The average thickness ranges between 8 and 10 m, rarely exceeding 20 m in total thickness. These deposits overlie most of the study region covering an area of approximately 500 km^2, parallel to the coastline from Cortegaça in the north to Quiaios in the south. This aquifer layer is highly permeable (20–30 m d^{-1}) and receives direct recharge from rainfall infiltration and irrigation returns. The natural flow pattern in the aquifer is from the

Table 11.1 Correlation between lithostratigraphic units and the corresponding hydrogeologic units and aquifer systems in the river Vouga basin.

Period	Stage	Lithostratigraphic units		Lower Vouga Basin Aquifer System Analysis	
				Hydrogeologic unit	Aquifer system
Quaternary	Holocene and Plio–Pleistocene	Alluvium, beach and eolian sands and sand dunes / Terrace and old beach deposits		Unconfined or locally semi-confined aquifer	Aveiro Quaternary aquifer
Cretaceous	Campanian–Maastrichtian	'Aveiro clay formation', C$_3$		Confining unit	Aveiro Cretaceous aquifer
	Coniacian–Santonian	Upper sandstone formation, C$_3$	Top	Aquitard	
	Upper Turonian-Lower Coniacian		Bottom	Confined aquifer	
	Upper Cenomanian–Turonian	'Furadouro sandstone', C$_2$		Confined aquifer	
	Cenomanian	Carbonate formation, C$_2$		Confined aquifer	
	Aptian/Albian–Lower Cenomanian	Lower sandstone formation, C$_{1A}$	Top	Confined aquifer	
			Middle	Confined aquifer	
			Bottom		

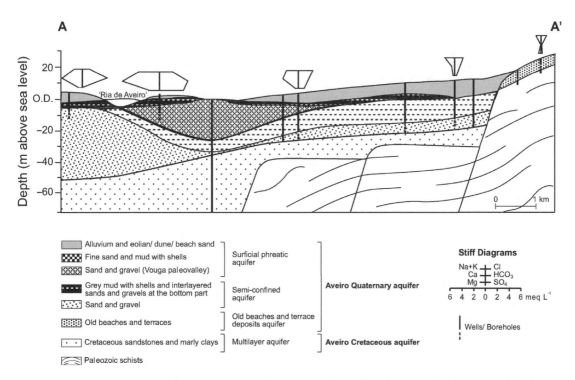

Fig. 11.2 W-E cross-section of the Aveiro Quaternary aquifer with indication of the principal hydrochemical facies for the three different hydrogeological units.

east towards the sea with smooth gradients ranging between 0.0012 in the north and 0.0036 in the south part of the aquifer. The aquifer discharges to the sea, to the rivers that flow in the area and to the underlying aquifer by vertical leakage through the underlying aquitard.

11.3.1.2 Semi-confined aquifer

This occupies an area of 650 km² and is composed of coarse highly permeable sediments (10–20 m d⁻¹) in close association with the river basins of the region (Vouga, Águeda, Cértima and Boco rivers). This aquifer is semi-confined by a low permeability organic mud and fine silt layer which acts as an aquitard and limits natural recharge to the aquifer. The deposition of this organic mud layer is closely linked to the palaeogeographic evolution of the Aveiro lagoon and the river Vouga estuary/delta and is very limited close to Estarreja (NE of the study area). In this area, the aquifer is unconfined, receiving direct recharge from rainfall infiltration and irrigation return. Recharge in the rest of the semi-confined area is by vertical leakage from the shallow aquifer and rivers, which may be hydraulically connected to the aquifer. The aquifer discharges to the sea, to the Aveiro lagoon and rivers, and to other aquifer layers where the hydraulic gradients are favourable. The natural gradient in the semi-confined aquifer layer is about 0.0014, flowing from the east towards the sea.

11.3.1.3 Old beaches and terrace deposits aquifer

The terrace deposits associated with rivers are older alluvial deposits (Plio-Pleistocene), visible in areas where erosion processes have deepened the river valleys and left the terrace deposits topographically above the present day alluvium. These deposits are predominantly coarse sands, gravel and pebbles, which occur mainly along the eastern part of region, where they overlie either Cretaceous or Triassic formations.

The thickness of the terrace deposits in the study area ranges between 10 and 20 m, and the average permeability ranges from 5 to 10 m d^{-1}. This unit receives recharge primarily by direct infiltration of rainfall, and secondarily by infiltration of excess irrigation.

Unlike the other surficial Quaternary deposits, the natural gradient in these terrace deposits is from west to east (gradient of 0.004 'units') where they discharge to the principal rivers that flow in the area.

11.3.2 Cretaceous hydrogeological units

The Aveiro Cretaceous multilayer aquifer is part of a thick sequence of Cretaceous sediments underlying Quaternary deposits (Fig. 11.3). Cretaceous sediments in the region are mainly siliciclastic and the aquifer has developed an extensive groundwater flow system, confined over two-thirds of its extension by a low permeability marly clay formation of Upper Cretaceous age, which limits modern recharge to the aquifer. This unit is not present in the eastern part of the

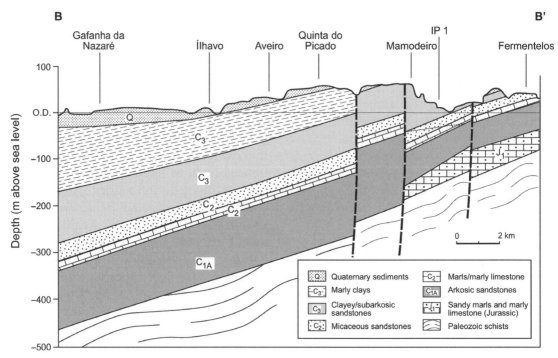

Fig. 11.3 W-E cross-section of the central part of the Lower Vouga River basin (the principal aquifer layers are the Quaternary units and the Cretaceous C_{1A}, C_2 and the bottom of C_3 units).

study area where the aquifer is unconfined. The principal hydrogeological units are described below, from youngest to oldest formations.

11.3.2.1 Marly clay formation ('Aveiro clay formation'), C_3

This unit consists of low permeability marly clays, occasionally with thin sandy clay interlayered beds, and is considered to be an aquiclude. It confines the western two-thirds of the aquifer extension, limiting recharge, but protecting the underlying permeable Cretaceous formations from contamination episodes.

11.3.2.2 Upper sandstone formation ('Verba sandstone formation'), C_3

Mainly clays and clayey sandstones form the top part of the Upper Sandstone Formation, which generally do not yield substantial volumes of water. The high clay content of this unit reduces the transmissivity and increases the groundwater residence times considerably. This |formation has often been considered to be an aquitard and very few small-yield boreholes around the village of 'Verba', for domestic use, are completed in this formation. The productivity of these wells is always very limited and the groundwater pumped usually has higher salinities than the typical baseline values for the other parts of the aquifer.

11.3.2.3 Upper sandstone formation ('Oiã sandstone formation'), C_3

Stratigraphically below the 'Verba sandstone formation', the bottom part of the Upper Sandstone Formation is significantly more permeable than the overlying unit. This unit consists of medium-grained sandstone with some interbedded clayey layers. The sand is predominantly quartz, but because this unit was deposited in a fluvial/deltaic sedimentary environment, the lithological variability may be very large even in short distances. However, from the hydrogeological point of view, this is considered an aquifer unit and may yield significant volumes of water to wells.

11.3.2.4 Micaceous sandstone formation ('Furadouro sandstone formation'), C_2

This sandstone formation is undoubtedly the aquifer layer with the highest transmissivities, and boreholes completed along a full section of the 'Furadouro Sandstone Formation' may yield as much as 40 or 60 L s^{-1}. The thickness of the unit ranges from 10 to 30 m with a median thickness of about 20 m, and the sand is predominantly very clean quartz with a wide range in granulometry, with increasing grain size towards the top of the unit. Most boreholes drilled in the region are completed in this formation, and the screens placed within it comprise the great majority of the groundwater pumped. The high micaceous content of the 'Furadouro Sandstone Formation' may sometimes create some groundwater-quality constraints related to the very fine mica particles in suspension and requires the correct placement of well screens.

11.3.2.5 Carbonate formation ('Mamarrosa limestone formation'), C_2

The lithology of the Carbonate Formation varies significantly within the study region but usually shows hydrogeological characteristics that permit its exploitation over the majority of the area. However, in the central part of the region, the permeability of this unit is significantly reduced and does not yield much water. The Carbonate Formation outcropping in the recharge area is also of low permeability.

11.3.2.6 Lower sandstone formation ('Palhaça sandstone formation'), C_{1A}

This Lower Sandstone Formation is uniform and exploited by several boreholes in the eastern part of the study area, where it yields considerable volumes of water. However, in the western part of the region it presents significant lithological variability with depth and has been divided into three different units from the hydrological point of view. The top part, which underlies the Carbonate Formation, is usually of high transmissivity and has good water quality. Most boreholes are completed in this unit. A reddish-brown clay layer separates this top part from an intermediate unit, where the clay content and degree of sandstone cementation increase and the well yield is reduced. The water quality decreases due to increasing mineralisation. Fewer boreholes are completed in this unit and it usually presents higher hydraulic heads than the upper overlying layer.

11.4 Background geochemistry

The study of the baseline quality in the two Aveiro groundwater bodies requires both information on the petrography and mineralogy of the aquifer matrix and on the rainfall chemistry, as rainfall is the major source of groundwater.

11.4.1 Petrography and mineralogy

The equilibrium geochemical reactions as well as kinetic properties in an aquifer are in great part determined by the petrography and mineralogy of the aquifer matrix. Both aquifer systems have hydrogeological units with distinct grain sizes, mineralogical compositions and clay contents, forming multilayer aquifer systems with different hydrogeological and hydrogeochemical properties (Table 11.2).

The Aveiro Quaternary aquifer is formed predominantly of siliceous material. Mineralogical studies by Rocha (1993) using x-ray diffraction and transmission electronic microscopy analysis show that Quaternary units are composed mainly of quartz, plagioclase, K-feldspar, calcite and dolomite. Gypsum, anhydrite, opal and zeolites amongst others may occur as accessory minerals. Illite, smectite and kaolinite are the most abundant clay minerals.

The principal mineral assemblages of the different Cretaceous units are largely identical although the relative abundance varies from layer to layer. They are composed mainly of quartz, phyllosilicates, calcite, dolomite, K-feldspar and plagioclase. Gypsum, anhydrite, jarosite (from the oxidation of pyrite); melanterite, pyrite and goethite may occur as accessory minerals. Kaolinite, smectite and illite are the most abundant secondary clay minerals.

11.4.2 Rainfall chemistry

Rainwater chemistry was monitored monthly from rainfall collected in bulk samplers at three monitoring sites: Aveiro (AVR), Albergaria-A-Velha (ALB) and Oliveira do Bairro (OBR) within the study area, over

Table 11.2 Principal mineral assemblages for the different aquifer layers.

Unit	Lithology	Mineralogy* (<38 μm)	Clay minerals* (<2 μm)
Q	Sands and sand dunes	Quartz – Plagioclase – Potassium feldspar – Calcite – Dolomite (Gypsum – Anhydrite – Opal – Zeoliths)	Illite – Smectite – Kaolinite
C_3	Marly clays	Phyllosilicates –Quartz – Potassium Feldspar – Plagioclase – Dolomite – (Calcite)	Illite – Kaolinite – (Smectite)
	Upper sandstone formation	Quartz – Potassium Feldspar – Phyllosilicates – Dolomite – (Plagioclase – Calcite)	Kaolinite – Smectite – Illite
		Quartz – Plagioclase – Potassium Feldspar – Phyllosilicates – (Dolomite – Calcite)	Kaolinite – Illite
C_2	Micaceous sandstone	Quartz – Phyllosilicates – Potassium Feldspar – Plagioclase – Dolomite – Pyrite – (Calcite)	Illite – Kaolinite – Sodium Smectite
	Carbonate formation	Calcite – Phyllosilicates – Potassium Feldspar – (Plagioclase – Dolomite – Quartz)	Kaolinite – Illite – Smectite and Sodium smectite
C_{1A}	Lower sandstone formation	Quartz – Potassium Feldspar – Plagioclase – Phyllosilicates – (Dolomite – Calcite)	

*Minerals are written in descending order of their relative abundance. Minerals written in brackets occur as accessories.

Table 11.3 Rainwater annual averaged composition at the three monitoring sites of the study area (all data in mg L^{-1}). Also shown are some characteristic chemical molar ratios for the rainwater.

	Cl^-	SO_4^{2-}	NO_3^--N	Br^-	Na^+	K^+	Ca^{2+}	Mg^{2+}	Si^{4+}	pH
AVR	5.29	3.86	0.124	0.024	4.28	0.28	1.35	0.48	0.23	5.37
ALB	3.57	3.68	0.104	0.023	3.13	0.34	2.98	0.59	0.19	6.15
OBR	3.36	4.87	0.156	0.023	3.14	0.36	3.09	0.60	0.29	6.00
	Na/Ca	K/Na	Mg/Na	Mg/Ca	K/Mg	Na/Cl	Mg/Cl	Si/Na	Br/Cl	
Seawater	44.65	0.02	0.12	5.43	0.18	0.85	0.10	0.0002	0.0016	
AVR	5.54	0.04	0.11	0.59	0.36	1.25	0.13	0.04	0.0020	
ALB	1.83	0.06	0.18	0.33	0.37	1.35	0.24	0.05	0.0029	
OBR	1.78	0.07	0.18	0.32	0.37	1.44	0.26	0.08	0.0030	

a period of five years from 1997 to 2001 (Condesso de Melo 2002). The average concentrations of major elements in the rainfall of the region and some characteristic molar ratios are compared to seawater in Table 11.3. Rainfall in the studied region is slightly acidic, with a solute content that demonstrates the marine aerosol influence as well as the result of the dissolution of atmospheric gases and aerosols.

The concentration of major elements in local rainfall follows a general pattern $Cl^- > SO_4^{2-} > Na^+ > Ca^{2+} > Mg^{2+} > K^+ > NO_3 - N$. Sodium and chloride are the dominant ions in Aveiro rainfall where it is basically a Na–Cl solution. These two ions are closely correlated in coastal regions due to their common origin in sea salt aerosols. In the other two sites, 20 km further inland, sodium becomes less abundant, calcium doubles in concentration, becoming almost as abundant as sodium, whilst chloride contents decrease approximately 35%. Thus, rainfall composition changes from Na–Cl near the coast to Na–Ca–SO_4 type inland, showing that the influence of marine aerosol on local rainfall composition decreases significantly a short distance from the coast. Rainfall samples from the Oliveira do Bairro site show an increase in sulphate, exceeding the chloride concentrations, probably reflecting an anthropogenic contribution related to industrial activities.

The average nitrate concentrations are low and vary between a minimum of 0.10 mg L^{-1} to a maximum of 0.16 mg L^{-1} of NO_3–N. Measured Br/Cl molar ratios at the Aveiro site (Br/Cl = 0.0020) are also close to marine aerosol composition, while for the other two sites, located in agricultural and forest areas further away from the sea, this ratio increases relative to seawater (Br/Cl = 0.0030), probably reflecting the addition from the biomass during burning and forest fires.

11.5   Groundwater chemistry

11.5.1   Groundwater sampling and analytical methods

Groundwater samples were collected from the two studied aquifer systems during different field campaigns. The Quaternary aquifer was sampled during field campaigns carried out in January and October 2001, and

later in April 2002. Groundwater samples were collected from 74 public supply and private boreholes and analysed for major, minor and trace elements, total organic carbon (TOC) and stable isotope contents (δ^2H, $\delta^{18}O$ and $\delta^{13}C$). The third campaign included the determination of methane and tritium (^3H) from pre-selected sites.

The Aveiro Cretaceous aquifer was sampled during October 1996, September 1997, August 2000 and December 2000, and groundwater samples for laboratory analysis were collected from 90 deep boreholes. The analysis included the determination of major, minor and trace elements, radiocarbon, TOC and stable isotope contents determined at pre-selected sites. For the present discussion, the isotopic database for the Cretaceous aquifer was increased, with 23 sets of stable isotope and radiocarbon groundwater analyses (δ^2H, $\delta^{18}O$, $\delta^{13}C$ and ^{14}C) published by Carreira et al. (1996).

All the groundwater sampling points discussed in this chapter are shown in Fig. 11.1. For the groundwater sampling, a multiport flow-through cell connected in-line to the sampling points was used. Water samples were taken from the discharge point during pumping, once stabilisation of the principal field parameters: pH, temperature (T), specific electrical conductance (SEC), redox potential (Eh) and dissolved oxygen (DO) were observed. Redox measurements, using a platinum electrode and referenced to ZoBell's solution, are reported relative to the standard hydrogen electrode. On-site measurements also included the determination of alkalinity (quoted as HCO_3^-) by acid titration.

The inorganic determinations in the samples were performed either by the Activation laboratories in Ontario (Canada) or the British Geological Survey (Wallingford, UK), which also carried out the stable isotope, TOC and methane analyses. Deuterium, oxygen-18 and carbon-13 results are reported as parts per thousand (‰) with respect to Vienna Standard Mean Ocean Water (VSMOW) standard and Vienna Pee Dee Belemnite (VPDB), respectively, using the standard δ (delta) notation (Gonfiantini 1978). The analytical precision for stable isotope analysis is ±0.2‰ for $\delta^{18}O$, ±2.0‰ for δ^2H and ±0.3‰ for $\delta^{13}C$.

Electroneutrality was used as a quality control for all the determinations, and ionic mass balances with errors between –5% and +5% were considered to be acceptable. Ionic charge imbalances were in all cases less than 5%.

The groundwater samples for radiocarbon analysis were prepared following transformation to graphite at the NERC Radiocarbon Laboratory in East Kilbride (Scotland) and then analysed by ^{14}C AMS at the University of Arizona NSF facility (USA). In keeping with international practice, the radiocarbon activities are expressed as a percentage of modern carbon (pMC) and reported ages are calculated as conventional radiocarbon years BP (before AD 1950), both expressed at the ±1σ level for overall analytical confidence. Groundwater ages have been derived applying the correction procedure of the ^{14}C value using the measured $\delta^{13}C$ values following the IAEA model (Salem *et al.* 1980).

11.5.2 Groundwater hydrogeochemical evolution

A general description of the chemical composition of the studied groundwater bodies is initially presented as cross-sections downgradient, followed by an explanation of the possible origin and natural evolution of the groundwater in the system, focusing on the controlling geochemical processes. Finally, the redox environment in the study units is discussed and a description on the occurrence of trace elements in the aquifer is presented. All the data are summarised in Table 11.4.

11.5.2.1 Aveiro Quaternary aquifer

The chemical analyses of groundwater samples collected in the Aveiro Quaternary aquifer revealed a groundwater body impacted by different types of diffuse and point-source contamination. The former sources of contamination (mainly agricultural) produce high nitrate (>11.3 mg L^{-1} NO_3–N), sulphate (>40.0 mg L^{-1} SO_4^{2-}) and potassium (>5 mg L^{-1} K) groundwaters. The latter (including industrial areas, septic tanks, petrol stations) impact groundwater with significant concentrations of persistent organic compounds (MNB, benzene, aniline), some inorganic compounds (Na, SO_4) and other minor and trace elements (Cd, Ni, Pb, Cr, As, Hg). However, parts of the aquifer (mainly, forested and dune covered areas) remain where the groundwater is pristine, which help to define the groundwater baseline.

The range of analytical results obtained for the whole groundwater body have been plotted in box plot diagrams and compared to the composition of diluted seawater normalised to the median chloride of the whole dataset (Fig. 11.4[a]) and to the median concentrations of each hydrogeological unit (Fig. 11.4[b]). Analysis of the results confirms the subdivision of the Aveiro Quaternary aquifer into three hydrogeological units with different hydrogeochemical characteristics and that different geochemical processes are contributing to the variations in major ion concentrations.

Table 11.4 Overall summary statistics and baseline values for the Aveiro Quaternary aquifer.

Parameter	Units	Min.	Max.	Median	Mean	97.7 % ile	n
pH		4.6	7.6	6.1	6.2	7.4	144
T	°C	13.9	21.2	17.3	17.2	20.2	144
DOC	mg L^{-1}	2.0	12.5	4.2	4.8	11.9	144
SEC	µS cm^{-1}	110	1474	456	508	1077	144
DO	mg L^{-1}	0.0	9.2	0.7	1.9	8	144
Eh	mV	−38	451	369	300	442	144
Na	mg L^{-1}	7.4	193	26.0	34.2	94.6	144
K	mg L^{-1}	1.0	40.3	7.8	11.8	38.6	144
Ca	mg L^{-1}	1.3	120	42.6	47.0	109	144
Mg	mg L^{-1}	1.9	48.3	6.0	8.5	27.4	144
Si	mg L^{-1}	1.7	28.0	5.4	6.4	13.5	144
Cl	mg L^{-1}	8.7	349	37.4	45.7	118	144
SO$_4$	mg L^{-1}	<0.03	192	41.5	46.6	134	144
HCO$_3$	mg L^{-1}	0	424	84.5	115	310	144
NO$_3$–N	mg L^{-1}	<0.05	47.1	4.4	9.3	39.6	144
NO$_2$–N	mg L^{-1}	<0.05	5.27	<0.01	0.07	0.20	144
PO$_4$–P	mg L^{-1}	<0.04	2.60	<0.02	0.08	0.89	144
Ag	µg L^{-1}	<0.2	<0.2	<0.2	<0.2	<0.2	144
Al	µg L^{-1}	<2	1110	10.00	49.10	275.1	144
As	µg L^{-1}	0.07	43.69	0.48	2.11	14.81	144
Au	µg L^{-1}	<0.002	0.005	<0.002	0.0004	0.005	144
B	µg L^{-1}	15.27	1300	78.13	158.1	756.4	144
Ba	µg L^{-1}	0.94	192.2	31.07	38.41	80.62	144
Be	µg L^{-1}	<0.10	2.59	<0.10	0.05	0.68	144
Bi	µg L^{-1}	<0.01	0.02	<0.01	<0.009	<0.01	144
Br	µg L^{-1}	<0.06	1.13	0.14	0.15	0.42	144
Cd	µg L^{-1}	<0.01	3.27	0.02	0.07	0.18	144
Co	µg L^{-1}	<0.005	10.10	0.17	0.48	1.56	144
Cr	µg L^{-1}	<0.50	1.02	<0.50	<0.41	0.70	144
Cs	µg L^{-1}	0.002	4.09	0.11	0.28	1.33	144
Cu	µg L^{-1}	<0.20	17.43	3.27	4.84	15.01	144
F	µg L^{-1}	<0.02	0.49	0.04	0.06	0.27	144
Fe	µg L^{-1}	<5.0	9520	163.5	1013	8760	144
Ga	µg L^{-1}	<0.01	0.28	0.01	0.01	0.06	144
Ge	µg L^{-1}	<0.01	0.15	0.02	0.02	0.07	144
Hf	µg L^{-1}	<0.002	0.17	0.003	0.009	0.074	144
Hg	µg L^{-1}	<0.20	<0.20	<0.20	<0.20	<0.20	144
I	µg L^{-1}	2.37	265.0	19.30	35.45	140.3	144
In	µg L^{-1}	<0.001	0.003	<0.001	<0.0009	0.001	144
Li	µg L^{-1}	<1	25.70	2.24	3.38	18.53	144
Mn	µg L^{-1}	0.13	418.6	29.71	57.07	209.6	144
Mo	µg L^{-1}	<0.1	13.19	0.12	0.45	2.79	144

Table 11.4 Continued.

Parameter	Units	Min.	Max.	Median	Mean	97.7 % ile	n
Nb	µg L^{-1}	<0.005	0.02	0.01	0.003	0.02	144
Ni	µg L^{-1}	<0.30	6.32	<0.30	0.64	4.84	144
Os	µg L^{-1}	<0.002	<0.002	<0.002	<0.002	<0.002	144
Pb	µg L^{-1}	<0.10	3.37	0.25	0.50	2.25	144
Pd	µg L^{-1}	<0.01	<0.01	<0.01	<0.01	<0.01	144
Pt	µg L^{-1}	<0.01	<0.01	<0.01	<0.01	<0.01	144
Rb	µg L^{-1}	1.72	65.95	9.67	15.53	48.20	144
Re	µg L^{-1}	<0.001	0.02	0.001	0.001	0.01	144
Ru	µg L^{-1}	<0.01	<0.01	<0.01	<0.01	<0.01	144
Sb	µg L^{-1}	<0.01	1.39	0.05	0.15	0.88	144
Sc	µg L^{-1}	<1	19.90	3.59	4.28	10.07	144
Se	µg L^{-1}	<0.2	7.83	0.52	0.74	2.72	144
Sn	µg L^{-1}	<0.1	0.48	<0.1	<0.09	<0.1	144
Sr	µg L^{-1}	13.40	754.5	171.2	214.6	703.5	144
Ta	µg L^{-1}	<0.001	0.017	<0.001	0.0003	0.009	144
Te	µg L^{-1}	<0.01	0.05	<0.01	<0.008	0.01	144
Ti	µg L^{-1}	0.70	18.58	2.04	3.17	12.36	144
Tl	µg L^{-1}	<0.005	0.45	0.04	0.07	0.39	144
U	µg L^{-1}	0.003	4.27	0.09	0.35	1.91	144
V	µg L^{-1}	<0.05	7.96	0.32	0.94	5.78	144
W	µg L^{-1}	<0.02	0.94	<0.02	0.01	0.12	144
Y	µg L^{-1}	0.01	41.61	0.59	2.96	29.87	144
Zn	µg L^{-1}	<0.5	2010	12.51	48.93	86.16	144
Zr	µg L^{-1}	<0.01	0.82	0.03	0.08	0.56	144

The surficial phreatic aquifer has dominant sodium chloride facies and lower salinities in areas not impacted by contamination, reflecting the local rainfall composition and the short residence times. This type of groundwater may be observed in the dune area north of the town of Ovar (northernmost part of the study area) and in alluvial areas adjacent to the main rivers. Towards the south of Ovar, the groundwater chemical composition is dominated by calcium and bicarbonate ions as a result of calcite dissolution and cation-exchange processes. The low chloride content for the whole aquifer (39 mg L^{-1}) and the relatively low salinities (median values for SEC are about 400 µS cm^{-1}) confirm the general presence of freshwater in the aquifer. However, locally there may be some boreholes pumping waters with chloride contents and salinities, greater than 100 mg L^{-1} and 900 µS cm^{-1} respectively, in areas where over-pumping produces saltwater up-coning.

Calcium-sulphate type waters are the dominant type in the old beaches and terrace deposits aquifer and reflect the impact of agricultural activities. Increased sulphate concentrations coincide with increased potassium and in general with the presence of nitrate (average value for NO_3–N in the

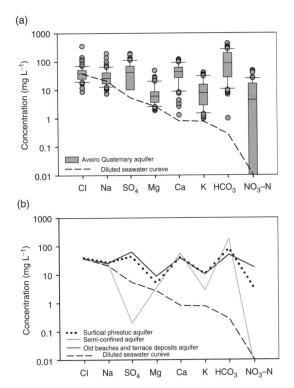

Fig. 11.4 Range of concentrations for the major elements in the Aveiro Quaternary aquifer; (a) box plots summarise the distribution of data set for the Aveiro Quaternary aquifer as a whole; (b) comparison between the median concentration of major elements in each hydrogeological unit.

aquifer is 18 mg L^{-1}). Groundwater specific electrical conductivity is usually greater than 400 µS cm^{-1} with a maximum value over 1100 µS cm^{-1}, which is significantly higher than background values for the aquifer. The semi-confined aquifer has calcium-bicarbonate type waters. However, in the northern part of the study area, close to the village of Torreira, there are some boreholes that showed a significant increase in chloride concentration and total hardness higher than typical aquifer background. These higher

salinity boreholes are usually located in areas close to the sea or to the Aveiro lagoon where groundwater discharge is almost insignificant, and where saltwater intrusion may occur locally.

For a better interpretation of the hydrochemical results, several chemical species and parameters for each hydrogeological unit are plotted downgradient along W–E cross sections in Fig. 11.5. Groundwater chemical evolution shows an increase in pH, calcium, bicarbonate (and strontium, but not shown) concentrations on the semi-confined aquifer when compared to the two shallow aquifers due to the prevalence of closed system conditions with respect to CO_2. Prevailing closed system conditions in the semi-confined aquifer may lead to calcite equilibrium (saturation indexes for calcite ~ 0) and is the justification for the higher $\delta^{13}C$ values observed in this hydrogeological unit (Fig. 11.6).

Sodium and chloride concentrations remain more or less constant along the three flowpaths and are not in agreement with the observed increase in SEC downgradient, which is in great part produced by the other major ions in solution. Silica also shows an incremental increase from the unconfined units to the semi-confined unit which may indicate not only an increase in the residence time (confirmed by tritium data) but also that silicate weathering is an active geochemical process in the aquifer.

To summarise, it could be said that despite its vulnerability and signs of human impact, the natural baseline composition of the Aveiro Quaternary aquifer system is of fresh calcium-bicarbonate type waters, with chloride concentrations of less than 40 mg L^{-1} and total salinities less than 400 µS cm^{-1}. The areas with sodium-chloride type waters, but low mineralisation, correspond to recent

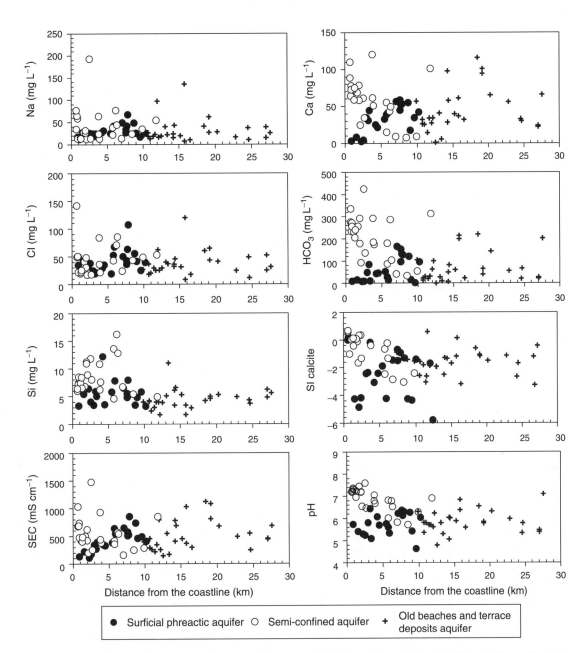

Fig. 11.5 Downgradient evolution of some of the principal chemical species in the three hydrogeological units of the Aveiro Quaternary aquifer. The results show a different geochemical behaviour between the two shallow unconfined units and the deeper semi-confined unit due to the system closed conditions.

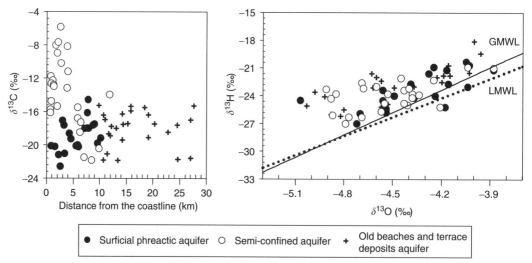

Fig. 11.6 Stable isotopic composition of the three hydrogeological units of the Aveiro Quaternary aquifer.

infiltrated groundwater, the composition of which reflects the rainfall composition and the influence of marine aerosol.

Redox patterns and their impact on groundwater chemistry

The redox characteristics of an aquifer may determine the geochemical and biological behaviour of several elements and influence their speciation, mobility, persistence and toxicity in the environment. Oxygen, as the principal oxidising agent, and organic matter, as a major reducing agent, have a significant role in controlling the extent of the redox processes and influence on the groundwater chemistry.

The three hydrogeological units of the Quaternary aquifer have distinct redox environments and active redox processes are responsible for the some of the most significant chemical variations observed in the semi-confined aquifer (Fig. 11.7). The two unconfined shallow aquifer units have oxidising waters with average Eh values over 300 mV and DO concentrations between 2 and 9 mg L^{-1}.

The semi-confined aquifer is distinguished from the two others because of the dominant anaerobic conditions (oxygen content in solution is below detection limit). The organic mud semi-confining layer acts as a reactive barrier enhancing redox processes and attenuating several potential contaminants. Organic matter decay starts under aerobic conditions but because of the lack of exchange with atmospheric O_2, oxygen is rapidly consumed, leading to anaerobic conditions and to the gradual decrease of Eh. The Eh values in the semi-confined aquifer indicate a reducing environment. Average Eh values for the aquifer are around 130 mV but in some deeper boreholes drilled close to the coast, groundwater Eh values may reach negative values (–40 mV).

Once groundwater becomes anoxic, available electron acceptors are consumed sequentially in the following order: NO_3–N > Fe(III) > SO_4 > CO_2 (methanogenesis). This leads initially to nitrate reduction and denitrification. Gradually, sulphate becomes the principal electron acceptor until concentrations of sulphate decrease below detection

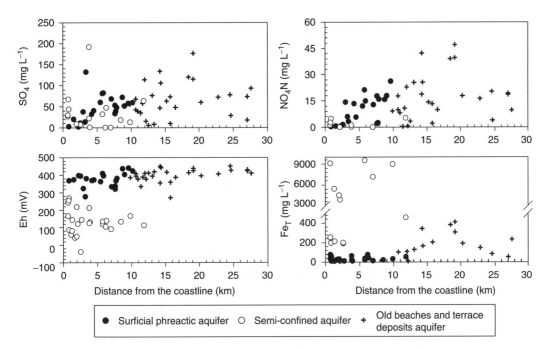

Fig. 11.7 Influence of redox conditions on the hydrochemistry of the Aveiro Quaternary aquifer. Samples collected in the semi-confined aquifer show clear redox patterns (lower Eh, sulphate and nitrate reduction, increase in total dissolved iron content).

limit and methanogenesis becomes an active process in the aquifer. Sulphate reduction in the semi-confined part of the aquifer is easily identified, not just by the low concentrations of sulphate in solution, often below detection limit, but also by the intense hydrogen sulphide smell.

Anoxic conditions induce nitrite and nitrate reduction and contribute to the protection of groundwater quality producing low nitrate waters. However, as soon as the groundwater becomes anoxic the concentrations of ferrous iron [Fe(II)] increase due to Fe(III) reduction and often create groundwater-quality constraints for human and industrial supply. Under reducing conditions, and pH values under 7.5, Fe(II) becomes soluble reaching concentrations over 6 mg L^{-1}. Like Fe(II), other redox sensitive chemical species such as Mn^{2+}, that were unstable

under aerobic conditions, become stable and their concentrations increase significantly in groundwater.

Groundwater results for the semi-confined aquifer provide evidence that biodegradation reactions have gradually consumed all the electron acceptors available in the aquifer (oxygen, nitrate and sulphate) and inhibitory to methanogenic activity. Evidence of methanogenesis is present in the aquifer and further studies confirmed concentrations of methane gas in the aquifer as high as 9 mg L^{-1} in pre-selected samples.

Minor and trace elements in the aquifer

An important number of minor and trace elements have been investigated in the aquifer to enrich the understanding of water chemical evolution. The concentrations of these elements are controlled in groundwater under natural conditions by atmospheric precipitation and weathering processes, and for some, their speciation, reactivity and mobility depends mostly on the groundwater redox conditions (Edmunds and Smedley 2000) and/or pH.

Most of the trace elements show a range of concentrations in each aquifer unit but the range of variation under baseline conditions does not often exceed one order of magnitude, and follows approximately the trend for the calculated diluted seawater curve. Due to the above described redox conditions, the semi-confined aquifer shows higher values for Fe, Mn, As and some other redox-sensitive chemical species. These higher concentrations are natural baseline values and may exceed current maximum admissible concentration (MAC) for human consumption.

Trace element concentrations in the shallow unconfined aquifers may also be affected by human activities. There is a potentially wide range of environmental distribution of trace elements in the region due to pesticide application, industrial activities, car exhausts, decorative materials, protective coatings, etc. that could be related to somewhat higher concentrations of trace elements such as zinc, copper, aluminium, cobalt, cadmium, lead, lanthanum and yttrium. However, the lower range of values determined should reflect mostly natural sources for these constituents.

11.5.3   Aveiro Cretaceous aquifer

The groundwater chemistry of Aveiro Cretaceous aquifer revealed an almost pristine groundwater body protected from contamination episodes by a thick confining layer of marly clays in two-thirds of its extent. The available geochemical data for the aquifer contributed for its division in three main hydrogeochemical zones with gradually different characteristics. The boundaries of these zones coincide approximately with redox boundaries in the aquifer system (Fig. 11.8). From the recharge area in the eastern part towards the coast we can identify:

1 *Modern groundwater* limited to the outcrop area of the aquifer under unconfined conditions (eastern part of the study area, 17–25 km away from coastline). These are recently recharged oxygenated groundwaters of predominantly $Ca-HCO_3$ hydrochemical facies, with pH values lower than 7.0 and water temperatures less than 20°C. Nitrate concentrations are generally low (median NO_3-N concentration is 2.5 mg L^{-1}) but some occasionally higher concentrations have been detected and reflect modern human impact. However, just two boreholes exceeded the limit of potability. Groundwater ages vary from recently recharged up to 7000 years BP waters that may occur in the western limit of this part of the aquifer, just 10 km from the outcrop area. Salinity is low with chloride median concentrations around 30 mg L^{-1} and SEC less than 650 μS cm^{-1}.

2 *Holocene to pre-industrial groundwaters* occur in the intermediate (5 to 17 km from coastline), confined parts of the aquifer. Redox potential and dissolved oxygen progressively decrease along the flow path with groundwater changing from oxidising to reducing conditions. These are groundwaters with either $Ca-HCO_3$ or $Na-HCO_3$ hydrochemical facies, pH values higher than 6.2 but lower than 8.3 and groundwater temperatures around 21.3°C. PCO_2 decreases in this part of the aquifer from 0.04 to 0.001 atm along the flow path. Background chloride

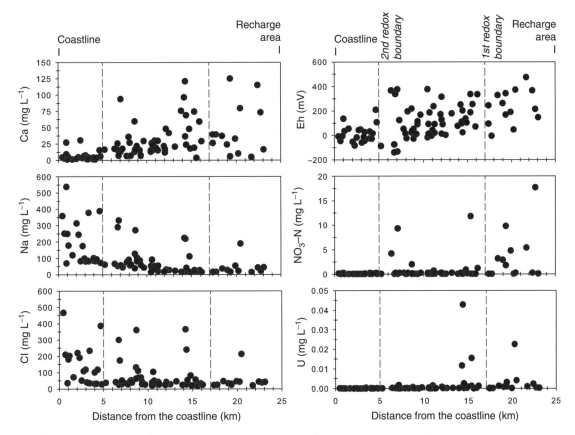

Fig. 11.8 Downgradient characterization of the hydrochemical evolution of the Aveiro Cretaceous aquifer with identification of major groundwater quality boundaries (dashed lines) due to redox and flow conditions.

concentrations increase to 40 mg L^{-1}, but SEC does not exceed 400 µS cm^{-1}. This is mainly unpolluted groundwater, enriched in ^{18}O compositions, and with modelled ages lying in the range of 3,500–18,000 years BP.

3 *Late Pleistocene–early Holocene groundwaters* have been identified in the deeper and confined part of the aquifer (0–5 km from coastline), well protected by the thick overlying aquitard. They are Na–HCO$_3$ or Na–Cl type waters, with pH in the range from 7.1 to 8.9, and average groundwater temperatures higher than 22°C, eventually reaching values higher than 30°C in the deepest part of the sedimentary basin (west of the town of Vagos). The negative Eh values reflect anaerobic (reducing) conditions and the water is unpolluted. The specific electrical conductance indicates that the confined groundwater in the Aveiro aquifer is fresh, with background chloride concentrations around 115 mg L^{-1}. Fresh groundwater with SEC less than 500 µS cm^{-1} and Cl$^-$ less than 40 mg L^{-1} is found down to a depth over 300 m below OD at less than 900 m from the present coastline.

The studied groundwaters have their origin, in part, associated with the infiltration of rainwater of Atlantic origin (which is essentially diluted seawater) and the gradual flushing of older formation waters (old seawater trapped in the sediments). Infiltrating rainwater percolates through the unsaturated zone to reach the underlying groundwater body, reacting with silicate and carbonate minerals in a system initially open to CO_2. Because carbonate dissolution kinetics are much faster than the silicate weathering reactions, a fresh groundwater dominated by Ca^{2+} and HCO_3^- initially prevails. Average Ca^{2+} concentrations in this part of the aquifer are about 33 mg L^{-1} while Na^+ concentration is less than 14 mg L^{-1} (Fig. 11.8).

As the water moves downward along the flow path into the deeper and confined part of the aquifer, Ca^{2+} is taken up from groundwater by cation exchange surfaces that form part of the aquifer matrix (mainly clay minerals), in return for Na^+. In the areas closer to the coast, average calcium concentrations are quite small, around 4 mg L^{-1}, while Na^+ becomes the dominant cation with concentrations higher than 110 mg L^{-1}. With the exception of the areas within the outcrop area, where calcium is the most abundant cation, sodium is the dominant solute in the aquifer. Its concentration varies from a minimum of 14 mg L^{-1} to a maximum of 400 mg L^{-1} with increasing residence time, and the same trend is observed for sodium/calcium molar ratios.

The natural hydrogeochemical evolution reflects these processes trending from $Ca–HCO_3$ type waters near the recharge area to $Na–HCO_3$ or $Na–Cl$ type waters in the areas close to the coast. The evolution of the groundwater towards an evolved $Na–HCO_3$ water type indicates that ion exchange and calcite dissolution are dominant geochemical processes in the aquifer.

At present, the baseline groundwater in the Aveiro Cretaceous aquifer is fresh and of remarkably low salinity (below 40 mg L^{-1} Cl^-), considering that it was, at least in part, of marine origin. The low chloride concentrations (minimum 16 mg L^{-1}, which is about 4 times the mean rainfall value) over most of the aquifer are consistent with the main input to groundwaters being derived from atmospheric inputs after allowing for evapotranspiration, and the remaining Cl^- is little more than 1–2% of that probably derived from (marine) formation water.

The dilute nature of the Aveiro groundwaters thus indicates previous freshening of the coastal aquifer since at least the late Pleistocene or even earlier. Higher hydraulic gradients (~0.004) during the last glacial maximum (LGM) or even before the LGM, imposed by a sea level lowered by approximately 130–140 m compared to present day, would have accelerated the complete refreshing of the aquifer, with fresh water flushing the original formation water.

The analysis of chemical patterns of dissolved species along evolutionary groundwater flow paths was also performed using elemental molar ratios. For the Aveiro Cretaceous aquifer, Br/Cl ratios were useful for the reconstruction of groundwater origin (Fig. 11.9). Late Pleistocene and early Holocene waters show approximately constant values for the Br/Cl ratio and samples plot on a mixing line between the rainwater and seawater ($[Br^-]/[Cl^-] = 0.0016$). Recent recharged waters have a Br/Cl ratio slightly enriched in bromine over the palaeo waters that could reflect the anthropogenic influence. The concentrations of sodium are not

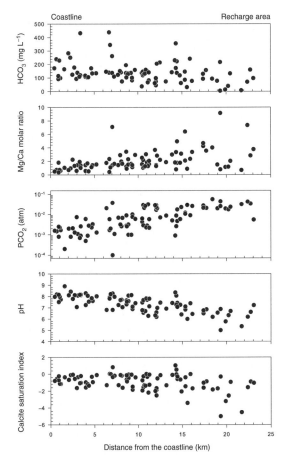

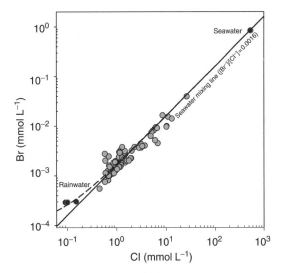

Fig. 11.10 Relation between Br⁻ and Cl⁻ content in the groundwater samples. Also plotted are Br/Cl molar ratios in seawater and rainfall as well as the seawater mixing line (in black) and the rainwater line (dashed).

Fig. 11.9 Groundwater pH, PCO₂, Mg/Ca molar ratio, HCO₃, and calcite saturation index evolution downgradient the Aveiro Cretaceous aquifer system.

totally compensated by chloride, as would be expected if seawater and aerosol inputs were the only sources of sodium, and Na/Cl molar ratio is not consistent with a simple fresh-/seawater mixing process. Sodium-feldspar (albite) dissolution is likely to be one of the geochemical processes responsible for liberating extra Na⁺ to solution, leading to Na/Cl ratios higher than one and high ratios of Na⁺ relative to Ca²⁺.

It is well known that aquifer silicate minerals can be a source of solutes through dissolution, and a sink via the processes of adsorption and precipitation. Taking into account the mineralogy of the aquifer matrix and the long residence times, alumino-silicate weathering and carbonate mineral dissolution are likely to be the other processes responsible for cation variation in the Aveiro aquifer. The observed decrease of PCO₂ along the flow path confirms that mineral reaction is taking place (Fig. 11.10).

The increase of Mg/Ca ratios in the initial 10 km of the aquifer flow path is probably the result of dolomite dissolution or the incongruent dissolution of impure calcite. The observed depletion of calcium and magnesium relative to bicarbonate ([Ca + Mg]/[HCO₃] < 0.5) in the last 15 km of the aquifer is due to the cation exchange phenomenon.

For a mainly silicate aquifer system, ground water silica concentrations (~5.5 mg L^{-1}) are remarkably uniform for the whole aquifer. Silica is being released into solution by alumino-silicate weathering but with a proportion being retained as secondary minerals (kaolinite and smectite); any excess of silica is likely to be removed initially as colloidal silica to maintain the apparent equilibrium with chalcedony or quartz.

The molar ratio of Si to Na$^+$ decreases along the flow path, from values around 0.40 in the recharge area to values of 0.02 in the deeper part of the aquifer. Considering just the sodium due to silicate reaction and neglecting Na derived from seawater (either rain or connate water), a ratio of Si to Na$^+$ still lower than two is derived. The weathering of primary minerals to kaolinite or gibbsite cannot generate that value; therefore a 2:1 silicate such as smectite must be forming. This is consistent with the clay minerals observed in the mineral assemblage.

11.5.3.1 Redox reactions in the Cretaceous multilayer aquifer

The natural redox conditions in the Aveiro Cretaceous multilayer aquifer change along the flow path, becoming gradually anoxic and affecting the solubility and transport of some major and minor solutes in groundwater. Decreasing Eh, NO$_3$ and U groundwater concentrations along the main aquifer flow path were used to indicate the presence of a redox boundary, which was observed close to the unconfined-confined boundary (Fig. 11.8). Oxidising conditions may generally be recognised by Eh values ≥300 mV and reducing groundwaters below 100 mV, with some of the most evolved groundwaters in the deeper part of the aquifer with negative measured Eh values.

Concentrations of dissolved oxygen in the unconfined part are commonly less than 6 mg L^{-1}, lower than atmospheric values and indicating some reaction following infiltration. Deeper confined groundwaters become gradually anaerobic due to the presence of traces of organic matter and/or Fe(II) and sulphide minerals at depth. Values of DO less than 0.2 mg L^{-1} have been determined throughout the aquifer in confined sections.

Nitrate concentrations in groundwater are only detected in the unconfined part of the aquifer. To the west of the redox boundary, NO$_3$–N concentrations are below detection limit, which confirms the existence of the redox boundary. This coincides with the drop in Eh to around +100 mV. Uranium, another good indicator of redox conditions, is also below the detection limit beyond the aquifer outcrop area.

Total dissolved Fe concentrations (≤ 0.45 µm fraction) are low or below detection limit (<5 µg L^{-1}) in unconfined oxygenated waters, but increase in the anaerobic confined aquifer to more than 1 mg L^{-1}. The same trend is observed for Mn, which reaches concentrations greater than 0.05 mg L^{-1} under anaerobic conditions. These high concentrations of Fe and Mn coincide with sulphate concentrations in the aquifer less than 50 mg L^{-1}, and an increase in As concentrations to greater than 0.01 mg L^{-1}.

11.5.3.2 Minor and trace elements in the Cretaceous aquifer

A range of minor and trace elements have been investigated in the aquifer to establish baseline compositions and to further assist the understanding of water chemical evolution. Results are shown in Fig. 11.11 and in Table 11.5. High concentrations of fluoride, iron, manganese and barium are evident in

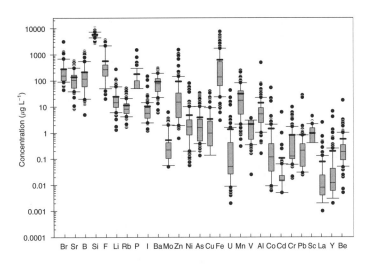

Fig. 11.11 Range of minor and trace element concentrations in the Aveiro Cretaceous aquifer groundwaters. The upper and lower quartiles of the data define the top and bottom of 'box', and horizontal line segments inside the box portray the median and the mean. The lower and upper inner fence values are the lowest and largest observed values, respectively, provided that they are less than 1.5 times the interquartile range.

some parts of the aquifer, exceeding occasionally the MAC values for human consumption. In addition to fluoride, the other halogen elements also increase with residence time due to the mixing with old formation water trapped in the clay minerals present in the aquifer matrix. The combined use of isotopic data with inorganic analysis was used as an important geochemical constraint to clarify the origin of the observed concentrations (Fig. 11.12).

In contrast, lithium concentrations which may often be used as a residence time indicator (Edmunds and Smedley 2000) do not show any increase along the line of section and do not follow the geochemical behaviour of sodium. Boron is also relatively high especially in the most cation-exchanged groundwaters (close to the coast) possibly as a result of mixing with old formation waters and further water–rock interaction. However, boron concentrations in the Aveiro Cretaceous aquifer are still less than the EC maximum permissible concentrations in drinking water (300 μg L^{-1}).

Most of the other minor and trace elements are present in concentrations very close to the detection limit, which is indicative of the low reactivity of the Aveiro Cretaceous aquifer.

11.5.3.3 Groundwater-quality stratification

The use of conventional methods of groundwater sampling from pumping multi-screened boreholes in the Aveiro Cretaceous multilayer aquifer inevitably leads to water mixing between the different aquifer layers, making an understanding of the water-quality patterns more difficult, and potentially leading to misinterpretation. To minimise this problem, the present study included not just the study of groundwater quality spatial variability but also the analysis of groundwater vertical stratification arising from geological layering (Buckley 2001; Condesso de Melo 2002).

The results of a geophysical logging and depth sampling campaign were used to characterise, from the geochemical point of view, each aquifer layer and confirmed that the Micaceous Sandstone Formation and the top part of the Upper Sandstone Formation are

Table 11.5 Overall summary statistics and baseline values for the Aveiro Cretaceous aquifer.

Parameter	Units	Min.	Max.	Median	Mean, μ	97.7 %tile	n
pH		5.0	9.2	7.3	7.3	8.4	90
T	°C	16.7	30.8	20.7	20.9	27.1	88
DOC	mg L^{-1}	0.1	11.0	1.5	2.3	9.0	29
SEC	μS cm^{-1}	141	1850	431	573	1710	88
DO	mg L^{-1}	<0.1	15.6	1.1	2.1	11.2	86
Eh	mV	−140	474	91	116	390	88
Na	mg L^{-1}	13.5	538	54.5	91.7	377	90
K	mg L^{-1}	2.0	27.4	7.9	8.3	19	90
Ca	mg L^{-1}	1.2	125	17.6	26.7	114	90
Mg	mg L^{-1}	0.5	45.1	6.1	7.9	28	90
Si	mg L^{-1}	2.9	10.4	5.7	5.8	10	90
Cl	mg L^{-1}	15.9	941	40.6	88.9	386	90
SO$_4$	mg L^{-1}	0.8	325	46.1	61.1	234	90
HCO$_3$	mg L^{-1}	4	438	131	140	355	90
NO$_3$–N	mg L^{-1}	<0.5	17.7	<0.02	0.76	9.78	90
NO$_2$–N	mg L^{-1}	<0.4	0.01	<0.003	<0.03	0.01	64
NH$_4$–N	mg L^{-1}	<0.02	1.77	<0.01	0.02	0.21	85
PO$_4$–P	mg L^{-1}	<3	0.8	<0.2	<0.23	0.40	90
Ag	μg L^{-1}	<0.2	0.06	<0.04	<0.09	0.04	90
Al	μg L^{-1}	<2	493	5.27	13.84	69.62	90
As	μg L^{-1}	<0.76	34.10	1.58	3.70	26.97	90
B	μg L^{-1}	<10	1350	111.5	200.8	1200	90
Ba	μg L^{-1}	6	305	74.92	91.08	262.5	90
Be	μg L^{-1}	<0.1	17.39	0.18	0.55	2.55	90
Bi	μg L^{-1}	<0.17	0.24	<0.02	<0.06	0.03	90
Br	μg L^{-1}	44	3150	153.5	283.3	1225	90
Cd	μg L^{-1}	<0.04	6.23	<0.01	0.08	0.33	90
Ce	μg L^{-1}	<0.06	6.84	0.01	0.16	1.97	90
Co	μg L^{-1}	<0.02	52.49	0.12	1.42	15.66	90
Cr	μg L^{-1}	<0.5	9.98	0.09	0.68	5.96	90
Cs	μg L^{-1}	<0.04	2.49	0.20	0.33	1.53	90
Cu	μg L^{-1}	<0.28	38.65	0.97	3.34	22.65	90
F	μg L^{-1}	20	3700	270.0	563.9	3486	90
Fe	μg L^{-1}	<6	7530	135.5	625.4	4526	90
Ga	μg L^{-1}	<0.04	0.28	0.02	0.02	0.12	90
Ge	μg L^{-1}	<0.05	0.45	0.14	0.16	0.35	90
I	μg L^{-1}	1.20	147	5.67	10.36	59.21	90
La	μg L^{-1}	<0.01	2.51	0.01	0.07	0.72	90
Li	μg L^{-1}	1.31	274	14.37	24.68	106.9	90

Table 11.5 *(Continued)*

Parameter	Units	Min.	Max.	Median	Mean, μ	97.7 %ile	n
Mn	μg L^{-1}	0.30	242	17.25	31.19	141.6	90
Mo	μg L^{-1}	<0.21	6.2	0.22	0.45	3.10	90
Ni	μg L^{-1}	<0.3	78.52	1.67	4.75	44.59	90
Pb	μg L^{-1}	<0.1	27.33	0.21	0.81	1.80	90
Rb	μg L^{-1}	1.64	51.67	8.22	11.36	44.23	90
Sb	μg L^{-1}	<1.22	0.19	<0.05	<0.38	0.15	90
Sc	μg L^{-1}	<1	4.39	<0.8	0.19	2.45	90
Se	μg L^{-1}	<2.3	24.05	0.82	1.63	10.04	90
Sr	μg L^{-1}	8.21	576	104.4	139.1	498.2	90
Th	μg L^{-1}	<0.13	0.40	0.004	<0.03	0.04	90
Ti	μg L^{-1}	<0.04	14.1	0.08	0.80	3.42	90
U	μg L^{-1}	<0.07	42.8	0.05	1.35	15.26	90
V	μg L^{-1}	<6	3.6	<6	<3.48	1.72	90
Y	μg L^{-1}	<0.01	6.6	0.01	0.19	1.30	90
Zn	μg L^{-1}	<1.23	1560	14.98	91.69	1053	90
Zr	μg L^{-1}	<6	179	<0.01	0.45	16.77	90
^{18}O	‰	−5.6	−4.1	−4.7	−4.7	−4.2	78
^{2}H	‰	−33.0	−20.5	−25.6	−25.7	−21.7	78
^{13}C	‰	−25.0	−7.3	−12.9	−13.4	−8.9	78
^{14}C	pmc	0.9	73.3	7.1	16.5	62.1	24

the best aquifer layers, providing the lowest salinities for the aquifer. The Micaceous Sandstone Formation is indeed contributing most of the water supplied by the aquifer, in a ratio for the transmitted flow, which could be as high as 9 : 1 of the abstracted volume. This implies that samples taken from the discharge during pumping conditions can be assumed to correspond in great part (~90%) to that from the Micaceous Sandstone Formation. For this reason, and although each aquifer layer has a characteristic chemical and isotopic composition, the chemical composition of the flow samples is dominated by the chemistry of the Micaceous Sandstone Formation.

However, the data arising from the study of the aquifer vertical stratification also showed that the Micaceous Sandstone Formation is highly vulnerable to groundwater contamination because of its high transmissivities and depleted hydraulic heads. In the boreholes where leaks at joints in the blank casing were producing inflow of high salinity water from the base of the Quaternary aquifer, it was observed that this water was leaving the borehole through the screens placed in front of the Micaceous Sandstone Formation. This problem was detected in several boreholes used for monitoring purposes in the region and may have some influence on the determined groundwater baseline quality of the aquifer.

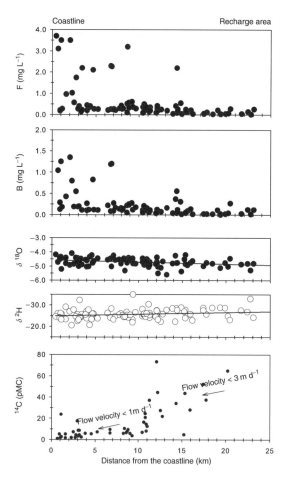

Fig. 11.12 Use of isotopic data as a geochemical constraint for baseline definition. The increase observed in B and F concentrations are due to water-rock interaction in groundwaters with the longest residence times.

11.6 Summary of the baseline quality

A statistical summary of the chemical characteristics of each aquifer is given in Tables 11.4 and 11.5, which outline conveniently the most important features of experimental distribution of data population, providing a selection of representative central and

extreme values, and a measure of the spread of the observations in the data-set.

The baseline quality of the groundwater in the Aveiro Quaternary and Cretaceous aquifers is represented by a wide range of concentrations as shown plotted on cumulative frequency plots (Fig. 11.13). It varies spatially along the main flow paths within the aquifer systems, and also varies with depth in the different aquifer units. Isotopes have proven to be a useful tool to distinguish in these ranges of concentrations that are due to natural geochemical evolution, from what is anthropogenic impacted.

The Aveiro Quaternary aquifer is in some areas impacted by human and industrial activities which influence groundwater hydrochemistry, requiring the use of a pre-selection method for the samples to calculate baseline and discard the anthropogenic influence. In shallow unconfined groundwaters with rapid turnover times (<1 yr) and in areas where there is very little or no agricultural practices, the natural baseline composition of the Aveiro Quaternary aquifer may be defined as predominantly of Na–Cl facies with low to moderate salinities (Cl < 40 mg L^{-1}; SEC < 300 µS cm^{-1}). This type of groundwater occurs in the dune areas in the north and in the Quaternary outcrops located at the very eastern part of the study area. The composition reflects: (1) the local rainfall composition; (2) CO_2 dissolution during rainfall infiltration; and (3) concentration due to evapotranspiration in the upper parts of the aquifer (some eight times that of rainfall composition). The short groundwater residence time limits the possibility of important contributions from other geochemical processes. Nitrate concentrations are generally low, but some occasional higher concentrations have been detected and reflect

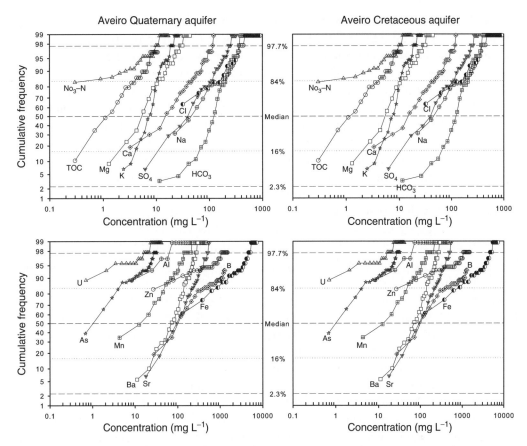

Fig. 11.13 Cumulative frequency diagrams for concentrations of major, minor and trace elements in the Aveiro Quaternary and Cretaceous aquifers.

modern human impact (>5 mg L^{-1}). Median concentrations for minor and trace elements are within the range of baseline values.

Groundwaters of Ca–SO$_4$ facies, with moderate salinities (SEC median values around 450 µS cm^{-1}) also occur in the shallow unconfined aquifer and are related to intensive agricultural activity. The relative increase in calcium can be attributed to calcite dissolution and cation exchange, the major baseline geochemical processes. High sulphate concentrations (>40 mg L^{-1}) are usually associated with high nitrate concentrations (>5 mg L^{-1}) and cannot be considered as baseline. Both are derived from the extensive application of fertilisers and manure used in agriculture. The small observed concentrations of minor and trace elements should reflect mostly natural sources for the constituents but eventually, some higher concentrations of trace elements such as zinc, copper, aluminium, cobalt, cadmium, lead, lanthanum and yttrium may be related to human impact (Table 11.5).

Ca–HCO$_3$ facies with low to moderate salinities (Cl < 40 mg L^{-1}; SEC < 500 µS cm^{-1}) are observed in the semi-confined aquifer, where three major geochemical processes contribute to groundwater baseline composition: (1) calcite dissolution; (2) cation exchange and (3) nitrate and sulphate reduction. Prevailing reducing conditions in the aquifer are responsible for nitrate, nitrite, and sulphate reduction producing baseline concentrations below detection limit. Sulphate reduction by organic matter contributes also to the high bicarbonate baseline concentrations (>200 mg L^{-1}). The aquifer reducing capacity produces high Fe^{2+} (>1 mg L^{-1}) and Mn^{2+} (0.1 mg L^{-1}) in solution which also represent baseline, despite posing limitations for human consumption. The presence of methane results from natural geochemical processes and should be considered as part of the baseline composition. The small concentrations observed for trace elements (Table 11.4) should reflect mostly natural sources for the constituents.

The present day groundwater composition in the confined part of the Aveiro Cretaceous aquifer represents still baseline quality. The spatial trends are a consequence of the geology, geochemical reactions, rainfall chemistry and palaeoclimate.

Identification of hydrochemical facies in the Aveiro Cretaceous aquifer system was based on the distribution of water types and used to divide the Aveiro Cretaceous aquifer into three main zones with gradually different hydrogeochemical characteristics along the flow path. A redox boundary was defined, corresponding approximately with the unconfined-confined boundary in the aquifer. This boundary also separates pristine groundwaters observed in the confined part of the aquifer from the waters with signs of human impact in the recharge area.

The studied groundwaters have their origin in part associated with the infiltration of rainwater of Atlantic origin (which is essentially diluted seawater) and the gradual flushing of older formation waters.

For an aquifer with residence times in excess of 30 ka, the Aveiro Cretaceous aquifer slow overall water–rock interactions, resulting from the mainly siliciclastic composition of the aquifer sediments and from a very low cation exchange capacity. Silicate weathering, calcite dissolution and cation exchange in the initial phases of seawater flushing are considered to be the processes responsible for the gradual groundwater chemical composition and were confirmed by the use of a reaction-path geochemical model (see Postma *et al.* Chapter 4 this volume).

The improved understanding of both aquifer systems resulting from these hydrogeological and hydrogeochemical investigations provides an important basis for present day and future water management in the region. The results will be of help to the Regional Water Authorities and Water Supply Companies in making decisions about the protection of the groundwater resources. This study also provides a consistent description of current water quality and baseline concentrations in the aquifers. This information may be used to establish the aquifer reference values when defining a groundwater-quality monitoring network for the aquifer: the major factors that affect the observed water-quality conditions and trends were identified, described, and explained and this is very important for monitoring purposes. Some of the natural baseline concentrations (e.g. Fe, Mn, F) exceed the EC MAC for human consumption, but they are related to natural geochemical processes and not the result of the impact of human activity.

The subdivision of the aquifer systems has shown that there are distinct water chemistries and vulnerabilities, calling attention to the necessity of recognising the natural heterogeneity (spatial and vertical) of the aquifer when defining background values and defining water bodies for management purposes. For management purposes it is pointed out that changes in the baseline chemistry are slow and monitoring for baseline properties need only be carried out, for example, on an annual basis and would benefit from the integrated use of isotopic methods as constraints for baseline characterisation and evolution.

References

Barbosa, B.P. (1981) Carta geológica de Portugal na escala de 1/50 000. Notícia explicativa da folha 16-C – Vagos. Direcção-Geral de Geologia e Minas, Serviços Geológicos de Portugal, Lisboa.

Buckley, D.K. (2001) Geophysical logging of 7 observation boreholes near Aveiro, Portugal. British Geological Survey Report CR/01/023, Keyworth, 20 pp.

Carreira, P.M.M., Soares, A.M.M., Marques da Silva, M.A. *et al.* (1996) Application of environmental isotope methods in assessing groundwater dynamics of an intensively exploited coastal aquifer in Portugal. In: *Isotopes in Water Resources Management.* Vol. 2, IAEA Symposium 336, March, 1995, Vienna, pp. 45–58.

Carreira Paquete, P.M.M. (1998) Paleoáguas de Aveiro. PhD thesis, Universidade de Aveiro, Portugal, 377 pp.

Condesso de Melo, M.T. (2002) Flow and hydrogeochemical mass transport model of the Aveiro Cretaceous multilayer aquifer (Portugal). PhD thesis. Universidade de Aveiro, 203 pp.

Condesso de Melo, M.T., Carreira Paquete, P.M.M. and Marques da Silva, M.A. 2001. Evolution of the Aveiro Cretaceous aquifer (NW Portugal) during the Late Pleistocene and present day: evidence from chemical and isotopic data. In: Edmunds, W.M. and Milne, C.J. (eds) *Palaeowaters in Coastal Europe: Evolution of Groundwater Since the Late Pleistocene.* Geological Society, London, Special Publications, 189, pp. 139–154.

Edmunds, W.M. and Smedley, P.L. (2000) Residence time indicators in groundwater: The East Midlands Triassic Sandstone aquifer. *Applied Geochemistry* 15, 737–52.

Ferreira, P.L.O. (1995) Hidrogeologia do Quaternário da região Norte da ria de Aveiro. Tese de Mestrado. Universidade de Aveiro, Portugal, 102 pp.

Gonçalves, A.C. and Vozone, M.L. (1999) Avaliação da qualidade de água para o consumo humano no concelho da Murtosa. Projecto 5°. Ano de Licenciatura. Departamento de Geociências, Universidade de Aveiro.

Gonfiantini, R. (1978) Standards for stable isotope measurements in natural compounds. *Nature* 271, 534–6.

Lauverjat, J., Martins de Carvalho, J. and Marques da Silva, M.A. (1983) Contribuição para o estudo hidrogeológico da região de Aveiro. *Boletin Soc. Geol.* (Portugal) 24, 295–304.

Leitão, T.B.E. (1996) Metodologia para a reabilitação de aquíferos poluídos. Tese de Doutoramento. Faculdade de Ciências da Universidade de Lisboa.

Marques da Silva, M.A. (1990) Hidrogeología del sistema multiacuífero Cretácico del Bajo Vouga – Aveiro (Portugal). PhD thesis, Universidad de Barcelona, Spain, 436 pp.

Oliveira, T.I.F. (1997) Capacidade de troca catiónica no Cretácico de Aveiro e sua influência no quimismo da água. MSc thesis, Universidade de Aveiro, Portugal, 132 pp.

Peixinho de Cristo, F. (1985) Estudo hidrogeológico do sistema aquífero do Baixo Vouga. Direcção-Geral dos Recursos e Aproveitamentos Hidráulicos, Divisão de Geohidrologia, Coimbra, 57 pp.

Peixinho de Cristo, F. (1992) Plano regional de ordenamento do território (PROT) do Centro Litoral. Recursos Naturais. Sistemas Aquíferos. Comissão de Coordenação da Região Centro. Ministério do Planeamento e da Administração do Território.

Rocha, F.T. (1993) Argilas Aplicadas a Estudos Litoestratigráficos e Paleoambientais na Bacia Sedimentar de Aveiro. PhD thesis, Universidade de Aveiro, Portugal, 399 pp.

Salem, O., Visser, J.M., Deay, M. and Gonfiantini, R. (1980) Groundwater flow patterns in the western Lybian Arab Jamahitiya evaluated from isotope data. In: *Arid Zone Hydrology: Investigations with Isotope Techniques*. International Atomic Energy Agency, Vienna, pp. 165–79.

Saraiva, M.P.S., Barradas, J. and Marques da Silva, M.A. (1983) Aquífero Cretácico de Aveiro – subsídios para a sua caracterização hidrogeológica. Hidrogeologia y Recursos Hidraulicos, VII, AEHS, Madrid, pp. 41–9.

Teixeira, C. and Zbyszewski (1976) Carta geológica de Portugal na escala de 1/50 000. Notícia explicativa da folha 16A – Aveiro. Direcção-Geral de Minas e Serviços Geológicos, Serviços Geológicos de Portugal, Lisboa.

Zbyszewski, G. (1963) Considerações acerca das possibilidades de se realizar uma nova captação de água para o abastecimento de Aveiro. Grupo de Estudos dos Recursos Hídricos Subterrâneos da Beira Litoral, Montemor-o-Velho. Relatório interno, 6 pp.

Zbyszewski, G., Alves, A.M. and Chaves, J.B. (1972) Contribuição de algumas sondagens de pesquisa e captação de água para o conhecimento hidrogeológico da região de Aveiro. In: *Proceedings I Congresso Hispano-Luso-Americano de Geologia Económica*, Lisboa, pp. 793–805.

12 The Neogene Aquifer, Flanders, Belgium

M. COETSIERS AND K. WALRAEVENS

The Neogene deposits in Belgium form a Tertiary marine sand aquifer which has undergone extensive freshening and is strongly exploited for drinking water supply and other uses. As a result of the depositional conditions the sediments are calcareous in the west, but almost without calcite in the east. The natural baseline quality of groundwater is given by a range of values, which are controlled by rainfall inputs, the marine depositional environment and the lithological composition. The variations in baseline chemistry and the dominant geochemical processes operating in the Neogene Aquifer have been studied. The natural variations of 91 constituents in the groundwaters of the area are represented. The hydrochemistry of the groundwater is variable, but in general the water is weakly mineralised. Geological differences, like calcareous versus carbonate deficit, are reflected in the groundwater quality. The dominant control on groundwater chemistry is related to dissolution of silicates and carbonates, and to redox reactions. In the calcareous western part calcite dissolution causes an increase in Ca and HCO_3 concentrations and pH. In the east silicate dissolution is the main reaction. These regional variations must be taken into account when defining the local baseline for the aquifer. Redox reactions cause a zoning with depth in the occurrence of dissolved species including O_2, Fe, Mn, SO_4 and As.

12.1 The Neogene Aquifer as a major groundwater resource

Tertiary marine sand aquifers are widespread in Western Europe. These aquifers have undergone continuing freshening on their emergence as continental sediments. Some are still showing a strong influence of their marine origin, with appreciable salinities, and evidence of freshening cation-exchange processes, such as the Ledo–Paniselian Aquifer in northern Belgium (Walraevens et al. 2007). Preserved salinities may sometimes be so high, that the aquifer's exploitation is impaired. Others, such as the Neogene Aquifer, have been extensively freshened, and are being exploited on a wide scale. Besides, the marine origin of these sand aquifers is often reflected by original calcite content. Those, that have preserved their calcite, show groundwater qualities that may be comparable to limestone and chalk, in the case when calcite dissolution is the main process determining groundwater chemistry. An example is the more freshened part of the marine Valréas Miocene sandstone in Provence, France, with CaHCO$_3$-type groundwater (Huneau et al. 2001). In other cases, when the marine sands have been decalcified by extensive flushing, their silicic nature shows up.

The North Sea Basin has been the scene of shallow-marine detritic deposition throughout the Tertiary era (Ziegler 1982), with coastal and deltaic to continental deposits at the basin's edges. The continental deposits often exhibit low to absent calcite contents. This is found also in the case of the Miocene Ribe Formation Aquifer of Jylland, Denmark, where the inert quartz sand leads to very limited water–rock interaction (Hinsby et al. 2001).

The Neogene Aquifer is situated on the southern edge of the North Sea Basin, on the transition from continental conditions in the southeast, over deltaic and coastal to shallow-marine conditions in the northwest. It presents a beautiful example of an extensively freshened Tertiary marine sand aquifer, which is calcareous in its western part, but almost without calcite in the east.

The Neogene Aquifer is located in the northeast of Flanders, in the provinces of Antwerp and Limburg at the border with the Netherlands. The Neogene Aquifer constitutes the most extensive aquifer of Flanders, containing important drinking water resources. As a consequence, the aquifer is affected by significant water exploitation, especially in the region of Antwerp. In the province of Antwerp the PIDPA water company has about 26 pumping stations and produced in 2004 about 70 million m^3 drinking water from groundwater. The VMW drinking water company has large groundwater pumping stations in the Neogene Aquifer in the provinces of Limburg and Brabant. The VMW

has a permit to extract more than 18 million m^3 groundwater per year. Additionally large pumping occurs for industrial uses, resulting in a total of more than 270 million m^3 groundwater extraction. Groundwater is withdrawn mainly from the Miocene layers (Berchem and Diest Sands), while Pliocene sands are less pumped to avoid surface effects such as pollution. The Pliocene sands also contain relatively more clay leading to smaller extraction rates. As the thickness of the Miocene sands can be more than 100 m, large pumping rates are possible. Groundwater extraction from these Tertiary sands is mainly important in Flanders. In the Netherlands, more to the north, these sediments occur at greater depth and the salinity increases, which make this aquifer less suitable for groundwater abstraction. Most of the aquifer area is considered as very vulnerable (De Coster et al. 1986; Van Dyck et al. 1986). Agriculture is the most important land use and light industries are locally present. In the north of the region coalmines and metallurgic industries were operating in the past.

In this chapter the variations in baseline chemistry and the dominant geochemical processes operating in the Neogene Aquifer are assessed. The hydrochemistry of the groundwater is variable, but in general the water is weakly mineralised. Geological differences, especially the presence or absence of carbonates, are reflected in the groundwater quality. Natural reactions between infiltrating water and the sediments mainly determine groundwater quality in the Neogene Aquifer. The dominant control on groundwater chemistry is related to dissolution of silicates and carbonates and to redox reactions. Towards the northwest, calcite dissolution occurs, in contrast with the southeastern zone, where the sediments are decalcified.

12.2   Geology and hydrogeology of the Neogene Aquifer

The geology of the study area is represented on a map in Fig. 12.1(a). The sediments of the Campine Basin dip gently towards the north-northeast with a slope of about 1–2% (Wemaere & Marivoet 1995) and are disturbed by different faults. The Neogene deposits rest on the Boom Clay, which reaches a thickness of 60 m up to 130 m (Laga 1973). The late Oligocene Voort Sands and Eigenbilzen Sands are included with the Neogene Aquifer system. These deposits are composed of dark green glauconitic, clayey sands (Laga 1973; Marechal and Laga 1988; PIH 1995). The lower Miocene is represented by the Bolderberg and Berchem Sands. The Bolderberg Sands consist of a lower marine unit, with dark green glauconitic sands containing shells, and an upper continental unit, with white limonitic and lignitic sands (Laga 1973; PIH 1995). In the west the Bolderberg Sands are replaced by the marine Berchem Sands. The Berchem Sands can reach a thickness of about 75 m and are characterised by green, strongly glauconitic, slightly clayey sands with layers enriched with phosphate and shells. After the deposition of the lower Miocene sands, an erosive stage took place for about 5 Ma and formed a deep SW–NE orientated erosion trench through the Boom Clay (PIH 1995). In the deep parts of the trench the Dessel Sands were deposited. These sands consist of fine, micaceous, glauconitic and calcareous sands (Vandormael 1992; PIH 1995). Above and outside the trench, the strongly glauconitic Diest Sands were deposited. Phosphate concretions and limonite sandstones, formed by the oxidation of glauconite, are present.

The Pliocene sediments were deposited in shallow-marine conditions and contain fossils and variable amounts of glauconite. The lower Pliocene occurs as two chronostratigraphic equivalent deposits, namely the Kattendijk and Kasterlee Sands. The Kasterlee Sands are characterised by grey-green, micaceous and glauconitic fine sand (Laga 1973; Wemaere and Marivoet 1995). To the north and northwest the Kasterlee Sands pass gradually into the Kattendijk Sands. The Kattendijk Sands are dark green to green-grey, fine, slightly clayey, glauconitic sands (Laga 1973; Wemaere and Marivoet 1995). In contrast to the Kasterlee Sands, the Kattendijk Sands do contain shells. The upper Pliocene deposits change from fluviatile depositions in the east (Mol Sands), over coastal sands (Poederlee Sands), to shallow-marine sands (Lillo Sands) in the west. The Mol Sands consist of white quartz sand, sometimes lignitic and with layers of micaceous clay (Laga 1973; Vandormael 1992; PIH 1995). More towards the west the coastal Poederlee Sands occur, consisting of slightly glauconitic fine sands with small lenses of clay. The marine Lillo Sands are grey to grey–brown sands, which are clayey in the lower part and contain several shell layers (Laga 1973; PIH 1995). The top of the Lillo Sands contains no fossils and is decalcified.

The top of the aquifer system consists of Pleistocene deposits of the Brasschaat Sands and Merksplas Sands and the heterogeneous Campine Complex. The Brasschaat

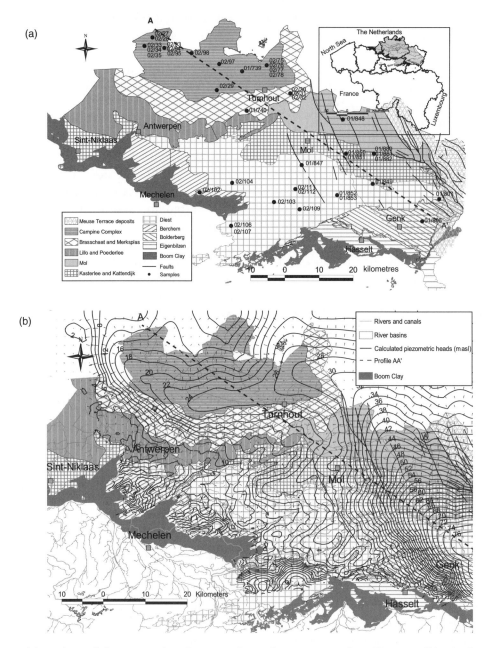

Fig. 12.1 (a) Geology of the area with indication of sampling points and profile A–A'; (b) calculated piezometric map for the Neogene Aquifer.

Sands are constituted of white–yellow, between fine and quartz sands with phosphate minerals vivianite and apatite (PIH 1995; Wemaere and Marivoet 1995). In the west the Merksplas Sands replace the Brasschaat Sands. The Merksplas Sands are constituted of coarse sands with shells at the base (PIH 1995) and some local clay lenses. The Campine Complex consists mainly of fine, locally coarse sand with clay lenses forming sometimes a clay layer (PIH 1995; Wemaere and Marivoet 1995). During the late Pleistocene and Holocene loess deposition occurred.

In the northern part of the province of Limburg, post-Oligocene faults occur, which were formed during the development of the Roer graben, the northwestern branch of the Rhine graben. The alpine earth movements, during the Upper Oligocene, caused a strong fault-connected subsidence of the Roer graben. This subsidence became stronger during the Mio- and Pliocene and is still active at present. The Feldbiss fault zone, which is partially located in Belgium, defines the western boundary of the Roer graben and the eastern boundary is defined by the Peel boundary fault (Demyttenaere and Laga 1988; Camelbeeck and Meghraoui 1996; Demanet et al. 2001).

In Table 12.1 a hydrogeological overview is given of the deposits of the Neogene Aquifer based on the HCOV code (Meyus et al. 2000). The Boom Clay is considered as the major aquitard in the region and it separates the Neogene Aquifer above from the lower-Rupelian aquifer below. Because the Neogene sediments contain some rather clayey layers, the aquifer can be subdivided into several parts. The Miocene Aquifer System is situated between the Lillo-Kasterlee semi-pervious layer and the Boom Clay Aquitard and contains the sandy formations of Kattendijk, Kasterlee, Diest, Berchem, Voort and Eigenbilzen (Beaufays et al. 1990). The Pleistocene and Pliocene Aquifer is delimited at the bottom by the combined effect of clay layers in the Lillo Sands in the western part and clay lenses or layers in the Kasterlee Sands in the east. It comprises the Brasschaat, Merksplas, Mol and Poederlee Sands (Beaufays et al. 1990) and the upper sandy layers of the Lillo and Kasterlee deposits. The Campine Clay Sand Complex forms the upper aquifer unit; it is isolated from the underlying formations by the Clay of Rijkevorsel (Beaufays et al. 1990). The Quaternary Aquifer System can be defined as the Quaternary terrace deposits of the Meuse and Rhine rivers and occurs only in the east. The combined Pleistocene and Pliocene Aquifer together with the Miocene Aquifer System are hereafter referred to as the 'Neogene Aquifer'. Patyn et al. (1989) concluded from hydrogeological observations that the Neogene sands, notwithstanding their lithological differences, behave as a single aquifer.

A regional groundwater flow model for the Neogene Aquifer was developed using the MODFLOW code. Groundwater recharge mainly takes place on the topographically elevated Campine Plateau in the east of the area (Fig. 12.1[b]). From the Campine Plateau groundwater flow is radial towards the Meuse Basin in the east and north and towards the Demer and Nete Basins, which are part of the larger Scheldt Basin, in the west and south. In the northwest of the study area a cuesta is present in the landscape where the clay layers of the Campine Complex are outcropping. This topographical elevation forms a second groundwater

Table 12.1 Schematisation of the Campine Aquifer System with indication of the reactive mineral phases (in italic).

Aquifer/Aquitard	West			East
Quaternary aquifer systems	Meuse and Rhine deposits			
Campine clay sand complex	Campine Clay Sand Complex			
Campine clay layer	Rijkevorsel Clay			
Pleistocene and Pliocene aquifer	Brasschaat Sands (*vivianite, apatite*) and/or Merksplas Sands (*calcite*)			
	Sandy top of Lillo (*calcite*)	Sands of Poederlee and/or Sandy top of Kasterlee (*calcite*)		Sands of Mol
Pliocene aquitard	Lillo Clay and/or top of Kattendijk Clay		Clayey layers in Kasterlee Sands	
Miocene aquifer	Kattendijk Sands and/or lower layer of Lillo Sands (*calcite, vivianite*)		Kasterlee Sands	
	Diest Sands (*vivianite*)			
		Dessel Sands (*calcite*)		
	Berchem Sands (*calcite, vivianite*)		Bolderberg Sands (*calcite*)	
			Voort Sands (*calcite*)	
			Eigenbilzen Sands	
Boom aquitard	Boom Clay			

divide between the Meuse and Scheldt river basins. Water infiltrating on the northern side of this cuesta will flow in northerly direction to the Meuse Basin and the Netherlands, while water recharged on the southern slope flows to the southern Scheldt Basin. In the south of the study area, where the Boom Clay is outcropping, a second cuesta dominates the topography. This older cuesta has several river breakthroughs and is more fragmented. Some small remnant hills, formed by the occurrence of iron sandstone layers in the Diest Sands, give rise to the occurrence of small recharge areas in the south. In general groundwater infiltrating in these southern recharge areas flows towards the north where it joins the Scheldt Basin. Between the northern cuesta of the Campine Complex and the southern cuesta of the Boom Clay a saddle shaped basin is present where groundwater outflow occurs by rivers and brooks.

The piezometric heads are characterised by a seasonal variation. In the aquifers of the Campine Complex and the Formation of Merksplas the amplitude of the fluctuations is about 2–3 m (Wemaere and Marivoet 1995). The Miocene Aquifer has similar seasonal fluctuations with amplitudes that rarely exceed 1 m.

12.3 Sampling and analytical methods

During December 2001 and January 2002, 32 samples were collected from 18 boreholes along a profile shown in Fig. 12.1. Sampling was performed after pH and electrical conductivity of the pumped water had stabilised. Thirty samples were taken from monitoring wells of the primary groundwater-monitoring network of the Flemish Environmental Administration (AMINAL). At these sampling points the screen length of the observation well was often not larger than 2 m, thus the pumped sample can be considered as representative for the depth at which it was taken and no mixing of waters with different chemistry occurred. The other two samples were taken from pumping wells belonging to industrial companies. In the frame of the Vlarebo project (Mahauden et al. 2002), eight additional samples were taken from monitoring wells of AMINAL. The location of these 40 sampled screens is indicated on Fig. 12.1. At some locations several screens on different depths were sampled.

The parameters pH, dissolved oxygen (DO) and redox potential (Eh) were measured in an anaerobic flow-through cell. Other on-site measurements included temperature and specific electrical conductivity (SEC). Where oxygen free measurement was not possible (samples taken at industrial companies), these parameters were measured immediately after sampling in the field. The TA-TAC titration was carried out in the field, since disturbance in carbonate equilibrium may occur by CO_2 contact or escape from the sample.

The samples were analysed at the Laboratory for Applied Geology and Hydrogeology (LTGH) of Ghent University and duplicates were sent to the British Geological Survey (BGS) laboratory. The samples taken for the LTGH were filtered in the field to remove colloidal or suspended particles. For the determination of total iron the sample was acidified with HCl to avoid the precipitation of iron in oxidising conditions. The organic cycle was fixated by means of chloroform in samples for the measurement of NO_3, NO_2, NH_4 and PO_4. To avoid the conversion of sulphide to sulphate, the sample was treated with zinc acetate. Cations were determined by atomic flame absorption spectrophotometry and trace elements were analysed by graphite furnace atomic absorption spectrophotometry. Molecular absorption spectrophotometry was used to determine the anions except for chloride and fluoride, which were respectively measured by chloridometer and ion-selective electrode. Stable isotope analysis was completed by BGS by mass spectrometry with results reported relative to the standards SMOW for δ^2H and $\delta^{18}O$ and PDB for $\delta^{13}C$. Ten samples were analysed for ^{14}C at the University of Bern.

12.4 Results

The results of the field measurements and the analytical results for the major ions are given in Table 12.2. A summary of the data for the isotopic measurements and trace elements is given in Table 12.3. This table shows the range and average of data as well as an upper concentration (defined as $+2\sigma$ or 97.7 percentile), which is used as a cut off for outlying data. The median is preferred to the mean as it is more robust and less affected by extreme values. Most groundwater samples from the Neogene Aquifer are characterised by low pH values (<6.5).

Table 12.2 Field parameters and major and minor element concentrations in the Neogene Aquifer.

Sample	Depth (m)	Geological formation	Date	T (°C)	pH	Eh (mV)	O_2 (mg L^{-1})	SEC (µS/ cm 25°C)	Na (mg L^{-1})
01/739	180	Berchem	Jul 01	12.7	7.2	−308	−2.9	489	83.8
01/740	68	Diest	Jul 01	10.4	7.3	−298	−7.9	329	15.2
01/806	20.13	Bolderberg	Dec 01	9.6	4.5	392	10	72	7.9
01/807	64.38	Bolderberg	Dec 01	5.4	5.6	377	5.9	12	5.3
01/847	25.52	Diest	Dec 01	9.6	5.9	202	0.1	62	9.3
01/848	157	Diest	Dec 01	10	6.9	265	0.1	62	5.4
01/849	42.12	Diest	Dec 01	8.9	6.2	245	0.1	61	5.2
01/850	10.49	Mol	Dec 01	10.7	6.1	375	0.2	161	16.8
01/851	25.49	Diest	Dec 01	9.8	4.8	291	0.2	66	4.2
01/852	8.98	Diest	Dec 01	9.9	5.6	205	0.3	117	13.8
01/853	48.98	Diest	Dec 01	9.6	5.9	285	0.2	127	27.1
01/880	15	Kasterlee	Dec 01	11.7	5.9	319	0.4	332	24.0
01/881	50	Kasterlee	Dec 01	10.7	5.4	302	0.3	145	16.6
01/882	70	Diest	Dec 01	10.5	6	235	0.2	114	13.9
02/102	24.8	Berchem	Feb 02	11	7.1	284	0.7	282	12.6
02/103	24.79	Diest	Jan 02	11.1	6	331	0.4	631	66.8
02/104	24.87	Berchem	Jan 02	10.2	5.7	307	0.4	132.2	7.4
02/106	9.21	Diest	Jan 02	9.5	6.7	269	5.8	150	21.0
02/107	34.87	Diest	Jan 02	10.4	6.8	272	1.4	133	7.7
02/109	58	Diest	Jan 02	9.6	5.8	299	0.8	84	8.1
02/111	10	Diest	Jan 02	10.6	5.5	269	0.8	109	13.8
02/112	20	Diest	Jan 02	10	5.4	250	0.7	120	15.9
02/27	11.5	Diest	Mar 02	10.6	5.2	211	0.1	228	9.3
02/28	117.5	Campine Complex	Mar 02	9.6	4.8	324	0.2	340	17.6
02/29	66	Diest	Mar 02	8.9	6.8	201	0.2	95	10.0
02/30	19.85	Campine Complex	Apr 02	8.9	5.6	311	0.2	77	13.6
02/31	81.76	Diest	Apr 02	10	6.5	171	0.2	87	6.9
02/2	209.74	Diest	Apr 02	12	6.6	283	0.2	130	27.7
02/33	15.9	Campine Complex	Jul 02	9.4	4.8	304	0.2	130	20.8
02/34	53.87	Merksplas	Jul 02	9.5	5.1	303	0.1	155	20.3
02/35	82.84	Diest	Jul 02	9.7	6.7	269	0.1	104	13.0
02/75	52.84	Brasschaat	Sep 02	9.8	5.8	162	0.1	117	9.5
02/76	69.81	Kasterlee	Sep 02	10	6.7	129	0.1	122	9.2
02/77	86.81	Diest	Sep 02	10.4	6.9	118	0.1	125	9.3
02/78	102.81	Diest	Sep 02	10.9	7	81	0.1	11	7.5
02/93	26.86	Campine Complex	Oct 02	10.1	5.3	216	0.1	94	10.9
02/94	53.84	Merksplas	Oct 02	10.4	5.4	247	0.2	67	9.0
02/95	70.82	Lillo	Oct 02	10.9	6.6	182	0.2	130	10.8
02/97	100	Diest	Jan 02	9.8	6.9	285	0.2	132	8.5
02/98	84.5	Diest	Jan 02	11.2	6.9	309	0.2	447	46.6

K (mg L^{-1})	Ca (mg L^{-1})	Mg (mg L^{-1})	Fe (mg L^{-1})	Mn (mg L^{-1})	NH$_4$ (mg L^{-1})	Cl (mg L^{-1})	SO$_4$ (mg L^{-1})	HCO$_3$ (mg L^{-1})	NO$_3$ (mg L^{-1})	NO$_2$ (mg L^{-1})	PO$_4$ (mg L^{-1})	SiO$_2$ (mg L^{-1})
11.5	23.1	29.6	29.6	0.01	0.79	12.8	2.8	439.2	0.96	0	0.13	15.5
2.0	100.0	4.0	4.0	0.01	0.01	32.0	79.8	212.9	2.1	0.01	0.06	16.7
2.0	6.5	2.4	2.4	0.04	0	11.4	10.9	12.2	6.81	0	0.1	12.2
1.2	5.3	0.5	0.5	0.02	0	4.0	0.0	26.8	1.1	0.02	0.04	15.3
4.7	18.2	9.4	9.4	0.19	0.31	31.6	36.3	95.2	0.77	0.15	0.21	21.9
1.2	13.9	1.8	1.8	0.25	0.11	8.4	3.1	71.4	0.23	0.03	3.07	24.0
1.1	11.5	1.3	1.3	0.07	0.05	8.4	13.1	45.8	0.63	0.05	0.18	15.3
7.2	11.2	2.7	2.7	0.1	0.19	27.3	73.2	1.8	3.16	0.01	0.04	10.6
1.8	9.8	1.1	1.1	0.07	0.09	2.7	9.3	64.1	1.42	0.06	0.3	20.9
4.3	9.2	4.8	4.8	0.32	2.32	26.1	55.4	131.2	1.52	0.23	0.12	14.3
3.4	23.8	2.2	2.2	0.17	0.11	9.0	82.6	108.6	1.68	0.08	0.33	17.5
51.8	55.6	7.7	7.7	0.03	0.01	33.8	103.8	94.6	84.9	0.02	0.61	7.6
20.6	21.6	2.9	2.9	0.06	0.06	15.9	98.2	12.2	0.89	0.02	0.04	10.3
4.9	11.0	1.1	1.1	0.2	0.05	15.4	81.4	26.2	1.56	0.02	0.02	13.4
14.5	61.4	12.7	12.7	0.44	0.34	33.1	32.6	206.2	0.85	0.01	0.73	27.2
60.9	83.4	15.9	15.9	0.1	0	102.7	104.2	292.8	1.35	0.02	0.16	9.8
5.9	32.6	7.2	7.2	0.38	0.9	16.8	14.5	150.7	0.59	0.17	5.18	56.8
6.8	47.4	4.3	4.3	0.15	0.12	12.6	35.1	164.1	1.84	0.02	0.03	34.6
5.1	49.6	4.7	4.7	0.07	0.03	12.8	47.8	126.9	0.49	0.01	0.54	30.0
3.7	16.9	2.1	2.1	0.23	0.07	9.6	8.3	97.0	0.69	0.02	0.06	25.1
10.0	6.7	2.4	2.4	0.18	0.03	29.2	45.8	48.8	0.07	0.06	0.23	25.8
10.8	6.4	1.2	1.2	0.1	0.07	17.9	63.9	56.1	0.14	0.11	1.59	25.4
8.1	78.9	5.4	5.4	0.04	0.58	9.4	4.0	280.6	1.01	0.02	0.43	23.1
6.3	55.0	21.7	21.7	0.42	0.16	33.2	271.1	14.7	0.79	0	0.23	16.7
1.8	63.1	2.1	2.1	0.14	0.27	7.5	0.0	236.1	1.09	0.1	1.26	27.1
1.6	19.9	3.8	3.8	0.28	1.59	27.3	57.5	48.2	1.16	0.02	1.07	26.8
1.7	66.2	8.6	8.6	0.15	0.77	5.0	0.0	286.7	1.01	0.16	1.01	23.9
7.9	25.2	8.7	8.7	0.19	1.69	8.7	16.6	186.7	0.96	0.14	0.32	18.6
3.9	32.5	25.3	25.3	0.47	0.12	24.6	222.1	18.9	0.75	0.01	0.07	18.7
3.8	59.4	8.8	8.8	0.36	0.21	21.3	219.3	18.9	0.75	0.01	0.07	21.8
2.7	71.5	2.9	2.9	0.39	1.41	13.7	101.3	169.6	1.28	0.09	1.45	19.4
1.7	16.8	4.5	4.5	0.36	0.17	13.5	43.7	56.7	0.58	0.09	1.45	20.9
1.5	33.9	6.8	6.8	0.54	0.21	10.8	45.4	121.4	1.08	0.04	0.67	20.1
1.4	64.7	5.7	5.7	1.14	0.42	8.0	2.9	244.9	1.07	0.03	0.35	26.0
2.1	70.3	6.4	6.4	0.39	0.64	6.1	0.0	262.3	0.57	0.03	1.48	24.1
1.5	13.0	2.1	2.1	0.27	0.32	19.0	25.6	42.7	0.58	0.06	0.32	37.3
1.4	8.7	1.1	1.1	0.15	0.19	9.4	6.5	47.6	0.49	0.03	0.47	32.1
2.1	89.5	2.5	2.5	0.14	0.57	9.0	0.0	311.1	1.26	0.06	1.03	29.7
2.6	62.5	6.1	6.1	0.05	1.09	10.1	6.5	227.5	0.98	0.23	0.77	28.7
18.0	85.7	11.6	11.6	0.04	0.56	22.5	0.0	419.7	0.83	0.02	0.3	24.5

Table 12.3 Trace concentrations in the Neogene Aquifer. The results are below detection limit for the elements In, Pt, Rh (DL: 0.01 µg L^{-1}); Ag, Bi, Ir, Os, Ru, Ta and Te (DL: 0.05 µg L^{-1}); Hg (DL: 0.1 µg L^{-1}) and Pd (DL: 0.2 µg L^{-1}).

Parameter	Unit	Min.	Max.	Median	Mean	97.7‰	n
δ^2H	‰	−48.8	−31.6	−44.9	−43.3	−33.9	32
δ^{18}O	‰	−7.76	−4.82	−7.10	−6.91	−5.32	32
δ^{13}C	‰	−23.6	−7.9	−13.9	−14.2	−7.9	32
Al	µg L^{-1}	7	4707	17.5	145	669	40
As	µg L^{-1}	1	52	6	12.5	49	40
B	µg L^{-1}	<20	1104	23.0	71	369	40
Ba	µg L^{-1}	9.02	203	53	60	143	40
Be	µg L^{-1}	<0.05	2.3	<0.05	0.16	1.3	40
Cd	µg L^{-1}	<0.05	8.8	0.51	1.2	7.7	40
Ce	µg L^{-1}	<0.01	65	0.13	2.0	11	40
Co	µg L^{-1}	<0.02	19	1.5	2.9	9.5	40
Cr	µg L^{-1}	<0.5	8.4	1.5	1.7	6.3	40
Cs	µg L^{-1}	<0.01	0.24	0.01	0.03	0.16	40
Cu	µg L^{-1}	0.5	7.8	1.8	2.4	7.4	40
Dy	µg L^{-1}	<0.01	2.5	0.01	0.14	0.86	40
Er	µg L^{-1}	<0.01	1.2	0.01	0.10	0.95	40
Eu	µg L^{-1}	<0.01	1.1	0.01	0.04	0.30	40
Fe	Mg L^{-1}	0.06	66	8.7	13	38	40
Ga	µg L^{-1}	<0.05	0.1	<0.05	<0.05	<0.05	40
Gd	µg L^{-1}	<0.01	3.3	0.02	0.16	1.2	40
Ge	µg L^{-1}	<0.05	0.3	<0.05	0.07	0.26	40
Hf	µg L^{-1}	<0.02	0.04	<0.02	<0.02	0.04	40
Ho	µg L^{-1}	<0.01	0.4	0.01	0.03	0.18	40
La	µg L^{-1}	<0.01	44	0.08	1.76	15	40
Li	µg L^{-1}	1.0	77	7.0	10.9	54	40
Lu	µg L^{-1}	<0.01	0.4	0.01	0.02	0.10	40
Mn	mg L^{-1}	0.0	1.1	0.16	0.21	0.59	40
Mo	µg L^{-1}	<0.1	8.3	0.05	0.46	2.5	40
Nb	µg L^{-1}	<0.01	0.01	<0.01	<0.01	0.01	40
Nd	µg L^{-1}	<0.01	35.3	0.10	1.2	8	40
Ni	µg L^{-1}	0.20	25	2.5	4.2	17	40
Pb	µg L^{-1}	<1	16	1.7	2.7	8.8	40
Pr	µg L^{-1}	<0.01	9.7	0.02	0.31	1.9	40
Rb	µg L^{-1}	1.5	52	3.6	7.9	40	40
Re	µg L^{-1}	<0.01	0.02	<0.01	<0.01	0.02	40

Table 12.3 *(Continued)*

Parameter	Unit	Min.	Max.	Median	Mean	97.7‰	n
Sb	µg L^{-1}	<0.05	0.51	<0.05	<0.05	0.17	40
Sc	µg L^{-1}	1.17	12.3	4.9	4.8	9.7	40
Se	µg L^{-1}	<0.5	2.5	<0.5	0.61	2.3	40
Sm	µg L^{-1}	<0.05	5.1	<0.05	0.20	1.4	40
Sn	µg L^{-1}	<0.05	0.48	<0.05	0.09	0.42	40
Sr	µg L^{-1}	18	505	137	209	504	40
Tb	µg L^{-1}	<0.01	0.61	<0.01	0.03	0.20	40
Th	µg L^{-1}	<0.05	0.50	<0.05	<0.05	0.20	40
Ti	µg L^{-1}	<10	18	<10	<10	17	40
Tl	µg L^{-1}	<0.01	0.21	<0.01	0.02	0.11	40
Tm	µg L^{-1}	<0.01	0.15	<0.01	0.02	0.12	40
U	µg L^{-1}	<0.05	1.53	<0.05	0.07	0.36	40
V	µg L^{-1}	<1	9.00	<1	<1	4.5	40
W	µg L^{-1}	<0.1	4601	0.25	125	845	40
Y	µg L^{-1}	<0.01	13.7	0.14	1.10	9.4	40
Yb	µg L^{-1}	<0.01	2.08	<0.01	0.11	0.74	40
Zn	µg L^{-1}	9.7	1580	35	82	326	40
Zr	µg L^{-1}	<0.5	1.8	<0.5	<0.5	1	40

All groundwaters are fresh with low mineralisation. The electrical conductivity is generally low; only three samples have a SEC value higher than 400 µS cm^{-1} at 25°C. Many groundwaters have low concentrations of DO which indicates a reducing environment. The Eh values are rather high for a reducing environment but other ions like nitrate, iron, manganese and sulphate clearly confirm the reduced environment. DO and Eh show a weak correlation. It is not well understood why the measured Eh values are high.

There is a clear difference between the eastern and western part of the Neogene Aquifer. In the eastern zone the sediments are completely decalcified, this part of the aquifer is further called the siliceous part. In the western zone sediments still contain some calcite and therefore this part is further called the calcareous part of the aquifer. Groundwater types were derived by means of the Stuyfzand classification system (Stuyfzand 1986). Groundwater in the Neogene Aquifer varies from $CaSO_4$, $MgSO_4$, $FeSO_4$ and $NaSO_4$ types to $CaHCO_3$, $NaHCO_3$ and $FeHCO_3$ types. Sulphate and iron water types occur in the upper part of the siliceous part of the aquifer and change into $CaHCO_3$ water types further along a flow line (Fig. 12.2). At the base of the aquifer $NaHCO_3$ water types occur. Coetsiers and Walraevens (2006) found also $MgHCO_3$ water types in the deeper parts of the aquifer.

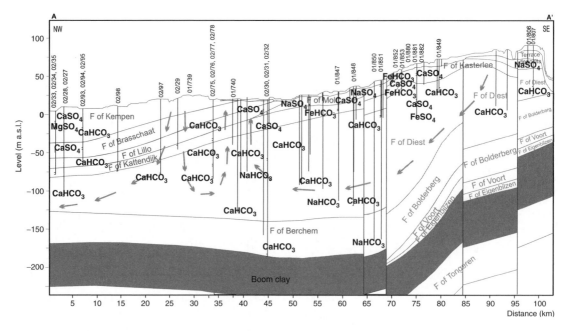

Fig. 12.2　Profile A–A′ (see Fig. 12.1) with indication of water types.

12.5　Hydrogeochemical processes controlling groundwater quality

On Fig. 12.3 the evolution of different species along profile A–A′ is given. Additional data were obtained from drinking water companies and the government and are also represented in Fig. 12.3. The main hydrogeochemical processes occurring in groundwaters are discussed more fully in Chapter 1.

12.5.1　Mineral dissolution reactions

In the siliceous part of the aquifer, where the sediments are almost completely decalcified, calcite dissolution has little influence on the groundwater chemistry. Where the calcareous formations of Lillo (in the northwest) and Berchem (in the southwest and in deeper parts of the aquifer) are present, calcite dissolution influences groundwater quality.

High quantities of dissolved CO_2 are produced in the soil atmosphere, and H_2CO_3 is dissolved in the water. As a consequence of calcite dissolution an increase in pH and Ca and HCO_3 concentration occurs (Fig. 12.3). The major part of the aquifer is undersaturated with respect to calcite and dolomite. SI for calcite varies between –7.78 and 0.07 and has a median value of –2.58. Calcite dissolution is thus likely taking place. The positive correlation between TAC and Ca indicates the occurrence of calcite dissolution in the aquifer (Fig. 12.4).

In the siliceous part of the aquifer silicate dissolution is the main reaction determining groundwater quality. Silicate dissolution via hydrolysis reactions involving dissolved CO_2 is a very slow process that adds silica and cations (Na, K and Ca) to the groundwater. The dissolution of amorphous SiO_2 (diatoms and radiolaria abundant in the

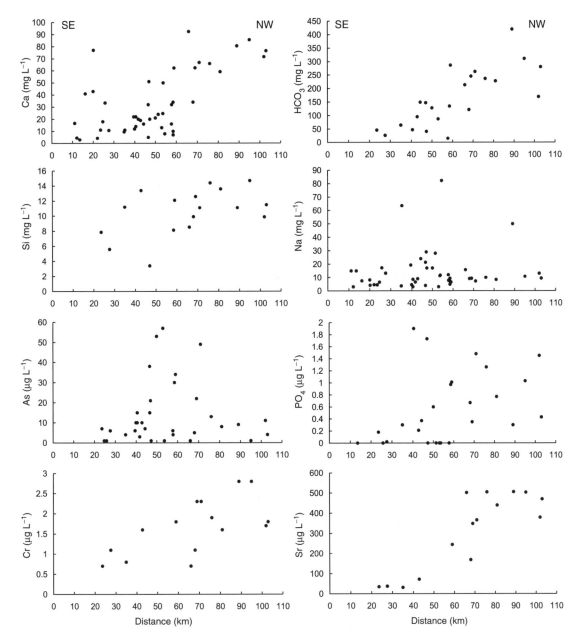

Fig. 12.3 Evolution of parameters (Ca, Na, HCO₃, Si, As, Cr, Sr, PO₄) along profile A–A′.

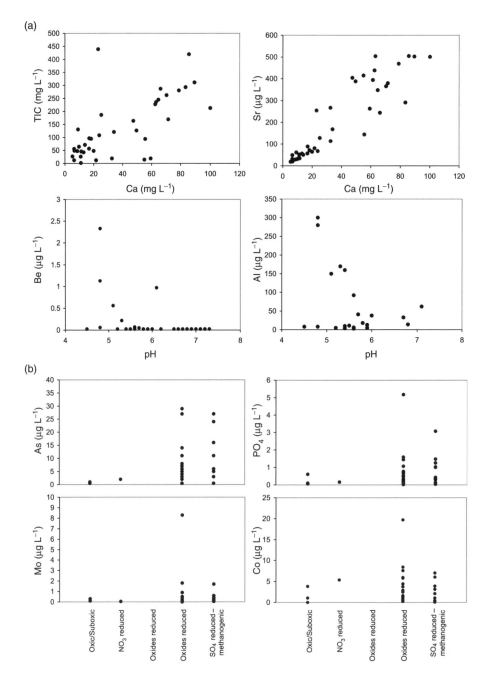

Fig. 12.4 Cross-plots illustrating the relation between Ca and TIC, Ca and Sr, pH and Be and pH and Al and concentrations of PO_4, As, Mo and Co along the different redox zones in the Neogene Aquifer.

marine formations), can also lead to elevated Si concentrations. Groundwaters from the Neogene have low TDS and Si increases along the flow line (Fig. 12.3). The silicate weathering leads to the formation of new secondary minerals, clays (illite, kaolinite and montmorillonite) and Fe-oxides (Appelo & Postma 2005).

Sodium contributes in all groundwaters significantly to the cations and its presence is not balanced by chloride as would be expected when seawater was the source of sodium. Sodium is mainly derived from weathering of Na-feldspar or any member of the plagioclase solid solution series between albite and anorthite (Ca-feldspar) (Appelo & Postma 2005). Plagioclase weathering releases in addition Ca^{2+}. The increase in cation concentration is accompanied by an increase in dissolved bicarbonate ($CO_2 + H_2O \rightarrow H^+ + HCO_3^-$).

Quartz is by far the most dominant component of the deposits but is strongly resistant to weathering and contributions to groundwater are the result of silicate weathering. The stability of the silicate minerals in a groundwater system can be evaluated by calculating the saturation state of the groundwater for a given mineral. For example, for albite, the dissociation reaction is

$$NaAlSi_3O_8 + 4H^+ + 4H_2O$$
$$\rightarrow Na^+ + Al^{3+} + 3H_4SiO_4$$

The single arrow is used to indicate the irreversible character of the reaction. With PHREEQC the saturation indices [Log (IAP/K)] for anorthite, albite, K-feldspar and kaolinite were calculated. Anorthite and albite are used as end-members to describe the plagioclases (Appelo & Postma 2005). Strong undersaturation is observed in the upper part of the aquifer, while the degree

of undersaturation is less in the deeper part of the profile. Dissolution of plagioclase is thus, from an equilibrium point of view, expected to take place. The groundwater has also a slight supersaturation with respect to the weathering product kaolinite.

An increase in phosphate occurs along the profile A–A' (Fig. 12.3) and can be explained by the dissolution of phosphate minerals and concretions, which are present in the Bolderberg, Berchem, Diest, Kattendijk and Brasschaat Sands.

12.5.2 Redox reactions

From the measurements of Eh it is difficult to define redox boundaries but the O_2 content indicates that most samples are taken from a reducing environment. The chemical elements C, N, O, S, Mn and Fe undergo redox reactions in the subsurface and from the low DO concentrations and the high dissolved Fe concentrations, it is apparent that reducing conditions prevail in most of the aquifer.

As water percolates beneath the water table oxygen is progressively consumed when organic matter is oxidised. When all DO in groundwater has been consumed, the oxidation of organic matter can still continue, but instead of free molecular O_2, the agents become NO_3, MnO_2, $Fe(OH)_3$ and SO_4. In a last stage of the redox sequence methane is produced. In the uppermost meters of the Neogene Aquifer nitrate is present in the groundwater but further downstream nitrate reduction has lowered these concentrations. As in the East Midlands Triassic Sandstone Aquifer in UK (Smedley & Edmunds 2002) nitrate is reduced rapidly after oxygen in the Neogene Aquifer and the nitrate reduction zone is very narrow (Coetsiers & Walraevens

2006). Deeper in the aquifer iron and manganese appear in the groundwater caused by the reduction of iron and manganese oxides and hydroxides. After that, sulphate is reduced, which can be coupled to the formation of sulphide minerals. In Fig. 12.4 the concentration of some trace elements is plotted against the redox environment. This redox environment was derived using the methodology described in Coetsiers and Walraevens (2006). The oxidation of organic matter causes an increase in PO_4 concentrations in the reduced groundwater environment (Fig. 12.4).

The oxidation of pyrite and other sulphide minerals has an important environmental impact. The oxidation of pyrite can occur by means of oxygen, free Fe^{3+} or nitrate. Raising and lowering of the water table, caused by pumping activities, provides optimal conditions for the weathering of pyrite. Arsenic and other heavy metals incorporated in pyrite are released when pyrite weathering occurs. The repeated drawdown and recovery pulls oxygen into the dewatered area and would in time cause oxidation of organic matter and the release of CO_2, oxidation of arsenic bearing pyrite and the formation of iron hydroxides. The Neogene Aquifer also contains large amounts of glauconite $[(K,Na,Ca)_{1.2-2.0}(Fe^{3+}, Al, Fe^{2+},Mg)_4(Si_7AlO_{20})(OH)_4 \cdot n(H_2O)]$, an iron containing hydroxide silicate, which is unstable in slightly aerobic conditions and brings Fe in solution when disintegrating to clay. The oxidation of glauconite is accompanied by release of SiO_2 (Nolan 1999).

12.5.3 Mixing with formation water and ion-exchange reactions

Although the Neogene Aquifer consists mainly of marine deposits, flushing of the sediments is already in an advanced stage, so that the marine influence has largely disappeared. The chloride concentrations are generally low in the deeper parts of the Neogene Aquifer. Only in some shallow wells elevated Cl concentrations were encountered, possibly indicating anthropogenic inputs. In sample 02/103 (25 m deep) the highest Cl concentration of 103 mg L^{-1} was measured, which is clearly caused by pollution. In the neighbourhood of this sampling site a well-known industrial pollution with chloride is present. In all other groundwater samples the Cl concentration is lower than 40 mg L^{-1}. The Na/Cl ratio is useful to determine whether an increase in Na is due to mineral weathering and ion exchange or related to mixing with older marine formation water as Cl behaves conservatively (Appelo & Postma 2005). The Na/Cl ratio in seawater measures 0.85. In the groundwater samples, this ratio varies significantly and reaches higher Na/Cl ratios than in seawater (up to 25; Fig. 12.5) indicating that water–rock interaction is more important than mixing with old formation water. Chloride concen-

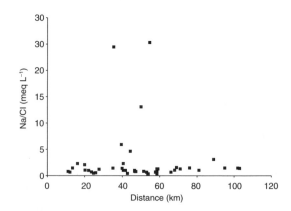

Fig. 12.5 Na/Cl ratio evolution along profile A–A'.

trations are generally low indicating that the aquifer has been well flushed of the original formation waters. The additional Na can be resulting from ion exchange due to freshening or silicate weathering. The samples with the highest Na/Cl ratios come from deep wells, where evidence of freshening is still present. The role of mixing with deeper connate water is not considered to be important along the flow line.

The Br/Cl ratio in the Neogene Aquifer is generally higher than that for seawater (Richter and Kreitler 1993), although Br concentrations are often below detection limit. The highest Br concentrations occur in the samples 03/35 and 02/103 (polluted). Boron concentrations in groundwater are generally lower than 0.4 mg L^{-1} except for sample 01/739 which has a concentration of 1.07 mg L^{-1}. This high B content occurs together with a low Cl concentration, which indicates that the B is probably not originating from seawater. The highest concentrations of Li and Rb are found in the polluted sample 02/103. Elevated Li concentrations occur in samples 01/881, 01/850 and 01/880 while elevated Rb concentrations are measured in the samples 01/739 and 02/98. These elevated Li and Rb concentrations show no correlation with Cl.

In the deeper parts of the aquifer cation-exchange reactions in freshening conditions take place and Ca from the groundwater is exchanged for Na, Mg and K adsorbed on the clay surfaces. The cations Na, K and Mg are successively being replaced in order of increasing affinity for the clay. This results in an increase of marine cations and a depletion of Ca in the groundwater. The chromatographic sequence of cation exchange is found in the subsequent surplus of Na, followed by K, and finally Mg, resulting in $NaHCO_3$ water types

(Walraevens and Cardenal 1999). Coetsiers and Walraevens (2006) also found $MgHCO_3$ water types in the deeper parts of the aquifer.

12.5.4 Trace elements

Strontium concentrations in the siliceous part of the aquifer are below 100 µg L^{-1} (Fig. 12.3). In the calcareous part of the aquifer the Sr concentration increases up to 500 µg L^{-1}. Possible sources for Sr are aragonite, calcite and/or gypsum (Appelo & Postma 2005). There is a good correlation of Sr, both with Ca (Fig. 12.4) and HCO_3 but not with SO_4, implying that the source of Sr is rather calcite dissolution than gypsum dissolution. Stuyfzand (1989) indicated shell fragments as a natural source for Sr in groundwater in the Netherlands.

High arsenic concentrations, above the WHO limit of 0.05 mg L^{-1}, are probably released with pyrite oxidation or reduction of iron hydroxides. Arsenic may substitute in pyrite (FeS_2) or can be found as the separate mineral arsenopyrite (FeAsS). High arsenic concentrations are also associated with the occurrence of glauconite in deposits (Huisman et al. 1997; Barringer et al. 1998). Fe(III) hydroxides have a high capacity and strong affinity for adsorbing As. The source of the high arsenic concentrations is thus natural, that is, derived from sediments by natural geochemical processes. The elevated arsenic concentrations are observed in the zones where Fe hydroxides are reduced and where sulphate is reduced (Fig. 12.4). Fe, Mn, As, Mo and Sb have similar chemical redox behaviour (Smedley and Edmunds 2002; Vissers et al. 2005) and are all found to increase at the iron reduction front (Fig. 12.4) caused by the dissolution of iron hydroxides. Vissers et al. (2005) also found Co,

Ni and U to increase at the iron reduction front. In the Neogene Aquifer only Co shows a similar trend (Fig. 12.4). For U the highest concentration occurs in the zone where nitrate is reduced in sample 02/103. The vanadium concentration is also highest in the zone where iron hydroxides are reduced. Elevated phosphate concentrations also occur in the Fe-reduced and SO_4^{2-}-reduced zone (Fig. 12.4) and are caused by the oxidation of organic matter.

High Al concentrations occur in the shallow parts of the aquifer, where a low pH is measured (Fig. 12.4). Be (Fig. 12.4) and the lanthanide elements, (Ce, Pr, Nd, Sm, Eu, Gd, Tb, Dy, Ho, Er, Tm, Yb and Lu), are detectable in the shallow parts of the aquifer where low pH values are measured. In other studies, Edmunds et al. (1992) found Ni, Co, Zn and Be to be dependent on the acidification level; Stuyfzand (1993) found also Li, Rb and Cd to be pH dependent and Vissers et al. (2005) also noted Tl, Ga, the rare earth elements and Y to be mobile under acidic conditions. At higher pH these trace elements become adsorbed. The dissolution of $Al(OH)_3$, ion exchange of aluminium or the weathering of primary silicates can have a buffering effect on pH. Several other trace elements are found at higher concentration in the calcareous part of the aquifer including Ni, Co, Cd, Ba, Sc and Cr. Some of the metal species are associated with sorption sites on Fe and Mn hydroxides and are most likely related to the mobilisation of these oxy-hydroxides in the reducing groundwaters.

Heavy metals, mainly Cd, Ni and Zn and in some cases also Pb and Cu, are high in the soil and shallow groundwater in some regions of the provinces of Antwerp and Limburg and represent regional pollution. Metallurgical industries (since the end of the nineteenth century) and mining activities (coal exploitation) are responsible for this pollution (Walraevens et al. 2003). The delimitation of the region in which the groundwater is influenced by this industrial pollution is not exactly known.

12.5.5 Temporal variations

There is little historical data available and no long time series were obtained from the Neogene Aquifer. Drinking water companies probably possess long time series but these were not made available for this study. Knowledge of the baseline conditions is therefore very difficult to assess directly. Such limited data also makes assessment of trends very difficult. Temporal variations were studied using data supplied by the PIDPA water company. Most boreholes had limited data for many solutes but it was clear that in most cases no significant trends were apparent. In some case however, an increase in As content was observed (Mahauden et al. 2002). In Fig. 12.6(a) the As-concentration measured between 1990 and 2000 in a well at Ravels is given. However the way in which the observed variations are related to changing extraction conditions is not known (Walraevens et al. 2003).

12.5.6 Depth variations

In the frame of this study some samples were taken from different wells (with different depth) at the same location for a couple of locations. In most cases, however, the screens of the different monitoring wells lie at a large vertical distance from each other so that no detailed information is available about depth variations. At some locations there is an increase of the HCO_3, Ca, Si and Mg

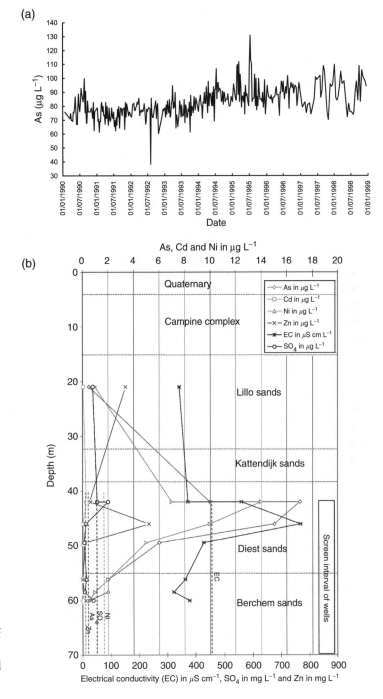

Fig. 12.6 (a) Evolution of the As content in the period 1990–2000 at Ravels; (b) vertical variation of groundwater composition in pumping wells (dashed lines) and piezometers at Brasschaat.

concentrations and pH with depth, while the SO₄, Cl, Fe concentrations decrease. Calcite dissolution, silicate dissolution and sulphate reduction accompanied by the precipitation of Fe-sulphides are mainly responsible for these changes in groundwater chemistry with depth; the decrease in Cl may be ascribed to decreasing anthropogenic influence.

The LTGH has conducted a policy-supporting research project, which attempts to provide more information concerning the natural background concentrations of trace elements in groundwater in Flanders (Mahauden et al. 2002). In a few cases, information was available allowing an insight to be gained into the variability of concentrations of inorganic pollution parameters in depth profiles. This was the case for some groundwater catchment areas of the PIDPA water supply company, where not only analyses were available from long-screened pumping wells, but additionally from several piezometers placed at different depths. The data for the Brasschaat catchment are shown in Fig. 12.6(b). The pumping wells are screened between 40 and 55–70 m depth. The eight piezometers have a screen length of 1 m, situated in the depth range of 20.9–60 m. The piezometers and pumping wells are situated in the Diest and Berchem Sands, except for the shallowest monitoring well (–20.9 m) which lies in the Lillo Formation.

The depth stratification of groundwater quality is clear: electrical conductivity (EC), sulphate, As and Ni abruptly increase at 42 m depth, reaching a maximum, except for EC and Zn, for which the maximum is reached at 46 m. At greater depth, all parameters decrease again. For Cd, only minor variations are found, although the maximum is also reached at 42 m. It is not clear whether pumping activities are influencing the observed depth stratification, or if it is reflecting the natural background variation in the aquifer. The elevated trace element concentrations occur all in the Diest Sands. The natural groundwater table is only a few meters below ground surface. Coetsiers (in preparation by 2007) calculated a drawdown caused by pumping activities of about 12 m at Brasschaat. The possible evidence for oxidation near the top of the exploitation screens is not consistent with data from other catchments. It is however clear that the groundwater produced by the pumping wells represents a mixture of different chemical compositions, originating from different depths.

12.5.7 Age of the groundwater

From the 38 samples taken for the purpose of this investigation 10 samples were selected for ^{14}C analysis. All samples have conventional ages less than 10,000 years and consequently infiltrated during Holocene times. Groundwater infiltrated during Pleistocene times would reflect the colder climate with lighter signatures of the stable isotopes δ^2H and $\delta^{18}O$. Such significant differences were

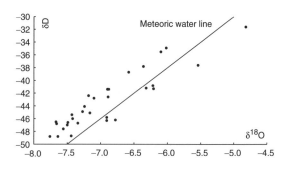

Fig. 12.7 Plot of δD versus $\delta^{18}O$ for groundwaters in the Neogene Aquifer.

not encountered supporting the hypothesis of the Holocene age of water. On a δ^2H–$\delta^{18}O$ plot (Fig. 12.7) the results plot close to the meteoric waterline indicating groundwater to be resulting from rain with the same composition as actual precipitation.

Carbon isotopes provide evidence of the degree of reaction with carbonate minerals, which may be loosely related to residence time. The concentration of $\delta^{13}C$ in groundwater is determined by the input of recharge water and by reactions with rocks (Mazor 1997). The $\delta^{13}C$ concentrations lie in between –23.6‰ and –7.9‰. These values suggest a combination of common plant material (–23 ± 3‰) and calcareous rocks (–2 to 0‰). In the northwest part of the aquifer, where calcite dissolution occurs, the $\delta^{13}C$ concentrations increase once the groundwater has passed the Lillo Formation. In this area samples with high HCO_3 concentrations have also high $\delta^{13}C$ values reflecting the greater extent of water–rock interaction.

12.6 Regional variations

There is a clear contrast in chemistry between the southeastern and northwestern parts of the aquifer. Lower SEC waters are found in the southeastern part of the aquifer where the sediments are almost completely decalcified or contain naturally low calcite concentrations. In the northwestern area the Kattendijk and Lillo Sands occur, which contain calcite. Here calcite dissolution occurs resulting in higher Ca and HCO_3 concentrations. This is also supported by elevated Sr concentrations, which are typically enriched in aragonite and calcite. The difference between the two zones becomes also clear in the pH values. The southeastern zone is characterised by typically low pH values (<6.5). Because the sediments in this zone are completely decalcified, the pH is determined by the infiltrating acid rain and the influence of organic CO_2 in the soil atmosphere. The Houthalen Sands, part of the Bolderberg Formation, still contain some remnants of marine limestone and at the base of the aquifer calcite dissolution can still occur. In the western zone the pH values are higher and determined by the dissolution of calcite.

In the southeast, where the Diest Formation crops out, the pH values are the lowest. The Diest Formation is rich in iron minerals, like pyrite and glauconite and the oxidation of this pyrite has an acidifying effect. The formation contains very little calcite so that the buffer capacity is almost zero. Glauconite also breaks down to clays by which Fe^{2+} is released into solution. In combination with acid rain and an acidifying atmospheric deposition the pH can further decrease in this non-buffered system. In the western part the sediments contain calcite and shells and pH is buffered.

High As- concentrations occur together with high Fe and Mn and are coupled to the oxidation of pyrite minerals and the reduction of iron hydroxides. Heavy metals, mainly Cd, Ni and Zn and in some cases also Pb and Cu, in the soil and groundwater can locally be elevated due to pollution from former metallurgical industry or mining activities. Locally elevated nitrate or chloride concentrations can be encountered in shallow wells indicating anthropogenic influences. In some shallow samples enhanced levels of V, U, Tl, Ti, Th, Sb, Rb, Cu, Cs, Cr, Co, Cd, Be, Al, F, rare earth elements and Br were encountered together with high Cd, Ni and Zn. In sites where increased concentrations are unrelated to low pH conditions pollution may be an explanation for the enhanced levels.

12.7 Summary and baseline chemistry of the Neogene Aquifer

Natural baseline is given by a range of values, which are controlled by rainfall inputs, the marine depositional environment and the geology of the area. Details of the natural variations and the overall statistical data for 91 constituents in the groundwaters of the area are represented. Mostly these are inorganic ions, which form the bulk of the interest of this study as far as the natural baseline is concerned, since the many possible micro-organics that are often sought in monitoring are by definition introduced substances.

The Neogene Aquifer can be divided in two parts, one decalcified while the other still containing calcite. The most important influences on water chemistry are redox reactions and mineral dissolution reactions involving silicates and calcite. In the calcareous western part calcite dissolution causes an increase in Ca and HCO_3 concentrations and pH. The aquifer is well flushed resulting in low amounts of total dissolved solids. Silicate reactions, although slower, have provided significant amounts of silica and cations including Na, K and to a lesser degree Ca. In addition redox reactions are important for many species.

The data have been presented in Tables 12.2 and 12.3 and the median value and 97.7 percentile provide a good estimate of the average and upper baseline concentrations in the aquifer. However this should be used in conjunction with maps, which show that regional variations do exist in the aquifer. The baseline chemistry changes spatially across the aquifer in relation to the extent of water–rock interaction and to geochemical controls imposed by the local geochemical environment. The baseline for the groundwater in the southeastern siliceous area is clearly different from that in the northwestern calcareous part of the aquifer. These regional variations must be taken into account when defining the local baseline for the aquifer.

Although most groundwaters are dominated by baseline concentrations, this is not the case for all solutes. Enhanced concentrations of N-species, pesticides and potentially a range of other solutes from urban and industrial usage, including some heavy metals, indicate the effect of anthropogenic inputs. The most obvious external inputs over the last few decades have been derived from agricultural pollution. Baseline nitrate concentrations are most likely to be around a few milligrams per litre as indicated by the limited historical data. Most analysed groundwaters had nitrate concentrations below detection limit. The presence of Fe^{2+} in the groundwater indicates a reducing environment in which nitrate reduction has occurred. Phosphate concentration can be locally elevated due to dissolution of vivianite present in the sediments or by the oxidation of organic material. Other elements, which may have been modified by agriculturally derived anthropogenic inputs, include K, Cl and P. Some trace elements including As and Sr are relatively high in parts of the aquifer but they are considered to represent natural baseline. A strong vertical variation in several parameters occurs due to redox changes with depth.

References

Appelo, C.A.J. and Postma, D. (2005) *Geochemistry, Groundwater and Pollution*, 2nd edition. Balkema Publishers, Rotterdam, 649 pp.

Barringer, J.H., Szabo, Z. and Barringer, T.H. (1998) Arsenic and metals in soils in the vicinity of the Imperial Oil Company Superfund site, Marlboro Township, Monmouth County, New Jersey: U.S. Geological Survey Water Resources Investigations Report 98-4016.

Beaufays, R., Bonnyns, J., Bronders, J. et al. (1990) Mathematical model of groundwater flow in the Neogene aquifer of northeastern Belgium. (In Dutch: Mathematisch model van de grondwaterstromingen in de Neogene meerlagige aquifer van noordoost België.) Report Study Centre for Nuclear Energy, SCK-CEN, Mol.

Camelbeeck, T. and Meghraoui, M. (1996) Large earthquakes in Northern Europe more likely than once thought. EOS 42, 405–9.

Coetsiers, M. and Walraevens, K. (2006) Chemical characterization of the Neogene Aquifer, Belgium. *Hydrogeology Journal* 14(8), 1556–1568.

Coetsiers, M. (2007) Investigation of the hydrogeological and hydrochemical state of the Neogene Aquifer in Flanders using modeling and isotope hydrochemistry

De Coster, D., De Smedt, P. and Van Autenboer, T. (1986) Groundwater vulnerability map of Limburg. (In Dutch: Kwetsbaarheidskaart van het grondwater in Limburg.) Report, AROL, Administration for Spatial Organisation and Environment, Administration for Environment, Brussels.

Demanet, D., Renardy, F., Vanneste, K. et al. (2001) The use of geophysical prospecting for imaging active faults in the Roer graben, Belgium. *Geophysics* 66, 78–89.

Demyttenaere, R. and Laga, P. (1988) Faults and isohyps maps of the Belgian part of the Roer Valley. (In Dutch: Breuken – en isohypsenkaarten van het Belgisch gedeelte van de Roerdal Slenk.) Belgian Geological Survey – Professional Paper 1988/4, 234, 20 pp.

Edmunds, W.M., Kinniburgh, D.G. and Moss, P.D. (1992) Trace metals in interstitial waters from sandstones: Acidic inputs to shallow groundwaters. *Environmental Pollution* 77, 129–41.

Hinsby, K., Harrar, W.G., Nyegaard, P. et al. (2001) The Ribe Formation in western Denmark – Holocene and Pleistocene groundwaters in a coastal Miocene sand aquifer. In: Edmunds, W.M. and Milne, C.J. (eds) *Palaeowaters in Coastal Europe: Evolution of Groundwater Since the late Pleistocene*. Special Publication, 189, Geological Society, London, pp. 29–48.

Huisman, D.J., Vermeulen, F.J.H., Baker et al. (1997) A geological interpretation of heavy metal concentrations in soils and sediments in the southern Netherlands. *Journal of Geochemical Exploration* 59, 163–74.

Huneau, F., Blavoux, B. and Bellion, Y. (2001) Differences between hydraulic and radiometric velocities of groundwaters in a deep aquifer: Example of the Valréas Miocene aquifer (Southeastern France). *Comptes Rendues. Académie Sciences. Paris, Earth and Planetary Sciences* 333, 163–70.

Laga, P. (1973) *The Neogene Deposits of Belgium*. Guide book for the Field Meeting of the Geologists' association London. Belgische Geologische Dienst.

Mahauden, M., Coetsiers, M. and Walraevens, K. (2002). Investigation of the natural concentrations of inorganic VLAREBO-parameters. Project TGO 99/19. Final Report. Laboratory for Applied Geology and Hydrogeology, Ghent University.

Marechal, R. and Laga, P. (1988) Proposal lithostratigraphy of the Paleogene. (In Dutch: Voorstel lithostratigrafie van het paleogeen.) Nationale commissies voor stratigrafie: commissie Tertiair.

Mazor, E. (1997). *Chemical and Isotopic Groundwater Hydrology*. Marcel Dekker, New York.

Meyus, Y., De Smet, D., De Smedt, F. et al. (2000) Hydrogeologische codering van de ondergrond van Vlaanderen (HCOV). *Water* 8, 1–13.

Nolan, B.T. (1999) Nitrate behavior in ground waters of the Southeastern United States. *Journal of Environmental Quality* 28, 1518–27.

Patyn, J., Ledoux, E. and Bonne, A. (1989) Geohydrological research in relation to radioactive waste disposal in an argillacous formation. *Journal of Hydrology* 109, 267–85.

PIH (Provincial Institute for Hygiene, Antwerp) (1995) Groundwater quality in the Province of Antwerp. (In Dutch: Grondwaterkwaliteit in de Provincie Antwerpen.) Provincial Council of Antwerp, Bestendige Deputatie.

Richter, B.C. and Kreitler, C.W. (1993) *Geochemical Techniques for Identifying Sources of Ground-Water Salinization.* C.K. Smoley, CRC Press, Florida.

Smedley, P.L. and Edmunds, W.M. (2002) Redox patterns and trace-element behaviour in the East Midlands Triassic Sandstone Aquifer, UK. *Ground Water* 40, 44–58.

Stuyfzand, P.J. (1986) A new hydrochemical classification of water types: Principles and application to the coastal dunes aquifer system of the Netherlands. In: *Proceedings of the 9th Salt Water Intrusion Meeting.* Delft, pp. 641–55.

Stuyfzand, P.J. (1989) Factors Controlling Trace Element Levels in Groundwater in the Netherlands. Water–Rock Interaction, WRI-6. In: *Proceedings of the 6th International Symposium of Water–Rock Interaction.* Malvern, 3–8 August 1989. Balkema, Rotterdam, The Netherlands, pp. 655–9.

Stuyfzand, P.J. (1993) Behaviour of major and trace constituents in fresh and salt intrusion waters, in the western Netherlands. Study and modeling of saltwater intrusion into aquifers. In: *Proceedings of the 12th Saltwater Intrusion Meeting.* Barcelona.

Van Dyck, E., Van Burm, P., De Vliegher, B. et al. (1986) *Groundwater Vulnerability Map of Antwerp. (In Dutch: Kwetsbaarheidskaart van het grondwater in Antwerpen.)* AROL, Administration for Spatial Organisation and Environment, Administration for Environment, Brussels.

Vandormael, C. (1992) *Groundwater Quality in Limburg. (In Dutch: Grondwaterkwaliteit in Limburg.)* Ministry of the Flemish Community, AMINAL.

Vissers, M.J.M., van der Veer, G., van Gaans, P.F.M. et al. (2005) The controls and sources of minor and trace elements in groundwater in sandy aquifers. In: Vissers, M.J.M. (ed.) Patterns of groundwater quality in sandy aquifers under environmental pressure. PhD thesis, Utrecht University, pp. 89–122.

Walraevens, K. and Cardenal, J. (1999) Preferential pathways in an Eocene clay: Hydrogeological and hydrogeochemical evidence. In: Aplin, A.C., Fleet, A.J. and Macquaker, J.H.S. (eds.) *Muds and Mudstones: Physical and Fluid Flow Properties.* Geological Society of London, Special Publication, 158, 175–86.

Walraevens, K., Mahauden, M. and Coetsiers, M. (2003) Natural background concentrations of trace elements in aquifers of the Flemish region, as a reference for the governmental sanitation policy. ConSoil 2003, Gent. Conference Proceedings. Theme A: Policies and Strategies on Soil and Groundwater, pp. 215–24.

Walraevens, K., Cardenal-Escarcena, J. and Van Camp, M. (2007) Reaction transport modelling of a freshening aquifer (Tertiary Ledo–Paniselian Aquifer, Flanders-Belgium). *Applied Geochemistry* 22, 289–305.

Wemaere, I. and Marivoet, J. (1995) Geological disposal of conditioned high-level and long lived radioactive waste, Updated regional hydrogeological model for the Mol site (The northeastern Belgium model). Study Centre for Nuclear Energy SCK-CEN, Mol.

Ziegler, P.A. (1982) *Geological Atlas of Western and Central Europe.* Shell Internationale Petroleum Maatschappij B.V., Elsevier, Amsterdam. 130 pp.

13 The Miocene Aquifer of Valréas, France

F. HUNEAU AND Y. TRAVI

The Valréas Miocene sandstone forms an important aquifer in Provence, France. This paper assesses the variations in baseline chemistry and the dominant geochemical processes operative in the aquifer, from the recharge area, close to Valréas, to the confined aquifer in the south of the basin. Much of the area is dominated by farmland and vineyards although local light industries are present around Valréas. The aquifer provides public and private water supplies to towns, farms and industry. The hydrochemistry of the groundwater is variable and a strong contrast between the recharge area (northern part of the basin) and the confined aquifer (south part of the basin) exists. The unconfined aquifer contains recent and sub-modern Ca–HCO$_3$ type waters, whereas the confined aquifer contains older Na–K–HCO$_3$ type waters. This contrast in hydrogeochemistry is accompanied by a strong contrast in the isotopic content of waters, which clearly shows the Pleistocene origin (>20,000 years) of confined groundwaters.

The geochemical environment of the aquifer is controlled by the presence of a marly Pliocene cover, which reaches thicknesses of more than 300 m in the south of the area. The flow of groundwaters, infiltrated in the northern area, beneath the impermeable clayrich Pliocene cover, where a redox boundary is present, has allowed increasing concentrations of many elements such as arsenic and iron. In addition, cation exchange reactions are responsible for the evolution of the water type, and an increase in Na and K along the flow line. A clear anthropogenic influence is detected in the recharge area where groundwaters contain high concentrations of NO$_3$, Cl and SO$_4$. These concentrations are mainly related to agricultural inputs to vineyards.

It is concluded that the properties of groundwaters in the Valréas Miocene basin are mainly determined by natural reactions between rainwater reacting with the bedrock and the poorly permeable sediments of the Pliocene cover. The evolution along flow paths is closely related to the intensity of water–rock interactions, which are also proportional to residence time of waters within the aquifer. The natural baseline is expressed as a range of concentrations, which can vary over several orders of magnitude for some elements. However, some elements are enhanced over the natural baseline: increases in nitrate and chloride, for example, have been modified by agricultural and industrial practices. The presence of relatively high concentrations of some trace elements (and more particularly arsenic) is considered to be due to entirely natural processes.

13.1 Introduction to the Valréas aquifer

The Valréas basin is located in southeastern France, near the Rhône Valley, at the northern boundary of Provence. This study area is centred on the town of Valréas and is included in the Department of Vaucluse. The basin is geologically a syncline surrounded by mountains of Cretaceous age such as Mont-Ventoux, the Lance Mountain, and the Tricastin Hills. It comprises an area of typical Miocene 'peri-Alpine molassic' dowland, with Quaternary and Pliocene cover. The Lez River and its small tributaries drain the whole basin and join the Rhône River close to Bollène.

The Valréas Miocene aquifer is a major groundwater resource in Provence. During the twentieth century it has been extensively used, mainly for agricultural purposes such as irrigation of vineyards. As a consequence, a dramatic lowering of potentiometric heads has been recorded and most of the wells are now no longer artesian. In recent years, the Valréas Miocene aquifer has been the object of new investigations. The increasing demand for drinking water supplies forces the authorities to find alternative resources to the alluvial aquifers, which were previously intensively exploited. Due to its structure and its nature (mainly cemented sandstones), the Valréas Miocene aquifer is believed to contain high-quality waters suitable for drinking water supply purposes and is considered as a strategic resource due to the potential for pollution of surface waters. The reservoir in its southern portion is perfectly confined under a marly cover of around 100–300 m and is thus protected from most modern pollutant sources.

However, much of the area is given over to wine production and the impact of irrigation has been significant, especially since the Second World War. Over-abstraction of groundwater may lead to unacceptable lowering of the water table. In addition, the intensive agricultural activity of the region has already caused diffuse pollution by nitrate in the recharge area of the system. Increasing nitrate concentrations in the recharge area are worrying and consequently increasing nitrate concentrations may reach the deep confined aquifer. As well as diffuse pollution, point source pollution from urban and rural wastes is occurring.

Knowledge of the groundwater quality and the potential influence of high abstraction rates are limited. The main threat would be that these young polluted waters from the recharge area and the surface of the system would reach the confined aquifer, which normally provides high-quality water. For these reasons, a good definition of the structure of the system and a detailed study of hydrogeochemical processes has been necessary to improve knowledge of residence time of water and to evaluate the dynamics of groundwater flow.

13.2 Description of the study area

An essential first step for assessing the natural baseline groundwater quality is an understanding of the system within which the groundwater is contained. This requires information on the geological and hydrogeological properties to provide the physical framework of the system being studied; definition of the mineralogy and geochemistry of the component minerals to explain the characteristic of groundwater chemistries; and finally the initial inputs to the system, principally rainfall chemistry to define the source term.

13.2.1 Geology

13.2.1.1 Structure

The Valréas Basin (Fig. 13.1) has undergone extensive deformation as part of the Alpine earth movements, which has produced a complex geological and hydrological system. Many fault systems affect the basin; they are mainly oriented SSW–NNE and known as post Alpine disturbances. These faults strongly affect the Cretaceous basement of the basin and can easily be observed on the peripheral limits of the basin (Baronnies, Tricastin, Massif d'Uchaux). The faulting also affects the lower levels of the Miocene sediments but observations of surface structures are most of the time difficult due to the presence of Quaternary and Pliocene cover over the Miocene deposits. The faults are likely to have an important control on groundwater flow and quality in this area. The Cretaceous deposits found in some places under the Miocene basin contain strongly mineralised waters that can potentially circulate upwards along the faults and influence the geochemistry of Miocene waters.

13.2.1.2 Stratigraphy

Miocene sediments are heterogeneous and mainly composed by calcareous cemented sandstone alternating with marly layers. The depth of sediments reaches 600 m (Fig. 13.2) in the central basin. The outcrop area of Miocene sediments occurs in the north-eastern part of the basin and the sediments are covered by Pliocene deposits coverage of approximately 150–300 m depth in the southwestern part of the basin.

13.2.1.3 Messinian crisis

A major palaeo-ria resulting from the drastic lowering of the Mediterranean Sea during the Messinian period cut the Miocene sediments in the southern part of the aquifer (Fig. 13.2). This palaeo-ria follows the line of the Aigues River. During the Pliocene, this

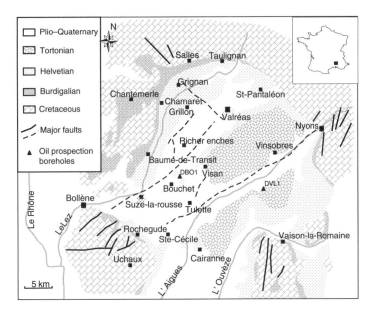

Fig. 13.1 Geology of the study area.

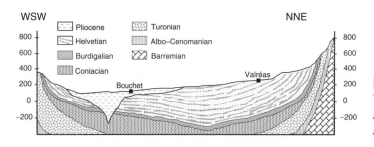

Fig. 13.2   Cross section through the Valréas Miocene basin (the Valréas aquifer is composed of Helvetian and Burdigalian sediments).

deep canyon was infilled with poorly permeable sediments, mainly marls and clays. This Pliocene cover confines the Miocene aquifer in the south of the basin.

13.2.2   Hydrogeology

Miocene sediments have an average hydraulic conductivity of approximately 10^{-6} m s^{-1}. This is particularly true in the central part of the basin were the productivity of the aquifer is very high. Close to the eastern and western boundaries and in the northern recharge area, hydraulic conductivity is slightly lower and is around 10^{-7} m s^{-1}. Groundwater flow occurs from the northeast to the southwest and reaches the Rhône River Valley through the Lez River Valley (Fig. 13.3). The main flow path is from the recharge area (North of Valréas town) to Suze; an additional flow path may be considered from Tulette to Suze. The central part of the basin, covered by the poorly permeable Pliocene sediments, used to be an important artesian area at the beginning of the century (Gignoux 1929). Artesian flow has now almost disappeared due to intensive abstraction of groundwater. Only a few deep boreholes located around the village of Bouchet are still artesian. The existence of close relations between surface drainage and the upper part of the aquifer is clearly identified along the Lez River valley (Huneau et al. 2001). Since the beginning of the exploitation of the aquifer, these relations have been strongly modified and it is now difficult to evaluate the intensity of the exchange between the Miocene aquifer and the Lez alluvial aquifer.

13.2.3   Lithology and mineralogy of the aquifer

The Valréas sandstone is a marine sediment, which accumulated slowly on the shore of

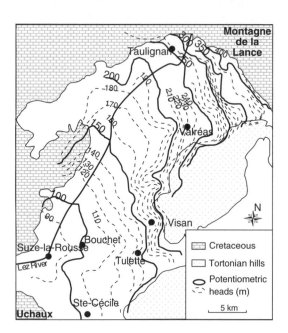

Fig. 13.3   Water level contours across the Miocene aquifer.

the Miocene sea. Detrital materials derived from erosion of the Alps are mainly sands and clay minerals, occurring as interbeded multi-layers. In some places, close to the Miocene shore, seashells can be found in the sandstone. The porosity of such complex sediment is difficult to evaluate, but it can be roughly estimated around 5–10%. The average composition of the sediment is shown in Table 13.1 and can be summarised as follows: 35% calcite, 35% quartz, 25% clay minerals plus additional fractions of dolomite, K-feldspar and plagioclase. Clay minerals have been identified and mainly comprise smectite and illite (Table 13.2).

13.2.4 Rainfall chemistry

Rainfall chemistry can be regarded as the primary input for explaining the baseline quality. For some elements it may well be the major source of solutes with very little being added to infiltration and flow. For this region, no rainfall stations exist which routinely measure rainfall chemistry and therefore a station located in Avignon is used. Avignon is 50 km south of Valréas in a similar position relative to the coastline but with a lower rainfall amount than Valréas. The average rainfall at the Valréas station is 760 mm yr^{-1}, and 660 mm yr^{-1} at the Avignon station.

Table 13.1 Composition (%) of some samples of Miocene and Pliocene sediments at different depths.

Boreholes	Depth (m)	Clays	Quartz	K-Feldspar	Plagioclase	Calcite	Dolomite
Su26	54–58	26	34	1	2	35	2
(Pliocene)	95–100	21	35	1	5	34	4
Tu20	18–22	26	32	0	2	26	14
(Pliocene)	58–62	5	36	2	3	50	4
	85–90	27	41	0	2	28	2
Ri5 (Miocene)	91–90	33	27	0	1	36	3
	154–159	24	37	5	6	24	4
	219–223	8	56	1	9	18	8

Source: Huneau (2000).

Table 13.2 Composition (%) of the clay minerals contained in Pliocene and Miocene sediments.

Boreholes	Depth (m)	Kaolinite	Illite	Chlorite	Smectite 10–14	Corrensite
Su26 (Pliocene)	54–58	3	30	4	63	Traces
	95–100	2	31	4	63	
Tu20 (Pliocene)	18–22	0	39	6	55	Traces
	58–62	2	37	6	55	
	85–90	1	36	5	58	Traces
Ri5 (Miocene)	91–90	2	30	4	64	Traces
	154–159	1	22	5	72	Traces
	219–223	1	8	2	89	

Source: Huneau (2000).

Table 13.3 Rainfall chemistry for Avignon station.

	Units	Rainfall	Rainfall × 4.3
pH	pH	4.6	
Na	mg L^{-1}	0.73	3.14
K	mg L^{-1}	0.22	0.94
Ca	mg L^{-1}	1.04	4.47
Mg	mg L^{-1}	0.12	0.51
Cl	mg L^{-1}	1.33	5.72
SO$_4$	mg L^{-1}	2.26	9.72
NO$_3$	mg L^{-1}	1.49	6.40
NH$_4$	mg L^{-1}	0.25	1.07
Rainfall	mm	900	

Note: Weighted average composition from October 1997 to April 1999.

Source: Celle (1999).

Chemical analyses for the station of Avignon are given in Table 13.3 for major elements. In addition, these have been multiplied by a factor of 4.3, which roughly accounts for the likely concentration due to evaporation under the prevailing climatic conditions (ratio corresponding to the rainfall amount divided by the amount of effective infiltration after evaporation). These values may be used as a guide for comparison with the groundwater. It is important to note that Cl is inert and groundwater concentrations may be largely rainfall derived. The high concentrations of NO$_3$ and SO$_4$ observed in Avignon most likely relate to the active industrial activity of the Rhône Valley. In the Valréas region, these concentrations are obviously lower and the nitrate content of rainfall is close to detection limit. Note that K and NO$_3$ can be taken up by vegetation and may therefore be found at lower concentrations in recharge waters than in rainfall.

13.3 Data for the Miocene Aquifer of Valréas

13.3.1 *Historical data on water quality*

Historical data on water quality are very sparse for the Valréas area. The most recent analyses were carried out by Roudier (1987) who was the first to evaluate the potential resource of the Miocene sediments as a water supply. Extensive searches as part of the present programme have failed to locate significant records of archive material of value to the present study, although odd, single analyses have been found in the DDASS (Sanitary and Social Direction of Vaucluse) data bank. This lack of quality data could be explained by a great number of private boreholes only devoted to agricultural use.

13.3.2 *New sampling programme*

A total of 110 samples were collected during the summer of 1999 from representative sites over the whole Miocene Basin, providing a good regional coverage of the aquifer with an emphasis on boreholes located along the main flow line from the recharge area to the Pliocene cover in the south. These form the main data-set used for deriving baseline conditions. They have been analysed for a full range of inorganic species, and in addition, field measurements (Eh, DO, pH, temperature, alkalinity and SEC) are used in the interpretation. All samples were filtered in the field and acidified with nitric acid (1% v/v) to stabilise trace elements. Additional samples were collected in glass bottles for stable isotopes (δ^2H, δ^{18}O and δ^{13}C). Boreholes were all pumped for an estimated two well bore volumes prior to sampling.

13.4 Hydrogeochemical characteristics of the groundwater

13.4.1 Summary statistics

A summary of the data is shown in Table 13.4 for the study area. This shows the range and average concentrations as well as an upper concentration (defined as mean $+2\sigma$ or 97.7 percentile), which is used as a cutoff for outlying data. The median is preferred to the mean as it is more robust and less affected by extreme values.

13.4.2 Water types and physicochemical characteristics

Groundwaters of the Valréas Miocene aquifer show a wide range of characteristics in terms of physicochemical parameters and element concentrations (Table 13.4). Although the waters are fresh, mineralisation varies from weakly to moderately mineralised (SEC from 310 to 1320 μS cm^{-1}). Temperatures are relatively high around 22–24°C in the confined aquifer at a depth of 350 m under Pliocene marls, but are generally in equilibrium with the atmosphere in the outcrop area (around 12–14°C). The waters are well buffered at circumneutral pH with a median of 7.4. Many groundwaters contain low concentrations of dissolved oxygen and have negative Eh values indicating that the aquifer varies from reducing to oxidising. However, it was not possible to measure this parameter at all sites in a flow-through cell and the median Eh of 218 mV is undoubtedly an overestimate.

Table 13.4 Field parameters, isotope data and range of major and minor element concentration in the Valréas Miocene aquifer.

Parameter	Units	Min.	Max.	Median	Mean	97.7%ile
T	°C	12.2	23.9	16.2	16.7	22.8
pH		7.00	8.76	7.41	7.46	8.12
Eh	mV	−202	372	219	136	349
DO	mg L^{-1}	0.47	10.02	4.45	4.37	9.02
SEC	μS cm^{-1}	310	1324	519	533	863
$\delta^{18}O$	%	−9.23	−6.44	−7.28	−7.47	−6.58
$\delta^{13}C$	%	−14.2	−5.44	−10.7	−10.2	−6.13
Ca	mg L^{-1}	12.1	196	92.3	93.9	158
Mg	mg L^{-1}	1.12	66.3	14.9	17.0	45.3
Na	mg L^{-1}	2.26	318	8.2	17.2	85.9
K	mg L^{-1}	0.2	9.5	1.5	1.8	7.3
Cl	mg L^{-1}	3.4	260	11.7	17.3	42.3
SO$_4$	mg L^{-1}	4.6	223	42.1	48.2	144
HCO$_3$	mg L^{-1}	203	483	299	310	409
NO$_3$	mg L^{-1}	0.0	428	4.8	17.3	93.2
F	mg L^{-1}	0.00	0.84	0.12	0.13	0.39
Br	mg L^{-1}	0.00	1.67	0.03	0.06	0.22
Sr	mg L^{-1}	0.07	5.87	0.55	0.72	1.74
Fe	mg L^{-1}	0.00	4.34	0.23	0.36	1.89
Si	mg L^{-1}	8.96	36.8	18.1	19.3	34.1

The groundwaters in the Valréas Miocene aquifer vary from Ca–HCO$_3$ to Na–K–HCO$_3$ types as shown on the Piper diagram in Fig. 13.4. Although Na and Cl concentrations are generally low in most groundwaters, concentrations can be relatively high locally and the waters may tend toward the Na–K–Cl type. This is particularly true in the area of Suze-la-Rousse where a deep borehole samples such waters. The village of Suze-la-Rousse is located along a major fault and upward groundwater circulation from the Cretaceous sediments may occur. This special point is not considered in the baseline box plot.

13.4.3 Major elements

The summary data are shown graphically as box plots and cumulative frequency plots

(Figs. 13.5 and 13.6). The boxplots display the range of data and are designed to show the distribution of data on a percentile basis. Cumulative probability plots are useful in visualising the distribution of data and can be of use in determining outlying data or discriminating pollution. These single values are useful for comparison between areas but it should be stressed that a range of compositions in any given aquifer represents the baseline. For the vast majority of elements the distribution represents geological and geochemical controls on the groundwater composition. The slope, curvature and overall shape of the lines are indicative of different processes.

Most major elements display trends on a cumulative frequency plot that tend to approach linearity, that is, a log-normal distribution. Although the median nitrate

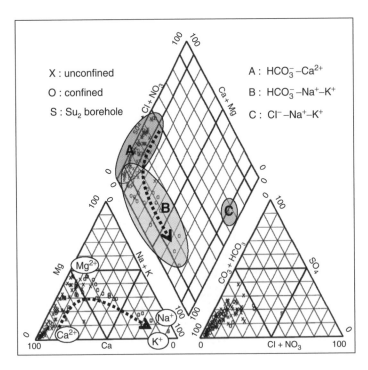

Fig. 13.4 Piper diagram showing groundwaters from the reference aquifer of Valréas.

concentration is close to 5 mg L^{-1}, the distribution of concentrations clearly shows the intense anthropogenic contamination of many boreholes in the recharge area of the aquifer. The median concentration of chloride (12 mg L^{-1}) indicates a source additional to rainfall input. Bicarbonate shows a relatively narrow range of concentration; SO$_4$, on the other hand shows a much wider range. Silicon shows a limited range of concentrations approaching a normal distribution as indicated by the similarity of the median to

the mean concentration. Iron and strontium also show a wide range of concentrations.

13.4.4 *Minor and trace elements*

Minor and trace elements are displayed on a box plot in Fig. 13.7. On the cumulative probability plot (Fig. 13.8), it can be seen that many samples are below the detection limit, shown by vertical lines at low concentrations. Bromide is close to the seawater dilution line determined using median Cl, even if additional source to rainwater may induce some variability. Fluoride is higher than expected compared with Br, but concentrations are still low (maximum value 0.84 mg L^{-1}), and approach a normal distribution as indicated by the similarity of the median (0.12 mg L^{-1}) to the mean (0.13 mg L^{-1}) concentrations (Table 13.4). Silicon shows a relatively narrow range (9–37 mg L^{-1}) with a quite normal distribution (median 18.2 mg L^{-1}, mean 19.3 mg L^{-1}).

The cations Sr and Ba have a wide range of concentrations; Sr reaches more than 5.8 mg L^{-1}, and Ba, although limited in many waters by high SO$_4$ (due to barite saturation),

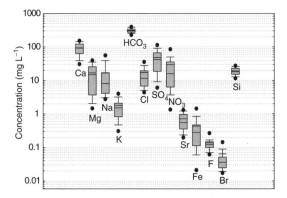

Fig. 13.5 Boxplots of selected major and minor elements.

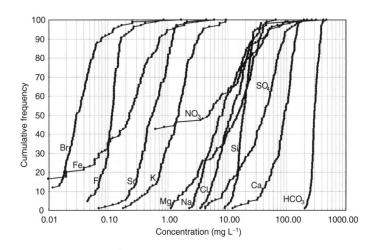

Fig. 13.6 Cumulative probability plots for selected major and minor elements.

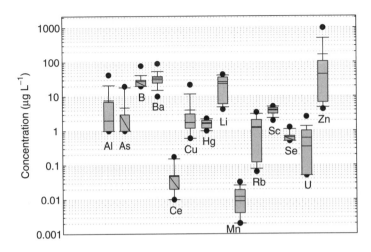

Fig. 13.7 Box plot showing selected minor and trace elements.

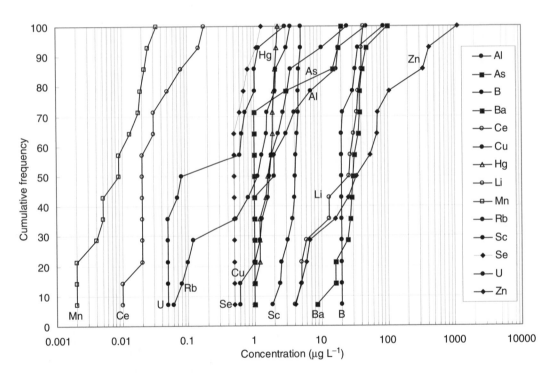

Fig. 13.8 Cumulative probability plots for selected minor and trace elements.

may reach 100 µg L^{-1}, nevertheless staying under the EU maximum admissible concentration (MAC) of 150 µg L^{-1}. Of the metals, Fe shows the highest concentrations varying from less than the detection limit up to 4.34 mg L^{-1} with a median value of 0.23 mg L^{-1}. In contrast, Mn stays relatively low (maximum 34 µg L^{-1}).

Of some concern is the presence of arsenic. Three samples had concentrations greater than 10 µg L^{-1} and up to 20 µg L^{-1}. In general, the high As is associated with reducing groundwaters. These three samples represent old confined groundwaters. Zinc shows a wide range of concentrations and reaches more than 1 mg L^{-1} in the confined aquifer. Such concentrations may be related to both water reducing conditions and the nature of the casing of some very old boreholes, which may include traces of zinc.

Most other metal species lie below the detection limit or have low median values, but may be locally of concern. Aluminium, mobile under relatively low pH conditions, shows relatively high concentrations (up to 48 µg L^{-1}), however, the median stays below 2 µg L^{-1}. Copper was found in some samples up to 24.7 mg L^{-1} (Table 13.4).

13.4.5 *Indicators of pollution*

In order to determine baseline concentrations, it is desirable to study pristine waters and, as far as possible, avoid polluted waters since these may alter baseline concentrations either directly by adding solutes or indirectly by promoting chemical reactions in the aquifer. It is generally difficult to obtain pristine waters, in part because waters may be sampled over a large screened interval where polluted waters are present at shallow depth or because diffuse pollution (especially agricultural pollution) is present in large parts of the unconfined aquifer.

Boreholes clearly affected by point source pollution have been avoided in the recharge area of the aquifer. Within this area, most of the sampling points show indicators of agricultural pollution such as high nitrate concentrations. Most of the time high nitrate concentration correlates with high Cl, K and SO$_4$ concentrations. The use of fertilisers by farmers in the recharge area can be responsible for such concentrations (Cl and K are present as traces in agrochemical or organic fertilisers). Baseline nitrate concentrations are most likely to be around a few mg L^{-1} as indicated by limited historical data. Furthermore, on the cumulative probability plot (Fig. 13.6), the main curvature occurs at around 4.8 mg L^{-1} and this is around the concentration expected for nitrate-N baseline concentrations. However, many waters are relatively reducing and denitrification may in certain places have lowered concentrations. A good understanding of the chemical controls may help to discriminate the anthropogenic influence.

Concerning the confined aquifer, most of the boreholes are still likely to be free from any anthropogenic influence. Waters from the confined aquifer of Valréas may thus be considered as pristine waters.

13.4.6 *Geochemical controls and regional characteristics*

This section deals with the dominant geochemical processes which influence groundwater chemistry including mineral dissolution/precipitation, redox reactions, ion exchange and residence time. Samples have been studied in detail along a potential flow path in order to understand the geochemical changes that take place with distance and time in the aquifer. These geochemical changes are then evaluated and put into a regional context.

13.4.7 *Chemical evolution along flow lines*

The main flow path studied trends approximately from the north of Valréas city to the village of Suze-la-Rousse, that is to say, from the northeast to the southwest. This flow line covers the recharge area of the aquifer and the confined aquifer in its southern portion. Selected data, including residence time indicators, are plotted against distance in Fig. 13.9.

The most obvious hydrochemical evolution is seen in the cation concentrations. The Ca–HCO₃ type in the recharge area changes to Na–K–HCO₃ type along the flow path. Calcium concentrations strongly decrease from 150 mg L⁻¹ to less than 50 mg L⁻¹ whereas sodium increases from 5 to more than 100 mg L⁻¹ and K from 0.5 to 3.5 mg L⁻¹. All groundwaters confined under the Pliocene cover are of Na–K–HCO₃ type.

13.4.8 Mineral dissolution reactions

Much of the chemistry is established at the start of the flow line due to dissolution reactions involving the calcite and dolomite cements of the sandstone. The production of CO_2 in the soil zone initially leads to slightly acidic waters due to production of carbonic acid, but this acidity will be neutralised through reaction with carbonate minerals. All groundwater samples along the flow path are at saturation with respect to calcite, controlled by the reaction:

$$CaCO_3 + H_2CO_3 \rightarrow Ca^{2+} + 2HCO_3^{-}$$

The first samples at the eastern edge of the flow line are undersaturated with respect to dolomite but reach saturation after approximately 10 km by dolomite dissolution:

$$CaMg(CO_3)_2 + 2H_2CO_3 \rightarrow Ca^{2+} + Mg^{2+} + 4HCO_3^{-}$$

The dissolution of calcite and dolomite provides the initial dominant control on water chemistry. The above reaction (congruent dissolution) is rapid, taking place in

the top few metres of the aquifer. Once saturation with respect to calcite is reached, no further calcite will dissolve rapidly. Congruent dissolution yields Ca and Mg concentrations, which are identical to those in the dissolving calcareous cement of sandstone. However, with passage of time in waters that are in equilibrium with the calcareous matrix, a second process (incongruent dissolution) may occur where the impurities (e.g. Mg) in the matrix are slowly released and a purer calcite left behind. In this process, the Mg/Ca ratio in groundwater progressively increases with time.

13.4.9 Major element controls and ion exchange reactions

The chemical evolution of major cations, pH, and alkalinity along the flow path in the aquifer suggest that the aquifer previously contained a more saline component than the present day recharge water, and therefore a freshening pattern is observed, with Ca–HCO₃ type water at the upstream end of the aquifer replacing Na–HCO₃ type water in the downstream part of the aquifer. The observed pattern is typical of freshening aquifers where ion exchange is the dominating geochemical process. Calcium displaces the original saline-derived cations, Na, K and Mg, from the ion exchange complex in a chromatographic pattern releasing first Na and K and then Mg to the aqueous phase. However, Mg fixation processes including dolomite precipitation may influence the freshening pattern evolution. The marine influence probably originates from the deposition of the marine Pliocene cover in the southern part of the Valréas basin, and thus the observed hydrochemistry along the flow line in the aquifer is probably the result of millions of years of geochemical evolution.

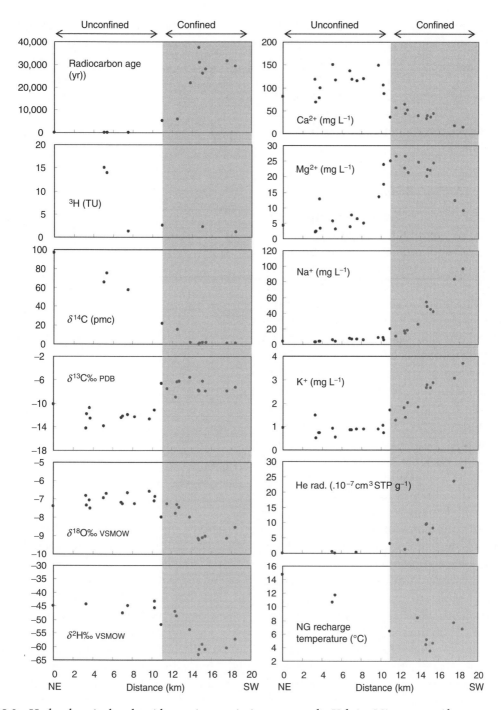

Fig. 13.9 Hydrochemical and residence time variations across the Valréas Miocene aquifer.

Modelling of the geochemical evolution along the flow line in the aquifer was carried out with the 1D reactive-transport code PHREEQC to identify and quantitatively describe the major controlling processes which have been responsible for the present day observed hydrochemistry (Postma et al. this volume).

13.4.10 Redox reactions

The parameters, redox potential (Eh) and dissolved oxygen (DO) concentrations, provide the primary indicators of the redox status of natural groundwaters. Unfortunately, it was not possible to measure these in all groundwaters due to sampling from an intermediate storage tank or because the diameter of the out flow was too large to connect to our flow-through cell. Nevertheless, a redox boundary can be recognised. This boundary coincides with the coverage of the Miocene sediments by the Pliocene marls, that is, to say approximately 10 km far from the recharge area (Fig. 13.10).

Water at recharge is saturated with DO at the partial pressure of the atmosphere (10–12 mg L^{-1} depending upon barometric conditions). Passing through the soil and the saturated zone, some of this O_2 will react as a result of microbiological processes and oxidation–reduction reactions. In the Miocene aquifer, however, almost all water reaching the water table still contains several mg L^{-1} O_2. Geochemical reactions such as the oxidation of Fe^{2+} in the sandstones progressively remove the O_2 along the flow lines. Once all the oxygen has reacted an abrupt change of water chemistry takes place (redox boundary). Other changes may occur at and downgradient of the redox boundary, especially denitrification and the probability that total

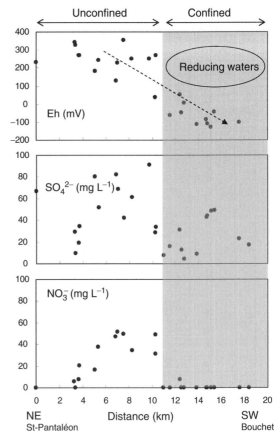

Fig. 13.10 Redox parameters and redox sensitive species across the flow line.

dissolved iron (Fe^{2+}) concentration will increase. Sulphate reduction and the production of sulphide (H_2S as HS^- in solution) also occur at greater depths.

This boundary is recognisable when considering redox sensitive species such NO_3, SO_4, Fe and Mn. Nitrate concentrations are relatively high northeast of the redox boundary but concentrations rapidly decrease to the southwest. The reduced nitrogen species NO_2 and NH_4^+ are present in some samples beyond this zone indicating denitrification and nitrate reduction is likely to have occurred.

Iron and Mn concentrations increase beyond the boundary, consistent with the change in redox condition. The redox sensitive trace elements As and Se are found to be slightly higher in the more reducing waters. This is particularly true for arsenic where concentrations reach more than 20 µg L^{-1}. However, the chemistry of these trace elements is relatively poorly understood in terms of mobility and speciation in the aquifer.

13.4.11 Mixing with older formation water

Evidence for the existence of old deep saline groundwaters has been found in the area of Suze-la-Rousse. Here, a major fault system allows mixing between Miocene groundwaters and Na–K–Cl type waters ascending from the Cretaceous basement of the basin. Only one borehole producing such waters has been discovered and the mixing seems to be extremely localised to this small area. No evidence of diffused upward leakage was found since the chloride concentration of confined groundwaters is extremely low around 5 mg L^{-1} and clearly from rainfall origin.

13.4.12 The age of the groundwater

Figure 13.9 shows the evolution of the age of groundwaters along the main flow line. Stable isotopes in addition to radiocarbon (^{14}C) have also been analysed (Fig. 13.11). ^{13}C ratios are enriched, whereas ^{18}O and ^{2}H values

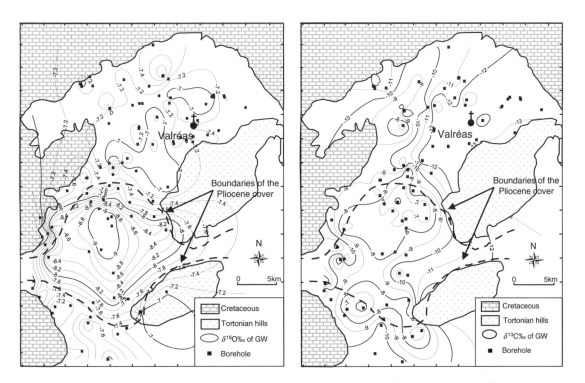

Fig. 13.11 Distribution of oxygen-18 and carbon-13 in groundwaters of the Miocene aquifer.

are depleted in comparison with modern recharge groundwaters from the north of the basin. The plot of $\delta^{13}C$ and $\delta^{14}C$ along a downgradient direction (Fig. 13.9) suggests an evolution of $\delta^{13}C$ with groundwater residence time. This phenomenon is related to isotope exchange between groundwater and the aquifer matrix. The same processes affect ^{14}C, which necessitates the correction of radiocarbon ages by means of an appropriate correction model, in this case the Fontes and Garnier model (1979) was used. Groundwater ages increase rapidly away from the recharge area and reach ages of more than 30,000 years in the confined aquifer. A strong isotopic contrast exists between the unconfined and the confined portions of the aquifer. A discontinuity can be observed between recent groundwaters from the recharge area ($^{18}O \approx -7\permil$ VSMOW; $^2H \approx -45\permil$ VSMOW) and old groundwaters confined under Pliocene sediments ($^{18}O \approx -9\permil$ VSMOW; $^2H \approx -60\permil$ VSMOW). This indicates that old groundwaters have been recharged under colder climatic conditions than at present time. The depletion in stable isotopes occurs around 18,000 years BP and points to the transition from the late-Glacial to the Holocene. Noble gas measurements were used to confirm and to quantify the climate signal recorded in the aquifer, the results proving the Pleistocene origin of confined groundwaters (Huneau 2000).

A confirmation of the long residence time of groundwaters within the aquifer has been shown using radiogenic helium concentrations dissolved in groundwaters. Confined groundwaters from the south of the basin show strong accumulation of helium indicating a residence time of at least 30,000 years (Huneau 2000).

13.4.13   Regional variations

The geochemical controls highlighted in the main cross-section can be applied to the chemical variations present regionally in the study area. There is a clear contrast in chemistry between the northeastern and the southwestern parts of the area, which can be separated along a line linking Richerenches to Visan. Ca–HCO$_3$ type waters are found in the whole north sector where Miocene sediments are not confined, but the groundwaters in the north generally show strong anthropogenic influences (high NO$_3$, SO$_4$ and Cl concentrations). Data from the southern part of the aquifer shows Na–K–HCO$_3$ water types, which are all confined under 100–300 m of Pliocene sediments. The low chloride concentrations of these waters indicate that no upward leakage from the Cretaceous basement occurs. The Mg/Ca ratio increases from the recharge area to the confined aquifer, related to the increasing of residence time of waters within the aquifer. The confined aquifer produces highly reducing waters (presence of H$_2$S) in the south of the basin.

13.5   Summary of baseline quality

The chemical data show that the primary important influences on water chemistry are mineral dissolution reactions, involving in particular calcite and dolomite. This has resulted in relatively high natural concentrations of Ca, Mg and HCO$_3$. Cation exchange reactions are also responsible for a significant amount of the increase in Na and K, these ions indicate a marine origin and show the existence of a freshening pattern within the aquifer. In addition, redox reactions are important for many species due to the confined nature of the aquifer as a consequence of the existence of a thick Pliocene cover in the south of the study area. Anthropogenic inputs over much of the aquifer are minimised by the protection afforded by the thick Pliocene deposits and the flow conditions. Therefore, in the southern part of the study area, for most elements the concentrations measured and the ranges found can be taken as representative of the natural baseline. On the contrary, the recharge area of the aquifer, which is located north from Valréas, tends to show strong pollution indices mainly resulting from agricultural activities. This conclusion is a major source of concern and should be taken into account by regional policy-makers.

The data have been presented in Table 13.4 and the median value and 97.7 percentile provide a good estimate of the average and upper baseline concentrations in the aquifer. However, this should be used in conjunction with maps that show that regional variations do exist in the aquifer. The baseline chemistry changes spatially across the aquifer in relation to the extent of water–rock interaction (hence residence time within the aquifer) and to geochemical controls imposed by the local geochemical environment (e.g. oxidation–reduction controls). The baseline for the groundwater in the recharge area is clearly distinct from that in the central part of the confined aquifer, the limit often being close to the median value. In addition, the effect of both facies changes in the aquifer and increasing residence time means that the baseline concentrations increase southward for many parameters. These regional variations must be taken into account when defining the local baseline for the aquifer.

It is clear that the baseline concentrations and ranges are different between unconfined and confined areas.

The effect of anthropogenic inputs is indicated by enhanced concentrations of N-species (Fig. 13.12), pesticides, and potentially a range of other solutes from urban and industrial usage (SO_4, Cl, K). The most obvious inputs over the last few decades are derived from agricultural pollution. Baseline nitrate concentrations are most likely to be around a few mg L^{-1} as indicated by limited historical data. The use of cumulative probability plots in the study area is difficult because much of the original nitrate might have been reduced in the confining parts of the aquifer. A cumulative probability plot has been drawn for nitrate on Fig. 13.6. The main curvature occurs at around 4.8 mg L^{-1} and this is around the concentration expected for nitrate-N baseline concentrations. Other elements which may have been modified by agriculturally derived anthropogenic inputs include K, Cl and SO_4. Some boreholes show evidence of increasing chloride in the recharge areas (Fig. 13.12). High chloride concentrations are always accompanied by high SO_4 and K concentrations and are directly related to agricultural inputs in vineyards. Potassium is not very mobile in the surface environment and most anthropogenic inputs will be consumed by biomass or through reactions involving clay minerals. Chloride, on the other hand, is very mobile and likely to modify the baseline significantly where evaporites are not present. Chloride concentrations may be enhanced locally in general around urban areas such as Valréas, Grignan or Taulignan. Although individual water samples may have been affected by such inputs, these are generally

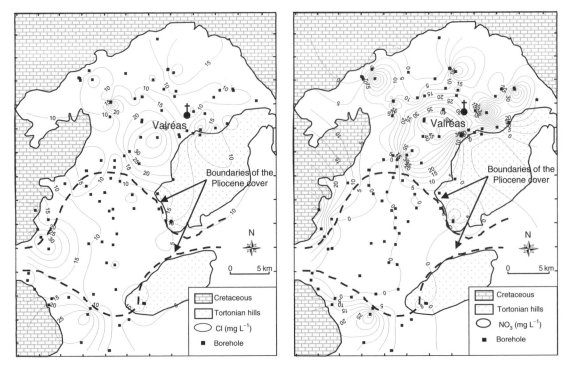

Fig. 13.12   Distribution of nitrate and chloride in Miocene aquifer groundwaters.

within the range of baseline concentrations produced through natural processes.

Some trace and minor elements including As and Fe are relatively high in parts of the aquifer, but they are considered to represent natural baseline. High concentrations of dissolved As are known from other sandstone groundwaters in Europe (Heinrichs and Udluft 1999; Shand et al. 2002). The range of variation in these elements is related to variations in source as well as geochemical environment, particularly changes in redox conditions within the aquifer.

References

Celle, H. (1999) Caractérisation des précipitations sur le pourtour de la Méditerranée occidentale, approche isotopique et chimique. Université d'Avignon, thèse, 222 pp.

Fontes, J.C. and Garnier, J.M. (1979) Determination of the initial ^{14}C activity of the total dissolved carbon: A review of the existing models and a new approach. *Water Ressources Research* 15, 399–413.

Gignoux, M. (1929) Forages artésiens et rivages pliocènes sur la rive gauche du Rhône entre Carpentras et Valréas. Etudes Rhodaniennes, Revue Géographiques Region. *Lyon* 5, 27–39.

Heinrichs, G. and Udluft, P. (1999) Natural arsenic in Triassic rocks: A source of drinking-water contamination in Bavaria, Germany. *Hydrogeology Journal* 7, 468–76.

Huneau, F. (2000) Fonctionnement hydrogéologique et archives paléoclimatiques d'un aquifère profond méditerranéen. Etude géochimique et isotopique du bassin miocène de Valréas (Sud-Est de la France). Université d'Avignon, thèse, 192 pp.

Huneau, F., Blavoux, B. and Bellion, Y. (2001) Differences between hydraulic and radiometric velocities of groundwaters in a deep aquifer: Example of the Valréas Miocene aquifer (Southeastern France). Comptes Rendues Académie Sciences Paris, *Earth and Planetary Sciences* 333, 163–70.

Roudier, P. (1987) Etude hydrogéologique et hydrochimique des nappes aquifères des bassins Miocènes de Valréas, Vaison-la-Romaine, Malaucène et Carpentras (Vaucluse). Université Lyon-1, thèse, 297 pp.

Shand, P., Tyler-Whittle, R. Morton, M. et al. (2002) Baseline Report Series 1: The Triassic Sandstones of the Vale of York. British Geological Survey Commissioned Report CR/02/102N.

14 The Miocene Sand Aquifers, Jutland, Denmark

K. HINSBY AND E. SKOVBJERG RASMUSSEN

The deeper part of the Miocene quartz sand aquifers of Jutland, Denmark, contain pristine freshwater of very high quality. Hydrochemistry, environmental tracers and groundwater modelling from previous studies show that the deepest of three major Miocene sand aquifers, the Ribe Formation, contain high-quality groundwater recharged during a few thousand years in the Holocene. There is no evidence of contamination from agriculture, point sources, urban environments or any other human impact. The trace elements As, Al, Ni and Zn that represent the trace elements most frequently found above guideline values in the Danish Groundwater Monitoring Program (in up to 16% of the wells) and all other solutes, are generally well within drinking water standards. The content of total dissolved solids are below 400 mg L^{-1} in most wells although natural chemical processes in adjacent aquitards lead to an increase in chloride, dissolved organic carbon (DOC), trace metals and rare earth elements such as, for example, As, Al, B, Cr, Se, U, Th, La and Yb in some downgradient wells. The content of total dissolved solids in these approach 1000 mg L^{-1} or more and may be brackish in some hydraulically isolated areas. In areas with elevated DOC values the drinking water standards may be breached for some of the trace elements, which probably form metallo-organic complexes with the humic substances.

The deep Miocene quartz sand aquifers are composed of fluvio-deltaic sands dominated by pure and inert quartz sands and water–rock interaction is limited. The redox conditions are anaerobic and sulphate concentrations are low (<10 mg L^{-1}), and although sulphate reduction may occur it is slow and insignificant due to the relatively low and recalcitrant organic carbon content in the sediment.

The hydrochemistry of major constituents is quite similar to the hydrochemistry of uncontaminated anaerobic carbonaceous Quaternary aquifers and evaporated precipitation in the area. The most significant geochemical processes that the groundwaters have undergone during recharge – either before recharge to the Ribe Formation or close to the recharge area in the aquifer itself – are dissolution of carbonates and reduction of dissolved oxygen possibly mainly by oxidation of both pyrite and organic carbon. These processes primarily increase the contents of Ca and HCO_3 (calcite dissolution) and Fe and SO_4 (pyrite oxidation). Hence, the hydrochemistry of the Miocene sand aquifers also approximate the natural baseline hydrochemistry of the anaerobic calcareous part of the Pleistocene sand aquifers above.

The study provides baseline information for a large number of mainly inorganic constituents, as an aid to future water-quality management of the aquifer and identification of temporal and spatial hydrochemical trends.

14.1 Introduction

The Miocene aquifers in Denmark consist of three major sand units deposited during the early Miocene. All three units constitute important aquifers with valuable water resources. In this chapter we compare groundwater hydrochemistry from wells screened in the lowermost pristine formation, the Ribe Formation quartz sand aquifer, with hydrochemistry from monitoring wells in Miocene sands in general as extracted by a query from the National Groundwater Monitoring Database. The investigated wells and data are used to estimate the natural baseline quality of Miocene sand aquifers in Denmark. To a certain extent the data can also provide information about the natural background level of some of the anaerobic Quaternary aquifers above. Data on natural background levels are crucial for the sustainable management of the subsurface water resource and the dependent ecosystems (Hinsby et al. 2008).

14.2 Geology and hydrogeology

14.2.1 *Depositional environment and palaeogeography of the Miocene sand aquifers*

The development of the Miocene depositional system, which developed the Miocene sand aquifers in Denmark, was controlled by reactivation of older Late Carboniferous-Early Permian fault systems (Michelsen and Nielsen 1993), tectonic uplift in the hinterland and by relative sea-level changes (Hansen and Rasmussen 2007; Rasmussen 2004a, b).

The main structural elements in the area were the WNW–ESE striking Sorgenfrei-Tornquist Zone which separated the Fennoscandian Shield from the main depositional basin of the North Sea area. This basin was subdivided into three minor parts: the Norwegian-Danish basin, the Ringkøbing-Fyn High and the North German Basin. The Ringkøbing-Fyn High was further segmentated by the NNE-SW trending Rødding Graben and the N-S striking Brande Trough (Fig. 14.1).

Uplift of the Fennoscandian Shield in the late Eocene resulted in high influx of clastic quartz-rich sediments into the northeastern North Sea Basin during the Oligocene. Resumed uplift in the Oligocene and Early Miocene times provides a continually high supply of sediments into the basin (Rasmussen and Dybkjær 2005). The most basinal distribution of the clastic deltaic system occurred in the Early Miocene associated with a distinct eustatic sea-level fall (Rasmussen 2004b).

The Miocene of the Danish area is composed of three stacked deltaic successions from below named the Ribe Formation, the Bastrup sands and the Odderup Formation (Figs. 14.2 and 14.3). Each deltaic complex shows a coarsening upward succession from fine-, medium- to coarse-grained sand with few intercalations of gravel layers. The gravel layers are found both in the lower part and in the upper part of the deltaic succession deposited by gravity flows and shoreface/beach sedimentation, respectively. The coarsening upward succession is from

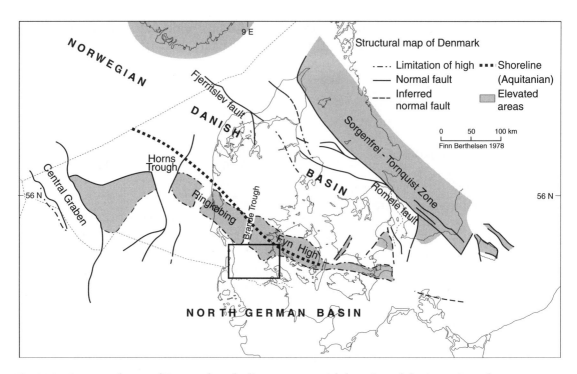

Fig. 14.1 Structural map of Denmark and adjacent areas with location of the investigated area.

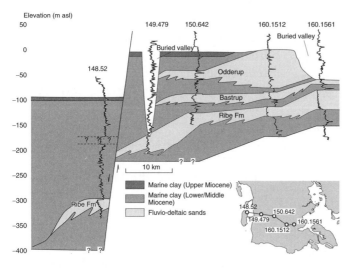

Fig. 14.2 Cross section showing the extent of the Miocene sand aquifers in southern Jutland, and the location of selected wells and two Pleistocene buried valleys. The buried valleys are located in the recharge and discharge areas in eastern and western Jutland, respectively. The valleys and the location of wells on the cross section are not to scale.

20 to 40 m thick. This is commonly capped by a coarse-grained sand or gravel layer deposited as a transgressive lag deposit. This deposit is generally a few metres thick. Channel fill deposits are also common and sometimes shows well developed point bar succession. Lignite-rich deposits are concentrated to the northernmost part. Here it is

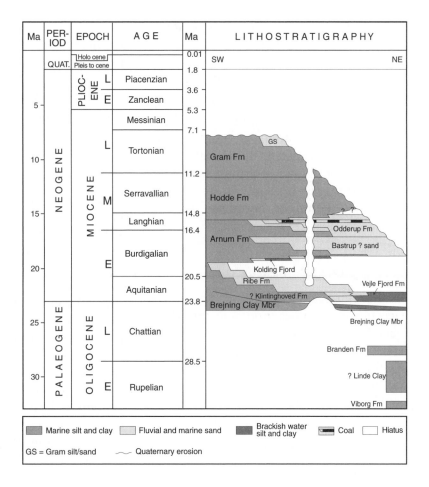

Fig. 14.3 Litho- and chronostratigraphy of the Oligocene and Miocene in Denmark.

associated with stacked fluvial deposits. More widespread formations of lignite deposits are concentrated to the uppermost deltaic succession, especially adjacent to former graben structures. The sand-rich deltaic successions are separated by marine organic-rich, clayey silt deposits. The thicknesses of these deposits are from a few metres to tenths of metres. During the Middle and Late Miocene the accelerated subsidence of the North Sea Basin resulted in a flooding of the area and deposition of marine clayey deposits was widespread, and probably covering the whole Danish area (Koch 1989; Rasmussen 2004b).

During the Miocene the shoreline moved back and forth across Denmark four times. The trend of the shoreline was NW–SE during the early and early middle Miocene (Fig. 14.1), but turned to a dominating N–S trend in the Middle Miocene. The climate was warm temperate to subtropical with summer rain (Koch 1989; Utescher et al. 2000). However, a cooler period occurred in the Early Miocene (Mai 1967; Utescher et al. 2000; Larsson et al. 2006) and a warm climate dominated in the early Middle Miocene (the so called Middle Miocene climatical optimum – Zachos et al. 2001).

14.2.2 Geology of the three major sand aquifers

14.2.2.1 Ribe Formation

The Ribe Formation consists of sand and gravel intercalated with mud deposited in a fluvio-deltaic depositional environment and massive sand deposited during falling sea level. The net thickness of the formation is around 40 m, but is locally up to 80 m in delta forsets within the central part of a delta complex.

14.2.2.2 Bastrup sand

The Bastrup sand consists of sand and gravel deposited in a delta environment and within incised valleys. The sand is often intercalated with marine – or flood plain mud. The net thickness rarely exceeded 30 m in the southern part, but up to 55 m of massive sand has been penetrated at some sites.

14.2.2.3 Odderup Formation

The Odderup Formation is dominated by sand and gravel deposited in fluvio-deltaic, shoreface and beach environments. In central and western Jutland up to three brown coal layers are interbedded in the Odderup Formation. The net thickness of the sand of the Odderup Formation is up to 40 m.

14.2.3 Hydrogeology

The main recharge to the Miocene sand aquifers occurs through Pleistocene sediments just outside the Main Stationary Line (MSL) of the last glaciation, that is, primarily in the areas with elevations of the water table above 30 m (see equipotential lines in Fig. 14.4). The advance of the Weichselian ice sheet stopped at the MSL in the eastern part of the investigated area at around 18,000 years BP. The flow direction in the Miocene sands is from the recharge area towards the west coast of Jutland, where it discharges to the North Sea partly through deep Pleistocene buried valleys located along the coast (Figs. 14.2 and 14.4).

Two types of flow systems exist within the study area: (1) shallow aquifers consisting of Pleistocene glacio-fluvial and interglacial sediments; and (2) a deep regional confined or semi-confined aquifer system consisting of Miocene fluvio-deltaic and marine sediments located at varying depths between less than 50 and up to more than 300 m below surface. The areal extent of the shallow aquifer systems is of the order of tens to thousands of km^2 and aquifer boundaries often coincide with those of surface water catchments. The areal extent of the deep aquifer systems is larger than 1000 km^2. The estimated recharge to the shallow aquifer system ranges from about 200 mm yr^{-1} to more than 400 mm yr^{-1} in the investigated area. Discharge from the shallow aquifers is to streams, wetlands, lakes, wells and to the deeper regional Miocene sand aquifer system. Discharge to surface water bodies accounts for 70–90% of the total recharge. The transmissivity of the Miocene sediments controls the flow paths and the depth of circulation in the shallow aquifer system. The hydraulic conductivity of the confining layers is low and estimated to be in the order of 10^{-8} m s^{-1} or less. The hydraulic conductivity of the Ribe Formation is quite high with an average or geometric mean of 7.7×10^{-4} m s^{-1} in the central parts of the investigated area according to both pumping tests and grain size analyses, respectively. The results of the

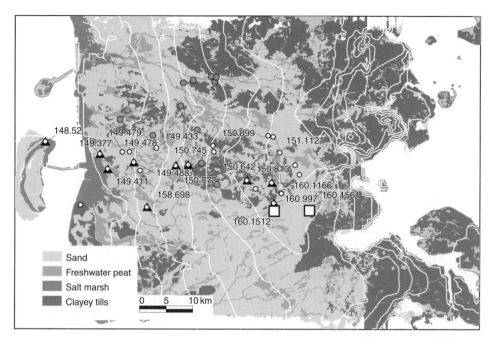

Fig. 14.4 Quaternary geology map with the location of the evaluated wells in southern Jutland and equipotential lines. Triangles are wells sampled in baseline, white circles were sampled in the Palaeaux project (Hinsby et al. 2001a), grey circles were sampled in a PhD project (Jakobsen 1995). Squares are wells investigated in the 'BurVal' project on Buried Quaternary Valleys (BurVal 2006; Hinsby 2006).

grain analyses show that the Ribe Formation sands are generally medium to very coarse grained and well sorted, although unsorted gravel does appear. The geometric mean mentioned above was obtained based on 20 grain size analyses at different levels in wells no. 149.433, 150.745 and 150.899. It is estimated that 10–20% of the recharge to the shallow aquifer system in the recharge area ultimately discharges to the Ribe Formation.

The Ribe Formation quartz sand aquifer is generally a confined aquifer at depths between 100 and 200 m below sea level (150–250 m below surface), with increasing depths from east to west, where it is faulted down to 300 mbsl (Fig. 14.2). The Miocene sands and clay sediments in the investigated area are

typically covered by 40–50 m of Pleistocene deposits. However, several Quaternary buried valleys cut through the area and have locally left valleys with an infill of up to a couple of hundred metres of Pleistocene sediments. Two such incised valleys have been identified in the investigated area, one in the western and one in the eastern part. The locations of the buried valleys are indicated schematically (not to scale) by the white incised areas on the cross section in Fig. 14.2. The buried valleys are believed to create hydraulic windows to the Ribe Formation both in recharge and discharge areas.

A thick succession of Eocene and Oligocene clays, of which some are swelling, forms the lower boundary of the Ribe Formation aquifer and an efficient seal to deeper saline waters (Dinesen 1961; Hinsby et al. 2001a). The Ribe Formation quartz sand aquifer is in the recharge area covered by the other Miocene units or directly by sandy Pleistocene sediments deposited during several glaciations, for example, in buried valleys. Downgradient towards the west coast of Jutland the Ribe Formation is confined mainly by Miocene marine clay aquitards except where the incised buried valley with infills of outwash sands and sandy and clayey tills cut down into the Ribe Formation close to the west coast of Jutland (Fig. 14.2; Hinsby et al. 2001a). Hence the topography of the pre-Quaternary surface and the depths to the Miocene sediments vary considerably in the investigated area. The depths to the high-quality groundwaters in the Miocene sands are generally greater than 50–80 m, and recently drilled wells show that in some areas these can be found to depths of more than 400 m (Rud Friborg, personal communication). The discharge from modern intensively cultivated lands, which cover most of the investigated area, and which clearly influence the shallow aquifers in the recharge area has not yet reached these deep aquifers as demonstrated in the following sections.

14.2.4   The water balance and groundwater/surface water interaction

The investigated area is located in the humid temperate zone and annual precipitation and evaporation is about 950 and 450 mm, respectively (Sonnenborg et al. 2003). A national hydrological model (integrated groundwater/surface water model) was completed in 2003 covering 43,000 km^2 (Henriksen et al. 2003). Denmark is divided into 11 areas in this model, each covered by

a regional hydrological model based on a 1 km^2 computational grid. The model covering the area investigated in this study shows that there are presently no significant impacts on surface water quantity and quality or other dependent ecosystems directly by the Ribe Formation groundwaters.

14.2.5   Pollution, climate and exploitable freshwater resources

The shallow groundwater resources in the investigated area, and in Denmark in general, are threatened by pollution from the surface (especially nitrate and pesticides). A new assessment by the national water resource model is estimating that the sustainable exploitable freshwater resource for Denmark is in the order of just 1 Mm^3 yr^{-1} or 200 m^3 capita^{-1} yr^{-1}, when the polluted resources and effects on dependent ecosystems are taken into account (Henriksen et al. 2003). This amount is similar to what is currently abstracted when permissions for irrigation are fully utilised (Henriksen and Sonnenborg 2003). The new estimate for Denmark is a serious reduction, compared to an earlier estimate of 1.8 Mm^3 yr^{-1} made 11 years ago, and it places Denmark among the 10–15% of UN countries with the lowest renewable water resources per capita when compared to a recent estimate of the global water resources (UNESCO 2003). The new lower estimate is an effect of a more detailed approach and a more thorough evaluation of the sustainability, which includes the consideration of pollution, climate variation (change) and effects on dependent ecosystems. In the new assessment the sustained exploitable groundwater resource is estimated to be only 6% of the actual recharge, when the state of the aquatic and terrestrial ecosystems (e.g. the ecology of surface waters in rivers and wetlands) and pollution of groundwater is taken into account. This emphasises the need for careful management and protection of the water resources.

The sandy soils in the western, less populated part of the country, which we investigate in this study, require considerable increased abstraction during the spring and summer due to irrigation of the cultivated areas; this results in critical low flow conditions primarily during summer time.

Climate models and scenarios indicate that Denmark will receive higher precipitation in this century, but that the dry season probably will be longer. Integrated hydrological modelling based on such scenarios indicate that the area investigated in this study may have to more than double the irrigation if the land use is not changed (Sonnenborg et al. 2006). Increased irrigation will also increase the recharge to the Ribe Formation, and hence the risk of contamination of the high-quality groundwaters, since much of the irrigation water is expected to be abstracted from the Ribe Formation. A sound knowledge of the natural baseline chemistry is therefore important in order to be able to recognise quality trends and human impact in time.

14.3 Background geochemistry and data for the Miocene sand aquifers

14.3.1 Rainfall chemistry

The rainfall chemistry often constitutes the major part of the dissolved components in shallow groundwater in uncontaminated areas. In such areas, chloride from seawater dissolved in rainwater is often the only source for this conservative/inert element in the hydrological cycle, and the ratio between chloride and sodium in precipitation is generally close to the ratio in seawater (e.g. Appelo and Postma 2005). Some other atmospherically derived elements have significant additional anthropogenic and/or natural contributions to the solutes in groundwater. For example, sulphate originates to some extent from global atmospheric pollution, as both wet (rainfall and snow) and dry (particles or gases deposited directly on vegetation or land surfaces) deposition, and contributes significantly to the total mineralisation. The atmospheric sulphur pollution and deposition contribute significantly to the acidification of groundwater in certain lithologies (Paces 1985). Such acidification processes are also described in western Denmark, where they create high dissolved aluminium concentrations in groundwater in some sandy non-calcareous parts of western Jutland (Hansen and Postma 1995). Other compounds such as the nitrogen species also have additional significant local and regional input by contamination from agriculture and combustion engines.

Table 14.1 compares the precipitation weighted rainfall chemistry of monthly measurements during a 5-year period (1970–74), based on data from a station about 40 km north of the recharge area (Goffeng 1973, 1977), to the groundwater chemistry below natural and agricultural areas (Postma et al. 1991), as well as data from a well in the Ribe Formation located in the recharge area, and 50 and 97.7 percentiles of data from the 17 groundwater wells in anaerobic Miocene sand aquifers investigated in research projects. Assuming that chloride inputs from dry

Table 14.1 Comparison of rainfall chemistry with groundwater chemistry in natural heath and agricultural areas at Rabis Creek, and in the Ribe Formation. Concentrations are in mg L^{-1}. The precipitation station at Askov is located about 40 km north of the recharge area with about the same distance to the North Sea.

| | Rainfall Askov[a] 1970–1974 ($n = 60$) | Rabis (Pleistocene sands)[b] | | Rabis (Pleistocene sands)[c] | | Ribe "recharge" well 150.813 (anaerobic) | Miocene sand[d] Median (anaerobic) ($n = 17$) | Miocene sand[d] 97.7% (anaerobic) ($n = 17$) |
		Natural (aerobic) ($n = 14$)	Natural (anaerobic) ($n = 7$)	Fertilised fields (aerobic) ($n = 50$)	Fertilised fields (anaerobic) ($n = 35$)			
pH	5.00	5.7	7.7	5.58	5.56	7.36	7.47	7.94
DO	11	7.7	0.39	8.3	0.23	0.0	0.03	0.04
Ca	1.2	4.5	34	20	18	48	48	80
Mg	0.52	2.6	2.3	11	8.8	4.6	7.9	12
Na	4.7	12	10	11	12	11	27	168
K	0.72	0.89	0.74	1.7	1.6	1.2	3.2	4.9
Cl	8.2	20	15	25	23	17	25	187
SO$_4$	2.4	13	16	32	57	6.8	7.5	17
HCO$_3$	–	5.5	97	6.0	6.0	215	188	226
NO$_3$–N	0.7	0.39	0.023	14	0.023	0.02	0.05	0.10
NH$_4$–N	1.2	0.0077	0.019	0.0093	0.019	0.17	0,22	0.42
P-total	–	0.011	0.23	0.012	0.024	–	0.18	0.94

[a] Precipitation weighted average of monthly analyses for the period 1970–1974 (Goffeng, 1973, 1977); [b,c] Medians for non-fertilised and fertilised fields from Postma et al. (1991) and the National Groundwater Monitoring Database. [d]Miocene sand groundwater > 120 m (of wells investigated in research projects), n is number of screens.

deposition are insignificant in the recharge area, which seems reasonable since the area is 50 km from the sea and has no large forests (which increase dry deposition), and that chloride in precipitation is the only significant source for chloride in groundwater in the deeper parts of the Miocene sand aquifers (here no human impact), the rainfall has to be evaporated by a factor of 2.1 (17/8.2 – Table 14.1). This factor compares extremely well to the present day ratio between precipitation (950 mm) and the actual groundwater recharge (450 mm) in the area as estimated by the national water resources model (Sonnenborg et al. 2003). This may indicate that both the chloride content in precipitation

and the water balance (precipitation and evapotranspiration) have not changed significantly during the past few thousand years in the investigated area. This conclusion is supported by the fact that the measured chloride content at the Askov station is very similar to the chloride concentration in precipitation of remote unpolluted areas of the world (Galloway and Gaudry 1984).

14.3.2 Geochemistry of the Miocene sands

The Miocene sands are mature sediments originating by weathering from the Fennoscandian basement rocks. The sands of the Ribe Formation are generally quite pure quartz sands and gravel with a very high content of silica (an average in 16 samples of 97%) and a relatively low content or absence of carbonate and organic carbon as indicated by the LOI (loss on ignition) value (Table 14.2). Unfortunately, the LOI is only measured at 1000°C, and the measurement therefore includes the loss of CO_2 from carbonates, organic carbon and water. It is not possible to estimate the relative importance of these without new measurements.

14.3.3 Hydrochemistry of groundwater recharging the Ribe Formation

The hydrochemistry of the Ribe Formation aquifer reflects the inert character of the Miocene quartz sands, and by far the major part of the dissolved ions are inherited from the overlying Miocene and Quaternary sediments. The hydrochemistry of the Ribe Formation groundwaters therefore also approximates to the baseline hydrochemistry of the overlying anaerobic parts of the Miocene and Pleistocene sand aquifers in the area (compare Table 14.1 and 14.3a). Hence, the 50 and 97.7 percentiles of the Ribe Formation and Miocene sand groundwaters also describe the natural baseline quality range for many major components in the overlying Pleistocene sand aquifers in the area. The major differences between shallow uncontaminated Pleistocene sands and the deeper Miocene sands are ascribed to carbonate dissolution and pyrite and organic carbon oxidation by oxygen and small amounts (<5 mg L^{-1}) of nitrate. Calcium, magnesium and pH, therefore, increase, while concentrations of redox sensitive elements such as iron and manganese vary according to the redox environment, when

groundwater flows from the shallow aerobic Pleistocene sands to the deeper lying anaerobic Miocene sands. This is seen in Fig. 14.5(b) where Fe and Mn concentrations are very low in the oxic waters in the recharge area and increase rapidly when the groundwater is reduced downgradient.

14.4 Hydrochemistry and quality of groundwater in the Ribe Formation

14.4.1 Historical data on water quality

The Ribe Formation has only been exploited for a few decades and historical data and possibilities of observing geochemical trends are very limited. Well number 149.488, which is a water supply well, has the most extensive dataset of the investigated wells with complete analyses of major ions, selected minor and trace elements, and a wide range of pesticides, but only from 1990, 1995, 1999 and 2003. No trends are observed in the major components or parameters such as Cl, SO_4, HCO_3, electrical conductivity and neither are any human impacts found. Since the groundwater in the Ribe Formation is of pre-industrial age the geochemical environment has a relatively homogeneous regional extent, no strong spatial or temporal trends are to be expected. An exception to this is where chloride, DOC and other elements and compounds from marine aquitards close to the aquifer, or from lignite layers in the aquifer itself, affect the groundwater chemistry.

However, since well number 149.488 is located at the edge of the buried valley in western Jutland (compare Figs. 14.2 and 14.4) careful monitoring is important, since such

Table 14.2 Geochemistry of Ribe Formation sands at different levels in three different wells. The two last columns show the maximum and median values of all samples analysed (up to 16).

Well depth (m)		149.433 145–152	149.433 161–164	150.745 191–194	150.745 203–204	150.899 152–155	150.899 171–174	Max. $n = 16$[a]	Median $n = 16$[a]
Al_2O_3	%	1.34	0.82	1.35	1.04	1.18	0.53	1.4	0.94
CaO	%	0.12	<0.1	0.11	0.093	0.35	<0.1	0.35	0.11
Fe_2O_3	%	0.40	0.28	1.61	1.55	0.77	0.95	1.6	0.31
K_2O	%	0.79	0.49	0.76	0.57	0.65	0.28	0.79	0.54
LOI	%	0.4	0.5	1.3	0.1	0.4	0	1.3	0.30
MgO	%	0.051	0.029	0.047	0.039	0.042	<0.03	0.053	0.039
MnO_2	%	0.011	0.0064	0.019	0.020	0.012	0.009	0.020	0.008
Na_2O	%	0.14	0.083	0.14	0.096	0.01	<0.05	0.14	0.092
P_2O_5	%	0.025	0.032	0.015	0.011	0.006	0.013	0.036	0.023
SiO_2	%	95	96	96	98	98	100	99.5	97
Sum	%	98	98	100	101	101	101	101	99
TiO_2	%	0.17	0.047	0.16	0.16	0.12	0.026	0.18	0.12
As	ppm	–	–	1	1.3	0.65	1.5	1.5	1.1
Ba	ppm	128	72	141	108	121	56	141	88
Be	ppm	<0.6	<0.6	<0.6	<0.6	<0.6	<0.6	n.d.	n.d.
Cd	ppm	–	–	0.075	0.082	0.041	0.080	0.082	0.077
Co	ppm	<6	<6	<6	<6	<6	<6	n.d.	n.d.
Cr	ppm	38	36	65	38	53	67	76	48
Cu	ppm	28	14	47	29	18	26	47	16
Ga	ppm	–	–	<10	<10	<10	<10	n.d.	n.d.
Hf	ppm	–	–	1.77	1.12	0.74	0.74	1.77	0.93
La	ppm	<6	<6	–	–	–	–	9.4	8.2
Mo	ppm	<6	<6	0.715	<0.6	<0.6	0.61	0.72	0.66
Nb	ppm	<6	<6	1.55	1.48	1.01	0.48	1.6	1.2
Ni	ppm	13.9	13.2	29	24	25	35	35	22
Rb	ppm	–	–	7.41	10	4.46	<2	10	7.4
Sc	ppm	<1.2	<1.2	<1.2	<1.2	<1.2	<1.2	1.5	1.4
Sn	ppm	–	–	1.6	11	0.75	1.6	11	1.6
Sr	ppm	19	49	20	18	22	32	56	22
Ta	ppm	–	–	<0.05	<0.05	<0.05	<0.05	n.d.	n.d.
Th	ppm	–	–	0.75	0.48	0.65	0.5	0.75	0.56
U	ppm	–	–	0.090	<0.028	<0.028	0.272	0.27	0.18
V	ppm	5.2	2.84	5.37	5.71	3.88	3.41	6.9	4.0
W	ppm	–	–	0.62	1.64	0.85	2.53	2.5	1.2
Y	ppm	4.08	<3	3.61	3.34	3.11	<3	4.1	3.3
Zn	ppm	<12	<12	31.4	<12	<12	<12	31	31
Zr	ppm	78	39	108	89	76	39	147	70

[a] Elements with no results in columns 3 and 4 have only the 4 results, which are shown in columns 5–8. Elements with no values in columns 5–8 have only 12 results. LOI = loss on ignition.

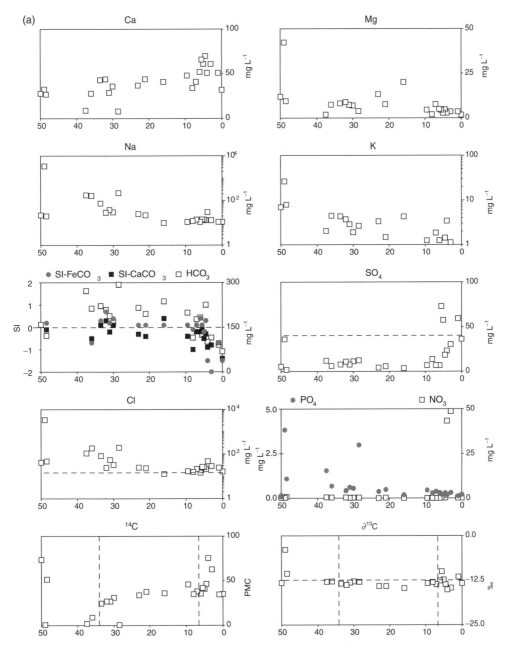

Fig. 14.5 (a) Hydrochemical evolution of selected ions, isotopes and saturation indices in the Ribe Formation and its recharge and discharge zones in southeastern and western Jutland, respectively. Flow is from right to left (east to west). SI-FeCO$_3$ and SI-CaCO$_3$ are saturation indices of siderite and calcite, respectively, computed by PhreeqC (Appelo and Postma, 2005).

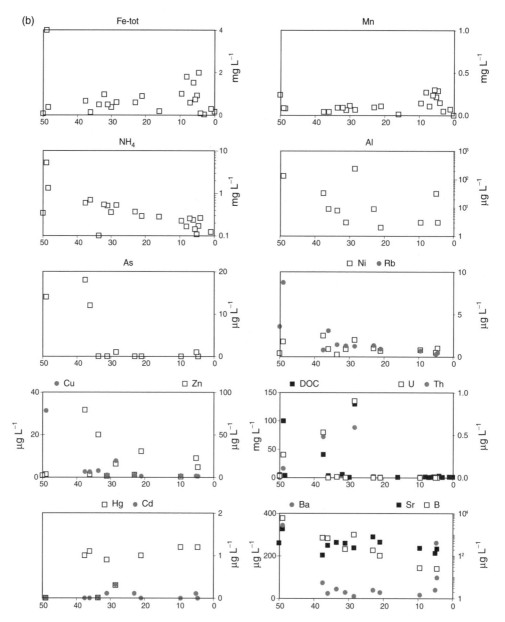

Fig. 14.5 (b) hydrochemical evolution of trace elements in the Ribe Formation and its recharge and discarge zones in southeastern and western Jutland, respectively. Flow is from right to left (east to west).

valleys in some cases create shortcuts for pollution from the surface (Kirsch and Hinsby 2006; Seifert et al. 2007). If abstraction from the Ribe Formation is strongly increased in this region the risk of pollution from the surface will increase considerably.

Besides the required monitoring by the water works, the Danish National

Groundwater Monitoring Network has been operating and collecting data from a total of more than 1000 groundwater monitoring wells since 1989 (e.g. Czakó 1994; Henriksen and Stockmarr 2000). The data are evaluated and presented every year at a national level by the Geological Survey of Denmark and Greenland (e.g. GEUS 2003) and in more local detail at the regional level (e.g. SA 2002). Data from the monitoring database for Miocene sand aquifers have been evaluated and compared to the results obtained from research projects in a later section of this chapter.

14.4.2 Downgradient hydrochemical evolution in the Ribe Formation

The downgradient evolution of the groundwater chemistry can be evaluated by comparing Figs. 14.4, 14.5(a) and (b) and Tables 14.3(a), (b) and 14.4. The geochemical environment in the Miocene sands is quite inert and reactions significantly influencing the distribution of the major elements are very limited, except where the chloride and total organic carbon from adjacent aquitards or coastal aquifers are mixed with the advancing freshwater. Hence the content of substances and elements is primarily inherited from the Quaternary and Miocene layers above the Ribe Formation in the recharge area (easternmost 10 km in Fig. 14.5[a] and [b]). The main reason for this is that equilibrium with the carbonates, calcite ($CaCO_3$) and siderite ($FeCO_3$), is established in the recharge area by dissolution of calcite and weathering of iron minerals, and subsequently possible precipitation of siderite as demonstrated by the saturation index (SI) of these two minerals (shown in HCO_3 diagram in Fig. 14.5[a]) calculated by PHREEQC (Parkhurst and Appelo 1999). Siderite probably

controls the Fe(II) concentrations in most wells in the Ribe Formation since the geochemical calculations by PHREEQC show a strong undersaturation of all iron sulphides.

Downgradient towards the coastline a small increase is observed in, for example, dissolved sodium, chloride, ammonia and boron as remnants of former more saline conditions and/or diffusive loss from surrounding marine clays. Small but significant effects on the hydrochemical evolution due to exchange with Ca on exchange sites in the Ribe Formation itself are indicated for NH_4 and Na. However, they do not generally change the water type (except for wells 148.52 and 158.698). Hence, for the majority of the major elements the composition of the Ribe Formation groundwaters is inherited from rainwater percolating through the Quaternary sands above undergoing carbonate dissolution and oxygen and nitrate reduction. No indicators of human impact (Hinsby et al. 2001b) have been found in the Ribe Formation groundwater.

The attenuation potential of the Ribe Formation may be limited due to the possible inert character of the quartz sands. If for instance, nitrate were to reach the deeper Miocene layers the attenuation of this contaminant may be limited due to a low reduction capacity of the sediment. However, the content of iron sulphides and organic carbon in the sediment is not known, and should be measured in order to be able to estimate the reduction and attenuation capacity of the Ribe Formation sands. The marine transgression during the Burdigalian-Langhian age in early-middle Miocene (Fig. 14.3) may have resulted in diffusion of reduced iron and sulphide from the marine Arnum clays to the Ribe Formation resulting in precipitation of pyrite also in the Ribe Formation. During this period the Ribe Formation sand aquifer must have been salinised, when the

Table 14.3(a)   Selected major and minor elements and parameters measured on groundwater samples collected in the Ribe Formation.

	Units	148.52	149.479	149.488	150.745	150.642	150.813	160.1512	50%[b]	97.7%[b]
n[a]		1	5	4	3	1	2	1	17	17
Elevation	m	0.7	25	28	38	42	75	37	–	–
Screen	mbs	296–304	192–232	200–224	180–202	168–186	180–209	154–162	>120	>120
WT	mbs	~–3	15	13	19	19	45	3.2	–	–
T	°C	14.2	10.9	11.2	10.4	10.4	9.2	10.2	–	–
pH		7.8	7.92	7.70	7.50	7.47	7.34	7.64	7.00	7.81
DO	mg L^{-1}	0.0	0.02	0.45	0.01	–	0.14	0.0	0.8	1.6
H$_2$S	mg L^{-1}	–	0.01	0.03	0.03	0.006	<0.002	<0.002	0.02	0.23
CH$_4$	mg L^{-1}	7.1	0.02	0.012	0.013	–	0.009	<0.001	0.05	0.1
EC	µS cm^{-1}	11800	731	475	420	–	450	–	240	450
Ca	mg L^{-1}	28	42	33	43	65	55	85 (43)	29	69
Mg	mg L^{-1}	38	8.2	6.15	13	11	6.1	5.1	2.9	14
Na	mg L^{-1}	2958	75	55	27	18	11	56	10	26
K	mg L^{-1}	25	4.3	3.6	4.3	3.7	1.42	2.8	1.3	3.5
Cl	mg L^{-1}	4275	84	55	24	22	17	59	14	26
SO$_4$	mg L^{-1}	19	3.2	7.1	4.5	6.1	6.1	5.1	6	20
HCO$_3$	mg L^{-1}	–	221	184	220	218	–	350 (200)	120	240
NO$_3$–N	mg L^{-1}	0.0	<1	<0.1	<0.1	–	0.02	<0.1	0.13	0.23
NH$_4$–N	mg L^{-1}	5.06	0.40	0.40	0.32	0.13	0.17	0.17	0.086	0.21
P	mg L^{-1}	1.5	0.19	0.20	0.13	–	0.15	<0.05	0.12	0.18
TOC	mg L^{-1}	59	–	1.80	1.80	–	–	–	–	–
DOC	mg L^{-1}	59	–	0.85	3.89	–	1.6	2.8	0.77	1.40
F	mg L^{-1}	<0.01	0.04	0.14	0.13	–	–	0.1	0.12	0.13
Si	mg L^{-1}	4.7	7.77	7.95	8.23	7.8	10.9	–	8.9	11
Fe	mg L^{-1}	4.0	0.47	0.41	0.46	0.61	0.97	0.45	2.1	4.2
Mn	mg L^{-1}	0.055	0.08	0.06	0.08	0.09	0.15	0.13	0.15	0.5

[a] n is number of samples. Where $n > 1$ the median is shown; [b] statistical data from all wells investigated in research projects.

sea transgressed and covered the area with marine sediments. However, the Ribe Formation was generally completely freshened again during the Pleistocene possibly partly through flushing with glacial meltwaters at high hydrostatic pressure at lower sea levels (up to 130 m at the glacial maximum) and later during the Holocene (Hinsby et al. 2001a).

14.4.3   Vertical hydrochemical profiles

The groundwater age generally increases with depth, while the risk of pollution and the contents of environmental tracers originating from the atmosphere decrease. Table 14.5 shows the contents of selected ions at different depths in the recharge and discharge zone of the Ribe Formation. No significant

Table 14.3(b) Trace elements measured on groundwater samples collected in the Ribe Formation (columns 1, 3–7) and in recharge (columns 7) and discharge areas (column 2).

	Unit	1 148.52[a]	2 149.479	3 158.698[a]	4 149.488	5 150.745	6 150.772	7 150.813	8 50%[b]	9 97.7%[b]
Elevation	m	0.7	25	8	28	38	35	75	17	17
W.Table	mbs	~–3	15	~–1?	13	19	14	45	–	–
Screen	mbs	292–304	192–232	192–224	200–224	180–202	174–195	180–209	>120	>120
T	°C	12.3	11.2	11.5[a]	11.0	10.4	11.6	9.2	–	–
pH		7.79	7.79	8.91	7.95	7.75	7.50	7.36	7.47	8.60
Eh	mV	–150	–140	–	–141	–122	–124	–111	–117	0.7
Ag	$\mu g\,L^{-1}$	–	–	–	–	–	–	–	0.05	0.095
Al	$\mu g\,L^{-1}$	133	–	240	3	9	2	3	5.4	213
As	$\mu g\,L^{-1}$	14	1.6	1	<1	<1	<1	<1	1.0	1.5
B	$\mu g\,L^{-1}$	5971	–	1006	210	185	102	27	185	933
Ba	$\mu g\,L^{-1}$	344	44	10	29	39	28	17	29	44
Be	$\mu g\,L^{-1}$	–	–	–	–	–	–	–	0.05	0.39
Cd	$\mu g\,L^{-1}$	<0.05	0.013	0.32	0.11	0.11	<0.05	<0.05	0.08	0.30
Cu	$\mu g\,L^{-1}$	31	3.1	7.8	1.1	1.1	0.6	0.8	1.1	7.3
Hg	$\mu g\,L^{-1}$	<0.1	0.0049	0.3	0.9	3	1	1.2	0.75	2.8
Li	$\mu g\,L^{-1}$	342	9.7	19	16	9	11	8	10	19
Mo	$\mu g\,L^{-1}$	0.6	0.33	1	0.3	0.3	0.3	0.2	0.30	0.92
Ni	$\mu g\,L^{-1}$	1.8	0.26	2	0.9	1	0.7	0.8	0.80	1.9
Pb	$\mu g\,L^{-1}$	<2	0.39	<2	<2	<2	<2	1.1	<2	<2
Rb	$\mu g\,L^{-1}$	8.8	1.5	1.3	1.3	1.4	0.93	0.7	1.3	1.4
Sc	$\mu g\,L^{-1}$	4.2	–	2.7	3.7	4.0	4.1	4.1	4.0	4.1
Se	$\mu g\,L^{-1}$	26	–	1.6	0.5	<0.5	<0.5	<0.5	0.5	1.5
Sn	$\mu g\,L^{-1}$	0.49	–	0.48	0.11	0.13	0.11	0.08	0.11	0.45
Sr	$\mu g\,L^{-1}$	1872	435	240	397	796	449	235	416	755
Th	$\mu g\,L^{-1}$	0.11	0.016	0.59	<0.05	<0.05	<0.05	<0.05	0.05	0.53
Ti	$\mu g\,L^{-1}$	314	–	488	<10	<10	<10	<10	10	444
U	$\mu g\,L^{-1}$	0.27	0.0088	0.9	<0.05	<0.05	<0.05	<0.05	0.05	0.80
V	$\mu g\,L^{-1}$	169	–	108	1	1	<1	<1	1	98
Zn	$\mu g\,L^{-1}$	3.4	50	16	1.5	3	31	0.8	9.4	47
Zr	$\mu g\,L^{-1}$	58	–	90	0.6	<0.5	<0.5	<0.5	0.5	82

[a] Artesian ('flowing') well; [b] statistical data from all wells investigated in research projects.

Table 14.4   Selected parameters, environmental tracers and groundwater age estimates in wells in the investigated area.

1 Well no.	2 Screen mbs	3 pH	4 O_2 mg L^{-1}	5 Cl mg L^{-1}	6 SO_4 mg L^{-1}	7 NO_3 mg L^{-1}	8 DIC[a] mg L^{-1}	10 CH_4 mg L^{-1}	11 DOC mg L^{-1}	12 ^{14}C pmc	13 δ^{13}C ‰	14 ^{18}O ‰	15 ^{3}H TU	16 ^{14}C[b] pmc	17 ^{14}C[c] pmc	18 Age[d] years
151.1121	31–51	7.11	2.00	48	23	43.2	86.7	–	3.4	76	–15.1	–7.82	–	125	170	Modern
160.1166	65–73	7.31	0.01	26	58	1.6	104	–	1.1	41	–12.3	–7.99	18	83	123	Modern
160.997	89–92	6.87	0.02	16	36	0.3	58	–	1.5	35	–13.3	–8.19	1.1	66	143	Modern
150.813	180–209	7.34	0.04	17	7	0.1	158	.026	1.6	47	–13.1	–7.58	<0.04	89	111	?
150.745	180–202	7.65	0.01	25	4	0.1	163	.013	1.9	34	–14.1	–7.49	–	60	73	1340
149.479	192–232	7.89	0.02	83	7	0.0	158	.020	3.0	23	–13.4	–7.41	–	43	54	2360
149.377	60–90	8.24	0.00	109	12	0.5	(196)	.019	41	1.1	–13.0	–8.23	0.25	2.1	2.5	12200
148.52	292–304	7.78	0.00	3400	36	0.3	(631)	6.9	156	0.7	–5.8	–9.23	–	3.0	3.4	11200

[a] As CO_2, calculated from field alkalinities by NETPATH (Plummer et al. 1994) except numbers in parentheses which are measured at the ^{14}C lab. at Århus University; [b] δ^{13}C corrected (Ingerson & Pearson 1964) [c] δ^{13}C corrected and maximum correction for O_2 oxidation of fossil ^{14}C 'dead' organic matter (Boaretto et al. 1998), and diffussion (Sanford 1997);[d] The calculated ^{14}C groundwater ages are corrected for chemical reactions and diffusion (Hinsby et al. 2001). Corrected ^{14}C values above 100 pmc in columns 16 and 17 indicate modern post-bomb waters.

pollution is observed in the wells below a depth of 47 m in the analysed wells. This is confirmed by the very low tritium activities (0.02 and 0.04 TU) in these wells (not shown). In contrast, sulphate concentrations in the two screens above this depth show a clear agricultural impact. The source of the sulphate is not known. It could be applied with the fertilisers or produced by the oxidation of pyrite in the sediments by oxygen and nitrate. However, the latter would result in increased carbonate dissolution, which is not indicated by the relatively low alkalinity compared to the Ribe Formation. The three wells at approximately the same depth 47–56 m below surface show comparable unpolluted geochemical composition. The well 160.1512 is screened in Miocene sands, while well 160.1526 and 160.1561 both are screened in Pleistocene sands. This supports the conclusion that the natural baseline quality of Pleistocene and Miocene sands in the area is quite similar as argued in an earlier section. Downgradient in the discharge zone the three samples collected in the top, centre and bottom of the 40 m long screen of well 149.479 in the Ribe Formation show no significant differences either in major ions (as shown) or in ^{14}C contents (not shown).

14.4.4 Groundwater ages, spatial and temporal hydrochemical trends

The groundwater chemistry and ages in the Ribe Formation were investigated and discussed by Hinsby et al. (2006). Based on ^{14}C dating results corrected for both geochemical reactions and diffusion, and groundwater flow modelling, it was concluded that the groundwaters in the Ribe Formation generally are a few thousand years old. Locally, groundwater of Pleistocene age (>10,000 years)

may occur in the discharge areas around the west coast of Jutland and in hydraulically isolated areas (Table 14.4).

As mentioned earlier, the geochemical results from analyses of groundwater samples from the Ribe Formation show no evidence of human influence as indicated by the very low nitrate and sulphate concentrations (Table 14.3(a), Figs. 14.5[a] and [b]). This is in accordance with the low activities of the environmental tracers tritium and ^{14}C (Table 14.4, Fig. 14.5[a]). Groundwater becomes progressively older towards the discharge area along the west coast of Jutland with ^{14}C values gradually decreasing from about 47 pmc (per cent modern carbon) in the recharge zone to around 1 pmc in several wells in the coastal area. The δ^{13}C values are quite constant around 12.5‰ except for a few wells with high organic contents (Figs. 14.5[a] and [b]), indicating that the groundwater chemistry has developed an equilibrium with calcite under closed conditions (e.g. Appelo 1994; Appelo and Postma 2005).

The very low ^{14}C contents of the wells 148.52 and 149.377 indicate that these groundwaters may be of Pleistocene age. This is corroborated by the relatively low ^{18}O content in well 148.52 and by noble gas measurements in well 149.377, which indicate a recharge temperature 5–6°C lower than the present annual average temperature (Hinsby et al. 2001a). However, the high DOC and/or sodium concentrations in well 148.52 influences the inorganic carbon pool either directly through production of CO_2 from fermentation processes and/or indirectly through the triggering of carbonate dissolution/precipitation processes (Appelo 1994). Evaluation of the groundwater chemistry at well 148.52 shows that these processes significantly affect the dissolved carbon pool in

Table 14.5 Comparison of selected major and minor ions in groundwater at different depth in the recharge area, and in the Ribe Formation in the discharge area (well 149.579).

Well	Elevation/ Depth (m)	Screen length (m)	Ca (mg L^{-1})	Na (mg L^{-1})	K (mg L^{-1})	Fe (mg L^{-1})	Cl (mg L^{-1})	HCO$_3$ (mg L^{-1})	NO$_3$ (mg L^{-1})	SO$_4$ (mg L^{-1})	P (mg L^{-1})
Recharge area											
160.1512-5	*31/6.5*	*2*	*70*	*16*	*2.8*	*3.8*	*27*	*140*	*<0.5*	*80*	*0.12*
160.1512-4	*10/27*	*6*	*94*	*13*	*1*	*1.2*	*24*	*200*	*<0.5*	*94*	*0.087*
160.1512-3	**-10/47**	**2**	**64**	**12**	**2.1**	**0.66**	**17**	**210**	**<0.5**	**6.3**	**0.096**
160.1512-2	-88/125	14	59	16	3.1	0.86	23	200	<0.5	8.2	0.091
160.1512-1[a]	-121/158	8	43	49	3.3	0.42	50	200	<0.5	8.0	0.013
			85	56	2.8	0.045	59	350	<0.5	5.1	0.005
160.1526-4	*-7/55*	*6*	*74*	*10*	*1.6*	*1.8*	*13*	*210*	*<0.5*	*18*	*0.12*
160.1561-3	*-12/56*	*2*	*60*	*10*	*1.8*	*0.59*	*15*	*200*	*<0.5*	*4.0*	*0.15*
Discharge area											
149.479	-167/92	40	43	74	4.3	0.46	81	220	–	3.2	–
	-187/212	40	42	77	4.4	0.51	86	223	–	3.2	0.17
	-207/232	40	38	82	4.5	0.43	84	221	–	3.1	0.12

[a] Analysis in upper row was sampled March 2003; analysis in lower row was sampled July 2003.

Note: Wells in italics are screened in Pleistocene sands, wells in normal font are screened in Miocene sands. The three wells in bold are screened at approximately the same depth in Miocene and Pleistocene sands.

the groundwater at the well. This is demonstrated by the high DIC and CH_4 concentrations, and the relatively enriched ('heavy') $\delta^{13}C$ values. These processes will lower the measured percentage of modern carbon (^{14}C) and increase the uncertainty on the groundwater age estimate, although the effect is partly taken into account by the $\delta^{13}C$ correction (Hinsby et al. 2001a).

Severe saltwater problems are relatively uncommon in the Ribe Formation within the study area although they do exist, and chloride concentrations increase slightly towards the coast. Locally, elevated chloride concentrations and even brackish waters associated with high dissolved organic matter content are encountered in western Jutland, and these occasionally result in problems for water supplies (Villumsen 1985). The sources of the chloride and DOC seem primarily to be from Tertiary marine and lagoonal sediments adjacent to the Ribe Formation (Jørgensen and Holm 1995; Rasmussen 2004a, b). Organic matter from the aquifer itself, although possibly present in significant quantities (up to ~0.3 wt%), does probably not contribute significantly to the occasionally observed high DOC values in the investigated area due to a low reactivity of the organic carbon. At other sites north of the investigated area the dissolved organic matter in the Ribe Formation groundwaters was found to be of terrestrial origin from adjacent lacustrine sediments (Grøn et al. 1996). Groundwater with elevated chloride and DOC contents also have elevated concentrations of many major and minor components (e.g. Na, NH_4, Fe), trace metals (e.g. Al, Ga, V) and rare earth elements (e.g. La, Yb, Th, U) originating from the marine sediments and for many of the minor and trace elements with a strong affinity to form metallo-organic complexes with humic

substances in both freshwater (Dupre et al. 1999) and saline waters (Moran et al. 2003; Weinstein and Moran 2005). This can be seen, for example, for U and Th in Fig. 14.5(b), and is also reflected in the concentrations of many of the trace metals and elements when comparing the medians and 90 percentiles in Table 14.5(b).

The main conclusions are that the Ribe Formation generally contains a Ca-bicarbonate type groundwater saturated or slightly undersaturated with calcite and siderite (Fig. 14.5[a]), and that the aquifer was flushed during the Quaternary and at present is more or less completely freshened. There is, however, a decrease in Ca and an increase in Na towards the west coast of Jutland, indicating that Ca is still replacing a small amount of Na on exchange sites in the aquifer sediments. Such processes are well known from freshening aquifers and may occur over long timescales of thousands of years (e.g. Appelo 1994) as in the Ribe Formation, and over short timescales of a few years at present day coastlines (e.g. Andersen et al. 2005).

The redox environment is mainly anaerobic because oxygen is reduced to low concentrations in the recharge zone in the layers above the Ribe Formation. The 'inert' (mature) character of the Ribe Formation sands, that on average contain 97% SiO_2 primarily as quartz, and less than 0.3 wt% recalcitrant organic carbon, generally prevent the development of methanogenic environments. Slow sulphate reduction, however, is indicated by the relatively low sulphate concentrations and redox potentials in the Ribe Formation as well as the decrease in Fe typically seen when sulphide is produced, as has been suggested in an earlier study based on, for example, H_2 measurements (Jakobsen 1995). On the other hand, no clear trend is observed and the low concentrations could, at least partly, be a result of much lower sulphur deposition in pre-industrial times (Graedel and Crutzen 1993). Assuming that sulphate contents in precipitation during the time of precipitation of Ribe Formation groundwater is similar to the present day contents in uncontaminated precipitation (Galloway and Gaudry 1984), and that dry deposition of sulphate is comparable to the wet deposition as has been found for recent conditions in Denmark (Jørgensen 1979), the recharge to shallow groundwater should contain about 6 mg L^{-1} sulphate. This amount compares quite well with the amount measured in the Ribe Formation groundwater. Note, however, that reduction of oxygen by pyrite may produce more than twice this amount in the aquifers above the Ribe Formation. However, sulphate reduction probably occurs before recharge to the Ribe Formation. Once the groundwaters have recharged the Ribe Formation the sulphate reduction rate is generally too slow to affect the dissolved inorganic carbon pool in the groundwater significantly. Only in the case of fresh- and saltwater (with very high DOC values) mixing as in well 148.52 does carbonate dissolution/precipitation, oxidation of organic carbon by sulphate and methanogenesis seem to affect the inorganic carbon pool in the Ribe Formation. In this well data indicate that there is plenty of reactive organic carbon in the surrounding aquitards affecting the geochemical processes and the groundwater composition in the aquifer, which is only 6–12 m thick at this location.

14.5   The natural baseline quality of Miocene sand groundwaters

Data has been abstracted from monitoring wells located in Miocene Quartz sands from the national groundwater monitoring database, based on different selection criteria, in order to evaluate and compare the data collected in the Ribe Formation in research projects to the general hydrochemistry of the Miocene sands. Fig. 14.6 show a cumulative frequency plot of data from water samples from the Miocene sands with nitrate concentrations below 10 mg L^{-1} NO_3 as a first approximation of natural unpolluted groundwater. However, the high sulphate concentrations of more than 40 mg L^{-1} and up to more than 100 mg L^{-1} clearly show a significant human impact in more than 30% of the samples in this subgroup. Current 'natural' baseline values for sulphate (without agricultural but with atmospheric pollution) should not exceed about 40 mg L^{-1}, which corresponds approximately to equal amounts from precipitation (including dry deposition) and maximum additional sulphate from oxidation of pyrite by oxygen in the sediment. The high sulphate concentrations above 40 mg L^{-1} must result from nitrate reduction by pyrite or by lowering of the water table (Postma et al. 1991; Thorling 1994; Larsen and Postma 1997). Even higher sulphate concentrations (>200 mg L^{-1}) can be obtained where the increased sulphate is a result of lowering of the water table below pyrite-containing layers. In this case, oxygen may be renewed by gas diffusion and/or barometric ventilation and be a continuous oxidant (Larsen and Postma 1997). In Table 14.6 we compare statistical data for the wells investigated in the research projects to a subset of data from the Miocene sands, which has both nitrate below 10 mg L^{-1} and sulphate below 40 mg L^{-1} as an approximation to true natural baseline quality groundwater. The composition of the groundwaters described by the statistical parameters for

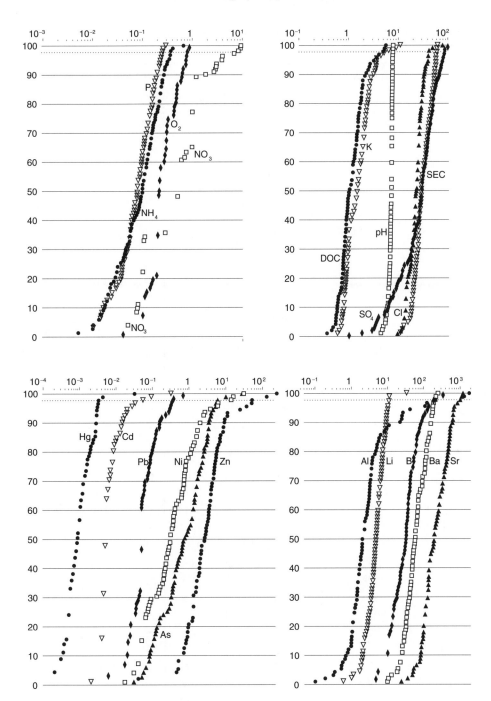

Fig. 14.6 Cumulative frequency plots of selected substances and elements. Data represent a subgroup of analyses from the groundwater monitoring database at the Geological Survey of Denmark and Greenland for Miocene sand aquifers with $[NO_3^-] < 10$ mg L^{-1}.

Table 14.6(a) Selected statistics on major ions, etc. in groundwater of the Miocene sand aquifers in Denmark.

			Wells investigated in research projects				Groundwater monitoring wells			
	Unit	EU/DK WHO[a]	n	50%	90%	97.7%	n	50%	90%	97.7%
pH		7–8.5[b]	17	7.47	7.94	8.60	66	7.28	7.60	7.71
Eh	mV		9	−117	−17	0.66				
DO	mg L^{-1}		9	0.02	0.03	0.038	60	0.29	0.76	0.98
SEC	µS cm^{-1}	250[c,d]	9	383	501	678	65	280	550	610
δ^{18}O	‰		7	−7.58	−7.47	−7.45				
^{14}C	pmc		8	33	41	45				
Ca	mg L^{-1}		17	48	74	80	68	53	95	122
Mg	mg L^{-1}	50[b]	17	7.9	10	12	68	4.8	9.2	14
Na	mg L^{-1}	175[b]	17	27	75	168	68	14	22	40
K	mg L^{-1}	10[b]	17	3.2	4.6	4.9	80	1.6	2.6	4.5
Fe	mg L^{-1}	0.1[b]	17	0.84	1.5	1.8	81	3.1	7.2	15
Mn	mg L^{-1}	0.02[b]	17	0.09	0.16	0.22	81	0.26	0.52	0.87
Cl	mg L^{-1}	250	17	25	79	187	82	22	33	43
SO$_4$	mg L^{-1}	250[b,c]	17	7.5	11	17	82	21	36	39
HCO$_3$	mg L^{-1}		17	188	219	226	69	171	320	363
NO$_3$–N	mg L^{-1}	~11[a]	8.0	0.05	0.09	0.10	82	0.18	0.23	0.88
NH$_4$–N	mg L^{-1}	~0.039[d]	17	0.22	0.38	0.42	81	0.14	0.28	0.41
P	mg L^{-1}	0.15[b]	7	0.18	0.56	0.94	69	0.11	0.2	0.24
Si	mg L^{-1}		17	8.7	9.7	11	67	8.9	11	12
DOC	mg L^{-1}	4[b]	8	2.9	30	74	68	1.3	2	4.3
H$_2$S	mg L^{-1}	0.05[b]	17	0.02	0.03	0.04	57	0.082	0.2	0.2
CH$_4$	mg L^{-1}	0.01[b]	7	0.01	0.07	0.11	78	0.035	0.05	0.16
F	mg L^{-1}	1.5	5	0.050	0.402	0.44	68	0.14	0.24	0.27
Br	mg L^{-1}		5	0.13	0.41	0.52	64	0.06	0.1	0.185
I	mg L^{-1}		5	0.003	0.009	0.011	29	0.0028	0.0084	0.011
Ag	µg L^{-1}	10[b]	5	0.050	0.080	0.095	39	0.05	0.1	0.1
Al	µg L^{-1}	100[b]	6	5.43	125	213	65	1.5	5.7	110
As	µg L^{-1}	5[b]	6	1.0	1.3	1.5	65	0.67	3.6	10
B	µg L^{-1}	500[d]	5	185	688	933	52	24	110	190
Ba	µg L^{-1}	700[d]	6	29	42	44	65	50	160	205
Be	µg L^{-1}		5	0.05	0.27	0.39				
Cd	µg L^{-1}	2[b]	6	0.08	0.22	0.30	65	0.005	0.01	0.015
Ce	µg L^{-1}		5	0.01	3.1	4.7				
Co	µg L^{-1}		6	0.02	0.07	0.11				
Cr	µg L^{-1}	20[b]	6	0.5	13	23	65	0.0575	0.65	3.2

Table 14.6(a) Continued

	Unit	EU/DK WHO[a]	n	50%	90%	97.7%	n	50%	90%	97.7%
			Wells investigated in research projects				Groundwater monitoring wells			
Cs	µg L^{-1}		5	0.01	0.01	0.01				
Cu	µg L^{-1}	100[b]	6	1.1	5.5	7.3	65	0.16	0.4	1.4
Dy	µg L^{-1}		5	0.01	0.21	0.32				
Er	µg L^{-1}		5	0.01	0.15	0.22				
Eu	µg L^{-1}		5	0.01	0.07	0.10				

[a] Guideline values – the lowest is listed. [b,c] and [d] indicate that the EU, DK or WHO guideline value, respectively, is indicated. The DK value is for water supplied to consumer's house.

Note: Comparison of data from research projects and the groundwater monitoring database at the Geological Survey of Denmark and Greenland.

Table 14.6(b) Selected statistics on trace elements in groundwater of the Miocene sand aquifers in Denmark.

	Unit	EU/DK WHO[a]	n	50%	90%	97.7%	n	50%	90%	97.7%
			Wells investigated in research projects				Groundwater monitoring wells			
Ga	µg L^{-1}		5	<0.05	0.66	0.97				
Gd	µg L^{-1}		5	<0.01	0.23	0.35				
Ge	µg L^{-1}		5	<0.05	0.16	0.21				
Hf	µg L^{-1}		5	<0.02	1.2	1.8				
Hg	µg L^{-1}	1[b]	6	0.75	2.0	2.8	64	0.001	0.003	0.005
Ho	µg L^{-1}		5	<0.01	0.05	0.07				
In	µg L^{-1}		5	<0.01	<0.01	<0.01				
Ir	µg L^{-1}		5	<0.05	<0.05	<0.05				
La	µg L^{-1}		5	0.01	1.3	2.0				
Li	µg L^{-1}		6	10	18	19	64	5.1	9.8	11
Lu	µg L^{-1}		5	0.01	0.03	0.037				
Mo	µg L^{-1}	70[d]	6	0.3	0.66	0.92	64	0.2	1.2	2
Nb	µg L^{-1}		5	0.01	0.04	0.055				
Nd	µg L^{-1}		5	0.01	1.7	2.5				
Ni	µg L^{-1}	20[b]	6	0.8	1.5	1.9	66	0.2	1.1	1.7
Os	µg L^{-1}		5	<0.05	<0.05	<0.05				
Pb	µg L^{-1}	5[b]	6	<2	<2	<2	65	0.05	0.1	0.23

(Continued)

Table 14.6(b)   Continued

	Unit	EU/DK WHO[a]	Wells investigated in research projects				Groundwater monitoring wells			
			n	50%	90%	97.7%	n	50%	90%	97.7%
Pd	µg L^{-1}		5	0.2	0.68	0.93				
Pr	µg L^{-1}		5	<0.01	0.39	0.58				
Pt	µg L^{-1}		5	<0.01	<0.01	<0.01				
Rb	µg L^{-1}		6	1.3	1.4	1.4				
Re	µg L^{-1}		5	<0.01	<0.01	<0.01				
Ru	µg L^{-1}		5	<0.05	<0.05	<0.05				
Sb	µg L^{-1}	2[b]	5	<0.05	<0.05	<0.05	53	0.02	0.13	0.13
Sc	µg L^{-1}		5	4.01	4.06	4.06				
Se	µg L^{-1}	10[b]	5	0.5	1.2	1.5	65	0.1	0.15	0.34
Sm	µg L^{-1}		5	0.05	0.33	0.48				
Sn	µg L^{-1}	10[b]	5	0.11	0.34	0.45	39	0.05	0.1	0.11
Sr	µg L^{-1}		6	416	622	756	64	253	610	1100
Tb	µg L^{-1}		5	0.01	0.052	0.074				
Te	µg L^{-1}		5	<0.05	<0.05	<0.05				
Th	µg L^{-1}		6	<0.05	0.32	0.53				
Ti	µg L^{-1}		5	10	297	444				
Tl	µg L^{-1}		5	<0.01	<0.01	<0.01	43	0.0485	0.23	0.23
Tm	µg L^{-1}		5	0.01	0.022	0.028				
U	µg L^{-1}	2[d]	6	<0.05	0.48	0.80				
V	µg L^{-1}		5	<1	65	98	64	0.44	0.58	2.80
Yb	µg L^{-1}		5	<0.01	0.14	0.21				
Zn	µg L^{-1}	100[b]	6	9.4	40	48	65	2.2	9.3	73
Zr	µg L^{-1}		5	<0.5	54	82				

[a] Guideline values – the lowest is listed; [b, c] and [d] indicate that the EU, DK, or WHO guideline value, respectively, is indicated. The DK value is for water supplied to consumer's house.

Note: Comparison of data from research projects and the groundwater monitoring database at the Geological Survey of Denmark and Greenland.

the different ions and so forth for the two data groups compare fairly well except for a few elements in the research wells with elevated concentrations. The differences are primarily ascribed to the wells with high contents of dissolved organic matter originating from marine aquitards, which have been investigated in the research proj-ect, but which are not part of the monitoring dataset. Hence, we conclude that both datasets describe the natural baseline quality of the Danish Miocene sand aquifers quite well, and that these may be used as the natural baseline concentrations (natural background levels) for similar aquifers in other parts of Denmark where detailed information on groundwater chemistry is missing.

14.6 Conclusion

The Miocene sands in Denmark found at depths greater than 60–100 m (e.g. The Ribe Formation) generally contain high-quality natural baseline groundwater. The main composition of this valuable freshwater resource is determined by natural reactions between rainwater and aquifer sediments. There is low total mineralisation controlled by calcite dissolution partly in the overlying Quaternary aquifers and aquitards. The conditions are anaerobic and sulphate reduction may occur, but it is generally slow and insignificant.

Elevated concentrations of trace metals and rare earth elements (e.g. As, Al, B, Cr, Se, U, Th, La and Yb) are found in a few saline high DOC groundwaters. Such water types are used for irrigation in the area, which may result in enhanced uptake of trace elements in the crops.

Carbon, oxygen and hydrogen isotope signatures as well as noble gas data indicate that groundwater in the Ribe Formation aquifer and in discharge areas around the coast, in some cases dates back to the last glaciation (older than 10,000 years). The Ribe Formation must have been salinised at least partly during the Middle to Late Miocene, when the North Sea covered most of the investigated area, about 10–20 million years ago. However, it was freshened again during the Pliocene and Pleistocene epochs over the past approximately 5 million years.

It is concluded that the deeper parts of the Miocene sand aquifers (especially The Ribe Formation) contain a valuable resource of pristine groundwater without any human impact. In some areas, where buried Quaternary valleys are incised into or close to the Miocene sands, the pristine baseline groundwaters may be threatened by contamination if exploitation is not properly developed, for example, the hydraulic gradients are reversed by over-abstraction. Proper monitoring and management is therefore important in order to avoid the advance of young polluted groundwater into the system. Data indicate that the potential for natural attenuation may be limited in the Ribe Formation. However, data on the contents and reactivity of electron donors such as organic carbon and iron sulphides are not available for the Ribe Formation and need to be measured before concluding on this issue.

The groundwater chemistry of the Ribe Formation describes the baseline of the Miocene sand aquifers and can to a large extent also be used as the baseline for the anaerobic and carbonate containing parts of the overlying Quaternary sands or those occurring in deeper parts of Pleistocene buried valleys. The Ribe Formation hydrochemistry for some major ions is furthermore quite similar to the hydrochemistry of shallow Quaternary aquifers and streams in natural heath areas in Jutland.

References

Andersen, M.S., Nyvang, V., Jakobsen, R. and Postma, D (2005) Geochemical processes and solute transport at the seawater/freshwater interface of a sandy aquifer. *Geochimica et Cosmochimica Acta* 69, 3979–3994.

Appelo, C.A.J. (1994) Cation and proton exchange, pH variations, and carbonate reactions in a freshening aquifer. *Water Resources Research* 30, 2793–2805.

Appelo, C.A.J. and Postma, D. (2005) *Geochemistry, Groundwater and Pollution.* 2nd edn., A.A.Balkema, Rotterdam, The Netherlands, 649 pp.

BurVal. (2006) Groundwater resources in buried valleys – a challenge for geosciences. Leibniz Institute for Applied Geosciences (GGA), Geozentrum Hannover, Germany, 303 pp. (www.burval.org).

Czakó, T. (1994) Groundwater monitoring network in Denmark: example of results in the Nyborg Area. *Hydrological Sciences Journal* 39, 1–17.

Dinesen, B. (1961) Salt mineralvand fra Danmarks dybere undergrund. *Geological Survey of Denmark, IV Series 4*, 6, 4–20.

Dupre, B., Viers, J. Dandurand, J. L. et al. (1999) Major and trace elements associated with colloids in organic-rich river waters: Ultrafiltration of natural and spiked solutions. *Chemical Geology* 160(1–2), 63–80.

Galloway, J.N. and Gaudry, A. (1984) The composition of precipitation on Amsterdam Island, Indian Ocean. *Atmospheric Environment* 18, 2649–2656.

GEUS (2003) Groundwater Monitoring (2003) Geological Survey of Denmark and Greenland, annual groundwater monitoring report (www.geus.dk).

Goffeng, G. (ed.). (1973) Hydrological Data – Norden. IHD stations – Basic Data 1970–1971. Ås-Print, Ås, Norway, 118 pp.

Goffeng, G. (ed.) (1977) Hydrological Data – Norden. IHD stations – Basic Data 1972–1974. Ås-Print, Ås, Norway, 1977, 149 pp.

Graedel, T.E. and Crutzen, P.J. (1993) *Atmospheric Change – An Earth System Perspective.* Freeman, New York, 446 pp.

Grøn, C., Wassenaar, L. and Krog, M. (1996) Origin and structures of groundwater humic substances from three Danish aquifers. *Environment International* 22, 519–534.

Hansen, B.K. and Postma, D. (1995) Acidification, buffering, and salt effects in the unsaturated zone of a sandy aquifer, Klosterhede, Denmark. *Water Resources Research* 31, 2795–2809.

Hansen, J.P.V. and Rassmussen, E.S. (2007). Structural, sedimentological and sea-level controls on sand distribution in a steep-clinoform asymmetric wave-influenced delta Miocene Billund Fm., Eastern Danish North Sea and Jylland. *Sedimentary Resarch.*

Hinsby, K., Condesso de Melo, M.T. and Dahl, M. (2007). Case studies supporting the derivation of natural background levels and groundwater threshold values for the protection of dependent ecosystems and human health. *Sci Total Environ.*

Henriksen, H.J. and Sonnenborg, A. (eds) (2003) Ferskvandets kredsløb. NOVA 2003 temarapport, GEUS, Copenhagen, Denmark.

Henriksen, H.J. and Stockmarr, J. (2000) Groundwater resources in Denmark – Modelling and monitoring. *Water Supply* 18, 550–557.

Henriksen, H.J., Troldborg, L., Nyegaard, P. et al. (2003) Methodology for construction, calibration and validation of a national hydrological model for Denmark. *Journal of Hydrology* 280, 52–71.

Hinsby, K. (2006) Environmental tracers, groundwater age and vulnerability. In: *BurVal. Groundwater Resources in Buried Valleys – A Challenge for Geosciences.* Leibniz Institute for Applied Geosciences (GGA), Geozentrum Hannover, Germany, pp. 141–148.

Hinsby, K., Harrrar, W.G., Nyegaard, P. et al. (2001a) The Ribe Formation in western Denmark: Holocene and Pleistocene groundwaters in a coastal Miocene sand aquifer. In: Edmunds, W.M. and Milne, C.J. (eds) *Palaeowaters in Coastal Europe: Evolution of Groundwater Since the Late Pleistocene.* Geological Society, London, Special Publication, 189, 29–48.

Hinsby, K., Edmunds, W.M., Loosli H.H. et al. (2001b) The modern water interface: Recognition, protection and development – advance of modern waters in European coastal aquifer systems. In: Edmunds, W.M. and Milne, C.J. (eds) *Palaeowaters in Coastal Europe: Evolution of Groundwater Since the Late Pleistocene.* Geological Society, London, Special Publication, 189, 271–288.

Hinsby, K., Purtschert, R. and Edmunds, W.M. (2008) Groundwater age and quality. In: Quevauviller, P. (ed.) Groundwater Science and Policy. RSC publishing.

Jakobsen, R. (1995) Sulfate reduction, Fe-reduction and Methanogenesis in Groundwater. PhD thesis, Department of Geology and Geotechnical Engineering, Technical University of Denmark.

Jørgensen, N.O. and Holm, P.M. (1995) Strontium isotope studies of "brown water" (organic-rich groundwater) from Denmark. In: *Proceedings of the International Association of Hydrogeologists*. Congress, Edmonton, pp. 291–295.

Jørgensen, V. (1979) Luftens og nedbørens kemiske sammensætning i danske landområder. *Tidsskrift for Planteavl* 1434, 633–656.

Kirsch, R. and Hinsby, K. (2006) Aquifer vulnerability. In: *BurVal. Groundwater Resources in Buried Valleys – A Challenge for Geosciences*. Leibniz Institute for Applied Geosciences (GGA), Geozentrum Hannover, Germany, pp. 149–156.

Koch, B.E. (1989) Geology of the Søby-Fasterholt area. Danmarks Geologiske Undersøgelse Serie A 22, 177 pp.

Larsen, F. and Postma, D. (1997) Nickel mobilization in a groundwater well field: Release by pyrite oxidation and desorption from manganese oxides. *Environmental Science and Technology* 31, 2589–2595.

Larsson, L.M., Vajda, V. and Rasmussen, E.S. (2006) Early Miocene pollen and spores from western Jylland, Denmark – environmental and climatic implications. GFF.

Mai, D.B. (1967) Die Florenzonen, der Florenwechsel und die Vorstellung uber den Klimaablauf im Jungtertiar der DDR. *Abhandlungen Zentral Geologisches Institut* H 10, 55–81.

Michelsen, O. and Nielsen, L.H. (1993) Structural development of the Fennoscandian Border Zone, offshore Denmark. *Marine and Petroleum Geology* 10, 124–134.

Moran, S. B.,Weinstein, S. E., Edmonds H. N. et al. (2003) Does Th-234/U-238 disequilibrium provide an accurate record of the export flux of particulate organic carbon from the upper ocean? *Limnology and Oceanography* 48 (3), 1018–1029.

Paces, T. (1985) Sources of acidification in Central Europe estimated from elemental budgets in small basins. *Nature* 315 (6014), 31–36.

Parkhurst, D.L. and Appelo, C.A.J. (1999) User's guide to PHREEQC (Version 2); a computer program for speciation, batch reaction, one dimensional transport and inverse geochemical calculations. US Geological Survey, Water Resources Investigations, Report 99-4259 Reston, VA.

Plummer, L.N., Prestemon, E.C. and Parkhurst, D.L. (1994) An Interactive Code (NETPATH) For Modeling NET Geochemical Reactions along a Flow PATH. Version 2.0 1994, US Geological Survey, Water-Resources Investigations Report 94–4169, Reston, Virginia, 130 pp.

Postma, D., Boesen, C., Kristiansen, H. and Parkhurst (1991) Nitrate reduction in a sandy aquifer: Water chemistry, reduction processes, and geochemical modeling. *Water Resources Research* 27, 2027–2045.

Rasmussen, E.S. (2004a) The interplay between true eustatic sea-level changes, tectonics, and climatical changes: What is the dominating factor in sequence formation of the Upper Oligocene-Miocene succession in the eastern North Sea Basin, Denmark? *Global and Planetary Changes* 41, 15–30.

Rasmussen, E.S. (2004b) Stratigraphy and depositional evolution of the uppermost Oligocene – Miocene succession in Denmark. *Bulletin Geological Society of Denmark* 51, 89–109.

Rasmussen, E.S. and Dybkjær, K. (2005) Sequence stratigraphy of the Upper Oligocene – lower Miocene of eastern Jylland, Denmark: role of structural relief and variable sediment supply in controlling sequence development. *Sedimentology* 52, 25–63.

SA (2002) Water Environment Monitoring 2001 – Groundwater. Technical Report, May 2002, Sønderjyllands Amt (In Danish), 49 pp.

Sanford, W.E. (1997) Correcting for diffusion in carbon-14 dating of ground water. *Ground Water* 35, 357–361.

Seifert, D., Sonnenborg, T.O, Scharling, P. and Hinsby, (2007) Use of alternative conceptual models to assess the impact of a buried valley on groundwater vulnerability. *Hydrogeology Journal*.

Sonnenborg, T.O., Christensen, B.S.B., Nyegaard, P. Henriksen, H.J. and Refsgaard, J. (2003) Transient modelling of regional groundwater flow using parameter estimates from steady-state automatic calibration. *Journal of Hydrology* 273 (1–4),188–204.

Sonnenborg, T.O., Christensen, B.S.B., Roosmalen, L. van and Henriksen, H.J. (2006) Klimaændringers betydning for vandkredsløbet i Danmark. GEUS report no. 2006/22, Geological Survey of Denmark and Greenland, Copenhagen, Denmark, 75 pp.

Thorling, L. (1994) Sulphate as age-indicator in groundwater. *Vand og Jord* 1, 113–115 (In Danish).

UNESCO (2003) World Water Development Report. www.unesco.org/water/wwap/wwdr.

Utescher, T., Mosbugger, V. and Ashraf, A.R. (2000) Terrestrial climate evolution in Northwest Germany over the last 25 million years. *Palaios* 15, 430–449.

Villumsen, A. (1985) Chemical limitations for increased exploitation of groundwater resources in Denmark. In: *Proceedings of the Jerusalem Symposium for the Scientific Basis for Water Resources Management*. September 1985, IAHS 153, 423–431.

Weinstein, S.E. and Moran, S.B. (2005) Vertical flux of particulate Al, Fe, Pb, and Ba from the upper ocean estimated from Th-234/U-238 disequilibria. Deep-Sea Research Part I-Oceanographic Research Papers, 52 (8), 1477–1488.

Zachos, J.C., Pagani, M., Sloan, L.C. Thomas, E. and Billups (2001) Trends, rhythms, and aberrations in global climate 65 Ma to present. *Science* 292, 686–693.

15 Tracer Based Study of the Badenian Bogucice Sands Aquifer, Poland

S. WITCZAK, A. ZUBER, E. KMIECIK, J. KANIA,
J. SZCZEPANSKA AND K. ROZANSKI

A hydrochemical study of a typical sandy aquifer in relation to flow and transport modelling supported by environmental tracer data is presented. The study was directed towards determining the natural background levels (baseline) for the implementation of the EU Groundwater Directive (GWD). Typical hydrochemical zones within the aquifer are shown to be related to timescales (ages of water) determined with the aid of numerical transport modelling and the interpretation of tritium by lumped-parameter models (box models). Both approaches yielded very wide distributions of groundwater ages in abstraction wells, which means that mixing of waters characterised by greatly different flow paths and possibly different hydrochemistry takes place. Such mixing pattern has a bearing on actually observed concentrations of important water constituents and may also significantly smooth the changes in concentrations expected as a result of either potential pollutant sources or changes in flow pattern caused by intensive abstraction. Though the investigated groundwater system is relatively small, the natural distributions of water constituents are very wide. Some modern waters present in the outcrop areas contain significantly elevated concentrations of Cl^-, SO_4^{2-} and NO_3^- due to anthropogenic influences. In the confined area, the ages of water range from about 0.2 ka to several ka, while the concentrations of Fe, Mn and NH_4 exceed the maximum permissible levels (MPLs) due to anaerobic conditions. In the deepest part of the aquifer, waters are of glacial age and of poor chemical status, with Na, NH_4, As and B exceeding the MPL values, though Fe and Mn contents are reduced due to favorable pH and Eh conditions. In general, the baseline concentrations of dissolved constituents in groundwater bodies (GWB) can be adequately characterised by 2.3 and 97.7 percentiles of the cumulative distributions, although more detailed information on different hydrochemical zones should be taken into account for the most effective management of groundwater systems.

15.1 Introduction

A number of sandy aquifers of different ages are exploited in Poland. In most cases they contain fresh waters of different ages: from modern (recent) in the unconfined recharge areas to older Holocene (ancient) and even Late Glacial (palaeowaters) in the confined parts. The Miocene aquifer (Bogucice Sands)

chosen for the present study is also characterised by such range of ages, and can be
regarded as a typical sandy aquifer slightly
influenced by anthropogenic pollutants in
the recharge area, and therefore suitable for
determining the natural background levels
(baseline) needed for the implementation of
the EU Ground Water Directive (GWD 2006).
Thirty-five wells, with 32 of them abstract-
ing water for municipal and private use,
were sampled for analyses of major, minor
and trace water constituents in order to
determine the natural background levels
(baseline). Aquifer material was sampled at
selected sites in the outcrop area for determining the mineralogy needed for hydrogeochemical modelling. Numerical flow and
transport modelling served for determining
the flow pattern within the aquifer and
aquitards. Environmental isotope tracers as
well as man made atmospheric trace gases
were used to recalibrate the numerical flow
and transport model and to determine

independently the ages of water (timescales)
and age distributions in individual wells.

15.2 Geology and hydrogeology

The Bogucice Sands aquifer is situated in the
Cracow area, southern Poland (Fig. 15.1).
The sediments of the aquifer and its confining cover were deposited in the Carpathian
Foredeep Basin during the Upper Badenian
when paleorivers from the mountains in the
south discharged at the deltaic shoreline.
Their lithofacies are irregular, due to variable gravitational segregation of the material
transported in the delta and secondary
changes caused by submarine landslides
(Porebski and Oszczypko 1999). The present
southern boundary of the outcrop area is
about 4 km to the north of the Carpathian
overthrust. Only thin and discontinuous
Quaternary layers of sand, loess and locally
boulder clay cover the outcrops. Permeable

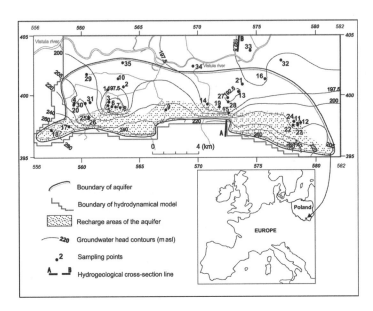

Fig. 15.1 Hydrogeological map of
the Bogucice Sands aquifer with the
positions of sampled wells.

sands and sandstones, which represent 20–80% of the aquifer with thicknesses of 100–310 m, deepen to the north, becoming confined by Badenian marine mudstones and claystones. The aquifer is underlain by impermeable clays and claystones of the Tertiary Chodenice beds (Fig. 15.2). The mineralogy of the aquifer material is highly heterogeneous as indicated by the following percentage concentration ranges of the main minerals: quartz 42.5–77.5%, calcite 2.0–24.0%, dolomite 0.0–2.4%, K-feldspar 0.0–7.0%, plagioclase 0.0–8.0%, muscovite 0.0–1.4%, biotite 0.0–2.2%, glauconite 0.0–1.9%, organic matter 0.0–2.0%, Fe-hydroxide 0.0–3.0% and carbonate cement 3.0–30.0%. Typical carbonate contents are 3–10% in sands and 25–30% in sandstones. They are represented by cements and calcareous debris of marine fauna. Minor components are represented by Al_2O_3 (0.7–9.7%), Fe_2O_3 (0.7–4.6%), CaO (3.2–6.7%), MgO (0.1–1.4%), MnO (0.03–0.09%), Na_2O (0.5–0.9%), K_2O (1.1–2.1%), P_2O_5 (0.03–0.09%) and SO_4 (0.04–0.07%), whereas the main trace components comprise Sr (34–100 ppm), Ba (20–340 ppm), Zn (30–70 ppm), Mo (12–27 ppm), Co (5–29 ppm), Ni (10–40 ppm), V (10–70 ppm) and Rb (7–28 ppm).

Presumably, during the Pliocene, the Badenian sediments still contained connate marine formation water. During the Pleistocene, the Vistula River valley was formed and infilled with fluvial and glacio-fluvial sands and gravels up to 10–15 m thick, and also with the loess up to 20 m thick in the south of the valley. During that period, the marine sediments of the delta were gradually flushed by freshwater.

Outcrops define the southern boundary of the Bogucice Sands aquifer, whereas other boundaries are arbitrarily chosen as those corresponding to the minimal transmissivity of 10 $m^2\,h^{-1}$. The mean hydraulic conductivity obtained from pump tests decreases in the direction of flow from 4.5×10^{-5} to 8.0×10^{-6} ms^{-1} with a total geometric mean of 1.5×10^{-5} $m\,s^{-1}$. Recharge occurs in the areas of outcrop and by downward leakage from the Quaternary sands near the outcrop areas, and in the areas of intensive withdrawal. The general flow direction is to the Vistula valley, with discharge by upward leakage through the confining layers. Mean precipitation, potential evapotranspiration and infiltration rates are 760, 500 and 140 mm a^{-1}, respectively (Kleczkowski et al. 1990). With the area of 176 km^2, estimated safe yield of

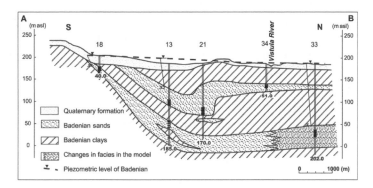

Fig. 15.2 Simplified geological cross section of the Bogucice Sands aquifer.

40,000 m³ d⁻¹ and typical well capacities of 96–4800 m³ d⁻¹, the aquifer is classified as belonging to the medium-size main aquifers in Poland, and supplies water for public and private use.

There is a distinct preference among individual users towards the use of surface water supplied by pipelines from a reservoir situated on the Raba River, about 30 km south of Cracow. This preference mainly results from the high hardness of the groundwater, which is inconvenient to individual users causing problems in central heating systems and washing machines. In spite of this preference, the yield of the aquifer is insufficient to meet local demands and, as a consequence, licensing conflicts arise between water supply companies and food industry on the amount of water available for safe exploitation. The positions of abstraction wells investigated within the project are shown in Fig. 15.1, with water contours clearly indicating three areas of intensive abstraction. Wells 18, 32 and 33 are unexploited due to unstable (18) or low output (32 and 33), whereas well 34 is exploited only sporadically.

15.3   Methods

The entire field procedures of hydrochemical sampling were similar to those described by Salminen et al. (2005). Unfiltered water was taken to 500 ml polyethylene bottles for major anion analysis by ion chromatography, filtered water was acidified by HNO_3 to pH < 2 and taken to new hardened polyethylene 100 mL bottles for ICP-MS and ICP-AES for analysis of the major cations, minor and trace components. Filtered water was taken in 100 mL umber glass bottles for DOC analysis. For pH and Eh field determinations,

two laboratory calibrated instruments were used immerged in pumped water until equilibrium and minimal difference were reached. If the readings still differed, the mean value was taken. Alkalinity was measured in the field by titration. Polyethylene bottles of 25 mL, 1.5 and 3×20 L were used to sample water for $\delta^{18}O$ and δ^2H, ^3H, and ^{14}C and $\delta^{13}C$ analyses, respectively. Special 2.5 L glass bottles were used to sample for N_2, Ne, Ar, SF_6 and freons. Sampled water was not allowed to have any contact with the atmosphere or to be degassed. The sampling bottles were also used as measuring vessels by the head-space method.

Environmental isotopes (^3H, ^{14}C, $\delta^{18}O$, δ^2H and $\delta^{13}C$) were measured only in selected wells. Tritium and ^{14}C were determined by liquid scintillation spectrometry, with electrolytic enrichment for tritium, and typical uncertainties in the order of 0.5 TU and 1.0 pmc, respectively. Stable isotope ratios of O, H and C were determined by mass spectrometry and expressed relative to V-SMOW and V-PDB standards with uncertainties of 0.1, 1.0 and 0.1‰, respectively. For some sites, SF_6, freons (F-11, F-12 and F-113), N_2, Ne and Ar were extracted from water in sample bottles and measured by gas chromatography methods as described by Zuber et al. (2005). The numerical hydrodynamic MODFLOW code (Guiger and Franz 2003) as well as transport MODPATH (Pollock 1988) and MT3D codes (Zengh and Wang 1999) with cells of 250 × 250 m were used to construct a model of flow pattern and to obtain tracer distributions and travel times within the flow system. As abstraction has been more or less constant for several decades, the modelling was performed for assumed hydrodynamic steady state, although the tracer data indicate that in areas of intensive withdrawal this assumption is not valid. The dispersion and the piston flow models combined in line with the exponential model (Maloszewski and Zuber 1996) were additionally used to obtain information on flow times to individual wells with available tritium records of several years, if the assumption of steady state was acceptable.

15.4 Environmental tracers and modelling of timescales

Environmental isotope data are shown in Table 15.1 together with the estimates of tritium ages obtained from simple box models and corrected ^{14}C ages obtained from the piston flow model. The δ^{18}O and δ^2H values of nearly all samples were scattered around typical values of modern infiltration in the Cracow area confirming recharge under moderate Holocene climate, whereas for sites 32–34 they were distinctly lower suggesting recharge under cooler pre-Holocene climate as confirmed by noble gas temperatures (NGT) of about 2°C determined for wells 32 and 33, in contrast to about 8°C in other wells, the latter corresponding to the mean long-term air temperature in the area (Zuber et al. 2005).

Spatial distributions of ^{3}H, shown in Fig. 15.3, indicate the presence of modern waters in the recharge areas and in some intensively exploited wells in the confined parts. In most cases, measurable tritium was accompanied by elevated concentrations of SF_6 and freons. The numerical flow and transport models were corrected with the aid of SF_6 data to obtain better fits for the areas with modern waters. The corrected transport model was confirmed by comparison with observed space and time distributions

of tritium for the period 2000–2002 (Zuber et al. 2005). For a number of sites, fairly good agreements were obtained, especially for wells 5–8 characterised by tritium in the range of 7–17 TU with decreasing values over time at individual sites, and for wells 11 and 19 characterised by tritium contents around 5 TU with increasing values. For some wells, no agreement was obtainable between the numerical transport model and tracer data, which means that the conceptual (hydrogeological) model is not adequately constructed for some areas. For instance, in well 25 situated in the unconfined recharge area, no tritium was found, which was in contradiction with numerical models. Most probably, this well is within a deep erosion structure undiscovered so far by geological and geophysical methods, which still contains pre-bomb era water. Similarly, for well 13, in spite of a very wide age distribution (residence time distribution) calculated with the aid of the MT3D code, the existence of very short travel times seen in Fig. 15.4, should result in measurable tritium contents, which is not the case. In addition, the mean modelled age of 502 years is too low in comparison with the estimated ^{14}C ages given in Table 15.1. In well 12, the tritium content was 24 TU in 1987 and about 1 TU in 2001 suggesting a change from modern to pre-bomb era water, whereas in the adjacent well 11, tritium contents of 6–7 TU were observed, with a slight tendency to increase. These examples show that, for some wells, it is difficult to obtain good fits of the box models to the tracer data due to unsteady flow conditions, and of numerical models due to insufficient knowledge on the aquifer structure.

The tritium data interpreted with the aid of box models yielded mean ages of about

Table 15.1 Tracer data of and tracer ages estimated with the aid of lumped-parameter models.

Well no.	Date	$\delta^{18}O$ (‰)	δ^2H (‰)	^{14}C (pmc)	$\delta^{13}C$ (‰)	^{14}C Age[a] (10^3 years)	Tritium (TU)	Model[b]	^3H Age (years)
Modern waters									
17	01.08.00	−10.0	−69	nm	nm	−	31.3±1.4	Modern,	
	18.05.01	−10.1	−70	66.4	−14.5	Modern	41.3±1.8	unsteady flow	
18	17.05.88	−10.5	−74	60.5	−14.5	Modern	85.5±4.0	DM	33
	01.08.00	−10.3	−72	nm	nm	−	40.3±1.8		
6	24.07.00	−9.8	−69	45.9	−13.5	Modern	17.2±0.8	EPM	58
	07.05.01	−10.0	−70	nm	nm	−	16.9±0.8	DM	16.5
	21.09.02	−9.7	−70	45.8	−13.4	Modern	14.7±0.7		
	14.07.03	−9.8	−71	nm	nm	−	14.6±0.7		
4	21.07.00	−10.1	−71	37.1	−11.8	0–1.2	19.0±0.9	EPM	79
	11.07.01	−10.2	−72	nm	nm	−	19.1±1.0	DM	57
5	24.07.00	−10.0	−70	55.0	−10.5	Modern	15.7±0.8	EPM	70
	07.05.01	−9.8	−69	nm	nm	−	15.4±0.7		
	21.07.03	−10.0	−72	nm	nm	−	12.5±0.7		
7	24.07.00	−9.8	−69	44.0	−12.8	Modern	11.2±0.6	EPM	124
	08.05.01	−9.9	−70	nm	nm	−	12.1±0.6	DM	128–175
	17.07.03	−9.8	−70	nm	nm		10.1±0.6		
8	24.07.00	−9.8	−70	43.4	−13.5	Modern	8.7±0.6	EPM	156
	08.05.01	−9.9	−69	nm	nm	−	8.3±0.5	DM	158
	21.09.02	−9.8	−71	44.1	−13.4	Modern	7.8±0.5		
11	25.07.00	−9.8	−70	nm	nm	−	6.4±0.5	DM	86
	10.05.01	−9.8	−70	64.6	−14.1	Modern	6.5±0.5		
	23.07.03	−9.6	−69	nm	nm		7.4±0.6		
19	30.09.01	−9.7	−68	nm	nm	−	4.5±0.7	EPM	163
	23.09.02	−9.6	−69	35.9	−12.7	Modern	6.6±0.5	DM	55
	25.07.03	−9.7	−69	nm	nm		5.6±0.4		
Holocene waters of pre-bomb era occasionally mixed with modern waters									
12	06.07.87	−10.0	−67	63.8	−11.0	Modern	24.0±1.5	^3H indicates water of	
	25.07.00	−9.9	−69	nm	nm	−	3.8±0.5	increasing age	
	10.05.01	−9.9	−70	63.5	−14.6	Modern	1.1±0.5		
15	28.07.00	−9.8	−68	nm	nm	−	0.2±0.5	Older without ^3H	
	24.05.01	−9.8	−68	55.0	−13.8	0–0.2	5.9±0.5	Younger with ^3H	
14	28.07.00	−9.7	−68	41.9	−14.1	0–1.1	4.7±1.5	Younger with ^3H	
	22.05.01	−9.7	−67	nm	nm	−	0.0±0.5	Older without ^3H	
	23.09.02	−9.6	−68	43.0	−14.2	0–1.2	0.1±0.5		
13	06.07.87	−10.6	−71	26.7	−12.1	0.5–2.0	2.2±1.5	^{14}C indicates water of	
	25.07.00	−10.7	−75	19.7	−12.3	1.2–2.7	0.0±0.5	increasing age	
	22.05.01	−10.7	−75	18.4	−12.2	1.5–3.0	0.0±0.5	initially with ^3H	

Table 15.1   *(Continued)*

Well no.	Date	$\delta^{18}O$ (‰)	δ^2H (‰)	^{14}C (pmc)	$\delta^{13}C$ (‰)	^{14}C Age[a] (10^3 years)	Tritium (TU)	Model[b]	^3H Age (years)
Holocene waters of pre-bomb era									
25	13.09.00	−10.0	−71	58.9	−13.8	0–0.1	0.0±0.5	No tritium	
	18.05.01	−10.1	−71	nm	nm	–	0.1±0.5		
	31.07.03	−9.7	−70	nm	nm		0.0±0.3		
9	27.05.87	−9.9	−68	51.2	−12.9	0–0.5	0.6±1.0	Traces of tritium due to	
	25.07.00	−9.6	−68	48.8	−13.1	0–0.4	0.9±0.5	small admixtures of	
	16.05.01	−9.7	−68	nm	nm	–	0.4±0.5	modern water	
	27.09.02	−9.6	−69	50.7	−13.3	0–0.5	1.4±0.5		
28	24.05.01	−9.7	−69	39.2	−13.2	0–1.0	0.1±0.5	No tritium	
2	27.05.87	−9.5	−69	40.1	−14.5	0–1.1	0.0±1.0	No tritium	
	19.07.00	−9.4	−66	41.0	−14.8	0–1.0	0.0±0.5		
27	24.05.01	−10.2	−72	32.3	−12.9	0–1.3	0.0±0.5	No tritium	
	26.09.02	−10.0	−71	34.0	−12.8	0–1.2	0.1±0.5		
16	28.07.00	−10.0	−70	31.8	−13.3	0–1.5	0.0±0.5	Occasionally traces of	
	22.05.01	−10.1	−70	nm	nm	–	0.4±0.5	tritium	
	23.09.02	−10.0	−69	32.5	−13.3	0–1.4	0.4±0.5		
3	25.07.87	−10.2	−69	30.5	−12.7	0–1.8	0.6±0.5	Occasionally traces of	
	21.07.00	−9.9	−70	30.3	−13.3	0–1.9	0.0±0.5	tritium	
	21.09.02	−10.0	−72	nm	nm	–	0.0±0.5		
1	13.04.88	−10.1	−71	22.5	−12.6	1.0–2.6	0.0±1.0	No tritium	
	19.07.00	−10.0	−71	23.6	−12.3	0.9–2.5	0.0±0.5		
35	30.07.03	−9.8	−70	18.2	−13.6	1.5–3.0	0.0±0.5	No tritium	
10	01.08.00	−9.9	−71	14.0	−12.3	2.0–3.5	0.0±0.5	No tritium	
16	28.07.00	−10.0	−70	31.8	−13.3	0–1.5	0.0±0.5	Occasionally traces of	
	22.05.01	−10.1	−70	nm	nm	–	0.4±0.5	tritium	
	23.09.02	−10.0	−69	32.5	−13.3	0–1.4	0.4±0.5		
3	25.07.87	−10.2	−69	30.5	−12.7	0–1.8	0.6±0.5	Occasionally traces of	
	21.07.00	−9.9	−70	30.3	−13.3	0–1.9	0.0±0.5	tritium	
	21.09.02	−10.0	−72	nm	nm	–	0.0±0.5		
1	13.04.88	−10.1	−71	22.5	−12.6	1.0–2.6	0.0±1.0	No tritium	
	19.07.00	−10.0	−71	23.6	−12.3	0.9–2.5	0.0±0.5		
35	30.07.03	−9.8	−70	18.2	−13.6	1.5–3.0	0.0±0.5	No tritium	
10	01.08.00	−9.9	−71	14.0	−12.3	2.0–3.5	0.0±0.5	No tritium	
Waters of glacial age									
32	08.11.02	−10.8	−76	0.1	−10.6	about 12[c]	0.0±0.5	No tritium	
33	13.04.88	−11.2	−79	0.8	−8.8	about 12[c]	0.5±1.0	No tritium	
	15.07.02	−11.1	−78	0.2	−8.3	about 12[c]	0.0±0.5		
34	21.07.03	−10.8	−79	0.3	−8.9	about 12[c]	0.0±0.5	No tritium	

[a] ^{14}C Piston-flow ages corrected for isotopic exchange with carbonate minerals (see text); [b] EPM stays for the piston flow model in line with expenential model, DM stays for the dispersion model; [c] assumed ^{14}C age of late glacial water (see text; nm: not measured.

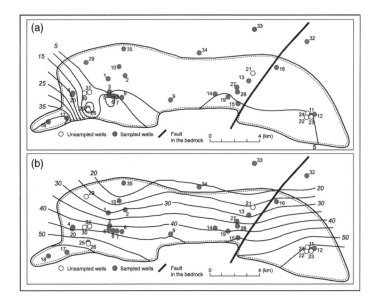

Fig. 15.3 Spatial distribution of (a) tritium and (b) ^{14}C observed in 2000–2001.

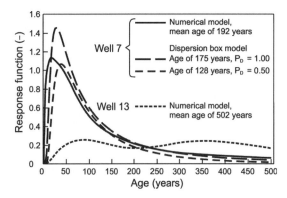

Fig. 15.4 Examples of age distributions in wells 7 and 13 calculated with the aid of numerical modelling (MT3D) in comparison with the distributions obtained from the dispersion box-model equally well fitted for two different pairs of parameters, that is, mean age and dispersion parameter (P_D), to the tritium data of well 7. No tritium was observed in well 13.

20–175 years with wide distributions resulting from the abstraction of water from different depths and different directions. Wide age distributions were also confirmed by modelling with the aid of the MT3D code, which

means that all abstracted recent waters represent mixtures of waters from nuclear and pre-nuclear bomb eras, that is, those recharged before and after 1952. Two examples of typical age distributions are shown in Fig. 15.4. For well 7, the dispersion box model yielded similar distributions of ages for two different pairs of fitting parameters (dispersion parameter and mean age) indicating the lack of a unique solution, which is typical, especially in the cases of short tritium records and high dispersivity (Maloszewski and Zuber 1996; Zuber et al. 2005). Similar difficulties in obtaining unique solutions can be expected in numerical modelling, though not necessarily apprehended by modelers. For well 13, the age distribution calculated with the aid of the MT3D code is even wider, more irregular than that in well 7. Very wide residence time distributions obtained from the interpretation of tritium data and the numerical transport modelling show that any trends in hydrochemistry expected due to changes in recharge areas (e.g. appearance or disappearance of pollution sources) will be considerably smoothed, and, therefore, difficult to observe in a reasonably prolonged period of observations.

The ^{14}C contents in waters containing tritium range from 37 to 66 pmc. Such low ^{14}C contents in modern waters may result from a number of reasons, among them from dilution by dead carbon due to reaction with primary or secondary carbonate minerals, mixing of bomb era water with distinctly older water, and from ^{14}C losses caused by isotope exchange with carbonates. The first process is usually accounted for by one of the methods leading to determining the so-called initial carbon content (e.g. Clark and Fritz 1997). The second process is very unlikely for the known geology of the recharge area of the Bogucice Sands. The third process can be significant if the exchange takes place with films on the surfaces of grains with δ^{13}C values modified by earlier exchange. The exchange does not then influence the δ^{13}C values of the dissolved inorganic carbon and, therefore, remains unaccounted for in the correction for the initial carbon content. As a consequence, the ^{14}C ages can be greatly overestimated as described for fissured carbonate aquifers by Maloszewski and Zuber (1991) and for a sandy aquifer with appreciable amounts of carbonate minerals by Aeschbach-Hertig et al. (2002). Some delay of all tracers, and especially of ^{14}C, is also possible in thin confined aquifers due to diffusional exchange between the aquifer and adjacent impermeable or semi-permeable sediments (Sudicky and Frind 1981; Sandford 1997). However, this process is considered to be of secondary importance for the Bogucice Sands.

Pre-bomb era Holocene waters dominate the confined part of the aquifer as indicated by the lack of tritium and ^{14}C contents in the

range of 10–43 pmc. In the discharge area (wells 32–34), pre-Holocene waters characterised by the most negative δ^{18}O and δ^2H values, low NGTs and ^{14}C contents of 0.1–0.3 pmc were identified. Assuming the continuity of recharge since the climate became distinctly warmer at the end of the last glacial period the recharge of glacial waters probably took place not earlier than in the time span of 14–11 ka BP. These values would then constrain the upper age in the aquifer. If the ^{14}C contents in the aquifer were governed only by radioactive decay with the half-life of 5.73 ka, they should follow lines 1 and 2 shown in Fig. 15.5 for ^{14}C contents measured in the recharge area. As a consequence of the constraint put on the upper age values, lines 3 and 4 in Fig. 15.5 bracket possible changes in ^{14}C contents caused both by radioactive decay and the exchange losses discussed above, with the effective half-life of about 2 ka. The uncertainty of the ^{14}C dating is

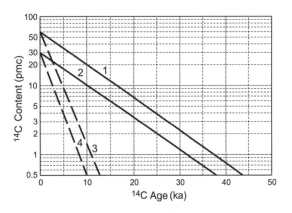

Fig. 15.5 Dependence of ^{14}C on age due to the radioactive decay for the initial ^{14}C contents observed in recent waters in the recharge areas (lines 1 and 2) and in the case of assumed additional decrease of concentration in the Bogucice Sands aquifer caused by exchange with abundant solid carbonates (lines 3 and 4).

admittedly very high in this case, mainly due to large heterogeneity of the solid carbonate contents (3–30%). Nevertheless, Fig. 15.5 provides a rough idea of possible timescales of groundwater within the confined part of the aquifer. Glacial age waters in wells 32 and 33 are unexploited; therefore, they are undoubtedly characterised by narrow distributions of travel times, which can reasonably well be estimated with the aid of MODPATH. The initial estimations yielded ages of the order of 2 ka which were in sharp conflict with the tracer data. However, subsequent recalibration of the numerical flow model for the drainage area yielded values in accordance with the tracer data.

15.5 Hydrochemistry and its relations to timescales

Geochemical modelling performed with the PHREEQC2 code (Parkhust and Appelo 1999) confirmed that rainfall chemistry and dry deposition contribute some major and minor groundwater components, but water–rock interaction is the main control of the hydrogeochemistry (Karlikowska et al. 2003), with additional influences of diffuse and local anthropogenic sources of pollution and external geogenic sources. Environmental tracers relate water-quality data to timescales. Recent waters with measurable ^{3}H and SF_6 contents are in most cases aerobic and of $Ca–HCO_3$ or $Ca–HCO_3–SO_4$ groundwater types. Bicarbonate results from the dissolution of carbonate cement and calcareous debris, whereas sulphate results from airborne contributions. Anaerobic conditions prevail in the confined aquifer containing pre-bomb era Holocene waters where the tendency to $Ca–Na–HCO_3$ type is observed

due to continued cation exchange. The oldest waters of glacial age are of $Na–HCO_3$ and $Na–HCO_3–Cl$ types due to prolonged cation exchange. They have distinctly elevated TDS contents, possibly due to diffusional exchange with pore water from adjacent impermeable sediments.

The distributions of hydrochemical parameters and constituents of groundwaters are known to be usually wide. For the Bogucice Sands, the number of sampled sites (35) was rather large considering the aquifer dimensions, but still too low to consider separately the statistical distributions for different aquifer zones and layers. As the anthropogenic influences are moderate and observed only in parts of the outcrop area, the pre-selection was applied only for Cl^- in one well where the presence of a downward seepage from the polluted Serafa River was evident. In Figs. 15.6 and 15.7, the total cumulative distributions of major, minor and selected trace components are shown, and the summary statistics of selected constituents are given in Table 15.2. Some toxic trace elements of natural origin are present but their concentrations are below MPLs.

It is evident from the inspection of Figs. 15.6 and 15.7 that too few sampled wells would undoubtedly yield less representative distributions. Therefore, in spite of some suggestions from law makers, a low number of sampling sites and/or a single value resulting from a large number of hydrochemical analyses (e.g. 97.7 percentile) should not be regarded as sufficient for describing adequately the natural baseline values (NBLs) and the chemical status of GWB. However, both the percentiles 2.3 and 97.7 can be regarded as a representative way for reporting the baseline values (NBLs) and supplying, in a simple way, information on

Table 15.2 Summary statistics of selected water constituents in the Bogucice Sands.

Parameter	Number of samples	Mean	Median	Mode	MPL	Percentiles 2.3	16	84	97.7
T	33	11.6	11.2	11.0		11.0	11.0	12.5	13.0
pH	35	7.56	7.44	7.08	6.5–9.5	7.00	7.08	8.18	9.04
Eh (mV)	35	118	94	76		-39	39	188	419
DO (mg L^{-1})	32	0.77	0.07	0.03^a		0.02	0.03	0.98	7.91
SEC (µS cm^{-1})	35	774	671	250^a	2500	250	452	1110	1696
Hardness (meq L^{-1})	35	5.19	5.58	0.07^a		0.07	2.00	8.21	11.20
TDS (mg L^{-1})	35	611	578	615		218	384	850	1308
Ca (mg L^{-1})	35	84	95	105		0.63	32	139	185
Mg (mg L^{-1})	35	12	13	0.3^a		0.30	5	20	24
Na (mg L^{-1})	35	68	21	3.1^a	200	3.1	6.5	139	**420**
K (mg L^{-1})	35	3.1	1.9	1.5		0.57	1.26	6.2	8.2
Cl (mg L^{-1})	35	54	26	1.1^a	250	1.1	6.3	130	246
SO$_4$ (mg L^{-1})	35	45	27	0.6^a	250	0.6	10	88	196
HCO$_3$ (mg L^{-1})	35	339	330	413		130	221	455	750
NO$_3$ (mg L^{-1})	34	3	0.02	0.02	50	0.02	0.025	1.9	**53**
NH$_4$ (mg L^{-1})	33	**0.84**	0.30	0.10	0.5	0.05	0.05	**2.1**	**4.5**
TOC (mg L^{-1})	33	1.13	0.91	0.30^a		0.3	0.6	1.4	4.6
DOC (mg L^{-1})	33	1.05	0.85	1.2		0.28	0.56	1.3	3.9
Si (mg L^{-1})	35	6.42	6.62	3.74^a		3.7	5.3	7.4	8.2
Fe (mg L)	35	**0.59**	**0.50**	0.015^a	0.2	0.015	0.12	**0.99**	**1.99**
Mn (mg L^{-1})	35	**0.078**	0.050	0.032	0.05	0.001	0.008	**0.16**	**0.25**
Sr (mg L^{-1})	35	0.31	0.27	0.19		0.015	0.16	0.54	0.86
F (mg L^{-1})	33	0.12	0.02	0.02	1.5	0.02	0.02	0.2	1.5
Br (µg L^{-1})	35	22	19	18^a		1	8	29	81
I (µg L)	35	8.9	6.8	0.35		0.35	3.8	15	37
Se (µg L^{-1})	35	0.23	0.18	0.18		0.10	0.18	0.22	1.00
B (µg L^{-1})	33	222	48	35	1000	5.8	16	362	**3130**
Ba (µg L^{-1})	35	64	46	7.3^a		7.3	21.6	115	245
As (µg L^{-1})	35	0.91	0.40	0.18	10	0.018	0.072	0.98	**13.0**
Li (µg L^{-1})	35	26.5	13.3	13.3		0.4	8.0	60	79
Pb (µg L^{-1})	35	0.165	0.061	0.020	10	0.014	0.020	0.19	1.40
Rb (µg L^{-1})	33	1.41	1.08	1.08		0.37	0.65	2.49	3.15
Sc (µg L^{-1})	30	4.71	4.54	4.50		2.81	4.07	5.55	6.54
Zn (µg L^{-1})	35	18.2	1.70	0.009		0.009	0.009	33.2	150
U (µg L^{-1})	33	0.131	0.012	0.012		0.002	0.004	0.364	0.827

^{a} The lowest mode shown if number of modes exceeds 1.

Values exceeding MPL according to Council Directive 98/83/EC are in bold.

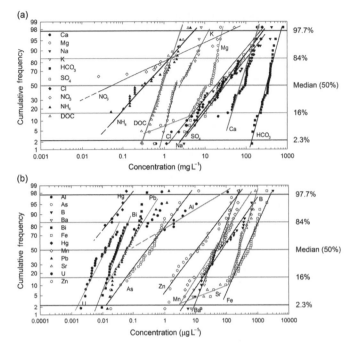

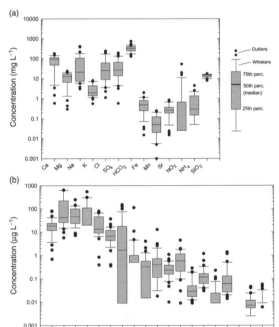

Fig. 15.6 Cumulative frequency diagrams of chemical constituents in the Bogucice Sands aquifer: (a) major and minor components; (b) selected trace components.

the uncertainty related to both the natural spread of values and possible anthropogenic influences (that range of percentiles corresponds in a good approximation to ±2 standard deviations).

The curves in Fig. 15.6 better represent the differences in modalities of particular constituents whereas boxes in Fig. 15.7 are better for a comprehensive presentation of all components. Wide ranges of natural distributions are in some cases broadened by anthropogenic and geogenic influences, which are clearly visible in Fig. 15.8, where horizontal space distributions of selected constituents are shown. However, the spatial variations shown in Fig. 15.8 are of approximate character due to a low number of wells available for sampling and their irregular distribution. No adequate data-sets are available to derive reliable time trends in the hydrogeochemistry of the aquifer caused by either variable input in

Fig. 15.7 Box-plots of chemical constituents in the Bogucice Sands aquifer: (a) major and minor components; (b) selected trace components.

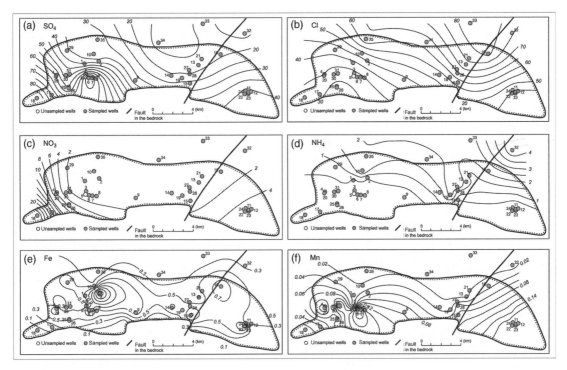

Fig. 15.8 Horizontal distributions of selected constituents in the Bogucice Sands aquifer with isolines in mg L^{-1}: (a) SO$_4$; (b) Cl; (c) NO$_3^-$; (d) NH$_4$; (e) Fe; (f) Mn.

the recharge area or intensive abstraction in some other areas.

The chemical status of the aquifer as a whole is good though several constituents exceed their MPLs in some wells, as indicated in Table 15.2 by bold numbers. Sulphate concentrations are below the MPL over the whole aquifer although they are elevated in and close to the recharge areas (Fig. 15.8[a]) mainly due to atmospheric deposition resulting from the nearby foundry and some contribution from the oxidation of pyrite. The anthropogenic impact was especially strong during the second half of the previous century. This is apparent from the following data related to percolating waters near the metallurgy plant in the Cracow area: the concentration of SO$_4^{2-}$ in 1973 to 1975 changed from 200 to 400 mg L^{-1}, dropped to

200 around 1982, reached 400 mg L^{-1} in 1988–1991, and subsequently dropped rapidly to ca. 50 mg L^{-1} in 1994 (Witczak et al. 1998). Similar changes, probably with slightly decreased concentrations, due to a larger distance from the plant, existed in the recharge area of the Bogucice aquifer. Elevated SO$_4^{2-}$ concentrations are observed in modern waters under aerobic conditions. However, no reduction of SO$_4^{2-}$ to H$_2$S is observed within the aquifer, therefore, the differences in the SO$_4^{2-}$ distribution most probably result mainly from differences in atmospheric concentrations. The distribution of Ca^{2+} is similar to that of SO$_4^{2-}$, although governed by different factors because reduced contents of Ca in the confined area, especially in wells with pre-Holocene water, result from exchange with sodium.

Chloride concentrations (median 26 mg L^{-1}) are generally in excess of those observed in local rainfall (c. 6.5 mg L^{-1} when corrected for enrichment by evapotranspiration). In the recharge areas, the excess concentrations are likely to be of anthropogenic origin resulting from different point sources of pollution, whereas far from these areas they are of geogenic origin (Fig. 15.8[b]). Distinctly elevated concentration occurs in well 5 (about 250 mg L^{-1}; omitted in the construction of the figures as a result of pre-selection) due to infiltration from Serafa River which is polluted by leakage from abandoned chambers where salt was exploited by the dissolution method. The solution chambers, which are situated south of the western part of the aquifer, undergo squeezing by the weight of municipal wastes being disposed in this area and release brines at the ground surface.

Distinctly elevated Cl concentrations (46–200 mg L^{-1}) in wells 13, 16, 21, 27 and 28 are probably related to the presence of faults in the bedrock, suggesting a source from ascending older waters of higher salinity. In all these wells, the values of the molar ratio of Na to Cl are distinctly below 1, but weight ratio of Cl/Br ranges from more than 300 to about 850, rather excluding the origin related to the remnants of formation water. The sodium concentration may be reduced in the case of upflow of older water induced by intensive abstraction due the exchange of $2Na^+$ for Ca^{2+}. However, elevated Ca^{2+} concentrations in the waters of these wells are not observed. A different situation exists in wells 32, 33 and 34, which are also characterised by elevated Cl concentrations, with MPL values of Na exceeded in wells 33 and 34. The molar Na/Cl ratios in these wells are 1.6, 4.6 and 2.9, respectively, confirming the Na excess of another origin to that of Cl.

The enrichment in Na results mainly from cation exchange of Ca with minerals of marine origin, which have not been completely washed out by freshwaters during the replenishment of the aquifer over a geological timescale. The possibility of exchange is confirmed by very low Ca concentrations in waters of these three wells (0.6–5.6 mg L^{-1}). The contribution from hypothetical remnants of formation water can be excluded by weight ratio of Cl/Br in the range 3300–5250.

Nitrate concentrations in some wells with recent waters (Fig. 15.8[c]) are distinctly elevated due to agricultural pollution; however, the MPL value is slightly exceeded only in well 18. In the confined areas, the concentrations of NO_3 decrease due to changes in the redox conditions and the N_2 contents increase, with the denitrification excess up to 6.7 mg L^{-1} as calculated from Ne and N_2 contents (Zuber et al. 2005). Ammonium concentrations in a number of wells of the confined area (Fig. 15.8[d]) exceed the MPL value due to purely geogenic influences. Similarly, the concentrations of Na, As and B exceed the MPL values in wells 33 and 34 in the deepest part of the aquifer. Well 33 is also characterised by an anomalous content of P (1.2 mg L^{-1}) in the form of PO_4^{3-}, which exceeds 7–60 times the concentrations in other wells.

Iron concentrations below the MPL value (0.2 mg L^{-1}) are observed only in wells containing the youngest waters under aerobic conditions (17, 18 and 23) and in some wells in the confined area containing modern water mixed with pre-bomb Holocene water as indicated for well 4 by the tracer data (see Table 15.1). Glacial age water in the deepest part of the aquifer (wells 33, 34 and 35) is also characterised by low Fe and Mn concentrations whereas the highest concentrations are observed between these two zones under anaerobic conditions; in most cases exceeding the MPL value (Fig. 15.8[e]). The Mn distribution within the aquifer is similar to that of Fe (Fig. 15.8[f]) although the MPL value of Mn (0.05) is exceeded in only a few wells. A similar distribution is also observed for strontium.

The existence of two groups of wells with low iron concentration is explained in Fig. 15.9. The first group of low Fe concentrations is related to the presence of aerobic conditions (high Eh) along the initial parts of flowpaths in the recharge area, whereas the second group results from low Eh values accompanied by high pH values in the deepest parts of the aquifer. A similar distribution also exists for Mn (Garrels and Christ 1965; Stumm and Morgan 1996), though the spatial distribution of Mn somewhat differs from that of Fe because in waters with active circulation its concentration does not depend to the same degree on Eh.

15.6   Conclusions

The Bogucice Sands, which can be regarded as a typical confined sandy aquifer, which is recharged at outcrop by downward seepage, and discharged by upward seepage in the farthest part. The aquifer is characterised by distinct hydrogeochemical zones related to timescales (ages) of water as shown in a simplified form in Fig. 15.10. In spite of anthropogenic influences observed in the recharge areas, the aquifer generally contains waters of acceptable quality, though some of them require treatment to reduce Fe and Mn concentrations. Waters characterised by youngest ages (c. <20 years) are influenced by

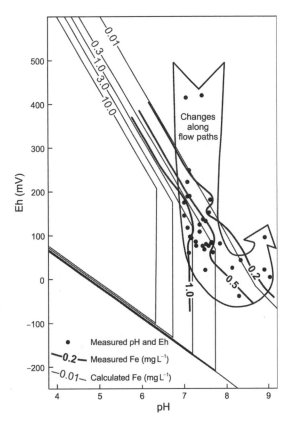

Fig. 15.9 The dependence of iron content on the Eh and pH values, adapted from Hem (1960) by Ratajczak and Witczak (1983) for HCO_3^- = 5.46 mmol L^{-1}, SO_4^{2-} = 0.6 mmol L^{-1}, ionic strength of 0.015, temperature of 10°C, and Eh and pH ranges typical for waters of active replenishment. Points represent data of the Bogucice Sands aquifer.

anthropogenic pollution. The deepest parts of the aquifer contain waters of glacial age, where the MPLs of Na, As, B and NH_4 are exceeded through natural anomalies.

Characterisation of chemical status of groundwater by single NBL (baseline) values (e.g. 90 or 97.7 percentiles of the cumulative distributions) is considered un-representative due to wide distributions of natural concentrations. The NBLs can be adequately

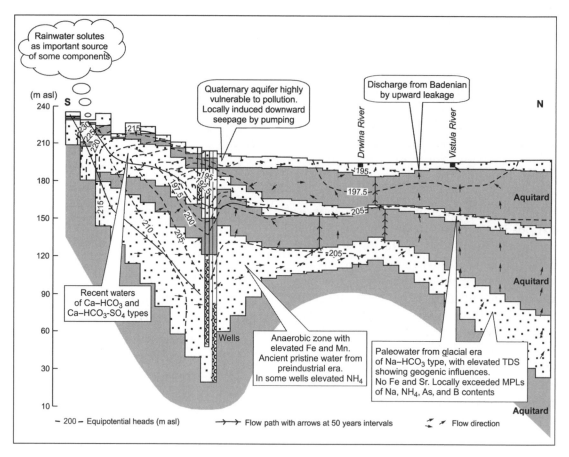

Fig. 15.10 Cross section of the conceptual model of the Bogucice Sands aquifer with flow paths and main characteristics of particular hydrogeochemical zones.

characterised as a range between 2.3 and 97.7 percentiles of the cumulative distributions.

If the number of sampled wells is low, or spatial distribution of wells is non-uniform, the NBL values may not be representative for whole aquifer or groundwater body. The problem becomes more acute if some anthropogenic influences are present, which are difficult to separate in a quantitative way from the natural background.

In aquifers containing water of greatly different ages different hydrochemical zones are usually observed. The management of waters in particular zones should be related to timescale. The description of the chemical status of such aquifers should account for differences in hydrochemical characteristics. Similarly, the pre-selection of indicator elements and the frequency of monitoring should be related to timescales.

Flow pattern and age distributions within the aquifer were obtained with the aid of numerical flow and transport modelling, which were additionally recalibrated with the aid of environmental tracers. However, for some wells, it was difficult to obtain a reasonable agreement between modelled and measured environmental tracer data. This discrepancy shows the importance of environmental tracer methods for obtaining a more detailed knowledge on the flow pattern and timescales of the investigated groundwater bodies.

Very wide residence time distributions obtained for the Bogucice aquifer from the interpretation of tritium data and the numerical transport modelling show that any trends in hydrochemistry expected due to changes in recharge areas (e.g. appearance or disappearance of pollution sources) will be considerably smoothed, and, therefore, difficult to detect.

References

Aeschbach-Hertig, W., Stute, M., Clark, J. et al. (2002) A paleotemperature record derived from noble gases in groundwater of the Aquia Aquifer (Maryland, USA). *Geochimica et Cosmochimica Acta* 66, 797–817.

Clark, I. and Fritz, P. (1997) *Environmental Isotopes in Hydrogeology*. Lewis Publishers, New York.

Garrels, R.M. and Christ, C.L. (1965) *Solutions, Minerals, Equilibria*. Harper Row, New York.

Guiger, N. and Franz, T. (2003) *Visual MODFLOW Pro v. 3.1*. Waterloo Hydrologic Inc., Canada.

GWD (2006) Directive 2006/118/EC of the European Parliament and the Council of 12 December 2006 on the protection of groundwater against pollution and deterioration. OJ L 327, 27.12.2006, p.19.

Hem, J.D. (1960) Restraints on dissolved ferrous iron imposed by bicarbonate, redox potential and pH. USGS Water-Supply Paper. 1459-B, Washington, DC.

Karlikowska, J., Ratajczak, T. and Witczak, S. (2003) Hydrogeochemical modeling of the processes forming the groundwater composition of the recharge zone of the Bogucice subbasin (MGWB 451). (in Polish). Zeszyty Naukowe Politechniki Slaskiej; no 1592, pp. 119–24.

Kleczkowski, A.S. et al. (1990) *The Map of the Critical Protection Areas (CPA) of the Major Groundwater Basins (MGWB) in Poland*. University of Mining and Metallurgy, Cracow.

Maloszewski, P. and Zuber, A. (1991) Influence of matrix diffusion and exchange reactions on radiocarbon ages in fissured carbonate rocks. *Water Resources Research* 27, 1937–45.

Maloszewski, P. and Zuber, A. (1996) Lumped parameter models for the interpretation of environmental tracer data. Manual on Mathematical Models in Isotope Hydrology. IAEA-TECDOC-910, IAEA, Vienna, pp. 9–58.

Pollock, D.W. (1988) Semi-analytical computation of path lines for finite difference models. *Ground Water* 26 (6), 734–50.

Porebski, S. and Oszczypko, N. (1999) Lithofacies and origin of the Bogucice Sands (Upper Badenian), Carpathian Foredeep (in Polish). In: *Proceedings of Polish Geological Institute*, CLXVIII, pp. 57–82.

Ratajczak, T. and Witczak, S. (1983) *Mineralogy and Hydrogeochemistry of Iron in Well Screen Encrustation in Quaternary Aquifers* (in Polish). Zeszyty Naukowe AGH, ser. Geologia, z.29, 229 pp.

Salminen, R., Batista, M.J., Demetriades, A., Lis, J. and Tarvainen T. (2005) Sampling In: Salminen, R. (chief ed.) *Geochemical Atlas of Europe. Part 1. Background Information, Methodology and Maps*. Geological Survey of Finland, Espoo, pp. 67–79.

Sandford, W.E. (1997) Correcting for diffusion in carbon-14 dating of ground water. *Ground Water* 35 (2), 357–61.

Stumm, W. and Morgan, J.J. (1996) *Aquatic Chemistry*. 3rd edn. J. Wiley & Sons, Inc., New York.

Sudicky, E.A. and Frind, E.O. (1981) Carbon-14 dating of groundwater in confined aquifers: implications of aquitard diffusion. *Water Resources Research* 17, 1060–4.

Witczak, S., Suder, M. and Wojcik, R. (1998) Monitoring of contaminant transport through the unsaturated zone using multi-layer sampling in large-diameter dug well (in Polish). In: *Municipal and Rural Water Supply and Water Quality*. PZITS, Poznan, pp. 111–23.

Zengh, C. and Wang, P.P. (1999) MT3DMS, a modular three-dimensional multi-species transport model for simulations of advection, dispersion and chemical reactions of contaminants in groundwater systems; documentation and user's guide. US Army Engineer Research and Development Center Contact Report SERDP-99–1. Vicksburg, MS.

Zuber, A., Witczak, S., Rozanski, K. et al. (2005) Groundwater dating with ^{3}H and SF_6 in relation to mixing pattern, transport modelling and hydrochemistry. *Hydrological Processes* 19, 2247–75.

16 The Cambrian–Vendian Aquifer, Estonia

R. VAIKMÄE, E. KAUP, A. MARANDI, T. MARTMA
V. RAIDLA AND L. VALLNER

In southern and central Estonia, the Cambrian–Vendian aquifer system contains relict saline groundwater of marine origin with total dissolved solids (TDS) up to 22 g L^{-1}. Cl$^-$ and Na$^+$ predominate over all other ions in this zone. In northern Estonia the Cambrian–Vendian aquifer system contains palaeo-groundwater, which recharged during the last glaciation. This is freshwater with TDS mainly below 1.0 g L^{-1}. The baseline chemical composition of the water is formed through the water–rock interaction during the last more than 10 ka. Generally, the groundwater is of good quality, but in some areas problems are associated with elevated Fe and Mn contents. Groundwater does not always fulfil the requirements of drinking water standards in respect of Cl$^-$ and Na$^+$ content. The most characteristic feature of the baseline quality of groundwater of the Cambrian–Vendian aquifer system in Northern Estonia is its lightest known oxygen isotopic composition (δ^{18}O values of c.–22‰) in Europe. This gives the possibility to use the isotopic composition of groundwater as an ideal tracer of possible changes in groundwater baseline quality.

In northern Estonia the groundwater of the Cambrian–Vendian aquifer system provides high-quality drinking water for communities and towns (including the capital city of Tallinn) but its industrial use is also important. The supply is very significant, amounting to 10–13% of the Estonian groundwater consumption. Overexploitation of freshwater groundwater resources in Tallinn and mine dewatering in northeast Estonia has resulted in the development of two basin-wide cones of depression. In turn, it has caused the changes in the direction and velocity of groundwater flow, which has led to 1.5- to 3.0-fold rise in the TDS content and concentration of major ions in groundwater. The main sources of dissolved load in the Cambrian–Vendian groundwater are the leaching of host rock and various geochemical processes that occur in the saturated zone. Leakage of saline water from underlying crystalline basement, is the second most important source of mineralisation. Intrusion of seawater with consequent implications for groundwater quality is at present time still not evident but should be considered in coming decades.

16.1 Background to understanding of baseline groundwater Quality in Estonia

16.1.1 Introduction

Estonia is a flat country with an area of 45,000 km^2, where plateau-like areas and regions of small hills alternate with low-lands. The average altitude is 50 m and only 10% of the territory has an elevation between 100 and 250 m above sea level. The Baltic Sea with the Gulf of Finland forms the main drainage basin. The climate is moderately cool and humid. Average annual precipitation ranges from 500 to 700 mm. The mean surface runoff from Estonia is 270 mm yr^{-1} (Perens and Vallner 1997).

An essential first step for assessing the natural (baseline) groundwater quality is an understanding of the hydrogeology. To define the source terms the input to the system must be known, that is, the chemistry of infiltrating water, derived mainly from rainfall. To explain the characteristics of groundwater chemistry within the aquifer, information on the geology, mineralogy, geochemistry, flow patterns and timescales are required.

16.1.2 Geological setting

16.1.2.1 Structure

Estonia is situated in the northwestern part of the East European Platform. Structurally, the main sedimentary basins, situated on the southern slope of the Baltic Shield, plunge southwards, sloping about 3–4 m km^{-1} (Fig. 16.1[c]). The crystalline Lower Proterozoic basement is overlain by Upper Proterozoic (Vendian) and Paleozoic (Cambrian, Ordovician, Silurian and Devonian) sedimentary rocks which are covered by Quaternary deposits (Perens and Vallner 1997).

In northern Estonia, the Vendian, Cambrian and Ordovician rocks are the only sedimentary rocks covering the crystalline basement, which lies approximately 150 m

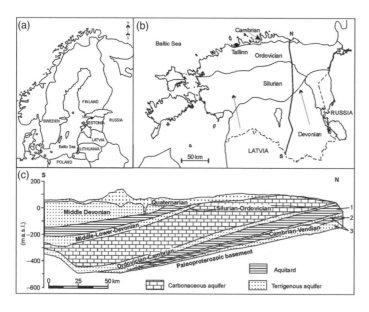

Fig. 16.1 (a) Schematic map showing the location of Estonia; (b) hydrogeological map of Estonia with the position of the line of cross section; arrows indicate the direction of groundwater flow and (c) the North–South cross section of major hydrogeological units of Estonia (1 – Lontova and 2 – Kotlin clays; 3 – Voronka (upper) and Gdov (lower) aquifers, confining the Cambrian–Vendian aquifer system).

below the surface (Fig. 16.1[c]). The crystalline basement comprises mainly gneisses and biotite gneisses (Koistinen et al. 1996) and its upper part (10–150 m thick) is fractured and weathered. Weathering profiles are predominantly composed of kaolinite, illite, chlorite and montmorillonite, depending on the original bedrock composition and the intensity of weathering.

16.1.2.2   Stratigraphy

Weathered basement rocks are overlain by water-bearing Vendian and Cambrian silt- and sandstones (with interlayers of clay), which form the Cambrian–Vendian aquifer system. In the east of Estonia, the Vendian sedimentary rocks are divided by a clay layer, the Kotlin Formation, into the Voronka and Gdov aquifers (Fig. 16.1[b] and [c]).

The terrigeneous rocks of the Cambrian–Vendian aquifer system occur all over Estonia, except in the Mõniste–Lokno uplift area in southern Estonia. The aquifer system is overlain by clays and siltstones of the Lükati–Lontova aquitard (Fig.16.1[c]), which has a strong isolation capacity, due to its low conductivity of 10^{-7}–10^{-5} m d^{-1} (Perens and Vallner 1997). However, in some places the aquitard and water-bearing bedrock formation have been penetrated by a relatively dense set of ancient buried valleys filled with loamy till and, in other places, glaciofluvial gravel in the lower parts of the valleys (Tavast 1997). These valleys are orientated northwest to southeast, approximately perpendicular to the north Estonian coastline.

16.1.3   Hydrogeology

There is an obvious difference between the cross sections of the Cambrian–Vendian aquifer system in western Estonia compared to those in the east. In the east, up to 53 m thick clays of the Kotlin Formation divide the aquifer system into two aquifers (Fig. 16.1[c]). The upper, Voronka aquifer, consists of quartzose sandstone and siltstone with a thickness of up to 45 m in northeastern Estonia. The hydraulic conductivity of the rocks ranges from 0.6 to 12.5 m d^{-1} with an average 2.6 m d^{-1}. The transmissivity decreases from 100 to 150 m^2 d^{-1} in northern Estonia to 50 m^2 d^{-1} (or less) in the south. Under natural conditions, the potentiometric levels along the coast of the Gulf of Finland are about 1.5–5.5 m a.s.l. The lower Gdov aquifer is formed by a complex of mixed sandstone and siltstone up to 68 m thick. It directly overlies the Pre-Cambrian basement and is confined by the overlying clay of the Kotlin Formation. In northern Estonia, the hydraulic conductivity of the water-bearing rocks is 0.5–9.2 m d^{-1} with an average of 5–6 m d^{-1}. Transmissivities in northeastern Estonia are in the range of 300–350 m^2 d^{-1} and decrease in a southerly and westerly direction to 100 m^2 d^{-1} or less. The potentiometric surface in the coastal area is about 3–5 m a.s.l. under natural conditions. West of the line where the Kotlin clays pinch out, the Cambrian and Vendian water-bearing rocks form the Lontova–Gdov aquifer. The Cambrian–Vendian aquifer system thins out towards the south and west. In northern Estonia, however, its thickness amounts to 90 m outcropping along the northern coast in the south of the Gulf of Finland. In northern Estonia, the aquifer system is mostly confined by 60–90 m thick clays of the Lontova Formation. However, in places the aquitard is penetrated by ancient buried valleys.

The Cambrian–Vendian aquifer system is underlain by Lower Proterozoic crystalline basement, whose fractures contain a small amount of water but is not exploited. The lower portion of basement serves as an impermeable base layer for all the overlying aquifer systems.

The Cambrian–Vendian aquifer system forms part of the regional flow system that recharges in southern Estonia, in the Haanja and Otepää heights, where groundwater levels are 180–280 m a.s.l. In these locations, the head declines with depth, indicating the existence of downward groundwater flow. On reaching the impermeable portion of the crystalline basement, this flow is directed towards the discharge areas in the depressions of the Baltic Sea and the Gulf of Finland. The length of deeper branches of the regional flow system can reach 250 km.

The Cambrian–Vendian aquifer system north of the recharge area belongs to a slow flow subzone (Vallner 1997). The calculated velocities of deep groundwater movement lie between 5×10^{-4} and 5×10^{-3} m d^{-1}. This indicates that during the last ~10 ka the deep groundwater could only have progressed several tens of kilometres, and complete water exchange along flow branches would not have been possible. Therefore it appears that, under natural conditions, groundwater recharged during the last glaciation has been preserved in the aquifer. This has been confirmed using isotopic tracers (Vaikmäe et al. 2001).

16.1.4 Palaeohydrological conditions during the Late Pleistocene

During the last glacial maximum (LGM) around 18 ka BP (21 calendar ka), the whole Baltic Sea area and northern Poland were covered by the Fennoscandian ice sheet.

Different reconstructions of the ice sheet have yielded different results. According to the ice model proposed by Denton and Hughes (1981), the ice thickness over the Baltic States area at that time was about 800–2500 m. A modified ice sheet model by Lambeck (1999) shows that the ice sheet thickness over the area at 18 ka BP was only about 600–800 m. The deglaciation of Estonia from the Haanja ice-marginal zone to the recession of ice from the Palivere zone took place in the 2000 years between ca 14.7 and 12.7 years BP (Kalm 2006; Rinterknecht et al. 2006). The rapid deglaciation produced huge volumes of meltwater as well as icebergs. Therefore, practically the whole area in front of the ice margin was covered with vast ice-dammed lakes during the last stages of ice sheet retreat. Starting around the Palivere stage, these lakes formed the eastern part of the Baltic Ice Lake.

Due to the stadial-oscillatory character of deglaciation, the level of the Baltic Ice Lake changed rapidly several times. At the end of the Younger Dryas cold stage the amelioration of the climate caused rapid retreat of the ice margin, and the Baltic Ice Lake drained catastrophically into the North Sea via the Öresund Strait. Its surface was lowered by 26–28 m within only a few years (Björck 1995). An open sound north of Billingen provided direct connection between the Baltic basin and the North Sea. According to Björck (1995) the final drainage is dated at c.10.3 ka BP.

In the context of groundwater formation, it is important to note that the main discharge area of the Cambrian–Vendian aquifers into the Gulf of Finland was submerged during all stages of the Baltic Sea in spite of both the several drastic changes in the Baltic Sea level and the high glacio-isostatic uplift rate in northern Estonia during and after the last deglaciation.

In contrast to the coastal areas of northwest Europe and the Mediterranean, where groundwater circulation during the LGM was in many cases activated by evolving shorelines due to sea-level lowering, the groundwater recharge and circulation in the Baltic region ceased or was strongly inhibited at this time due to ice cover and/or permafrost.

The results of isotope and geochemical investigations as well as noble gas analyses (Vaikmäe et al. 2001) showed that the water in the Cambrian–Vendian aquifer system in northern Estonia was recharged during the last glaciation. It has further been shown that for about 11 ka during the Late Pleistocene, the Scandinavian ice sheet in the outcrop area was underlain by meltwater (Jõeleht 1998). During this time the hydraulic head was controlled by the thickness of the ice. Although the aquifer has relatively high hydraulic conductivity, areas of low hydraulic conductivity surround it and therefore the hydraulic head in the outcrop area of aquifer system was probably close to floating point, for example, about 90% of ice thickness (Piotrowski 1997). Taking into account also the postglacial uplift and the present depth of the Cambrian–Vendian aquifer system (about –100 m below sea level), the hydraulic gradient was around 0.0031 (Jõeleht 1998). Thus, recharge probably occurred during the glaciation, most likely by subglacial drainage through the tunnel valleys (Vaikmäe et al. 2001).

16.1.5 *Hydrochemistry*

Earlier studies showed that Na–Ca–Cl–HCO_3 and Ca–Na–Cl–HCO_3 type waters with total dissolved solids (TDS) contents between 0.4 and 1 g L^{-1} dominate the Cambrian–Vendian aquifer system in northern Estonia. In northeastern, southwestern and southeastern Estonia, as well as on the islands of Saaremaa and Ruhnu, Na–Cl, Na–Ca–Cl and Ca–Na–Cl type waters with TDS contents from 2 to 22 g L^{-1} are widespread (Karise 1997). The aquifer has predominantly reducing conditions (Vallner 1997), and is usually rich in trace elements with concentrations increasing towards the east. East of Tallinn, for example, concentrations of iodide reach values between 120 and 280 μg L^{-1}. In some wells in northeastern Estonia, the concentrations of cadmium (Cd), lead (Pb) and lithium (Li) are slightly in excess of the EU drinking water standards. However, the concentrations of trace elements are highest at Värska in southeastern Estonia, where cadmium (Cd), lithium (Li), manganese (Mn) and lead (Pb) concentrations exceed the EU regulations for drinking water. In several regions of Estonia, high concentrations of bromide (Br) have been detected in the Cambrian–Vendian aquifer system, for example, values of 13 mg L^{-1} Br were observed in Kuressaare at depths between 540 and 555 m whilst concentrations in Värska ranged between 16 and 17 mg L^{-1} at 520–535 m and 51–56 mg L^{-1} at 540–600 m (Karise 1997).

The oxygen isotope compositions in the groundwater of most aquifer systems in Estonia range from –11.0 to –12.2‰ (Vaikmäe and Vallner 1989). However, groundwater in the Cambrian–Vendian aquifer system shows a heavily depleted oxygen isotope composition, with $\delta^{18}O$ values varying mainly from –18‰ to –22‰ (Vaikmäe et al. 2001).

In contrast, long-term mean annual $\delta^{18}O$ value in contemporary precipitation in

Estonia is −10.4‰ (Punning et al. 1987). Low (highly negative) $\delta^{18}O$ values in the Cambrian–Vendian aquifer are indicative of recharge in cold conditions, whilst low ^{14}C concentrations are indicative of long residence time of groundwater (Fig. 16.2). Low ^3H concentrations (<2 TU) in most of the studied well waters confirm that no detectable intrusion of modern water (including seawater) into the Cambrian–Vendian aquifer has occurred during the past approximately 45 years (Vaikmäe et al. 2001).

16.2 Data for the Cambrian–Vendian aquifer and interpretation

16.2.1 Historical and recent data on water quality

Extensive data collected during the last 55 yr exists in the Geological Survey of Estonia (GSE) and contains more than 1500 analyses from 967 wells for the Cambrian–Vendian aquifer system. The database contains information on the main components: TDS, Na^+, K^+, $Na + K$, NH_4^+, Ca^{2+}, Mg^{2+}, Fe^{2+}, Fe^{3+}, Fe_{tot}, Cl^-, SO_4^{2-}, NO_2^-, NO_3^-, CO_3^{2-}, HCO_3^-, pH, SiO_2, hardness (Perens et al. 2001). Data are also available from a number of published and unpublished (mainly various reports of the GSE) investigations. The aquifer has been studied at various times over the past decades (Mokrik and Vaikmäe 1988; Mokrik 1997; Perens and Vallner 1997; Groundwater State 1998; Vaikmäe et al. 2001). The results of these studies provide extensive information on the hydrogeology, geochemistry and lithology of the aquifer system. All these results were critically analysed for this study.

During 2001–2003 new samples were collected from representative sites of the

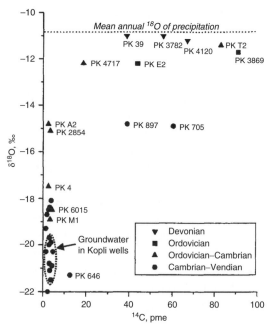

Fig. 16.2 Distribution of $\delta^{18}O$ values of groundwater from different aquifers in Estonia according to their ^{14}C concentrations. $\delta^{18}O$ values from two wells (PK 705 and PK 897) indicate the mixing with infiltrated modern water.

Cambrian–Vendian aquifer system in order to assess the major groundwater chemistry as well as the trace elements. The samples were analysed for a wide range of inorganic species, and field measurements including Eh, DO, pH, temperature and SEC, which are used to assist interpretation. All samples were filtered in the field and acidified with nitric acid (1% v/v) in order to stabilise the trace elements in solution.

16.2.2 Data handling

The two databases containing the historical and the new data were compared in order to evaluate the suitability of the historical database in the context of a baseline study. Where concentrations were below the detection limit of the method, half the value of the detection limit was used for the statistical data analysis and interpretation. Since the historical database goes back to the 1950s, the quality of the database had to be proved, using one of the following approaches: (1) check and evaluate the extreme values of different species by comparing the dry solid residue contents measured in the laboratory; (2) where the TDS were calculated (rather than measured), the cation–anion balance was used to assess the quality of sample. Where samples contained only a few measured major ions with extreme values, it was decided that, without knowledge of the background concentrations, it was better to omit these values from further calculations.

For data processing, interpretation and hydrogeochemical assessment of the results, MapInfo Professional 6.0 and AquaChem 3.7 were used. The summary statistics (maxima, minima, median and standard deviation values) showed that the ranges of major ion concentrations in the Cambrian–Vendian aquifer system for the samples from the historical GSE database and the new samples compare relatively well. The cumulative frequency plots were derived from those data showing the concentration ranges for selected elements/species (Fig. 16.3).

The median (50%) values and the 97.7 percentiles are used as reference values. These values are useful for comparison between different areas as well as for regulatory purposes. For the majority of elements, the range of distributions represents the geological and geochemical controls on the groundwater compositions whereby the slope and the shape of the plots are indicative of the different hydrogeochemical processes.

The cumulative frequency plots (Fig. 16.3) show that HCO_3, Ca and Mg concentrations vary within a small range and only about 10% of the data have values greater or smaller than the median. In the case of Na and Cl, about 20% of the samples have values greater or smaller than the median, but there are only a few samples with extremely high values. This is probably due to the small number of wells in the Cambrian–Vendian aquifer in South Estonia.

16.3 Geochemical controls and regional characteristics

16.3.1 Major element controls

The chemical type of groundwater in the Cambrian–Vendian aquifer system is determined by Na, Ca, Cl and HCO_3, with

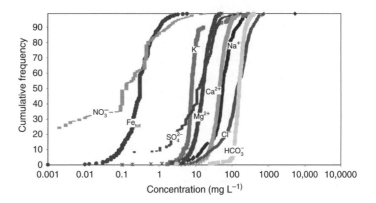

Fig. 16.3 Cumulative frequency diagrams of major ions in the Cambrian–Vendian aquifer.

Na and Cl the most abundant ions in the water. Magnesium is present at relatively low concentrations $(3.2–36.0$ mg $L^{-1})$ and the SO_4 concentration is mostly below the detection limit. The highest SO_4 concentrations are usually detected in wells located close to buried valleys, where groundwater freshening has contributed to the development of oxidising conditions.

The main source of dissolved solutes is leaching of the host rock as well as geochemical processes in the saturated zone. The underlying crystalline basement comprises saline groundwater in its upper weathered and fissured part and is hydraulically connected with the overlying Cambrian–Vendian aquifer system, forming a second important source of ions. The fractured basement and its clayey weathering crust contain Ca–Cl type groundwater, which is characterised by high TDS values $(2–20$ g $L^{-1})$. Intensive water abstraction in northern Estonia accelerates the groundwater exchange and also increases the area influenced by pumping. Chemical and isotopic groundwater studies indicate an increasing contribution of the leakage from the crystalline basement to the groundwater chemistry (Karro et al. 2004).

16.3.2 Downgradient evolution

The chemical evolution of groundwater in aquifers along flow paths depends on the age distribution with depth and distance, on geological conditions and the lithological composition of the water-bearing rocks and sediments. However, considering the very low velocity of groundwater in the Cambrian–Vendian aquifer system and taking into account both the palaeohydrological situation during the Late Pleistocene as well as the contemporary situation in

northern Estonia, where the groundwater geochemistry is influenced by intensive abstraction, the overall picture of the aquifer system is rather complicated and not yet well understood.

It is practically impossible to follow the evolution of the groundwater chemistry all the way back to the recharge area in Southern Estonia, because there are only three wells drilled in the aquifer in southern and central Estonia. Groundwater with high TDS values is pumped and used as a mineral water at two to three localities. In northern Estonia, however, the fresh-water of the Cambrian–Vendian aquifer system provides the main source of public water supply and the evolution of ground-water chemistry can be followed through many wells drilled within a distance of about 50 km south of the coastline. As a result of intensive groundwater abstraction, two extensive depression cones have formed in this area with centres around the capital city Tallinn and around the Kohtla–Järve mining industry region in northeast Estonia (Vallner 2003). The natural baseline characteristics of the aquifer can be further explained using a Piper diagram (Fig. 16.4).

Four major groundwater types can be distinguished in the diagram based on their chemical composition:

1 Na–Cl type, can be interpreted as a 'saline baseline' or relict formation groundwater of the Cambrian–Vendian aquifer system. It may be very old and probably formed long before the last ice age. TDS concentrations in waters of this type are higher that 2 g L^{-1}, and they are considered as mineral waters. In Värska (southeast Estonia), TDS concentrations in the groundwater of the deeper part of the aquifer system reach concentrations up to 18 g L^{-1}. The $\delta^{18}O$ values

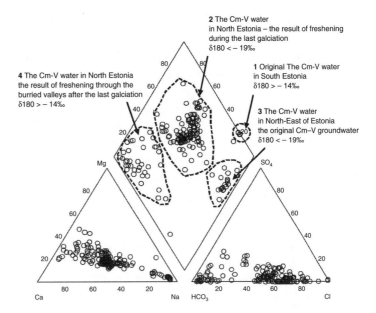

Fig. 16.4 The Piper diagram reflecting major chemical types of Cambrian–Vendian groundwater.

of this groundwater type are around –14‰. Relict groundwater of Na–Cl type is widely distributed in the Cambrian–Vendian aquifer system in southern and central Estonia but also in the Voronka aquifer in northeast Estonia at Narva–Jõesuu. It is also characterised by a very high Cl^- content, ranging from 1083 mg L^{-1} in wells at Narva–Jõesuu up to 10,919 mg L^{-1} at Värska. The Na^+ concentrations in these wells are 684 and 5222 mg L^{-1}, respectively. These characteristic features of the Na–Cl type groundwater are also well reflected in the cumulative probability plots (Fig. 16.3).

2　Ca–Na–HCO₃–Cl or Ca–Na–Cl–HCO₃ water. This is the 'fresh baseline' water of glacial origin, recharged during the last glaciation (Vaikmäe et al. 2001). The chemical composition has been formed through water–rock interaction during the last 10 ka or more. This water type has the largest spatial distribution, spreading from the north

coast to central Estonia, and also shows mixing with other water types (except with type 1) on the Piper diagram. The TDS concentrations vary from 300 mg L^{-1} to several g L^{-1}. Because of the lack of sampling wells in central Estonia, the exact border between the 'saline baseline' waters and 'fresh baseline' waters cannot be defined. The relative proportions of HCO_3^- and Cl^- define the different chemical water types. Thus, some waters are classified as different water types even though the actual differences in their chemical concentrations are small. In the probability plots, this group is represented by low variations in HCO_3^-, Ca^{2+} and Mg^{2+} concentrations and continuously increasing Cl^- and Na^+ concentrations (Fig. 16.3).

However, the most characteristic feature of this groundwater type is its strongly depleted stable isotope composition. The $\delta^{18}O$ values range between –19‰ and –22‰, indicating the formation of the water under cold climatic conditions (Fig. 16.5).

3 Na–Cl–HCO$_3$ groundwater type is interpreted as a mixture of glacial meltwater with some relict saline groundwater. This groundwater type is distributed predominately in northeastern Estonia, where the Kotlin clays divide the Cambrian–Vendian aquifer system into two aquifers and where the overlying clays reach their maximum thickness (Fig. 16.1). Therefore, intrusion of fresh meltwater into the aquifer during the last glaciation was probably lower in this area compared to the western part of north Estonia. However, freshening of the original, relict groundwater, by glacial meltwater is still evident from the δ^{18}O values of this water type, which range between –19‰ and –22‰. Moreover, the TDS values are lower (~1 g L^{-1}) compared to those of the relict saline water (Fig. 16.5).

4 Ca–HCO$_3$ type groundwater is found in northern Estonia, in areas around the ancient buried valleys, where intrusion of fresh groundwater from overlying aquifers and/or rainwater occurs. The intensity of such freshwater intrusion varies spatially and temporally, depending on the extent of groundwater exploitation near the valleys. The intensive groundwater abstraction in the late 1970s, for example, caused extensive cones of depression to develop around Tallinn and Kohtla–Järve (Vallner 2003). In these areas, the groundwater drawdown of 25 and 35 m, respectively, resulted in intensive freshwater intrusion into the aquifer through the buried valleys and caused the groundwater chemistry and its isotopic composition to change (Fig. 16.5). In the cumulative probability plots, this water type is represented by Ca^{2+}, Mg^{2+} and HCO$_3^-$ values above average and by Na$^+$ and Cl$^-$ values lower than average (Fig. 16.3). Due to the freshwater intrusion, the δ^{18}O values of this water type have been changed from values around –20‰ in early 1980s towards more positive values and today are around –15‰ (Fig. 16.5). In parallel the TDS concentrations have also diminished and today are between 200 and 500 mg L^{-1}.

16.3.3 The age of water in the Cambrian–Vendian aquifer

The aquifer system in southern and central Estonia contains relict groundwater with TDS concentrations up to 22 g L^{-1}. There are problems in determining the age of this groundwater, and age estimations have to rely on the data obtained from a groundwater

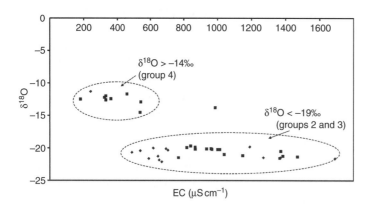

Fig. 16.5 δ^{18}O versus EC values in the Cambrian–Vendian groundwater.

flow model. The model shows that the lateral water movement in this area is directed predominantly towards the north at a velocity of 0.5–1 m a^{-1} (Vallner 2003). This indicates that the groundwater in this part of the aquifer may be very old, probably in excess of 100 ka.

The situation is much different in the northern part of the Cambrian–Vendian aquifer. Consideration of the data collected during the present study together with results from earlier studies lead to the convincing conclusion that the isotopically depleted groundwater in northern Estonia is of glacial origin (Vaikmäe et al. 2001). Results from radiocarbon dating and from noble gas analyses point toward the Fennoscandian ice sheet as a probable source for the groundwater in Estonia. This water is heavily depleted in ^{18}O. This ties in well with the palaeo-climatic and palaeo-environmental situation in the study area during the late Weichselian glaciation. However, so far there is no convincing answer to the question of how and when the meltwater of the ice sheet reached the aquifer system. Earlier studies have indicated that meltwater recharge into the Cambrian–Vendian aquifer system occurred at about 11–12 ka BP, after the retreat of the continental ice from Estonia and during the formation of the Baltic Ice Lake (Yezhova et al. 1996). However, the low ^{14}C concentrations (<5 pmC) suggests an age of the water of about 15–30 ka BP, which in turn implies that the meltwater intrusion took place much earlier, whilst Estonia was still covered by ice.

In contrast, Mokrik (1997) suggests that the freshwater in the Cambrian–Vendian aquifer complex in the north of Estonia was formed by cryogenic metamorphism. He proposes that during the Pleistocene

glaciation the aquifers went through several freezing–refreezing cycles down to depths of 50–200 m, which led to the freshening of the originally highly mineralised groundwater in the Cambrian–Vendian aquifer complex. This would imply that a permafrost zone had developed on the surface of the Baltic Shield about 50–100 km from the outcrop of the aquifer (Mokrik 1997). However, the existence of a thick permafrost layer under the Fennoscandian ice sheet in the region around Estonia seems unrealistic. According to Kleman and Borgström (1994), frozen conditions only existed in the central area of the Fennoscandian ice sheet, whilst most of subglacial areas reached melting temperatures due to frictional strain and geothermal heating. Thus, recharge of the aquifers by subglacial meltwater seems to be a more realistic explanation for the formation of isotopically light freshwater in the Estonian aquifers and this interpretation also agrees with the recent work of Boulton et al. (1995, 1996) and Piotrowski (1994, 1997). During the Late Weichselian, the base of the ice sheet in the Cambrian–Vendian outcrop area in Estonia was probably in a liquid state for about 11 ka (Jõeleht 1998) and the hydraulic head was controlled by the thickness of ice. Whilst the aquifer system has high hydraulic conductivity, areas of low hydraulic conductivity surround it and, therefore, the hydraulic head in the outcrop area of the aquifer system was probably close to floating point, for example, about 90% of ice thickness (Piotrowski 1997). Taking into account the postglacial uplift as well as the present depth of the Cambrian–Vendian aquifer system of about 100 m b.s.l., it appears that the hydraulic gradient was around 0.0031 (Jõeleht 1998). Thus, groundwater recharge to the aquifer probably occurred during the glaciation, presumably by subglacial recharge through the tunnel valleys.

16.3.4 Trends in water quality and changes with depth

Water quality parameters were investigated in northern Estonia (Karro and Marandi 2003; Karro et al. 2004; Marandi et al. 2004) where the Cambrian–Vendian aquifer system is major source of public water supply. In general, lateral changes in chemical composition and groundwater type are significantly larger than those in the vertical direction. In the Gdov aquifer (the deeper part of the aquifer system), the concentrations of Ca and HCO_3 in the groundwater decrease eastwards where Na and Cl ions dominate (Marandi et al. 2004). A similar trend was observed in the overlying Voronka aquifer (the upper part of the aquifer system). The content of Na and Cl varies from 26 to 405 mg L^{-1} and 46 to 700 mg L^{-1}, respectively, exhibiting the highest values in the eastern part of the country. The concentrations of Ca (6–188 mg L^{-1}) and HCO_3 (103–264 mg L^{-1}) show highest concentrations in the western part of the study area. The highest SO_4 concentrations were measured along the western margin of the area with concentrations between 19 and 22 mg L^{-1}. The TDS content in the analysed groundwater samples ranged from 0.1 to 1.5 g L^{-1} (Marandi et al. 2004).

Generally, the groundwater is of good quality in accordance with the limits set by the Estonian Drinking Water Standards (2001). However, in some areas, problems associated with elevated Fe and Mn concentrations occur. Highest Fe concentrations reach concentrations up to 6–7 mg L^{-1}, exceeding the drinking water limits by factor of 5–30 Groundwater does not always fulfil the requirements of drinking water standards also in respect of Cl and Na contents (Marandi et al. 2004).

Barium is the only toxic element that occurs in concentrations much higher than those permitted in water abstracted for drinking purposes. The modelling results of Marandi et al. (2004) showed that SO_4 and HCO_3 are the main anions that may control Ba precipitation, although HCO_3 has practically no influence when Ba concentrations are low (<10 mg L^{-1}). Increasing concentrations of Cl, in contrast, contribute to an increased solubility of Ba. The results also show that in the case of low SO_4 concentrations (<3 mg L^{-1}), the Ba content in the Cambrian–Vendian aquifer system can be higher than limit value of 2 mg L^{-1} set for drinking water.

The most serious consequence of intensive groundwater use in north Estonia is the formation of regional cones of depression around the capital Tallinn and Kohtla–Järve (northeast Estonian industrial area). A basin-wide model simulation showed that over-exploitation has caused the changes in the direction and velocity of groundwater flow (Vallner 2003). As a result, lateral and vertically rising groundwater flows support the transport of connate brackish water from the deeper parts of the aquifer system and from the underlying crystalline basement to the groundwater intakes and also promotes seawater intrusion (Yezhova et al. 1996; Mokrik 1997; Vallner 1999; Savitski 2001).

A case study was conducted on the Kopli peninsula in Tallinn in order to assess the possible causes for the increase in TDS (Karro et al. 2004). Groundwater production wells trapping the Cambrian–Vendian aquifer system on the Kopli Peninsula are situated close to the sea. The production wells in the pumping stations have depths between 40 and 107 m and drilled, more or less, through the full thickness of the aquifer system. For comparison, a groundwater monitoring well penetrating into the fractured basement, and two wells opening Lontova aquitard were included in the study.

Groundwater abstraction varied between years, ranging from 10 to 1300 m³ d⁻¹ during 1978–2002 and in 1992 resulted in a drop of the potentiometric surface by 17 m. At about the same time, groundwater extraction decreased due to declining industrial and agricultural production on the one hand and more sustainable groundwater use on the other. As a direct result, the potentiometric surface level of the aquifer system has steadily risen during the last 10 years and is now at –4.0 m a.s.l. (Savitski 2001).

Fresh groundwaters of Na–Ca–HCO₃–Cl type are characteristic of the upper part of the Cambrian–Vendian aquifer complex in the Tallinn region (Perens et al. 2001). In the deeper part of the aquifer and in the crust of the weathered crystalline basement,

groundwaters of Na–Ca–Cl–HCO₃ and Ca–Na–Cl type and with TDS content of 1.4–5.0 g L⁻¹ are common. In the upper part (–50 m asl) of the aquifer system, the groundwater has chloride concentrations of 100 mg L⁻¹, which increase with depth to 350 and 2500 mg L⁻¹ at 100 and 130 m, respectively (Savitski 2001; Boldõreva and Perens 2002).

The large-scale variations in TDS (0.49–4.6 g L⁻¹) and major element concentrations in the water of production wells are evident from data collected during 1978–2002 (Table 16.1). The maximum concentrations presented in Table 16.1 describe the water chemistry during 1994–1996, which clearly exceeded the permissible concentrations in major components as set by the Estonian Drinking Water Standard (2001) and by the Drinking Water Directive of the EU (Directive 1998). During subsequent years the TDS content of the groundwater decreased slightly. The most distinct temporal changes in water chemistry occurred in well 613 (Fig. 16.6), where the TDS reached concentrations of up to 1.5 g L⁻¹ increasing

Table 16.1 Statistical summary (range and mean values) of concentrations (mg L⁻¹) of major chemical compounds and TDS in studied wells, Kopli Peninsula. Analyses from 1978 to 2002.

Well no	TDS	Na⁺	K⁺	Ca²⁺	Mg²⁺	Cl⁻	SO₄²⁻	HCO₃⁻
598	707–1120	91.0–142.9	8.9–15.0	94.6–117.8	22.6–28.1	302.8–413.0	2.0–17.7	146.4–176.9
	804	114.0	10.7	110.4	25.6	364.1	3.7	158.6
613	624–1545	96.8–241.7	5.4–13.3	99.0–184.4	21.7–47.6	259.9–707.6	2.0–55.6	146.4–238.0
	1010	160.6	10.0	138.5	31.7	481.6	21.8	170.8
614	724–1100	89.6–172.2	7.0–20.0	102.8–149.2	23.7–33.0	330.1–546.7	0.0–19.4	121.9–189.2
	832	115.2	10.8	119.6	27.8	387.6	0.0	152.5
615	488–1170	68.6–222.2	7.8–15.0	71.9–127.3	19.2–40.1	196.1–546.7	0.0–56.4	61.0–189.2
	820	122.5	9.8	101.2	28.6	358.4	24.7	167.8
798	3616–4587	400.0–525.0	16.0–22.5	676.8–836.7	80.4–114.4	2109.5–2552.6	0.0–3.7	6.1–18.3
	4347	469.6	20.5	805.1	104.2	2439.7	2.0	12.2

Source: Karro et al. (2004).

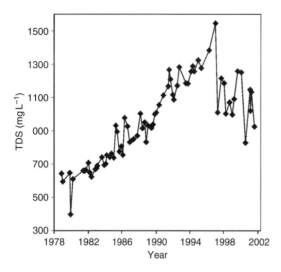

Fig. 16.6 Temporal changes of TDS in groundwater in the Cambrian–Vendian aquifer on Kopli peninsula, Tallinn (abstraction well no 613).

by a rate of 50 mg L^{-1} yr^{-1}. In the other wells of the Kopli peninsula, a less dramatic increase in TDS was observed.

Since major ions constitute the bulk of the mineral matter contributing to the TDS, most major elements (Na, K, Ca, Mg, SO_4, Cl) display the same trends as TDS (Table 16.1) with the exception of HCO_3. In comparison to other major ions, HCO_3 shows the smallest concentration range, and these relatively stable HCO_3 concentrations explain why, at high TDS values, bicarbonate remains insignificant in determining the groundwater chemical type. When TDS exceeds 0.9 g L^{-1}, Cl type waters dominate.

The Cambrian–Vendian aquifer system at Kopli peninsula is confined by the Lontova aquitard ($k = 10^{-7}$–10^{-5} m d^{-1}), which protects the aquifer from infiltrating modern water. Thus, human impacts are unlikely to have any effect on the groundwater chemistry.

Considering the hydrogeological situation in the study area, there are three major

processes which can be responsible for the increase in TDS values: (1) intrusion of present-day seawater; (2) pumping-induced upwards migration of deeper saline water from areas below the freshwater; and (3) a combination of these two processes.

Characteristic ion ratios and scatter diagrams for the most significant chemical components (Fig. 16.7) were used to explore the importance of individual components.

One of the main indicators of seawater intrusion into drilled wells is usually the increase in the conservative elements, especially Cl in the groundwater. However, within the Cambrian–Vendian aquifer system, the use of the Na/Cl ratio to identify modern seawater intrusion is somewhat limited as the original relict groundwater at least in southern and central Estonia has a marine origin with a corresponding Na/Cl ratio. Therefore, even in the fresh groundwater of glacial origin northern Estonia the Na/Cl ratio is very close to that of seawater.

The concomitant increase of Na and Cl in wells in the Kopli peninsula is quite clear (Table 16.1). However, the Na/Cl plot shows that the groundwater in the Kopli peninsula waterworks is depleted in Na relative to seawater (Fig. 16.7[a]) and that the values fall below the seawater dilution line (SDL). In well 615, the depletion of Na relative to seawater becomes more evident at higher Cl concentrations. The distribution of data below the SDL suggest mixing more saline water with dilute water.

Magnesium concentrations in Kopli lie approximately on the SDL when plotted against Cl, with small deviations to either side (Fig. 16.7[b]). This indicates that Mg may have a marine component, modified to some extent (mainly at lower Cl values) by dissolution/weathering reactions. Ca and K

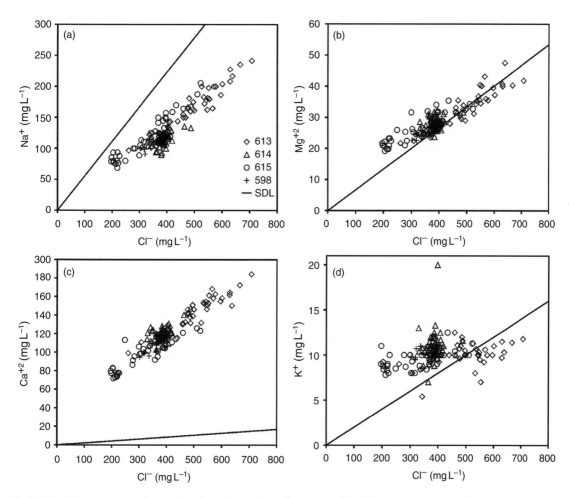

Fig. 16.7 The concentrations of various ions plotted against chloride concentration; SDL – seawater dilution line (Karro et al. 2004).

are enriched in the groundwater relative to seawater. K concentrations are fairly independent of Cl concentrations as shown in Fig. 16.7(c), while Ca exhibits a clear increase at high Cl values (Fig. 16.7[d]).

Karro et al. (2004) showed that leaching from the host rock and element release resulting from geochemical processes in the saturated zone provide the major sources of dissolved load in the Cambrian–Vendian groundwater of the Kopli peninsula. A sec-

ond important source of elements is the underlying crystalline basement, which is hydraulically connected with the overlying Cambrian–Vendian aquifer system and its upper weathered and fractured zone, contains saline groundwater. The fractured basement and its clayey weathering crust also host the Ca–Cl type groundwater, which is characterised by high TDS value (2–20 g L^{-1}). Intensive water abstraction in northern Estonia has accelerated the leakage from the crystalline basement as is seen in the results from chemical and isotopic groundwater studies (Vaikmäe et al. 2001; Karro et al. 2003).

Due to its glacial origin, the groundwater in the Cambrian–Vendian aquifer system in northern Estonia has a unique isotopic composition. Thus, the $\delta^{18}O$ values of the groundwater, as well as the ^{14}C and ^{3}H signature, provide the most sensitive indicators of modern seawater intrusion into aquifer system. The, $\delta^{18}O$ values in the palaeo-groundwaters vary between –19‰ and –22‰, while ^{14}C concentrations are lower than 5 pmC and ^{3}H concentrations lie typically below the detection limit (<1 TU) (Vaikmäe et al. 2001). According to Punning et al. (1991) the annual mean $\delta^{18}O$ value of water in the Gulf of Finland is about –7‰, the ^{14}C concentrations in the Baltic Sea are close to 100 pmC and ^{3}H concentrations range between 5 and 10 TU.

In 2001, groundwater samples from six wells of the Kopli peninsula were analysed for their isotopic composition, and a set of selected $\delta^{18}O$ and ^{14}C values as well as Br/Cl ratios are presented in Table 16.2 and in Fig. 16.2.

The $\delta^{18}O$ and ^{14}C values are typical for Cambrian–Vendian groundwater in northern Estonia, indicating the glacial origin of the water. Based on these results, it appears that intrusion of modern water (including seawater) into the aquifer as a consequence of heavy pumping has not occurred. This conclusion is also confirmed by the absence of ^{3}H in sampled wells in the costal areas of northern Estonia (Vaikmäe et al. 2001). The Br/Cl ratios (Table 16.2) are almost double those in seawater (0.0035) and suggest that older formation water (without traces of Cl from modern seawater) is present.

Table 16.2 Selected isotopic and chemical parameters in studied wells at Kopli peninsula, Tallinn.

Well no	$\delta^{18}O$ (‰)	^{14}C (pmc)	Br$^-$ (mg L^{-1})	Cl$^-$ (mg L^{-1})	Br$^-$/Cl$^-$
599	–20.4	2.5	0.63	99	0.0064
600	–19.8	5.7	2.15	494	0.0044
598	–21.5	3.4	3.62	403	0.0090
613	–21.5	3.0	3.84	481	0.0080
615	–21.6	2.8	3.18	521	0.0061
614	–21.5	5.4	3.96	409	0.0097

16.4 Summary

The groundwater in the Cambrian–Vendian aquifer system is generally not affected by present-day infiltration and the main controls on the baseline chemistry are the geochemistry of the bedrock sediment and processes of water–rock interaction. However, in places, the Lontova aquitard and water-bearing bedrock formation have been eroded forming a set of ancient buried valleys, filled mostly with loamy tills, although glacio-fluvial gravels also occur in the lower parts of the valleys. Where groundwater is intensively abstracted, these erosional valleys can provide the Cambrian–Vendian aquifer system with recharge of fresh meteoric water.

The aquifer can be divided in two main zones, which contain groundwater of totally different origin and contrasting baseline chemical composition. In southern and central Estonia, the aquifer system contains relict saline groundwater of marine origin with TDS values up to 22 g L^{-1}. The dominating solutes in this zone are Cl and Na and the age of the groundwater probably exceeds 100 ka.

In northern Estonia, the aquifer system contains palaeo-groundwater, which was recharged during the last glaciation, more than 10 ka ago, by subglacial drainage through the aquifers. The groundwater in this area is fresh with TDS mainly below 1.0 g L^{-1} with baseline chemical composition resulting from water–rock interaction during about the last 10 ka. Generally, the groundwater quality is good, although problems associated with elevated Fe and Mn concentrations are reported in some areas. However, the most characteristic feature of the baseline chemistry in northern Estonia is the oxygen isotopic composition of the groundwater, with the $\delta^{18}O$ values around $c.$–22‰ being the lightest signature reported anywhere in Europe. For comparison the long-term mean annual $\delta^{18}O$ values in contemporary meteoric water in Estonia are –10.4‰. Thus, the isotopic composition of the groundwater provides an ideal tracer for possible changes in the groundwater baseline quality.

The principal economic role of the Cambrian–Vendian groundwater in northern Estonia is the provision of high-quality drinking water for communities and towns (including the capital city of Tallinn) as well as to supply water for industrial use. The supply is very significant, amounting to 10–13% of Estonian groundwater consumption. Anthropogenic impacts on the aquifer are mainly related to overexploitation of freshwater resources in northern Estonia and mine dewatering in northeast Estonia and have resulted in the development of two basin-wide cones of depression, around Tallinn and Kohtla–Järve areas. This, in turn, has caused changes in the direction and velocity of groundwater flow and resulted in an increase in TDS and major ion concentrations in the groundwater. Where production wells are situated near the sea, lateral seawater intrusion into the groundwater intakes may occur. The long-term monitoring of the Cambrian–Vendian aquifer system, utilised for industrial water supply at the Kopli Peninsula in Tallinn, showed remarkable changes in the chemical composition of groundwater. The 1.5- to 3-fold increase in TDS and major element concentrations in the abstracted groundwater is probably a consequence of heavy pumping. The main sources of TDS in the Cambrian–Vendian groundwater are leaching from the host rock and release by different geochemical processes in the saturated zone. The second important source of ions is the underlying crystalline basement, which comprises saline groundwater in its upper weathered and fractured zone and is hydraulically connected with the overlying Cambrian–Vendian aquifer system. The fractured basement and its clayey weathering crust host Ca–Cl type groundwater, which is characterised by high TDS values $(2–20\,g\,L^{-1})$. Intensive water abstraction accelerated the groundwater exchange and also increased the area influenced by pumping, resulting in an increasing contribution from up-coning from the underlying crystalline basement, observed by chemical and isotopic studies of the groundwater. At present, there is no evidence for seawater intrusion into the aquifer system but it may occur in coming decades.

Acknowledgements

We are grateful to all individuals and organisations which contributed to the successful completion of this project. Historical database and various reports were provided by the Estonian Geological Survey. Baltic Ship Repair Factory Ltd., Environmental Surveys of North and South Estonia and the local water supply companies are thanked for granting access to their wells and for support in sampling. The assistance of W. M. Edmunds and P. Shand for editorial matters is greatly acknowledged. We would like to thank P. Shand and R. Purtschert for reviewing the paper, and for their helpful comments. This work was supported by the EC as part of the Fifth Framework Programme Project BASELINE under contract no EVK1-CT1999-0006, by the Estonian Science Foundation (grant no. 5925) and by the Estonian Ministry of Education and Research (research project no. 0332089s02).

References

Björck, S. (1995) A review of the history of the Baltic sea 13.0–8.0 ka BP. *Quaternary International* 27, 19–40.

Boldõreva, N. and Perens, R. (2002) Groundwater monitoring at national observation network 2002 (in Estonian). Estonian Geological Survey Report GR-02-8, CD.

Boulton, N.S., Caban, P.E. and van Gijssel, K. (1995) Groundwater flow beneath ice sheets: Part 1 Large scale patterns. *Quaternary Science Reviews* 14, 545–62.

Boulton, N.S., Caban, P.E., van Gijssel, K. et al. (1996). The impact of glaciation on the groundwater regime on Northwest Europe. *Global and Planetary Change* 12, 397–413.

Denton, G.H. and Hughes, T.J. (1981) *The Last Great Ice Sheets.* John Wiley & Sons, New York, 484 pp.

Directive. (1998) Council Directive 98/83/EC of 3 November 1998 on the quality of water intended for human consumption. *Official Journal of European Communities* No L 330, 32–54.

Estonian Drinking Water Standard. (2001) The quality and monitoring requirements for drinking water and methods of analysis. SOMm RTL 2001/100/1369 (in Estonian).

Groundwater State 1997–1998 (in Estonian). (1998) Estonian Geological Survey, Tallinn. 112pp.

Jõeleht, A. (1998) Geothermal studies of the Precambrian basement and Phanerozoic sedimentary cover in Estonia and in Finland. Ph.D. thesis, University of Tartu.

Kalm, V. (2006) Pleistocene chronostratigraphy in Estonia, southeastern sector of the Scandinavian glaciation. *Quaternary Science Reviews* 25, 960–975.

Karise, V. (1997) Composition and properties of groundwater under natural conditions. In: Raukas, A. and Teedumäe, A. (eds) *Geology and Mineral Resources of Estonia.* Estonian Academy Publishers, Tallinn, pp. 152–6.

Karro, E. and Marandi, A. (2003) Mapping of potentially hazardous elements in Cambrian–Vendian aquifer system, Northern Estonia. *Bulletin of Geological Society of Finland* 75, 17–27.

Karro, E., Marandi, A. and Vaikmäe, R. (2004). The origin of increased salinity in Cambrian–Vendian aquifer system at Kopli Peninsula, Northern Estonia. *Hydrogeology Journal* 12, 424–35.

Kleman, J. and Borgström, I. (1994) Glacial landforms indicative of partly frozen bed. *Journal of Glaciology* 40, 225–64.

Koistinen, T., Klein, V., Koppelmaa, H. et al. (1996) Palaeoproterozoic Svecofennian orogenic belt in the surroundings of the Gulf of Finland. In: Koistinen, T. (ed.) *Explanation to the Map of Precambrian basement of the Gulf of Finland and Surrounding Area 1:1 Million.* Special Paper 21, Geological Survey of Finland, pp. 21–57.

Lambeck, K. (1999) Shoreline displacements in Southern Central Sweden and the evolution of the Baltic Sea since the last maximum glaciation. *Journal of the Geological Society of London* 156, 465–86.

Marandi, A., Karro, E. and Puura, E. (2004) Barium anomaly in the Cambrian–Vendian aquifer system in Northern Estonia. *Environmental Geology* 47, pp. 132–9.

Mokrik, R. (1997) The Palaeohydrogeology of the Baltic Basin. Tartu University Press, Tartu, 138 pp.

Mokrik, R. and Vaikmäe, R. (1988) Palaeo-hydrogeological aspects of formation of isotopic composition of groundwater in the Cambrian–Vendian aquifer system in Baltic area. In: Punning, J.-M. (ed.) *Isotope-Geochemical Investigations in Baltic Area and in Belorussia.* Estonian Academy of Sciences, Tallinn, pp. 133–43.

Perens, R. and Vallner, L. (1997) Water-bearing formation. In: Raukas, A. and Teedumäe, A. (eds.) Geology and mineral resources of Estonia, Estonian Academy Publishers, Tallinn, pp. 137–45.

Perens, R., Savva, V., Lelgus, M. et al. (2001) *The Hydrogeochemical Atlas of Estonia* (CD version). Geological Survey of Estonia, Tallinn.

Piotrowski, J. A. (1994) Tunnel-valley formation in northwest Germany – geology, mechanisms of formation and subglacial bed conditions for the Bornhöved tunnel valley. *Sedimentary Geology* 89, 107–41.

Piotrowski, J.A. (1997) Subglacial hydrology in Northwestern Germany during the last glaciation: Groundwater flow, tunnel valleys and hydrological cycles. *Quaternary Science Reviews* 16, 169–85.

Punning, J.-M., Toots, M. and Vaikmäe, R. (1987) Oxygen-18 in Estonian Natural Waters. *Isotopenpraxis* 17, 27–31.

Punning, J.-M., Vaikmäe, R. and Mäekivi, S. (1991) Oxygen-18 variations in the Baltic Sea. *International Journal of Radiation and Applied Instrumentation. Part E. Nuclear Geophysics* 5, 529–39.

Rinterknecht, V.R., Clark, P.U., Raisbeck, G.M. et al. (2006) The last deglaciation of the southeastern sector of the Scandinavian Ice Sheet. *Science* 311, 1449–52.

Savitski, L. (2001) Groundwater monitoring at the groundwater supply plant of Baltic Ship Repair Factory in 2001 (in Estonian). Geological Survey of Estonia, Report of Investigation, 7316.

Tavast, E. (1997) Bedrock topography. In: Raukas, A. and Teedumäe, A. (eds) *Geology and Mineral Resources of Estonia.* Estonian Academy Publishers, Tallinn, pp. 252–5.

Vaikmäe, R. and Vallner, L. (1989) Oxygen-18 in Estonian groundwaters. Fifth Working Meeting on Isotopes in Nature, Leipzig, 25–29 September, pp. 161–2.

Vaikmäe, R., Vallner, L., Loosli, H.H. et al. (2001) Palaeogroundwater of glacial origin in the Cambrian–Vendian aquifer of Northern Estonia. In: Edmunds, W.M and Milne, C. (eds) *Palaeowaters in Coastal Europe: Evolution of Groundwater Since the Late Pleistocene.* Geological Society, London, Special Publications, 189, pp. 17–22.

Vallner, L. (1997) Groundwater flow. In: Raukas, A. and Teedumäe, A. (eds) *Geology and Mineral Resources of Estonia.* Estonian Academy Publishers, Tallinn, pp. 137–45.

Vallner, L. (1999) Development of drinking water resources in coastal areas of Estonia. *Limnologica* 29, 282–5.

Vallner, L. (2003) Hydrogeological model of Estonia and its applications. *Proceedings of the Estonian Academy of Sciences. Geology* 52 (3), 179–92.

Yezhova, M., Polyakov, V., Tkachenko, A. et al. (1996) Palaeowaters of Northern Estonia and their influence on changes of resources and the quality of fresh groundwaters of large coastal water supplies. *Geologija* 19, 37–40.

17 The Cenomanian and Turonian Aquifers of the Bohemian Cretaceous Basin, Czech Republic

T. PACES, J. A. CORCHO ALVARADO,
Z. HERRMANN, V. KODES, J. MUZAK,
J. NOVÁK, R. PURTSCHERT, D. REMENAROVA
AND J. VALECKA

The Bohemian Cretaceous Basin covers an area of 14,600 km^2, of which 12,490 km^2 lie within the territory of the Czech Republic. The most important aquifers are the deepest artesian aquifer in Cenomanian sandstones and the unconfined aquifer in Middle Turonian sandstones. The two aquifers are separated by a Lower Turonian aquitard of claystones and marlstones. The hydrogeological characteristics of the basin differ in its western, central and eastern parts. The presented results characterise the western part with the best-developed aquifers. It is located on the west bank of Jizera River.

Two regions in the western part were selected for the study: (1) the section along the west bank of the Jizera River to the confluence with the Labe (Elbe) River; and (2) the Straz Tectonic Block in the northern section of the western part of the basin. The Straz Tectonic Block contains large deposits of uranium.

The median chemical composition and maximum concentrations of major components in groundwater in the Cenomanian aquifer are in mg L^{-1}: Ca 43/290, Na 8/200, Mg 6/76, K 3/20, HCO$_3$ 159/360, SO$_4$ 16/800, Cl 5/54, NO$_3$ 0.8/67. The maxima of major ions may indicate an admixture of fossil brine and an agricultural pollution. The hydrochemical water type changes from the recharge to discharge area. The most important processes causing the change are dissolution of calcium carbonate cement and Na–Ca ion exchange on clay minerals in matrix of the sandstone. The median chemical composition and maximum concentrations of major components in groundwater in uncontaminated Cenomanian aquifer in the Straz Tectonic Block not affected by mining of uranium are in mg L^{-1}: Ca 31/65, Mg 7/14, Na 3/30, K 1/4 HCO$_3$ 91/278, SO$_4$ 30/118, Cl 5/16, NO$_3$ 0.5/23. The median baseline chemical composition and maximum concentrations of major components in groundwater in the Turonian aquifer are in mg L^{-1}: Ca 100/407, Na 6/211, Mg 6/175, K 2/23, HCO$_3$ 260/530, SO$_4$ 47/1590, Cl 13/120, NO$_3$ 11/108. The Turonian aquifer in the Sraz Tectonic Block contains water with lower TDS between 100 and 200 mg L^{-1}. The hydrochemical water type is Ca–SO$_4$–HCO$_3$. Water is often slightly acidic with increased concentrations of Al and Fe. The concentrations of trace components in uncontaminated groundwater in all four locations are close to or below the detection limits of analytical methods. The uranium deposit in the Cenomanian aquifer of the Straz Tectonic Block has been exploited by underground mining and by underground leaching with sulphuric acid. The mining and leaching ended in the mid-1990s; however, the contaminant mining waste solutions remain in the Cenomanian sandstones. There is a potential danger of spreading the contamination in a southwest direction through the Cenomanian sandstones and across the semi-permeable Lower Turonian aquitard to the superjacent Middle Turonian aquifer. The factors that influence the groundwater quality in the Cenomanian aquifer are (1) water–rock geochemical interaction; (2) mixing with fossil brine present in the subjacent Permo-Carboniferous sedimentary and older crystalline basement; (3) flux of CO$_2$ of magmatic origin and (4) mining of uranium deposits by acid leaching in the Cenomanian aquifer. The dominant control of chemistry in the part of the Cenomanian aquifer is water–rock interaction. Contamination of groundwater due to uranium mining is limited to the Straz Tectonic Block. The flux of magmatic CO$_2$ occur along several deep faults in the basement of the Cretaceous sediments. Such areas were avoided when the statistical parameters of the representative quality of groundwater were calculated. The factors that influence the groundwater quality in the Turonian aquifer are (1) water–rock geochemical interaction; (2) mixing with groundwater from the Cenomanian aquifer; (3) infiltration of surface water and irrigation water and (4) dispersion of contaminants from the uranium leaching in the Cenomanian aquifer across the Lower Turonian aquitard.

Activities of ^{39}Ar in the northern part of the aquifer close to the recharge area indicate the age of groundwater to be 600 years. Water downstream of the recharge area of the Cenomanian aquifer has ^{39}Ar activities below the detection limit. This indicates that the residence time is older than 1000 years. The ^{14}C activities indicate residence times between a few hundreds up to about 17,000 years. The ^{85}Kr results for groundwater in the Turonian show that the groundwater from the Turonian aquifer contains a large proportion of modern water. The calculated ^{14}C model ages indicate a mean residence time of less than 1000 years. The groundwater is probably an admixture of 25–45% of old groundwater in recent groundwater. The mean residence times of the younger groundwater component vary between 4 and 11 years. The age of the old component was determined with the tracer ^{39}Ar, resulting in ages up to 200 years.

17.1 Introduction

The Bohemian Cretaceous Basin is the most extensive, continuous sedimentary basin of the platform cover of the Hercynian Bohemian Massif of Central Europe. It covers an area of 14,600 km^2, of which 12,490 km^2 lie within the territory of the Czech Republic (Fig. 17.1). There are four sandstone aquifers in the basin: aquifer A in the Peruc–Korycany Formation (Cenomanian), aquifer B in parts of the Bila Hora Formation (Lower

Fig. 17.1 The location of the Bohemian Cretaceous Basin in Europe.

Turonian and basal part of Middle Turonian), aquifer C in the Jizera Formation (Middle Turonian and basal layer of the Upper Turonian), and aquifer D in the Teplice to Merboltice Formations (Upper Turonian, Coniacian and basal layers of Santonian).

The most important aquifers are the deepest artesian A aquifer in Cenomanian sandstones and the unconfined C aquifer in Middle Turonian sandstones. The two aquifers are separated by a Lower Turonian aquitard of claystones and marlstones over most of the territory of the Bohemian Cretaceous Basin. The best development of the Cenomanian and Turonian aquifers is in the eastern part of the western region on the west bank of the Jizera River (Fig. 17.2) and in the northern part of the western region

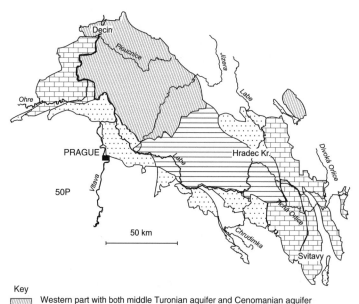

Key

▨ Western part with both middle Turonian aquifer and Cenomanian aquifer separated by the lower Turonian aquitard.

▤ Almost impervious central part.

▦ Eastern part where lower Turonian sediments have the character of an aquifer and join with the middle Turonian aquifer and western part with well developed Cenomanian aquifer.

⋯ Southern marginal part with Cenomanian aquifer.

Fig. 17.2 Extent of the aquifers in the Bohemian Cretaceous Basin in the territory of the Czech Republic.

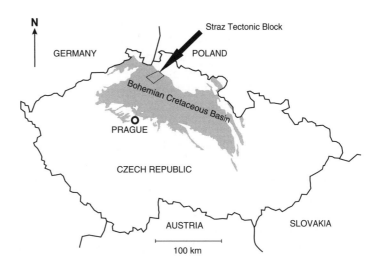

Fig. 17.3 Location of the Straz Tectonic Block within the Bohemian Cretaceous Basin.

formed by the Straz Tectonic Block (Fig. 17.3). Both parts of the Bohemian Cretaceous Basin have been extensively used for water supply since the 1930s. However, the hydrogeology and hydrochemistry of the Straz Tectonic Block is adversely affected by extensive mining and leaching of uranium deposits in the Cenomanian sandstones. Details of the geology and hydrology of the Bohemian Cretaceous Basin were described by Hercik et al. (1999, 2003). Over 4000 wells were drilled and investigated in the Bohemian Cretaceous Basin. New chemical data on groundwater quality were obtained from selected wells and modelling of the possible future deterioration of water was performed during the present studies.

17.2 Geology and hydrogeological characteristics

The Bohemian Cretaceous Basin is elongated, with its axis trending NW–SE and trending to the NNW–SSE near its eastern end. The basin is 290 km long in its axial

direction and up to 100 km across. Basin limits are generally erosional, less commonly tectonic. The areas beyond the basin margin are composed of Hercynian, crystalline rocks, which also continue beneath the basin fill. The surface of the Cretaceous Basin is mostly covered by a thin, discontinuous cover of Quaternary sediments, particularly colluvium, loess and sands and gravels of river terraces. Only the northwestern marginal part of the basin is covered by Tertiary basalts, andesites and tuffs of the Ceske Stredohori Mountains and Tertiary sedimentary rocks of the North Bohemian Basin.

The preserved thickness of the fill of the Bohemian Cretaceous Basin usually varies between 200 and 400 m with the greatest thickness of 1100 m in the western part of the basin. Sedimentary fill accumulated from the late Albian to the Santonian. It is composed of detrital sediments of variable grain size, with subordinate carbonate rocks. The basal Cenomanian sequence of the Peruc–Korycany Formation, up to 100–120 m thick but locally missing, is dominated by psammites. The rest of the fill is characterised by the presence of two facies: basinal facies and proximal facies. The basinal facies is represented by calcareous claystones to marlstones. The proximal facies comprises psammites of variable grain size, with variable contents of cement.

An important tectonic unit of the basin is the Straz Tectonic Block, located at the northwest margin of the basin (Fig. 17.3). The basal Cenomanian sandstones in the Block contain large bodies of concentrated and disseminated uranium deposits. The concentrated bodies were mined by underground operation (Hamr mine), while the disseminated mineralisation was leached using concentrated sulphuric acid. The underground mining was closed in 1993 and the acid leaching of the ore ended in 1996. The Bohemian Cretaceous Basin is subdivided into four contrasting parts (western, central, eastern and southern marginal part) shown in Fig. 17.2 in relation to the occurrence of groundwater. The groundwater bodies in the individual parts are associated with different geological settings and lithological character of the basin fill. The central part of the basin functions as an impermeable block between the western and eastern parts. Surface morphology affects the recharge and direction of groundwater circulation. The erosional base of the whole basin is the Labe River, however, smaller rivers drain some parts of the aquifers before water reaches the Labe River. The Jizera River is important in draining the central part of the basin and the Ploucnice River drains the northern part, namely the Straz Tectonic Block.

Four essential groups of aquifers may be recognised. The A aquifer is a basal aquifer, largely associated with the Peruc–Korycany Cenomanian sandstones, locally also called the Bílá Hora Formation (called AB aquifer); the B aquifer in the Bílá Hora Formation; the C aquifer in the Jizera Formation; and the D aquifer in the Teplice and Březno Formations. A detailed description of the geological and hydrogeological characteristics is given in Hercik et al. (2003). The most significant for water-supply purposes is the C aquifer, which contains 52% of the groundwater resources of the basin. Less significant aquifers are aquifer A (20%), B (14%) and D (14%).

The western part of the basin is characterised by extensive sandstone strata. The sandstones form thick aquifers with continuous groundwater bodies over the whole area. The structural setting plays a subordinate role in groundwater circulation. The aquifers lie mostly subhorizontally. The structural control if any is dominated by the tectonics of the SW–NE (Erzgebirge) strike. The basal aquifer A is present in the whole of the western part of the basin with the exception of several areas of high basement elevation and areas from which it was eroded. The groundwater body in the aquifer – except for peripheral zones – is confined. The permeability of aquifer A is mostly intergranular with the highest transmissivity of $160 \, \text{m}^2 \, \text{day}^{-1}$ (average of $64 \, \text{m}^2 \, \text{day}^{-1}$). Aquifer A is asymmetrically divided into two parts by the course of the Labe River. Each part has its own flow regime with the discharge to the Labe River. The effect of faults on groundwater flow directions is insignificant.

Aquifer C in the western part of the basin is developed mostly on the northeastern bank of the Labe River, where it hosts the largest resources of groundwater in the whole Cretaceous Basin. Groundwater of aquifer C is mostly unconfined, being confined only in those areas where the overlying aquitard of the marlstones and claystones of the Teplice and Brezno Formations is present. The average thickness of the saturated zone of the aquifer is 150 m. Aquifer C has dual porosity and generally decreases towards the SE due to the lithofacies development of the Cretaceous complex. The highest average transmissivity coefficient is 762 m^2 day^{-1}. The lowest values of 48 m^2 day^{-1} were measured at the southern margin. The ground-water flow regime in aquifer C differs markedly from that in aquifer A, especially in the Jizera River basin, where its piezomeztric head is controlled by the drainage effect of the river.

Aquifer D is formed in the sandstones that extend under the volcanic rocks of the Ceske Stredohori Mountains and in the Luzicke Hory Mountains and also has dual porosity. The average transmissivity coefficient of this aquifer is 128 m^2 day^{-1} and it has an average thickness of 150 m in the block of the Ceske Stredohori Mountains and in the Luzicke Hory Mountains.

The central part of the basin differs from the western and eastern parts of the basin in having negligible recharge areas, and small thicknesses of aquifer A. The flat relief is dominated by impermeable calcareous claystones and marlstones of the Teplice and Brezno Formations. Groundwater in aquifer A is confined with an average transmissivity coefficient of 17 m^2 day^{-1}. Aquifer A is recharged by flow across the faults in the north, and drainage occurs along the course of the Labe River in the south. Groundwater percolation is very slow and is almost stagnant. The groundwater is enriched in magmatic CO_2 from faults in subjacent rocks. This gives rise to sparkling mineral water. The mineral water is used for balneological purposes in the spas of Podebrady and Bohdanec.

The hydrology of the eastern part of the basin is controlled by the structural setting.

Cretaceous strata were shaped by Saxonian folding into a system of flat, asymmetrical faulted folds of mostly NW–SE (Sudetic) trend. Structural elevations and faults are responsible for segmentation of this part of the basin into lower-order hydrogeological structures with independent groundwater circulation. The basal aquifer A has intergranular and fracture permeability and the thickness and facies development of the aquifer is variable. Therefore, the hydraulic parameters of the aquifer vary over a wide range. Aquifers B and C are associated with sandstones in their upper portions with upward-coarsening cycles and to silty–sandy silicified spiculitic marlstones to spongolites. These brittle rocks had very low permeability coefficients and have undergone deformation, giving rise to a system of open fractures, which makes the aquifers very permeable. The groundwater basins are small and are located in synclines separated by unsaturated anticlines. As a result, no unified groundwater flow direction is observed over the eastern part of the Bohemian Cretaceous Basin.

17.3 Groundwater chemistry

The baseline quality of groundwater has been determined in Cenomanian aquifer A and in the Turonian aquifer C of the western part of the Bohemian Cretaceous Basin. Two regions were selected for the study: (1) the eastern region of the western part drained at the confluence of the Jizera River and the Labe River and (2) the Straz Tectonic Block. Both regions are used for the exploitation of groundwater and there are many drilled wells to be sampled and monitored for groundwater quality.

The factors that influence the groundwater quality in the Cenomanian aquifer A are (1) water–rock geochemical interaction; (2) mixing with fossil brine present in the subjacent Permo–Carboniferous sedimentary and older crystalline basement; (3) flux of CO_2 of magmatic origin and (4) mining of uranium deposits by acid leaching in the Cenomanian aquifer. The dominant control in the part of the Cenomanian aquifer on the west bank of the Jizera River is water–rock interaction. Contamination of groundwater due to uranium mining is an important feature in the Straz Tectonic Block of the Cretaceous structure at the northern margin of the Bohemian Cretaceous Basin (Fig. 17.3).

The factors that influence the groundwater quality in the Turonian aquifer C are (1) water–rock geochemical interaction; (2) mixing with groundwater from the Cenomanian aquifer; (3) infiltration of surface water and irrigation water; (4) flux of CO_2 of magmatic origin and (5) dispersion of contaminants from the uranium leaching in the Cenomanian aquifer across the Lower Turonian aquitard.

The dominant controls in the part of the Turonian aquifer along the Jizera River are water–rock interaction and infiltration of surface water. Mixing with water from the Cenomanian aquifer and pollution from the uranium mining are important processes in the Straz Tectonic Block. The flux of magmatic CO_2 occurs along several deep faults in the basement of the Cretaceous sediments. Such areas are avoided when the statistical parameters of the representative quality of groundwater are calculated.

The source of groundwater in the Cenomanian (A) and Turonian (C) aquifers is meteoric water, as evidenced by the isotopic composition of oxygen and hydrogen. Only

locally, in aquifer A, is the water enriched in sulphate and chloride components of fossil origin and its total dissolved solids (TDS) are higher due to an influx of magmatic CO_2.

Three sets of data are used to represent the baseline quality of groundwater in the Cenomanian aquifer and its spatial and temporal variations. They are (1) regional set of 43 wells, which were monitored and sampled for the quality of water in 2001 and 2002 in the area on the west bank of the Jizera River; (2) a temporal sequence of data from three drill holes monitored from 1987 till 2002 in the Straz Tectonic Block (wells STPC99, STPC143 and STPC158); and (3) a set consisting of seven wells (VP7502, 7506, 7500, 7515, 7517, 7519 and the artesian well B at Karany Water-Supply plant) along the flow trajectory of the groundwater in the Cenomanian aquifer sampled during the present study in the area of the west bank of the Jizera River.

Similar sets of chemical data represent the baseline quality of groundwater in the Turonian (C) aquifer and its temporal and spatial variations. They are (1) a regional set of 61 wells sampled in 2001 and 2002; (2) a temporal sequence of data from three drill holes (STPT10, STPT31 and STPT33) monitored from 1987 till 2002 in the Straz Tectonic Block and (3) a set of four wells (VP7523, 7512, 7524 and 7520) along the flow trajectory of the groundwater in the Turonian aquifer sampled during the present study.

The baseline chemical composition of groundwater in the two aquifers in the region west of Jizera River has been calculated using data from selected wells that have been sampled twice a year by the Czech Hydrometeorological Institute. The sampled wells yield unpolluted water. Wells suspected

Table 17.1 The statistical characteristics of the composition of the unpolluted groundwater from Cenomanian (A) and Turonian (C) aquifers in the region west of the Jizera River: (a) physico-chemical characteristics; (b) major components; (c) trace metals and other minor elements and (d) dissolved organic carbon and hazardous organic components.

Parameter	Units	Cenomanian aquifer (no. of sampled wells 43)					Turonian aquifer (no. of sampled wells 61)				
		Average	St. Div.	Maximum	Minimum	Median	Average	St. Div.	Maximum	Minimum	Median
(a) Physico–chemical characteristics											
T	°C	11.9	4.4	26.8	6.1	10.3	10.4	2.4	23.5	7.2	9.8
pH lab		7.3	0.67	9.5	5.4	7.35	7.1	0.66	8.8	4.75	7.26
pH field		7.28	0.61	9.4	6.18	7.29	7.09	0.61	8.21	4.9	7.2
Dissolved O_2	mg L^{-1}	3	3.2	12	0.6	1.7	5.4	3.4	10.4	0.5	6
SEC	µS cm^{-1}	43.2	27.9	160	9.6	37	61.6	45.8	308	15	56
(b) Major components											
Ca	mg L^{-1}	53.9	50.7	290	3.7	43	100.7	71.9	407	2.2	100
Mg	mg L^{-1}	8.3	11.2	76	0.29	5.8	13.7	25.0	175	0.64	6.1
Na	mg L^{-1}	25.1	41.9	200	1.1	8.4	13.4	31.0	211	1.2	5.9
K	mg L^{-1}	4.3	4.3	20	0.5	3.1	3.7	4.5	22.8	0.88	2.1
Cl	mg L^{-1}	12.7	13.0	54	1.2	5.2	17.8	19.5	120	1.2	13
SO_4	mg L^{-1}	50.6	127.5	800	0.7	15.9	93.2	212.6	1590	<1	47
HCO_3	mg L^{-1}	174	92	360	6.1	159	231.9	138.9	530	4.3	260
NO_3	mg L^{-1}	5.05	15.3	67.2	<1	<1	18.8	24.6	108	<1	11
NO_2	mg L^{-1}	0.01	0.03	0.21	<0.005	<0.002	0.01	0.03	0.15	<0.005	<0.002
NH_4	mg L^{-1}	0.09	0.18	0.72	<0.05	0.07	0.12	0.44	2.84	<0.05	<0.05
PO_4	mg L^{-1}	0.03	0.11	0.66	<0.04	0.04	0.041	0.10	0.5	<0.05	0.04
F	mg L^{-1}	0.39	0.87	5.6	<0.05	0.15	0.31	0.53	2.3	<0.05	0.12
SiO_2	mg L^{-1}	4.6	3.6	18.8	0.3	3.8	5.2	2.9	12.8	0.64	4.2
CO_2	mg L^{-1}	20.1	27.3	180	<2.2	12	21.1	13.9	88	<2.2	20
(c) Trace elements											
Al	µg L^{-1}	41.8	160	810	<10	<10	1.1	40	240	<10	<10
As	µg L^{-1}	0.2	2.7	10	<1	<1	<0.7		4.2	<1	<1
B	µg L^{-1}	26.2	95.7	510	<25	<20	65	260	1800	<25	<20
Ba	µg L^{-1}	7.2	60	160	<50	13	2.2	70	268	<50	<50
Be	µg L^{-1}	0.2	0.4	2.3	<0.1	<0.1	0.06		7.2	<0.1	<0.1
Cd	µg L^{-1}	<0.06	0.2	0.8	<0.1	<0.1	<0.02		0.85	<0.1	<0.1
Co	µg L^{-1}	<0.2	2.3	8	<1	<1	<0.8		4.7	<1	<1
Cu	µg L^{-1}	<1	1.7	4	<2	<2	<0.55	10	37	<2	<2
Cr	µg L^{-1}	<1	1.7	6	<2	<2	<0.13		8	<2	<2
Fe	µg L^{-1}	2700	6700	31200	<60	<60	330	1190	8500	<60	<0.60
Hg	µg L^{-1}	<0.1		<0.1	<0.1	<0.1	<0.09		0.5	<0.1	<0.1
Li	µg L^{-1}	<6	73	33	<50	<50	18	60	0.263	<0.05	<0.05
Mn	µg L^{-1}	180	280	1430	<10	99	40	80	411	10	12
Mo	µg L^{-1}	<0.8	2.5	6.2	<2	<2	<2		3.5	<2	<2
Ni	µg L^{-1}	2	7.4	41	<2	<1	0.71		17	<2	<2
Pb	µg L^{-1}	<1.5	1.0	2	<2	<2	<2		7	<2	<2
Sb	µg L^{-1}	<0.5		<0.5	<0.5	<0.5	<0.47		0.6	<0.5	<0.5
Se	µg L^{-1}	<1		<1	<1	<1	<1		<1	<1	<1
Sr	µg L^{-1}	345.6	370	1600	10	200	630	970	5950	30	240
Zn	µg L^{-1}	<3	14	52	<10	<10	3.1	30	239	<10	<10

Table 17.1 (*Continued*)

Parameter	Units	Cenomanian aquifer (no. of sampled wells 43)					Turonian aquifer (no. of sampled wells 61)				
		Average	St. Div.	Maximum	Minimum	Median	Average	St. Div.	Maximum	Minimum	Median
(d) Organics											
DOC	mg L^{-1}	1.19	0.55	2.6	0.15	1.1	1.74	0.76	4	0.43	1.7
CHCl$_3$	µg L^{-1}	<0.1		<0.1	<0.1	<0.1	<0.062	0.14	0.8	<0.1	<0.1
CCl$_4$	µg.L^{-1}	<0.1		<0.1	<0.1	<0.1	<0.1		<0.1	<0.1	<0.1
Atrazine	µg.L^{-1}	<0.0014	0.0058	0.014	<0.005	<0.005	0.05	0.30	2.3	<0.005	<0.005
1,1,2-trichloro-ethylene	µg L^{-1}	<0.1		<0.1	<0.1	<0.1	<0.097	0.03	0.1	<0.1	<0.1
1,1,2,2-tetrachloro-ethylene	µg L^{-1}	<0.095	0.0301	0.1	<0.1	<0.1	<0.097	0.03	0.1	<0.1	<0.1
Phenol	µg L^{-1}	<0.1		<0.1	<0.1	<0.1			<0.1	<0.1	<0.1
Simazine	µg L^{-1}	<0.005	0.0000	<0.005	<0.005	<0.005	<0.0041	0.00	0.012	<0.005	<0.005
Isoproturon	µg L^{-1}	<0.02	0.0000	<0.02	<0.02	<0.02	<0.02	0.00	<0.02	<0.02	<0.02

to be contaminated as well as wells that yield water of extreme chemical compositions due to the input of CO_2 and fossil brines are not included in this study. The statistical parameters of the baseline quality are given in Table 17.1 to characterise the groundwater in the Cenomanian (A) and Turonian (C) aquifers of the Bohemian Cretaceous Basin.

Temporal variations in groundwater from the Cenomanian aquifer are shown for three wells in the Straz Tectonic Block (Figs. 17.4 and 17.5). The medians and 97.9 percentiles of chemical components of groundwater from the three wells are summarised in Table 17.2.

The monitoring of groundwater quality since 1985 indicates that the chemical composition of groundwater in the Cenomanian aquifer does not exhibit any systematic trend; however it is not constant. This is exemplified by the fluctuation in TDS for the whole period of observation in groundwater from wells VP7517 and VP7515 (Fig. 17.4). The only exception is water sampled from the drill hole VP7502. Groundwater from this drill hole exhibits an increasing trend in TDS

since the winter of 1998. This corresponds to the location of the drill hole in the recharge area of the aquifer A, which is vulnerable to surface contamination. The steady-state TDS in the Cenomanian aquifer in the Straz Tectonic Block is a significant feature of the baseline properties of the groundwater, indicating that this water is not influenced by the mining activities in the Straz uranium deposit. Hence, the selected drill holes can serve as the baseline data.

The monitoring of groundwater quality since 1985 in the Turonian aquifer indicates that the temporal trends in chemical composition are not uniform and the observed variations are not easy to explain. There are basically three types of temporal changes. In the first case, exemplified by drill hole STPT 10, there is a gradual increase in TDS from 60 mg L^{-1} in 1985 to present level of 210 mg L^{-1}. There are, however, periodic changes during this time period as shown on in Fig. 17.5. The second type is a steady-state chemical composition without a systematic trend but with periodic variations. The periods are regular or irregular with time. Such trend is

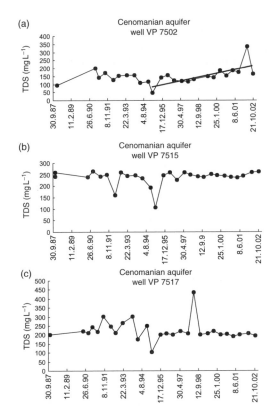

Fig. 17.4 Temporal trends and variations in wells tapping groundwater in the Cenomanian aquifer.

exemplified by water from well STPT 31. The third type is the steady-state chemical composition with an introduced anthropogenic pulses of pollution followed by some type of recovery either natural or manmade. This behaviour is demonstrated by the chemical changes in groundwater from the well STPT 33.

The third set of analyses of waters from aquifers A and C include samples from wells in the area located on the west bank of the Jizera River. The wells were selected in such a way, that they represent the north–south trajectory of groundwater flow in both the aquifers. The trajectory follows the piezometric gradient from the area of recharge to the area of discharge. The chemical data for water

from the Cenomanian aquifer and the distance from the recharge area to discharge area are given in Table 17.3. The trends in the concentrations of major cations and anions in groundwater along the flow trajectory in the Cenomanian aquifer are shown in Fig. 17.6.

Figure 17.6 illustrates the exchange of Ca for Na along the trajectory towards the discharge area at the confluence of the Jizera River with the Labe River. Bicarbonate decreases while sulphate and chloride both increase. The last artesian well Karany B discharges water which differs from the general chemical evolution. The mixing of old and young water probably affects its chemical composition. This is documented by a dating analysis discussed below.

The downgradient evolution in the Turonian aquifer was studied using four wells located from north to south along Jizera River. The position of the wells follows approximately the piezometric gradient from the area of recharge to the area of discharge. The chemical composition of groundwater and the distance from the recharge area to the discharge area are summarised in the Table 17.4. The chemical evolution is illustrated using major ions in Fig. 17.7. Contrary to the evolution of groundwater in the confined Cenomanian aquifer, groundwater in the unconfined Turonian aquifer is steadily enriched in all major cations and anions. The ion exchange between Ca and Na is not an obvious process. Instead, a continuous dissolution of carbonate and enrichment in sulphate and chloride ions take place.

17.4 The age of the groundwater in Cenomanian and Turonian aquifer

The non-detection of bomb ^3H and ^{85}Kr in the investigated wells of the Cenomanian

Table 17.2 Median values and 97.9 percentile of the chemical composition groundwater from three drill holes in the Cenomanian aquifer sampled in the period 1985–2002 in the Straz Tectonic Block outside the zone of contamination by acid leaching of uranium deposit in the Cenomanian sandstones.

Drill hole no.		STPC 99		STPC 143		STPC 158	
X		50°39'34"		50°39'12"		50°39'18"	
Y		14°46'47"		14°48'00"		14°47'53"	
n – number of samples		51		34		40	
		Median	Percentile 97.9	Median	Percentile 97.9	Median	Percentile 97.9
pH		7.57	8.06	7.05	8.00	7.20	7.90
Conductivity	$\mu S\,cm^{-1}$	15.0	173	16.6	27.4	16.0	27.0
TDS	$mg.L^{-1}$	129	228	148	179	139.5	218
Ca	$mg.L^{-1}$	25	49	34	46	32	37
Mg	$mg\,L^{-1}$	8.2	13.3	4.3	4.76	7.20	9.00
Na	$mg\,L^{-1}$	3.7	16.6	3.50	5.06	2.35	4.20
K	$mg\,L^{-1}$	1.1	2.27	1.25	3.34	1.00	1.40
CL	$mg\,L^{-1}$	3.7	12.0	6.45	13.9	5.15	9.45
SO_4	$mg\,L^{-1}$	28.0	71.2	42.0	76.5	30.0	58.7
HCO_3	$mg\,L^{-1}$	90.9	156	77.0	136	96.0	120
NO_3	$mg\,L^{-1}$	0.50	5.3	0.5	18	0.50	9.9
NH_4	$mg\,L^{-1}$	0.025	1.20	0.010	0.90	0.010	5.3
PO_4	$mg\,L^{-1}$	0.025	0.025				
HPO_4	$mg\,L^{-1}$	0.025	1.57	0.025	0.124	0.025	0.133
F	$mg\,L^{-1}$	0.50	0.60	0.50	0.52	0.50	0.74
Trace elements							
Al	$mg\,L^{-1}$	0.25	2.3	0.25	1.37	0.25	1.00
Cr	$mg\,L^{-1}$	0.00	0.025	0.00	0.00	0.00	0.00
Fe	$mg\,L^{-1}$	0.88	2.99	0.48	7.72	0.82	3.5
Mn	$mg\,L^{-1}$	0.090	0.22	0.033	0.056	0.069	0.53
Ni	$mg\,L^{-1}$	0.033	0.072	0.095	0.16	0.025	0.10
Ra	$Bq\,m^{-3}$	2080	3401	5800	6983	3850	17917
U	$mg\,L^{-1}$	0.0040	0.024	0.088	0.14	0.0030	0.037
Zn	$mg\,L^{-1}$	0.020	0.040	0.050	0.109	0.020	1.30

aquifer indicates that groundwater recharge occurred more than 50 years ago. ^{39}Ar activities were measured only in the northern recharge area of this aquifer, and are equivalent to a mean groundwater age of 600 years. Groundwater in the Cenomanian aquifer has ^{39}Ar activities below the detection limit. This indicates that groundwater from the wells located farther downstream of the recharge area should have residence times older than 1000 years. ^{85}Kr measurements indicate no admixture of modern water.

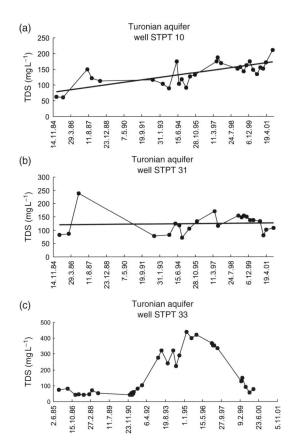

Fig. 17.5 Temporal trends and variations determined in wells tapping groundwater in the Turonian aquifer.

An increasing mean residence time of water in the Cenomanian aquifer in the direction of groundwater flow from north to south is confirmed by decreasing ^{14}C activities ranging from 54 to 6 pmC (Corcho Alvarado et al. 2004). The ^{14}C activities were converted into groundwater residence times using the model NETPATH (Plummer et al. 1991), which corrects for geochemical reactions occurring along the flow path. Calculations using this correction scheme indicate residence times from a few hundred years to about 17,000 years (Corcho Alvarado et al.

2004; Purtschert et al. Chapter 5B, this volume). Considering the uncertainties in the age determinations, it is reasonable to assume that the age of groundwater in the Cenomanian aquifer varies from 600 years in the recharge area to more than 20,000 years at the end of the flow path near the confluence of the Labe River and Jizera River. The well Karany B as a final point on the trajectory path of water flow has a screened interval several hundred metres shallower than the rest of the wells. An increased portion of younger water is thus very probable in this well. This is supported by the elevated ^{39}Ar and ^{14}C concentrations compared to the wells located further upstream. An admixture of between 40% and 60% of a younger water component in a very old groundwater is estimated based on the ^{39}Ar activity.

The ^{85}Kr results for groundwater in the Turonian aquifer (38.6 and 5.7 dpm cc^{-1} Kr for VP7524 and VP7512 respectively) show that the groundwater from the Turonian aquifer contains a large proportion of modern water. The calculated ^{14}C model ages (according to Ingerson and Pearson 1964) indicate a mean residence time of less than 1000 years for all investigated waters in the Turonian sandstones. This corresponds to the occurrence of high nitrate that is related to a recharge from agricultural fields around the sampled wells. Admixtures of between 25% and 45% of old groundwater in recent groundwater were calculated with the dispersion model (Zuber and Maloszewski 2001) for these two wells (Corcho Alvarado et al. 2004). The mean residence times of the younger groundwater component vary between 4 and 11 years. The age of the old component was determined with the tracer ^{39}Ar, resulting in ages up to 200 years.

Table 17.3 Chemical composition of groundwater from the Cenomanian aquifer along a trajectory from the recharge area to the discharge area on the west-bank of the Jizera River. Sampled during April 25–30, 2003.

Well No.		VP7502	VP7506	VP7500	VP7515	VP7517	VP7519	Karany B
Distance on trajectory	Kilometre	0	6.4	11.2	25.4	33.7	48.3	64.3
Physico-chemical characteristics								
pH lab	units	7.12	7.43	7.71	6.98	6.73	6.48	7
pH field	units	7.7	7.9	7.8	7.4	7.6	7.7	7.6
El. conductivity	$\mu S\,cm^{-1}$	400	318	327	301	258	321	489
Major components								
Ca	$mg\,L^{-1}$	64.5	59.3	48.1	35.0	23.0	22.1	34.6
Mg	$mg\,L^{-1}$	12.9	5.0	10.2	5.9	5.1	8.0	13.2
Na	$mg\,L^{-1}$	3.6	1.8	6.1	14.4	19.9	29.9	41.0
K	$mg\,L^{-1}$	4.4	1.0	4.6	2.4	3.0	5.2	9.3
Cl	$mg\,L^{-1}$	3.0	1.1	1.9	14.8	15.9	22.9	26.9
SO_4	$mg\,L^{-1}$	3.1	0.3	0.3	20.6	21.5	25.7	54.6
	$mg\,L^{-1}$	253	201	210	125	100	119	177
NO_3	$mg\,L^{-1}$	0.9	<0.30	<0.30	<0.30	<0.30	<0.30	0.9
NH_4	$mg\,L^{-1}$	0.3	<0.02	0.1	<0.02	0.1	0.2	<0.02
F	$mg\,L^{-1}$	0.2	0.1	0.3	0.1	0.1	0.1	0.4
SiO_2	$mg\,L^{-1}$	11.6	<2.0	2.5	8.0	8.0	7.5	6.9
Trace elements								
Al	$\mu g\,L^{-1}$	14	<10	25	90	95	43	<10
As	$\mu g\,L^{-1}$	0.6	0.6	0.6	0.5	1.6	<0.5	3.4
Be	$\mu g\,L^{-1}$	<0.02	<0.02	<0.02	0.2	0.0	0.0	0.1
Cd	$\mu g\,L^{-1}$	<0.04	<0.04	<0.04	<0.04	<0.04	<0.04	<0.04
Co	$\mu g\,L^{-1}$	<0.5	<0.5	<0.5	<0.5	0.7	<0.5	<0.5
Cr	$\mu g\,L^{-1}$	<0.5	<0.5	<0.5	0.5	19.7	1.4	<0.5
Cu	$\mu g\,L^{-1}$	1.1	3.4	0.3	0.4	17.4	0.7	1.1
Fe	$\mu g\,L^{-1}$	<0.05	<0.05	<0.05	<0.05	0.2	<0.05	0.4
Li	$\mu g\,L^{-1}$	36	5	15	16	23	43	84
Mn	$\mu g\,L^{-1}$	468	90	74	104	94	54	33
Mo	$\mu g\,L^{-1}$	2.7	3.4	2.1	<0.5	1.4	<0.5	<0.5
Ni	$\mu g\,L^{-1}$	1.2	<0.5	<0.5	<0.5	9.7	2.0	<0.5
Pb	$\mu g\,L^{-1}$	<0.4	0.4	<0.4	1.2	0.5	<0.4	<0.4
Rn	$Bq\,L^{-1}$	6.4	9.8	2.2	3.9	2.8	3.6	5.1
V	$\mu g\,L^{-1}$	<10	<10	<10	<10	<10	<10	<10
Zn	$\mu g\,L^{-1}$	<10	11	10	11	<10	<10	58
Organics								
DOC	$mg\,L^{-1}$	1.9	0.6	0.6	0.5	0.4	0.4	1.6

(a)

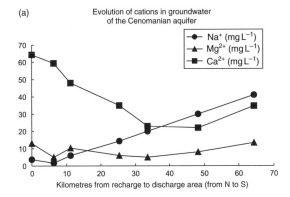

(b)

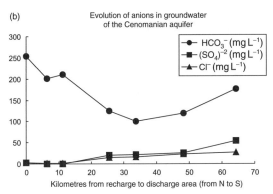

Fig. 17.6 The trend of cations and anions in groundwater along the flow trajectory in the Cenomanian aquifer.

These ages have a tendency to increase with the depth of the screens in the wells. Older ages are found in the wells with deeper screens: VP7523 and VP7512, indicating probable age stratification in this aquifer. The ^{14}C activities in bicarbonate vary between 63 and 79 pmC indicating a very recent origin for the groundwater in the Turonian aquifer, which may still contain an input of ^{14}C from thermonuclear weapon tests. The age and the degree of insulation from mixing with surface water are important differences between the groundwater in the Cenomanian and in the Turonian aquifer.

17.5 Major element controls

Groundwater in the Cenomanian aquifer of the western part of the basin obtains its chemical composition in the northern recharge area along the tectonic northern boundary of the Cretaceous sediments. The only important process is water–rock geochemical interaction. The absence of nitrate indicates that modern surface water does not enter the Cenomanian aquifer. Groundwater is much older even in the recharge area than water in the Turonian aquifer. In spite of the greater age, groundwater in the Cenomanian sandstones has low TDS concentrations varying from 100 to 200 mg L^{-1} in the recharge area. It is characterised by Ca–HCO$_3$ water type in the area on the west bank of the Jizera River. This type changes to Ca–HCO$_3$–SO$_4$ type in the Straz Tectonic Block. The concentrations of sodium increase in the direction of groundwater flow to the south along the Jizera River. No such trend is observed in the Straz Tectonic Block drained by the Ploucnice River. The chemical type Ca–HCO$_3$ passes to type Ca–Na–HCO$_3$ and the TDS concentration reaches 400 mg L^{-1} at the confluence of the Jizera River with the Labe River. A gradual substitution of Na for Ca ions occurs along the flow trajectory. This exchange takes place on clay minerals of the matrix of sandstones. The source of calcium and bicarbonate is from calcite in the cement. The groundwater in the Cenomanian aquifer changes its chemical composition to Na–Ca–HCO$_3$ or Na–Ca–HCO$_3$–Cl–SO$_4$ type, with TDS concentrations approaching 500 mg L^{-1}. It is possible that a small quantity of fossil chloride and sulphate brines in the basement Permo–Carboniferous sediments

Table 17.4 Chemical composition of groundwater from the Turonian aquifer on the west-bank of the Jizera River along the trajectory from recharge to discharge area. Sampled during April 25–30, 2003.

Well no.		VP7523	VP7512	VP7524	VP7520
Distanceon trajectory	Kilometre	0	8.4	14.2	25.2
Physico-chemical characteristics					
pH lab	units	7.76	8.05	8.09	7.75
pH field	units	7.14	7.41	7.09	7.18
Conductivity	$S\ cm^{-1}$	509	428	693	862
Ca	$mg\ ^{L-1}$	108.3	82.2	130.9	165.7
Mg	$mg\ ^{L-1}$	2.18	6.81	16.04	17.76
Na	$mg\ ^{L-1}$	1.66	2.92	4.99	5.38
K	$mg\ ^{L-1}$	1.04	1.13	1.61	2.56
Cl	$mg\ ^{L-1}$	5.82	5.57	18.8	44.3
SO_4	$mg\ ^{L-1}$	26.9	15.1	77.4	151
HCO_3	$mg\ ^{L-1}$	286.79	247.13	311.2	311.2
NO_3	$mg\ ^{L-1}$	<0.30	5.71	22.2	6.36
NH_4	$mg\ ^{L-1}$	<0.02	0.04	<0.02	0.13
F	$mg\ ^{L-1}$	0.07	0.09	0.11	0.23
SiO_2	$mg\ ^{L-1}$	7	6.5	9.2	6.9
Trace elements					
Al	$\mu g\ L^{-1}$	50	29	24	77
As	$\mu g\ L^{-1}$	<0.5	0.5	1.1	<0.5
Be	$\mu g\ L^{-1}$	<0.02	<0.02	<0.02	<0.02
Cd	$\mu g\ L^{-1}$	<0.04	<0.04	<0.04	<0.04
Co	$\mu g\ L^{-1}$	<0.5	<0.5	<0.5	<0.5
Cr	$\mu g\ L^{-1}$	<0.5	<0.5	<0.5	3.3
Cu	$\mu g\ L^{-1}$	0.4	0.9	0.7	1
Fe	$\mu g\ L^{-1}$	<0.05	<0.05	<0.05	<0.05
Li	$\mu g\ L^{-1}$	2	2	4	23
Mn	$\mu g\ L^{-1}$	5	10	<5.0	15
Mo	$\mu g\ L^{-1}$	<0.5	<0.5	<0.5	<0.5
Ni	$\mu g\ L^{-1}$	<0.5	<0.5	<0.5	6.1
Pb	$\mu g\ L^{-1}$	<0.4	0.7	<0.4	<0.4
Rn	$Bq\ L^{-1}$	3.2	3.1	23	14
V	$\mu g\ L^{-1}$	<10	<10	<10	<10
Zn	$\mu g\ L^{-1}$	12	13	<10.0	10
Organics					
DOC	$mg\ L^{-1}$	1.6	1.1	1.7	1.9

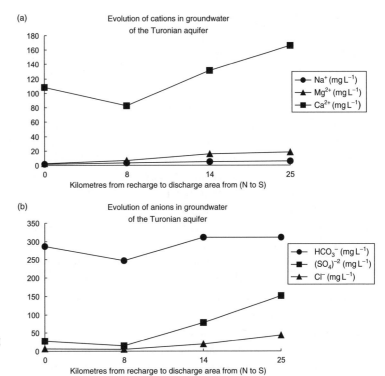

(a) Evolution of cations in groundwater of the Turonian aquifer

Na⁺ (mg L⁻¹)
Mg²⁺ (mg L⁻¹)
Ca²⁺ (mg L⁻¹)

Kilometres from recharge to discharge area from (N to S)

(b) Evolution of anions in groundwater of the Turonian aquifer

HCO₃⁻ (mg L⁻¹)
(SO₄)⁻² (mg L⁻¹)
Cl⁻ (mg L⁻¹)

Kilometres from recharge to discharge area from (N to S)

Fig. 17.7 The trend of cations and anions in groundwater along the flow trajectory in the Turonian aquifer.

under the Cenomanian aquifer (Jetel 1970) contributes to the increase.

The major element control on the baseline quality of groundwater in the Turonian aquifer is also water–rock interaction. The sources of solutes are from the Turonian sandstones with calcareous cement. The occurrence of a small quantity of feldspar is sufficient to supply small amounts of sodium and potassium. Bicarbonate is derived from calcite together with calcium. The other major anions, Cl^-, NO_3^- and SO_4^{2-} are usually found in very small concentrations. Increased concentrations of the anions occur in agricultural areas with intensive fertilisation. Groundwater of $Ca–HCO_3$ type with TDS of 150 mg L^{-1} is present in the divide area on the western bank of the Jizera River. Towards the south and southeast, in the direction of groundwater flow, TDS increases to

500–600 mg L^{-1}. The chemical type usually remains unchanged or passes into $Ca–Mg–HCO_3–SO_4$ type. Sulphate is supplied from dissolution of sulphate minerals in claystones and marlstones covering the sandstone aquifer.

The determination of the baseline quality of groundwater has avoided areas with obvious influence of admixtures of exotic components such as fossil brines or industrial contaminants derived from uranium leaching.

17.6   Summary of the baseline quality

The Bohemian Cretaceous Basin contains four aquifers in sandstone strata. The aquifers are denoted by letters A, B (locally joint aquifer AB), C and D. Their geological and hydrogeological properties are described in detail by Hercik et al. (2003). Two of the aquifers A and C are the most extensive water-bearing strata and the most important for the groundwater supply in the Czech Republic. These are the Cenomanian aquifer (A) developed in the Peruc–Korycany Formation and the Turonian aquifer (C) developed in the Jizera Formation. They are separated by a Lower Turonian aquitard of claystones. The origin of groundwater in both of the aquifers is meteoric. Two areas were selected to represent the baseline quality of groundwater. The first area is located in the eastern part of the western region of the Bohemian Cretaceous Basin on the west bank of the Jizera River (left bank tributary to the Labe River). The second area is located within the Straz Tectonic Block where the Turonian aquifer serves as a source of drinking water while the Cenomanian aquifer contains a uranium deposit. The deposit was mined by underground mining (Hamr mine) and by leaching of disseminated ore with concentrated sulphuric acid.

Water in the Cenomanian aquifer is recharged in the outcropping sandstones forming the northern rim of the Cretaceous Basin. The recharge to the Turonian aquifer is at the northern outcrops of sandstones and from the surface along the flow trajectory. Groundwater in the Cenomanian aquifer is old and obtains its chemical composition in the northern recharge area along the Luzice Fault. The only important process controlling groundwater chemical composition is water–sandstone geochemical interaction. Dominant processes are dissolution of matrix residual and cement minerals and ion exchange on clay minerals in matrix. The Cenomanian aquifer on the west bank

of the Jizera River contains high-quality groundwater. The water body is confined by the Lower Turonian aquitard. This aquifer is not vulnerable to agricultural pollution. This is manifested by an absence of nitrate. Many of the monitored drill holes in the Cenomanian aquifer are artesian. The chemical composition of groundwater does not exhibit any temporal trends, however, small variations in total dissolved solids displayed fluctuation over the monitoring period. The median baseline chemical composition and maximum concentrations of major components in groundwater in the Cenomanian aquifer in the west bank of the Jizera River are in mg L^{-1}: (median/maximum): Ca 43/290, Na 8/200, Mg 6/76, K 3/20, HCO_3 159/360, SO_4 16/800, Cl 5/54, NO_3 0.8/67. The concentrations of trace components are close to or below the detection limits of the analytical methods. The maxima of major ions may indicate a small admixture of fossil brine and a minor agricultural pollution. The hydrochemical water type changes from the recharge to discharge area. The most important processes are dissolution of calcium carbonate cement and Na–Ca ion exchange on clay minerals in the matrix of the sandstone.

The median baseline chemical composition and maximum concentrations of major components in groundwater in uncontaminated Cenomanian aquifer in the Straz Tectonic Block which are not affected by mining of uranium are in mg L^{-1} (median/maximum): Ca 31/65, Mg 7/14, Na 3/30, K 1/4, HCO_3 91/278, SO_4 30/118, Cl 5/16, NO_3 0.5/23. The trace components are close to or below the detection limits of the analytical methods. The determination of the baseline quality is important in the Straz Tectonic Block since it serves as a reference state in the evaluation of the impact of the uranium mining on groundwater quality and consequent restoration measures on local groundwater body.

The Turonian aquifer is unconfined and the quality of groundwater is often vulnerable from surface, namely agricultural, contamination. Groundwater in the recharge area has low TDS concentrations ranging from 100 to 200 mg L^{-1}. It is characterised by $Ca–HCO_3$ water type in the area on the west bank of the Jizera River. TDS increases with the distance of percolation to concentrations above 600 mg L^{-1}. The median baseline chemical composition and maximum concentrations of major components in groundwater in the Turonian aquifer in the west bank of the Jizera River are in mg L^{-1} (median/maximum): Ca 100/407, Na 6/211, Mg 6/175, K 2/23, HCO_3 260/530, SO_4 47/1590, Cl 13/120, NO_3 11/108. The concentrations of trace components are close to or below the detection limits of the analytical methods. The medians and maxima of all the components are higher than in the older and deeper groundwater in the Cenomanian aquifer. This not only reflects the more variable Turonian sedimentary sequence with claystones and marlstones substituting partly for sandstones but also reflects the infiltration of contaminated water from the surface. The chemical composition of groundwater is not as stable as in the Cenomanian aquifer. Some wells exhibit an increasing trend in TDS. Single anomalies with an increase in TDS followed by a decrease to the original level were determined by monitoring in other wells. Most monitored wells, however, produce groundwater with a fairly uniform chemical composition with only small fluctuations of their chemical components.

The Turonian aquifer in the Straz tectonic Block contains water with low TDS between 100 and 200 mg L^{-1}. The hydrochemical water type is $Ca–SO_4–HCO_3$. Water is often slightly acidic with increased concentrations of Al and Fe. Analyses selected for the definition of local baseline quality are probably close to the pristine composition of local groundwater. The uranium deposit in the Cenomanian aquifer of the Straz Tectonic Block has been exploited by underground mining and by underground leaching with sulphuric acid since 1968. The mining and leaching ended in the mid 1990s; however, the contaminant mining waste solutions remain in the Cenomanian sandstones. The major components of the contaminant solutions are free sulphuric acid, nitric acid, ammonium and aluminium. Very low pH (often 1–2) in the solutions increases concentrations of other trace elements. Pumping of the contaminant solutions with dissolved uranium from the Cenomanian aquifer prevents their dispersion outside the leaching fields. The pumping maintains the piezometric head in Cenomanian aquifer under the unconfined water table in the Turonian aquifer. In spite of this, there is a potential danger of spreading the contamination in a southwest direction through the Cenomanian sandstones and across the semi-permeable Lower Turonian aquitard to the superjacent Middle Turonian aquifer. This process has been modelled mathematically. The results show that pollution outside the mining area will probably occur after 200–500 years (Novak et al. 1998). The determination of the baseline quality in the Straz Tectonic Block is important because it will serve as a reference state in the evaluation of the impact of the uranium mining on local quality of groundwater.

References

Cech, S. and Valecka, J. (1991) Important trans-gressions and regressions in the Bohemian Cretaceous Basin. Report P.040.1981, Archive of the Czech Geological Survey, Prague (in Czech).

Corcho Alvarado, J.A., Purtschert, R. and PaCes, T. (2004) Establishing timescales of groundwater residence times based on environmental tracer data: A study of the Turonian and Cenomanian aquifers of the Bohemian Cretaceous Basin, Czech Republic. In: 32nd International Geological Congress, Florence, Italy.

Hercik, F., Herrmann, Z. and Valecka, J. (1999) *Hydrogeology of the Bohemian Cretaceous Basin.* Czech Geological Survey, Praha, 115 pp., ISBN 80-7075-309-9 (in Czech).

Hercik, F., Herrmann, Z. and Valecka, J. (2003) Hydrogeology of the Bohemian Cretaceous Basin. *Czech Geological Survey*, 91 pp., Praha. ISBN 80-7075-604-7.

Ingerson, E. and Pearson, F.J.,Jr. (1964) Estimation of age and rate of motion of ground-water by the 14C-method. In: Miyake, Y. and Koyama, T. (eds) *Recent Researches in the Fields of Hydrosphere, Atmosphere and Nuclear Geochemistry.* Maruzen, Tokyo, pp. 263–83.

Jetel, J. (1970) Hydrogeology of the Permo-carboniferous and Cretaceous in the profile Melnik-Jested. Sbornik Geol. Ved. HIG 7, 7–42, Praha.

Novak, J., Smetana, R., Strof, P. et al. (1998) Mining of uranium by acid leaching and its environmental consequences. In: Memmi, I., Hunziker, J.C. and Panichi, J. (eds) *A Geochemical and Mineralogical Approach to Environmental Protection.* Proceedings of the International School Earth and Planetary Sciences, Siena, pp. 215–22, .

Valecka, J., Kaas, A. and Prazak, J. (1984) Explanatory text to the map of isolines of the absolute height above sea level of the base of the Lower Turonian in the Middle-mountainous balance block (S3) Report P. 23.1984, Archive of the Czech Geological Survey, Prague (in Czech).

Zuber, A. and Maloszewski, P. (2001) *Lumped-Parameter Models. Environmental Isotopes in the Hydrological Cycle.* UNESCO/IAEA.

18 Quality Status of the Upper Thracian Plio-Quaternary Aquifer, South Bulgaria

M. MACHKOVA, B. VELIKOV, D. DIMITROV,
N. NEYKOV AND P. NEYTCHEV

The quality of groundwaters in the Upper Thracian Plio-Quaternary aquifer (South Bulgaria) was studied in order to determine the baseline concentrations and main factors influencing them. Detailed sampling of groundwaters was implemented and an extensive set of major and trace elements and other hydrochemical characteristics were determined to elucidate the present hydrochemical conditions of water bodies. Standard statistical methods and software were used to analyse modern and historical data. Despite the fact that the area is quite densely populated and industry and agriculture are well established, most of the studied waters were of good quality with respect to many elements. The observed trends for different hydrochemical parameters are due to the influence of both natural and anthropogenic factors. Baseline values cannot be clearly distinguished, as they are often masked by the influence of the anthropogenic factors.

Along with general decreasing trend of groundwater levels, decreasing trends for pH and Ca and increasing trends in Na and Cl are recognised. An analysis of spatial variability of hydrochemical components indicates the role of the main river channel as a dividing line in the aquifer. The lack of exchange of waters between the northern and southern parts of the aquifer highlights two different spatial distributions of the hydrochemical parameters on either side of the river.

18.1   Introduction

The Plio-Quaternary aquifer is situated in the Upper Thracian lowlands of southern Bulgaria (Fig. 18.1). The lowlands form part of the Maritza River basin with elevation ranging from 50 to 400 m, gradually decreasing to the east and southeast. The predominant part (more than 70%) is situated between 50 and 300 m. The Maritza River basin is the largest on the Balkan Peninsula, having an area of 53,000 km^2 to the river mouth and 21,084 km^2 (and length 321 km.) to the Bulgarian state border. The highest Bulgarian mountains surround the basin. The river is sourced from a spring below the highest peak in the Balkans: Musala at 2925 m asl. The Plio–Quaternary aquifer is the most important groundwater formation with a surface area of about 2000 km^2. The groundwater in the aquifer is the major water source for public water supply and industries. About 95% of water supply comes from this aquifer, with only

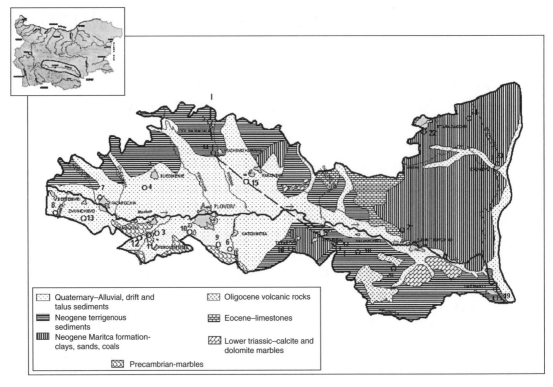

Fig. 18.1 Geological map of the study area (Upper Thracian Plio-Quaternary aquifer). Numbers refer to sample sites discussed in the text.

5% being derived from surface water. Major water sources for the irrigation system are surface reservoirs, river intakes and groundwater (Japan International Cooperation Agency 1999; Ministry of Environment and Waters of Bulgaria 2000). The area is relatively densely populated with well-established industry and agriculture.

The present chapter summarises some results of an EC 5th Framework project 'BaSeLiNe' (Baseline 2003). The chapter considers the methodology and presents results for evaluating the baseline hydrochemistry of the Upper Thracian Plio–Quaternary aquifer in southern Bulgaria where significant anthropogenic influences exist, but where the water overall is still in a state of good quality. The baseline chemistry is here defined as 'the range in concentration (within a specified system) of a given element, species or chemical substance present in solution which is derived from natural geological, biological or atmospheric sources' (Shand et al. 2002; Edmunds and Shand 2004).

18.2 Geology

The region forms part of the northern edge of the Alpine orogenic belt, the latest major tectonic event of the Tertiary. It is bordered to the west and east by the Ihtiman horst-anticline and the Chirpan horst, respectively, and is bounded to the north and south by longitudinal sets of faults (Geological Map of Bulgaria 1990). The region is a complex shallow graben, filled with sediments of Tertiary to Quaternary age (Fig. 18.1).

Precambrian and Palaeozoic plutonic and metamorphic rocks crop out mainly in the mountain areas. These consolidated rocks form the basement to the area and through weathering and erosion provided sediments for the younger formations. Mesozoic formations are exposed in small areas in the southeastern parts of the region forming narrow strips along the major faults. Tertiary sedimentary rocks cover most of the topographically flat area. Oligocene volcanic rocks include acid and intermediate volcanics and acid pyroclastics., and are present in some southern zones. The Upper Eocene is present in two formations: a lower continental coal-bearing molasse and an upper sequence of sandstone, calcareous marl and limestone of marine origin. Pliocene terrigenous sediments belonging to Ahmatovo and Elhovo formations are composed of conglomerate, sand, clayey sand and clay. The thickness of the sequence ranges from 10 to 500 m depending on the depth to the dislocated basement blocks. Its average thickness is about 150 m. Quaternary unconsolidated deposits are distributed along the mountain foot and along the Maritza River and its tributaries. Results of previous exploration boreholes showed that the depth to the basement rocks ranges from 5 to 10 m in small tributaries to more than 100 m downstream in the Maritza River basin.

18.3   Hydrogeology

Quaternary deposits distributed in the plains and hilly parts of the region form an unconfined aquifer. The Pliocene unconfined aquifer in places is hydraulically connected to the Quaternary sediments, mostly in the western and southeastern parts, forming a common Quaternary-Pliocene aquifer with an average thickness locally reaching up to 200 m (Hydrogeological Map of Bulgaria 1967). The alluvial sediments in some places are partially isolated due to clay layers forming locally confined conditions. The Plio-Quaternary aquifer has high productivity. Permeability coefficients range from 50 to 400 m d^{-1} and transmissivity varies between 200–500 and 2000 m^2 d^{-1}, locally reaching 4000 m^2 d^{-1} (Antonov and Danchev 1980). Lower transmissivities ranging from 100 to 400 m^2 d^{-1} were found in the Pliocene. Values less than 100 m^2 d^{-1} were found in the areas of Pliocene outcrop to the north and northeast. The groundwater hydraulic gradient is 0.003 for the northern part of the aquifer, while it is 0.005 in the south. The annual average groundwater discharge to the rivers is about 13 m^3 s^{-1} (Ministry of Environment and Waters of Bulgaria 2000). Groundwater is recharged from several sources including the Maritza River and its tributaries during high flow periods, precipitation and irrigated water. Groundwater tables range from the surface up to 15 m depth. Most frequently the depth of water table is 2.0–4.0 m. The highest groundwater levels often occur during the spring, mainly in March–May. This period coincides with the time of snowmelt, as well as the rainy period and high flows in the rivers. The minimum water levels occur during the summer–autumn period, mostly in August–October, corresponding to the driest period of the year (Machkova and Dimitrov 1999).

18.4   Sampling programme

A total of 22 samples were collected from boreholes during the period 4–10 November 2002. All boreholes sampled were pumping stations, providing relatively good spatial coverage over the Plio-Quaternary aquifer in the studied region (Fig. 18.1). The pumping stations are exploited continuously and water is used for different purposes, mainly for drinking and industrial water supply. The same pumping stations are used as observation points in the National Groundwater Monitoring Network (NGWMN). With few exceptions the samples collected represent a mixture of groundwater from several aquifers, since the boreholes sample water from different stratigraphic horizons (Machkova et al. 1994). When interpreting the water-quality data it should be taken into account that the regional variations might be caused by
• Intercalation of the Plio–Quaternary aquifer with clayey layers and the existence of different stratigraphic horizons.
• Boreholes which were designed and constructed 30–40 years ago. They have different depths and different casing configurations.

As a consequence, the variations in hydrochemical composition may not necessarily be related to the hydrogeochemical reactions along the groundwater flow path. These problems should therefore be borne in mind when interpreting regional variations in the hydrochemical data. Physical and hydrochemical characteristics of the samples measured on site included water temperature ($T°C$), pH, dissolved oxygen (DO), redox potential (Eh), specific electrical conductance (SEC), alkalinity (as HCO3), dissolved CO_2, NO_3–N, NO_2–N, NH_4–N and PO_4–P. Major and trace elements as well total organic carbon (TOC) were analysed in the chemical laboratory of the University of Mining and Geology in Sofia. The British Geological Survey analysed a representative set of 18 samples for trace elements. The chemical determinations from these two sources were combined and used for interpretation in this study.

18.5   Historical chemical data

The set of historical chemical data includes information on water samples from 22 boreholes. These form part of the NGWMN. Chemical data from NGWMN cover the period 1980 to the present, however sampling frequency was quite variable, being mostly seasonal. They are the same wells as used for the sampling programme in November 2002. There are limited historical data available for some of the chemical species (such as PO_4, DO, As, Cu, Ni, Pb), although good information exists for major and trace components (Tzankov et al. 1993; Pencheva and Fouillac 2004).

18.6   Interpretation of the new sampled data

All plots and tables are based on the newly obtained chemical data. Statistical procedures were applied using the 'Basic Statistics' procedures (STATISTICA 2000). Trilinear plots were made based on the new data and on the historical chemical information.

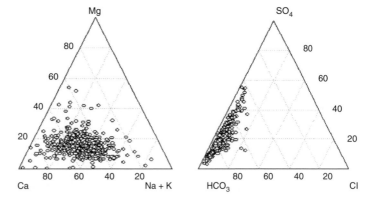

Fig. 18.2 Tri-linear plots showing the relative proportions of major cations (left) and anions (right) (Upper Thracian aquifer, annual averages).

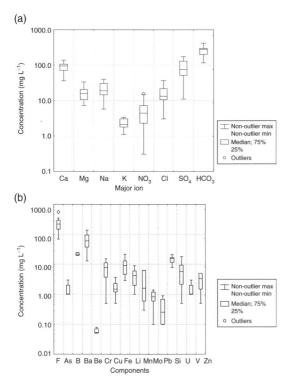

Fig. 18.3 (a) Box plot showing the range of major elements in the Upper Thracian groundwaters (2002); (b) range of minor ions and trace elements in the Upper Thracian groundwaters (2002).

Where concentrations were less than the detection limit, a value equal to the detection limit was applied.

18.6.1 *Hydrochemical characteristics*

The hydrochemical characteristics are summarised in Table 18.1(a) and (b) for the samples collected in 2002. The annual averages for the historical data are shown graphically in a tri-linear plot (Fig. 18.2). The box plots (Fig. 18.3[a] and [b]) and cumulative frequency plots (Fig. 18.4[a] and [b]) are based on the new data. The relative proportions of major cations are very variable, but the anions are dominated by HCO_3 and SO_4 reflecting the low concentrations of Cl (Fig. 18.2).

The waters have low concentrations of dissolved solids. SEC varies from 325 to 1116 µS cm^{-1}, the median value being 644 µS cm^{-1}. The temperature is slightly high with a median value of 14.7°C. The pH is around neutral (7.29) and DO concentration is variable from 2 to 10 mg L^{-1} with a median of 6.3 mg L^{-1}. The groundwater in the Upper Thracian aquifer is thus mainly of Ca–HCO$_3$

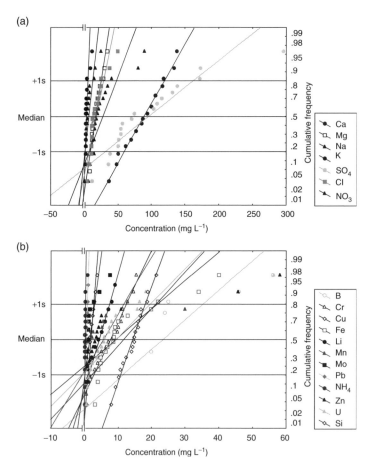

Fig. 18.4 (a) Cumulative probability plots of major ions in the Upper Thracian groundwaters (2002); (b) cumulative probability plots of minor ions and trace elements in the Upper Thracian groundwaters (2002).

to Ca–HCO_3–SO_4 types and in some places of Ca–Na–HCO_3 type.

18.6.2 Major components and characteristics

Table 18.1(a) and the box plots (Fig. 18.3[a]) highlight the concentrations and ranges of major element concentrations. The median concentrations of Ca (91.2 mg L^{-1}), Mg (15.2 mg L^{-1}) as well as HCO_3 (278 mg L^{-1}), for the 2002 data are relatively high, and typically approach log-normal distributions (Fig. 18.4[a]). Such distributions are considered to result from water–rock interactions.

The cumulative frequency plots for Na, SO_4 and Cl show a more complex distribution, which could signify additional sources of these components from pollution.

18.6.3 Minor and trace elements

The concentrations of the minor and trace elements are given in Table 18.1(b) and presented on box plots (Fig. 18.3[b]) and cumulative frequency plots (Fig. 18.4[b]). The trace cations Sr (169–1821 μg L^{-1}; median 416 μg L^{-1}) and Ba (12.9–268 μg L^{-1}; median 57.9 μg L^{-1}), are derived from various

Table 18.1a Statistical summary of groundwaters in the Upper Thracian Plio-Quaternary aquifer (major components and characteristics).

Elements	n	Unit	Mean	Median	Min.	Max.	2.3%-Qu	25%-Qu	75%-Qu	97.7%-Qu	Variance	St. Dev.
T	18	°C	15	15	12	17	12	14	15	17	1	1
pH	18		7.2150	7.2900	6.7400	7.670	6.7400	7.0700	7.3800	7.670	0.1	0.2810
Eh	18	mV	341	320	271	525	271	284	357	525	5955	77
DO	18	mg L^{-1}	6	6	2	10	2	4	7	10	6	2
DO	18	%	57	63	19	94	19	37	75	94	553	24
SEC	18	µS cm^{-1}	650	644	325	1116	325	445	786	1116	46280	215
Ca	18	mg L^{-1}	89	91	36	138	36	70	106	138	832	29
Mg	18	mg L^{-1}	17	15	7	34	7	11	21	34	61	8
Na	18	mg L^{-1}	27	18	6	90	6	14	31	90	534	23
K	18	mg L^{-1}	3	2	1	11	1	2	3	11	6	3
Cl	18	mg L^{-1}	17	13	3	49	3	11	22	49	127	11
SO$_4$	18	mg L^{-1}	97	74	11	296	11	51	131	296	4698	69
HCO$_3$	18	mg L^{-1}	263	278	116	421	116	201	296	421	7695	88
NO$_3$-N	18	mg L^{-1}	6	4	0.3100	24	0.3100	2	7	24	39	6
NO$_2$-N	18	mg L^{-1}	0.0069	0.0050	0.0050	0.028	0.0050	0.0050	0.0050	0.028	0	0.0056
NH$_4$-N	18	mg L^{-1}	0.0863	0.0440	0.0060	1	0.0060	0.0120	0.0770	1	0	0.1650
PO$_4$-P	18	mg L^{-1}	48	20	20	367	20	20	30	367	6934	83
TOC	18	mg L^{-1}	0.2000	0.2000	0.2000	0.200	0.2000	0.2000	0.2000	0.200	0.0	0.0
CODMn	18	mg L^{-1}	1	1	1	2	1	1	1	2	0.1	0.3243

Table 18.1b Statistical summary of groundwaters in the Upper Thracian Plio-Quaternary aquifer (minor and trace components).

Elements	n	Unit	Mean	Median	Min.	Max.	2.3%-Qu	25%-Qu	75%-Qu	97.7%-Qu	Variance	St. Dev.
F	18	µg L^{-1}	230	210	70	550	70	150	300	550	14365	120
Al	18	µg L^{-1}	1	1	1	5	1	1	1	5	1	1
As	18	µg L^{-1}	2	1	1	5	1	1	2	5	2	1
B	18	µg L^{-1}	24	20	20	56	20	20	24	56	101	10
Ba	18	µg L^{-1}	76	58	13	268	13	37	95	268	3764	61
Co	18	µg L^{-1}	0.0550	0.0200	0.0200	1	0.0200	0.0200	0.0200	1	0.0	0.1258
Cr	18	µg L^{-1}	8	8	1	16	1	4	11	16	23	5
Cu	18	µg L^{-1}	2	2	1	4	1	1	2	4	1	1
Fe	18	µg L^{-1}	12	9	3	40	3	5	13	40	114	11
Li	18	µg L^{-1}	4	4	1	9	1	2	6	9	8	3
Mn	18	µg L^{-1}	223	2	0.3000	2510	0.3000	1	6	2510	395280	629
Mo	18	µg L^{-1}	1	1	0.1000	8	0.1000	1	1	8	3	2
Ni	18	µg L^{-1}	0.3278	0.2000	0.2000	2	0.2000	0.2000	0.2000	2	0.2	1
Pb	18	µg L^{-1}	0.3722	0.2500	0.1000	1	0.1000	0.1000	1	1	0.1	0.3064
Pd	18	µg L^{-1}	0.2000	0.2000	0.2000	0.200	0.2000	0.2000	0.2000	0.200	0.0	0.0
Rb	18	µg L^{-1}	1	0.2200	0.0700	14	0.0700	0.1500	1	14	10	3
Re	18	µg L^{-1}	0.3222	0.0500	0.0100	3	0.0100	0.0100	0.1000	3	1	1
Sc	18	µg L^{-1}	3	3	1	4	1	2	3	4	0.3	1
Se	18	µg L^{-1}	1	1	1	3	1	1	2	3	1	1

mineral sources, especially carbonates and gypsum, responsible for the high concentrations and complex distribution. Similar complex distributions were found for Fe, Zn and Mn which have median concentrations of 8.6, 8.3 and 1.55 µg L^{-1}, respectively. The variation is very high for Mn with concentrations up to 1.09 mg L^{-1} in 'Septemvri' (Fig. 18.1, point 8) and 2.51 mg L^{-1} in 'Dimitrovgrad' (Fig. 18.1, point 21), both in the alluvial terrace of the Maritza River.

A number of trace elements were present including (brackets show median concentrations) As (up to 2.5 µg L^{-1}, but with a low median concentration of 0.1 µg L^{-1}), B (20 µg L^{-1}), Cr (7.65 µg L^{-1}), Cu (1.45 µg L^{-1}), Li (4.0 µg L^{-1}), Mo (0.8 µg L^{-1}), Pb (0.25 µg L^{-1}), Rb (0.22 µg L^{-1}), Re (0.05 µg L^{-1}), Sb (0.06 µg L^{-1}), Sc (2.49 µg L^{-1}), Se (1.05 µg L^{-1}), V (1.0 µg L^{-1}). Uranium had a relatively high median concentration (5.68 µg L^{-1}) with a maximum of 56.3 µg L^{-1} in the borehole 'Parvomai' (Fig. 18.1, point 17).

18.6.4 Pollution indicators

In general, the aquifer is free from organic pollutants. The presence of NH_4–N (median concentration of 0.044 mg L^{-1}; maximum in sample 6 'Katunitsa' 0.73 mg L^{-1}) and NO_2–N (0.005 mg L^{-1}), as well as the high local concentrations of Mn and occasionally, Fe indicate locally reducing conditions with derivation of these elements from natural origin in some points in the alluvial terraces of the rivers. In most cases, however, the redox potential is high enough so that aerobic conditions prevail and resulting low metal concentrations.

This aquifer is strongly affected by human inputs and in many places some element concentrations are far above the baseline.

For instance high nitrate is an indicator of pollution derived mainly from the application of fertilisers (e.g. point 3: 107 mg L^{-1}; point 15: 70 mg L^{-1}; point 16: 65 mg L^{-1}; point 20: 62 mg L^{-1} as NO_3) and in some regions they exceed the maximum permissible levels for drinking water. TOC is quite low, often below the detection limit. At the same time, the median value of chemical oxygen demand, another indicator of organic pollution, is 0.5 with concentrations increasing over time, as discussed below.

18.6.5 Main correlations and hydrogeochemical process

Only positive correlations are above the significance level in the dataset studied, for example, between HCO_3^- and pH, DO, SEC, Ca, Mg and Na. In this case, the predominant hydrogeochemical processes include carbonate and silicate dissolution reactions:

$$AECO_3 + H_2O + CO_2 \leftrightarrow AE^{2+} + 2HCO_3^-$$

(here AE = alkaline earth element), and

$$2CO_2 + 2NaAlSi_3O_8 + 11H_2O \cdot AlSi_2O_5(OH)_4 + 2Na^+ + 2HCO_3^- + 4H_4SiO_4$$

that is, hydrolysis with incongruent dissolution and with formation of residual kaolinite.

The degree of saturation (Paces 1983; Velikov 1985; Velikov and Panayotova 2001) was less than 1 with respect to calcite and dolomite. Only in one sample (point 9, Plovdiv) was the saturation for calcite slightly over 1 (1.09) showing minor oversaturation. The role of the alumosilicate rock–water interactions can be illustrated by the transformation index anorthite–kaolinite (Paces 1983; Velikov and Panayotova 2001), which has values ranging from –7.22 to –5.05.

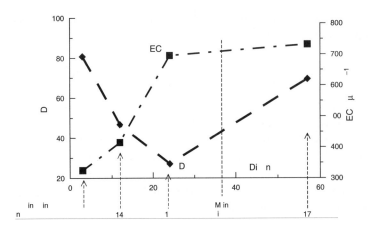

Fig. 18.5 Changes in DO and SEC along the profile I–I' shown in Figure 18.1

The relatively high Na/Cl ratio (being mostly 2–5, but higher for some points as, for example, 7.42 for point 12) indicates that some cation exchange on clay minerals in the aquifer has occurred. For Rb, ion exchange in the clay zones could be also of importance taking into account the correlations of Rb with Na and K, and also of Mo and Mn. The good correlations of Sr with Ca, Mg, SO_4, HCO_3 and Ba, show that the source of Sr is likely to be not only from carbonate minerals, but also from gypsum and possibly celestite or barite.

The reducing conditions in some parts of the aquifer near the Maritza River are due to degradation of organic materials with insufficient DO. Approximate illustrations of the decreasing DO and increasing SEC levels, when approaching the river are shown on Fig. 18.5. It can be seen that getting closer to the main river channel and down to the lower part of the basin, the waters become more mineralised.

18.6.6 *Temporal and spatial variations*

Some typical temporal variations over the last 20 years are shown in Fig. 18.6. The black straight line is the linear trend regression line estimated by the least squares method. However, the least squares estimate (LSE) is not robust against outlying observations in the data. A single outlier not obeying the trend of the rest of the data can give a totally misleading summary of the relationship. Because of this vulnerability to outlying observations, many robust regression estimators have been introduced (Rousseeuw and Leroy 1987). The three-groups Tukey's regression line avoids these difficulties by using medians within the groups to achieve resistance (Hoaglin et al. 1983). The dashed line presented in the plots is the Tukey's resistant regression line. The simple regression lines based on the least median of squares, trimmed least squares and S estimators, which are other positive breakdown point regression estimators (Rousseeuw and Leroy 1987), are not shown on these plots because of similar behaviour to the Tukey's line.

As a general trend, decreasing groundwater levels occur in the aquifer (Machkova and Dimitrov 1999; Ministry of Environment and Waters of Bulgaria, MoEW 2000). The hydrochemical characteristics have different

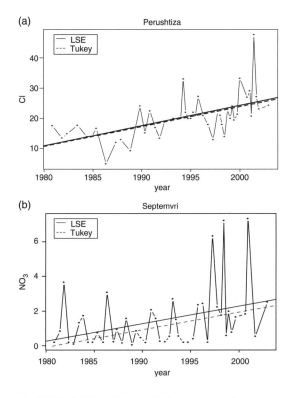

Fig. 18.6 (a) Trend line of Cl concentrations at sample point 3; (b) trend line of NO₃ concentrations at sample point 8.

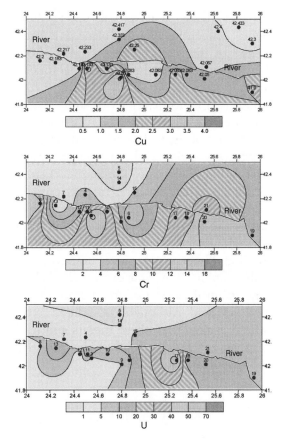

Fig. 18.7 Spatial distribution of some trace element concentrations ($\mu g\ L^{-1}$) in the Upper Thracian groundwaters.

trends at different points in the aquifer, more commonly showing a decrease in pH and Ca, a weaker decrease in SO_4 and a predominant increase in Na and Cl concentrations. Slight increasing trends in COD, NO_3^- and HCO_3^- were also found. Trend lines for NO_3 and Cl evaluated through classical LSE and robust Tukey methods are given as an example in Fig. 18.6(a) and (b). Although in most cases remaining far below the maximum permissible concentrations, the increasing trends are of concern and need to be investigated further in order to avoid unfavourable human impacts on the groundwater quality.

Another approach to visualise spatial variations of the hydrochemical variables is to plot and contour the concentrations. The kriging method (Golden Software 2000) was used to produce a grid and plot the contours. The maps, drawn using the 2002 data (Fig. 18.7), highlight the role of the main river channel as a dividing line in the aquifer. The lack of exchange of waters between the northern and southern parts of the aquifer highlights two different spatial distributions of the hydrochemical parameters.

Some elements (such as Cu, Cr, U) show higher concentrations set against a background of low values. This clearly shows the influence of anthropogenic factors on groundwaters status.

Increased concentrations of some metals are observed. These are observed over different parts of the region, which means that the factors causing those concentrations exist in many places of the aquifer. One of the factors is the location of poly-metallic ore deposits in the southern mountainous part of the region. There is a similar situation with some trace elements, especially uranium; Pliocene deposits of uranium exist in the aquifer. They were exploited for about 20 years until 1993, leaching uranium using sulphuric acid. Although also related to the natural background, these metals and trace elements can also be mobilised by industrial and agricultural activities.

18.7   Conclusions

The study of the groundwater in the Upper Thracian aquifer shows that the dominant control on water chemistry in the aquifer is water–rock interaction, especially carbonate and aluminosilicate dissolution, and also ion exchange. For some parameters, increased concentrations are due to a mixed influence of natural and anthropogenic factors. Although some of the waters are influenced by anthropogenic activities, for many elements it is still of good quality. Baseline trends could not be clearly distinguished, as they are in most cases masked by the influence of the anthropogenic factors.

The chemical types of water indicate that the samples are probably mixtures of waters, derived from different stratigraphic horizons.

In addition, the infiltration of fresh river waters close to the main river channel and major tributaries is important. The role of the main river channel as a hydraulic boundary between groundwater bodies located along its both sides is quite visible, especially with respect to some components (such as Cu, Cr, U, Sr), whose variations are caused by anthropogenic factors. Interesting in this respect are the distributions of pH and uranium, the latter having been leached in situ by sulphuric acid in the past.

References

Antonov, H. and Danchev, D. (1980) *Groundwaters in the Republic of Bulgaria*. Technica, Sofia.

Baseline. (2003) Natural baseline quality in European aquifers: a basis for aquifer management. Final Report to EC, Contract EVK1-CT1999-0006.

Committee of Geology (1967) Hydrogeological Map of Bulgaria 1:200,000.

Committee of Geology (1990) Geological Map of Bulgaria 1:100,000, Sofia.

Edmunds, W.M. and Shand, P. (2004) Geochemical baseline as a basis for the European Groundwater Directive. In: Wanty, R.B. and Seal, R.R. (eds) *Water–Rock Interaction 11*. Balkema, Rotterdam, pp. 393–7.

Golden Software. (2000) SURFER.

Hoaglin, D.C., Mosteller, F. and Tukey, J.W. (1983) *Understanding Robust and Exploratory Data Analysis*. John Wiley and Academia, Praha.

Japan International Cooperation Agency. (1999) *Study on Integrated Management for the Maritza River in the Republic Bulgaria*, Sofia.

Machkova, M. and Dimitrov, D. (1999) *Manual of the Bulgarian Groundwater*. MOEW-NIMH, Sofia.

Machkova, M., Tzankov, K., Velikov, B. and Dimitrov, D. (1994) *Contamination of Surface and Groundwaters in the Upper-Thracian lowland, Bulgaria. Water Down Under 94.* Adelaide, Australia.

Ministry of Environment and Waters of Bulgaria (MoEW). (2000) *General Master Plans for the Water Usage in the River Basin Districts*, Sofia.

Paces, T. (1983) Rate constants of dissolution derived from the measurements of mass balance in hydrological catchments. *Geochimica et Cosmochimica Acta* 33, 1855–63.

Pencheva, E. and Fouillac, C. (2004) WATMETAPOL, NATO SfP973739 Prj Bulgarian Academy of Science, Geological Institute; Ministry of Environment and Water, Ministry of Regional Planning and Public Works, Ministry of Health, Ministry of Agriculture and Forests.

Rousseeuw, P. and Leroy, A. (1987) *Robust Regression and Outlier Detection*. John Wiley and Sons, New York.

Shand, P. et al. (2002) Baseline Report Series: 1. The Triassic Sandstones of the Vale of York.

British Geological Survey Report, CR/02/102 N, pp. 48.

STATISTICA. (2000) StatSoft.

Tzankov, K., Machkova, M., Dimitrov, D. et al. (1993) *Groundwater Hydrochemical Reference Book for the Period 1980–1991.* MOEW&NIMH, Sofia.

Velikov, B. (1985) Estimation hydrochimique des interactions eau-roches carbonates. Congres International de Technique Hydrothermale 21st, Varna, v.2, pp. 170–9.

Velikov, B. and Panayotova, M. (2001) Thermodynamic calculation in elimination of water pollutants at hydrogeochemical barriers. *Chinese Journal Chemistry* 19, 222–6.

Whitfield, M. (1974) Thermodynamic limitations on the use of the platinum electrode in Eh measurements. Limnology Oceanography 19, 857–65.

19 The Mean Sea Level Aquifer, Malta and Gozo

J. MANGION AND M. SAPIANO

19.1 Introduction

The Maltese archipelago consists of three inhabited islands: Malta, Gozo and Comino, and a number of uninhabited islets scattered around the shoreline of the major islands. Its location is approximately 96 km south of Sicily and 290 km north of Tunisia and is located at latitudes 35°48' and 36°05' north and longitudes 14°11' and 14°35' east. The total surface area of the archipelago is approximately 316 km^2; with Malta and Gozo, the two largest islands, occupying 246 and 67 km^2, respectively.

The islands are typical examples of coastal carbonate aquifers, and thus representative of the coastal limestone areas of Southern Europe, where limestones, often deeply karstic, form the main and possibly only aquifers. In these coastal regions, the maintenance of high-quality waters as far as possible is also important, in view of both seasonal and perennial anthropogenic pressures.

19.2 Regional setting

The islands lie on the eastern edge of the North African continental shelf, geologically known as the Pelagian Block. This corresponds to an oceanic area in the central Mediterranean spanning from the shores of Tunisia in the southwest, to the shores of Sicily in the north and ending in an abrupt escarpment at the edge of the Ionian Sea. Late Cretaceous and Tertiary movements gave rise to a series of horst and graben structures running in a NE–SW direction and having a predominant regional dip to the northeast.

19.2.1 Local setting

The geology of the Maltese islands comprises a succession of Tertiary limestones and marls with scarce Quaternary deposits. Essentially, the islands are geologically made up of a core of clays and marls, the Blue Clay and the Globigerina Limestone formations stacked between two limestone formations known as the Upper and the Lower Coralline Limestones. The oldest formation, the Lower Coralline Limestone is of Oligocene Age whilst the Maltese succession ends in the Miocene, with the top of the Upper Coralline Limestone being chronologically dated to the Upper Messinian age possibly extending into the early Pliocene.

From a structural point of view, the Maltese islands can be subdivided into three regions, primarily consisting of two elevated blocks separated by the two major NE–SW fault lines present in the islands, namely the Ghajnsielem–Qala fault in the north and the Victoria fault in the south. Between these two faults a structural graben stretching between southern Gozo, Comino and northern Malta separates the two upthrown blocks.

In a significant part of the island of Malta, south of the Victoria faultline, the Upper Coralline Limestone and the Globigerina/ Lower Coralline Limestone formations are stacked vertically. The Lower Coralline Limestone in this region occurs mainly at sea level and is thus in lateral and vertical contact with seawater. The Upper Coralline

Limestone formation outcrops mainly on the western side of the island, perched over the Blue Clay formation (Fig. 19.1).

The downthrown region of the archipelago, north of the Victoria fault, is divided by a NE–SW fault system into a succession of horst and graben like structures. This structure with parallel compartments separated by faults leads to the formation of relatively small aquifer blocks, which are independent from one another from a hydrogeological point of view.

In the island of Gozo, north of the Ghajnsielem–Qala Fault, the Lower Coralline Limestone occurs at sea level and is overlain

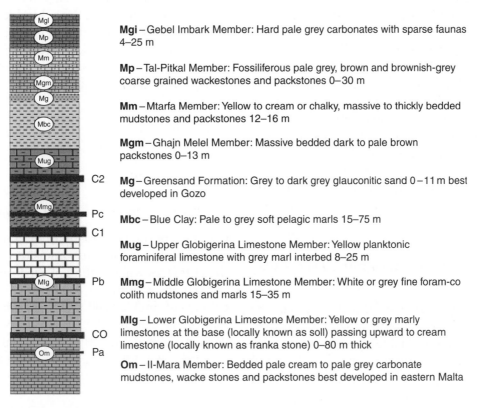

Mgi – Gebel Imbark Member: Hard pale grey carbonates with sparse faunas 4–25 m

Mp – Tal-Pitkal Member: Fossiliferous pale grey, brown and brownish-grey coarse grained wackestones and packstones 0–30 m

Mm – Mtarfa Member: Yellow to cream or chalky, massive to thickly bedded mudstones and packstones 12–16 m

Mgm – Ghajn Melel Member: Massive bedded dark to pale brown packstones 0–13 m

Mg – Greensand Formation: Grey to dark grey glauconitic sand 0–11 m best developed in Gozo

Mbc – Blue Clay: Pale to grey soft pelagic marls 15–75 m

Mug – Upper Globigerina Limestone Member: Yellow planktonic foraminiferal limestone with grey marl interbed 8–25 m

Mmg – Middle Globigerina Limestone Member: White or grey fine foram-co colith mudstones and marls 15–35 m

Mlg – Lower Globigerina Limestone Member: Yellow or grey marly limestones at the base (locally known as soll) passing upward to cream limestone (locally known as franka stone) 0–80 m thick

Om – Il-Mara Member: Bedded pale cream to pale grey carbonate mudstones, wacke stones and packstones best developed in eastern Malta

Fig. 19.1 Lithologic column of the Maltese rock formations showing the occurrence of the main conglomerate beds (after Scerri 2002).

by the Globigerina Limestone formation over much of the whole of the island. The Upper Coralline Limestone outcrops over the peaks and on high ground, perched over the impervious Blue Clay formation in the northern and central regions of the island.

The lithological different natures of these formations together with their geological position give occurrence to two broad aquifer types: the upper (perched) aquifers in the Upper Coralline Limestone and the lower (mean sea level) aquifers in the lower limestone units (the Lower Coralline Limestone, and where fractured in the Globigerina Limestone). Due to the depressed structure of the central region of the archipelago, the Upper Coralline Limestone also hosts small sea level aquifers in the Northern region of Malta.

The Upper and Lower Coralline Limestones are thus considered to function as the main aquifer formations in the islands. The Globigerina limestone functions only locally as an aquifer formation, only where it is fractured and/or is located at sea level, and is commonly expected to allow groundwater flow exclusively through fractures and fissures. The Blue Clay formation is normally impermeable and underlies the perched aquifers.

19.2.2 Factors which may influence the groundwater quality

On the basis of historical data, it can be noted that the quality of groundwater in Malta is highly variable, with the main influences arising from seawater salinity and from nitrate, mainly of agricultural origin. Groundwater sustained in the sea level aquifer formations in Malta has generally high levels of chloride and other seawater-related parameters, typical of a groundwater body which occurs in direct lateral and vertical contact with seawater. In such a scenario any alteration to the flow of groundwater caused by both natural and anthropogenic factors will result in the intrusion of saline waters.

Furthermore, in an island with a high population density and a rapidly developing economy, the aquifers are subject to intense pressures and impacts that have led to a gradual depletion in qualitative status over the years. Nitrate contamination is a source of particular concern since concentrations in the aquifers currently exceed accepted limiting values and worse, exhibit upward trends.

As is typical for carbonate aquifers, groundwater in Malta has a relatively high degree of hardness whereas fluoride levels are close to limiting values for drinking water in places where groundwater flow occurs through the phosphorite conglomerate beds.

19.3 Background

19.3.1 Geology

19.3.1.1 Lower Coralline Limestone

Outcrops of the Lower Coralline Limestone are limited to deep valley incisions and to the cliff faces where 140 m of the formation is exposed in southern Malta and Gozo. It consists of thickly bedded detrital limestone that can be distinguished into four members (Pedley, 1978) namely the Maghlaq member (lowest), the Attard member, the Xlendi member and the Il-Mara member.

Maghlaq member (Chattian). The lowest member of the Lower Coralline Limestone consists of pale yellow beds of foraminiferal biomicrites composed of tests of benthonic forams that constitute 40% of the rock. Typical microfossils recorded by Felix (1973), are the *Miliolid foraminifera*, *Austrotrillima paucialveolata*, *Praerhapydionina delicata*, *Borelis haueri* and *Peneroplis Thomasi*. The Maghlaq member passes transitionally upwards into the Attard member and extends uniformly beneath the whole territory.

Attard member (Chattian). The Attard member consists of pale grey biosparites (wackestone and packstone) associated with large *Archaeolithothamnium algal rhodolits* and *Archeolitothamnium intermedium ranier*. Rhodolite colonies consisting of Archeolitothamnium intermedium Ranier and Lithotamnium make up between 15% and 80% of the rock while bethonic foraminifera may form up to 11% of the rock. These assemblages are locally associated with bryozoa and strombid gastropods, indicating a depositional environment typical of open shoal-conditions with depths of less than 25 m.

Xlendi member (Chattian). The Xlendi member consists of cross-stratified coarse-grained limestone (packstone) with abundant aminiferal fragments, and giant foraminifera including Amphistegina, Spiroclypeus and Heterostegina. At Ghar il-Qamh in Gozo it reaches a maximum thickness of 22 m exposing low-dipping foreset bedding and predominantly west-facing cross-stratification.

Mara member (Chattian). This is the highest member of the Lower Coralline Limestone and consists of pale yellow massive bedded biosparites and biomicrites. The basal contact is transitional and overlies the Xlendi member. Towards the upper levels, the

Table 19.1 Geochemistry of the Lower Coralline Limestone formation.

Component	% Composition sample 1	% Composition sample 2
MgO	0.45	0
Al_2O_3	0.83	0.83
SiO_2	2.34	2.17
P_2O_5	0.56	0.54
K_2O	0.36	0.38
CaO	94.4	93.4
TiO_2	0.08	0.09
Fe_2O_3	0.78	0.74

member becomes richer in Lepidocyclina, bryozoan fragments and the echinoids Echinolampas.

The Lower Coralline Limestone is considered chemically as an almost pure limestone deposited in a shallow water patch-reef palaeoenvironment of deposition. The principal geochemical components of these limestone and their respective fractions were identified by x-ray fluorescence spectrometry (XRF) and are shown in Table 19.1. While the main components, expressed as oxides, are CaO, SiO_2 and Al_2O_3; phosphate content, on average is about 0.6%.

19.3.1.2 Globigerina Limestone formation

The Globigerina Limestone is the most widespread formation in Malta and Gozo. It consists of yellow to pale grey, fine-grained limestones composed entirely of tests of globigerinid foraminifera. There are several phosphorite horizons and hard grounds, two of which are ubiquitous and can be traced throughout the islands. These horizons, often less than 0.5 m thick and consisting of dark brown to black collophanite pebbles,

Table 19.2 Geochemistry of the Lower Globigerina Limestone member.

	Na_2O	MgO	Al_2O_3	SiO_2	K_2O	P_2O_5	CaO	TiO_2	Fe_2O
Mean %	0.04	0.71	1.18	4.0	0.19	0.21	49.71	0.08	0.66

have enabled the subdivision of this formation into three distinct members (House et al. 1961) namely, the Lower, Middle and Upper Globigerina members.

Lower Globigerina Limestone (Aquitanian). This member consists of yellow to pale cream indurated, globigerinid packstones becoming wackestone one metre away from the basal contact with the Blue Clay. At outcrop it exposes a characteristic honeycomb weathering and a reddish-yellow colour. Fossils, where present, include the molluscs Chlamys and Flabellipecten, the echinoids Schizaster and Eupatagus, pteropods such as Cavolina, and thallasinoidean burrows.

At the base, the member is transitional with the Lower Coralline Limestone whilst it ends at the top in a phosphorite conglomerate layer less than 1 m thick. This phosphorite bed, referred to by Pedley as the lower phosphorite horizon *Mc1*, is planar at the top and irregular at the base. It consists of rounded phosphatic pebbles that have been cemented by chemical and concretionary action of phosphates. Phosphatised clasts of molluscs and corals are locally common together with shark teeth *Carcharodon megalodon*, *Odontaspis*, *Isurus* and *Hemeprestes*, and tusks of marine mammals.

The basal boundary of the Globigerina Limestone formation is taken to be that first bed above the Scutella bed that does not contain Scutella echinoids. The Lower Globigerina is thus considered to be transitional with the Mara member.

Geochemical analysis of the Lower Globigerina Limestone member undertaken by Cassar (2002) is shown in Table 19.2. It is noted that P_2O_5 content of the samples ranges from 0.1 to 0.71. Although no data on fluoride content is available it is estimated to be of the order of 0.03%.

Middle Globigerina Limestone (Aquitanian – Burdigalian). The Middle Globigerina Limestone member lies unconformably over the first phosphorite horizon in distinct beds of white to pale grey finely laminated marly limestones. At the base it is more whitish in colour and finely laminated with occasional seams of brown chert nodules. Towards the top, it becomes grey in colour and marly with sparse clay lenses.

The Middle Globigerina Limestone ends with a phosphorite horizon referred to as Mc2 consisting of 0.5 m of reworked phosphorite pebbles, mollusc casts, corals, echinoids, shark teeth and casts of the nautiloid *Aturia aturi*. Below this bed, *thallassinoidean* burrows extend up to 0.75 m and are commonly filled with reworked phosphorite pebbles.

The Middle Globigerina Limestone is really a marl. Geochemical analysis (Table 19.3) indicate that phosphate and presumably fluoride are present though at subdued levels.

Upper Globigerina Limestone member (Langhian). The Upper Globigerina Limestone consists of a tripartite sequence comprising a lower division of pale yellow

Table 19.3 Geochemistry of the Middle Globigerina Limestone member.

Component	% Composition sample 1	% Composition sample 2
SiO_2	8.2	9.6
Al_2O_3	2.8	2.2
TiO_2	0.15	0.12
Fe_2O_3	1.2	1.0
MnO_3	0.02	0.01
CaO	44.7	44.5
MgO	1.9	2.3
K_2O	0.43	0.42
Na_2O	0.14	0.27
P_2O_5	0.33	0.45
SO_3	1.3	0.5

biomicrites, an intermediate pale grey marl and an upper cream coloured limestone. It lies conformable over the Upper Phosphorite Conglomerate bed Mc2 in eastern Gozo while it lies locally over an erosion surface west of Gharb. Schizaster euryonotus, the gastropod Epitonium, the Pteropod Maginella and thallasinoidean burrows are frequently found in the upper division.

Although no geochemical analyses of the Upper Globigerina Limestone member are available, geochemical data on the matrix of the Upper Main phosphate conglomerate bed are considered to be representative of this member. Geochemical data (average percentage composition) from Pedley and Bennet (1985) are shown in Table 19.4.

19.3.1.3 Phosphorite conglomerate beds

Hardgrounds result from cementation of pelagic sediments by high-Mg calcite, aragonite and in some instances, glauconite and calcium phosphate. They constitute pene-contemporaneous lithification and represent the earliest stage of diagenesis for some pelagic carbonates. A prolonged contact between sediments and seawater is necessary for hardground generation, and are therefore favoured by very low primary rates of sedimentation, and sediment bypass under the action of submarine currents, which remove the finer sediments. Areas of phosphorite generation are characterised by upwelling of seawater rich in nutrients and organic activity is therefore high. Phosphorite precipitation occurs by direct replacement of skeletal carbonate allochems and carbonate mud or indirectly as a result of biochemical concentration within living faunal tests. The first to be lithified are the existing burrows, which are filled with relatively more permeable coarser material. Partial or complete coalescence of the lithified burrows accompanied by winnowing produces a discrete nodular fabric or a continuous lithified surface. Periodic overturning by seabed dwelling invertebrates promotes nodule development (Scerri 2002).

The oldest phosphorite conglomerate pebble beds recorded in the Maltese islands occur at different levels within the Il-Mara Member of the Lower Coralline Limestone formation in east Malta. In this region this

Table 19.4 Indicative geochemistry of the Upper Globigerina Limestone member.

	Na_2O	MgO	Al_2O_3	SiO_2	K_2O	P_2O_5	CaO	F	Fe_2O_3
Matrix	1.2	1.34	0.38	7.14	0.3	6.43	41.14	0.85	2.25

member is thickly developed. In central and western Malta this member is usually thin and frequently occurs as a condensed sequence with no phosphorites.

19.3.1.4   Blue Clay (Serravallian – Tortonian)

The Blue Clay is a bluish grey formation composed of plastic colour-banded kaolinitic marls and clays or olive green marls and clays with no apparent banding. The darker bands are almost pure clay while the lighter bands have a lower clay content and can be described as clayey marls. Commonly found are rich assemblages of fossil macro-fauna Flabellipecten and Amusium with goethite casts of bivalves, gastropods, fish-teeth, solitary corals and echinoids. Being highly impermeable the Blue Clay forms an aquiclude at the base of the Upper Coralline Limestone where it supports the perched aquifers.

19.3.1.5   Greensand (Tortonian)

This consists of greyish-green, friable, sandy marls becoming dark coloured greenish-brown, and richer in glauconite near the top. Fossil content is abundant and predominantly consists of the giant foraminifera *Heterostegina depressa*, casts of the bivalves *Glycemeris, Cardium, Chlamys, Ostrea* and *Throcia* and the echinoids *Clypeaster altus*, and *Schizaster eurynotus*.

The formation is 11 m thick in Gozo (Gelmus) and ends with a sharp contact at the base of the Upper Coralline Limestone. West of a N–S line running from Marsalforn to Mgarr ix-Xieni (Pedley et al. 1976) this contact is ubiquitous marking an extensive hiatus between the Upper Coralline Limestone and the Greensand.

19.3.1.6   Upper Coralline Limestone (Messinian)

The Upper Coralline Limestone displays several lateral variations and has been studied by several authors. Pedley et al. (1976) subdivided the formation into four lithostratigraphical units with 12 distinct beds following the identification of a basal Upper Coralline erosional surface and a widespread brachiopod marker bed–the *Terebratula-Aphelesia* bed – that allowed accurate interpretation of the lower members.

19.3.1.7   Quaternary

This is very restricted and appears only as valley infills, raised beach deposits and modern beach sands.

19.3.2   Hydrogeology

The Lower Coralline Limestone formation represents the most important aquifer formation of the Maltese islands, sustaining the major sea level groundwater bodies, which by far are the primary sources of freshwater for the islands. As the formation is predominantly composed of algal-fossiliferous limestone with sparse corals, it has a moderate, irregular and channel-like permeability. In fact, the high permeabilities of coral reefs are absent and are replaced instead by an irregular permeability more characteristic of algal reefs. This heterogeneity is further accentuated by the presence of scattered patch-reefs in lateral contact with lagoonal and fore-reef facies.

The primary porosity of the formation is highly variable and varies from 7% to 20%. The different density indicates that a large part of the primary pore space is not interconnected, a fact which is also stressed by the

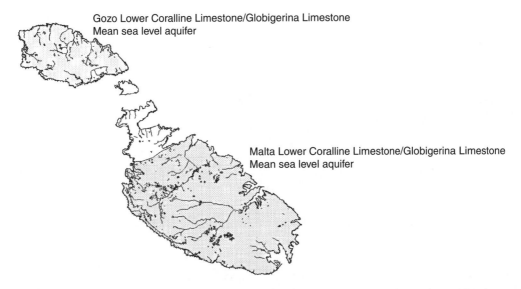

Gozo Lower Coralline Limestone/Globigerina Limestone
Mean sea level aquifer

Malta Lower Coralline Limestone/Globigerina Limestone
Mean sea level aquifer

Fig. 19.2   Extent of the Lower Coralline Limestone aquifer systems in the Maltese islands.

fact that the primary permeability is rather low. The effective porosity of the formation is mainly connected with fracture permeability, since otherwise the pores are very poorly interconnected. Flow and dewatering of pore spaces rely on secondary permeability by tectonic fracturing and dissolution. The fractures range from micro-fissures to karst solution cavities, frequently aligned in one direction. Secondary permeability is thus fissure dependent and is estimated to range between 10% and 15% whilst the average hydraulic conductivity as measured from pumping tests is $400 \times 10^{-6} \mathrm{m\ s^{-1}}$. The transmissivity of the formation is estimated to vary between 10^{-4} and $10^{-3} \mathrm{m^2\ s^{-1}}$.

The largest and by far the most important of these sea level bodies of groundwater is the mean sea level aquifer system occurring in the island of Malta. This aquifer stretches across an area of 216 km², primarily south of the Victoria fault. However, intense fracturing along the fault plane allows horizontal communication of groundwater; and thus the effective boundary of this aquifer system is considered as the 'sealing' Pwales fault, which is located further to the north. The body of groundwater sustained by the Malta mean sea level aquifer yields an estimated 66% of the total groundwater abstracted in the country and under optimum conditions the aquifer system is estimated to be capable of storing up to 1.5 billion m³ of groundwater (Fig. 19.2).

The second largest mean sea level aquifer system is found in the island of Gozo, north of the Ghajnsielem–Qala fault, stretching over an area of 50 km². The groundwater body sustained in this aquifer system is the major source of groundwater on the island of Gozo and an exclusive source of drinking water. Other sea level aquifer systems occur in the northern region of Malta, north of the Pwales fault, but being very limited in extent, these aquifers are considered as having only a relatively 'local' importance.

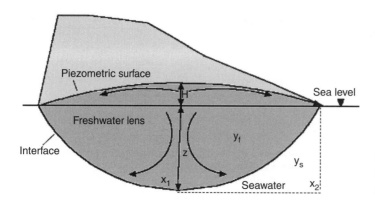

Fig. 19.3 Schematic representation of the mean sea level aquifer system.

The groundwater body occurring in these sea level aquifer systems is in lateral and vertical contact with seawater. Due to the density contrast of freshwater and saltwater a Ghyben–Herzberg system is developed. The outcome is a lens shaped body of freshwater that is dynamically floating on more saline water, having a convex piezometric surface and conversely a concave interface, both tapering towards the coast where there is virtually no distinct definition between the two surfaces. Maximum hydraulic heads of 4–5 m amsl were measured for the groundwater body sustained in the Malta mean sea level aquifer in the 1940s when the system was still largely unexploited. The lens sinks to a depth below sea level roughly 40 times its piezometric head at any point, fading into more saline water across a transition zone, the thickness of which depends on the hydrodynamic characteristics of the aquifer formation, and the dynamic effects of pumping. The limits of this transition zone are commonly defined by the surface of the 1% and 95% seawater content (Fetter, 1988), based on the total dissolved solids or chloride content (Fig. 19.3).

In practice, these mean sea level aquifer systems are very sensitive to point-form saline upconing due to their hydrogeological characteristics and the relatively small piezometric head. Wells drilled to some depth below sea level are prone to localised upconing of saline water in response to the drawdown in the piezometric head caused by abstraction, with the direct result being an increase in the salinity of the abstracted water.

19.3.3 Groundwater age

The age of the groundwater body in the mean sea level aquifer was investigated using isotopic techniques. Groundwater samples from the mean sea level aquifer revealed a low ^{3}H content, averaging around 1.15 TU. This indicates that this groundwater body is characterized by a relatively long residence time (in excess of 20 years on average).

Further investigations on the ^{3}H content in these groundwater samples were carried out, in which the results obtained were correlated with the thickness of the unsaturated zone at the sampling site and the nitrate content of the respective sample, nitrate being considered as a tracer of recently infiltrated water (Fig. 19.7). Invariably, the correlation showed that the thicker the unsaturated zone the lower the ^{3}H content thus confirming a slower and longer travel time from the surface. Moreover, a significant

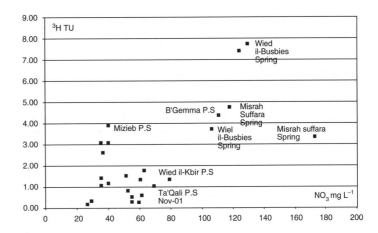

Fig. 19.4 Correlation of ^{3}H versus NO$_3$ content in groundwater samples (after Bakalowicz 2002).

correlation between ^{3}H and nitrate was registered. This correlation is due to a mixing between long- and short-residence time groundwater, respectively rich in ^{3}H and nitrate respectively. Lower nitrate values generally matched low ^{3}H content, although evidence of localised nitrate enrichment was also encountered. (Fig. 19.4).

The results obtained thus indicate that groundwater in the mean sea level aquifer is characterised by the mixing of waters with two different infiltration processes: (1) a very slow infiltration, through the matrix porosity, which is the dominant recharge process of the aquifers and (2) a fast infiltration, through cracks and fractures, which is a local and discrete flow occurring probably only during the main rain events and which most probably is responsible for the direct leaching of pollutants from the surface into the saturated zone.

The aquifer is thus primarily recharged through the slow percolation of groundwater in the unsaturated zone. However, bedding planes, erosional surfaces and hiata in the sedimentary facies are also important features in the circulation of groundwater. Mineralisation and groundwater chemistry

will thus be influenced by the spatial occurrence of these discontinuities.

19.3.3.1 Data availability

Generally, it is not always easy to identify waters where there are no traces of human impact, and this is particularly so in the aquifer system under consideration. Potentially, the only possible way to correct this would be to establish water quality levels prior to the start of the groundwater abstraction in the early 1900s – accepting the limitation of the probably poorer analytical techniques used at the time.

19.3.4 *Historical data*

The first available data on the quality of groundwater in the mean sea level aquifer dates back to analysis carried out in 1865, at the British War Department in Woolwich, UK (Table 19.6). These analyses were undertaken on samples collected from the first shaft drilled in the mean sea level aquifer when this was still in an undisturbed condition and thus reflect its natural background concentration. Of particular relevance is the

Table 19.5 Geochemical data on the Lower Main Phosphorite conglomerate bed.

	Na_2O	MgO	Al_2O_3	SiO_2	K_2O	P_2O_5	CaO	F	TiO_2	Fe_2O_3
Pebbles (outer rim)	0.34	1.09	0.23	5.37	0.28	13.3	47.0	1.09	0.01	2.33
Whole pebbles	0.75	1.29	0.25	4.56	0.2	16.8	42.4	1.35	<0.01	2.31
Matrix	0.11	1.31	0.41	7.86	0.25	3.72	50.3	0.41	0.01	1.14

Source: Scerri (2002).

Table 19.6 Water quality analyses – Malta 1865.

"The well at the Marsa, in the farm of Armier, from which water was raised by a steam pump"	mg L^{-1}
Total solids	761
Carbonate of lime	301
Nitrate of lime	14
Sulphate of lime	–
Carbonate of magnesia	53
Sulphate of soda	50
Chloride of soda	341
Ammonia	–
Organic and other volatile matters	1

Source: Sutherland (1867).

fact that although this analysis was performed at a time when the groundwater body was still practically unexploited, its salinity expressed as TDS is already relatively high (761 mg L^{-1}). and the chloride of soda value corresponds to a Cl concentration of 207 mg L^{-1}.

19.3.5 Current datasets

Regular monthly chemical analysis on groundwater abstraction sources, operated by the Water Works Department (now Water Services Corporation), date back to the early 1960s. However these analyses concern only a limited number of core parameters, such as chloride, nitrate, pH, TDS, hardness and turbidity.

More detailed groundwater analyses are also performed from time to time, particularly on samples collected from pumping stations in the sea level aquifers. The range of parameters monitored include colour, odour, pH, free chlorine, TDS, TSS, CO_2, total alkalinity, total hardness, Ca, Mg, Na, K, Fe, Pb, Cl, NO_3, NH_3, NO_2, P_2O_5, SO_4, F, Cu, O_2, Zn as well as microbiological content.

The results from these analyses immediately indicate that the major problem to be faced when dealing with groundwater quality of this aquifer system, is the chemical disturbance arising from the almost generalised influence from the bounding saltwater. In fact, a high number of monitoring stations exhibit a relatively high chloride content. The nitrate content of the groundwater is also high, exceeding the 50 mg L^{-1} threshold in a number of monitoring stations. On the other hand, the concentrations of heavy metals in the groundwater are far below the admissible drinking water standards; and all analysis for pesticides have resulted below the established EC threshold limits.

19.4 Hydrochemical characteristics

19.4.1 Groundwater typology

As water flows through an aquifer it interacts with the chemistry of the geological strata and builds up a typical chemical composition. In coastal and island aquifer systems, which are in direct contact with seawater, variations in the chemical composition of the groundwater due to the intrusion should also be taken into consideration. In fact, through this study, in order to fully evaluate the chemical composition of the groundwater and attribute the variations of certain parameters to the sea-water intrusion and possibly also to other forms of anthropogenic pollution, it has proved necessary to investigate the interdependence between the various chemical species, particularly those characteristic of saline conditions.

The Na and Cl content of the groundwater, showed an excellent correlation and their ratio can be used as an indicator of direct or indirect seawater entry. The same conclusion could also be reached with regards Mg and Cl. Moreover, plotting the conductivity versus the chloride content, results in the expected direct correlation between these parameters; effectively demonstrating that the intruding saline water has by far the major responsibility for the high conductivity observed in the groundwater.

The correlation between SO_4 and Cl is also good. However, when analysing this correlation in more detail, a number of deviations have been identified with groundwater from certain sources exhibiting an anomalously high SO_4/Cl ratio. These results suggest that the sulphate contained in the

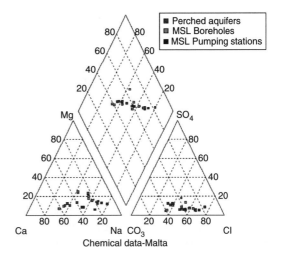

Fig. 19.5 Piper plot for the quality parameters of groundwater samples.

groundwater is not derived exclusively from the intruding seawater, but in certain places (particularly regions of the aquifer with a relatively low depth to groundwater) the influence exerted by other anthropogenic sources such as fertilisers used in agricultural activities, could be significant. Moreover, most of the monitoring stations in which a high SO_4/Cl ratio was identified, also registered high values for other chemical elements or parameters.

The typology of the groundwater in the mean sea level aquifer systems was further analysed through ternary (Piper) plots (Fig. 19.4) where, as expected, samples from these aquifer systems exhibited a general shift towards the apex characteristic of NaCl waters. Thus the Piper plot indicates that as a general rule, groundwater in the mean sea level aquifer is shifting from the fresh recharging carbonate waters position towards the saline water position primarily as a result of the mixing processes with intruding sea-water in the aquifer (Fig. 19.5).

19.4.2 Major physico-chemical characteristics

The groundwater in the mean sea level aquifer is also characterised by a high, anthropogenically derived, nitrate content which reaches levels ranging between 25 and 100 mg L^{-1}. The nitrate-containing recharge should be diluted by the intruding seawater, which has a very low nitrate content. In order to estimate the importance of any such dilution phenomena, the NO_3/Cl relationship was investigated on a source-by-source basis. No clear relationship could be established, with different monitoring points exhibiting totally different relationships. The following examples have been chosen to demonstrate the observed variations in the relationship between these parameters.

1 In a number of monitoring stations, an increase in the chloride content with time, was complemented by a decrease in the nitrate content. In such cases, the dilution of the freshwater by intruding saltwater can explain the decrease in the nitrate content.

2 In other monitoring stations, the chloride and nitrate concentrations have a similar behavior over time. This effect could be due to a variation in the area of influence of the pumping well in response to pumping rate variations. It is suggested that a smaller area of influence resulting from a decreased pumping rate would result in a smaller susceptibility for pollutants from the surface to reach the well.

3 In certain monitoring points, the evolution of the nitrate concentration was observed to be independent of the changes registered in the chloride concentration. One possible explanation for this is a dilution by intruding highly saline water; the dilution ratio however, is too low to influence the nitrate content. The salinity of the intruding water could also be variable, depending on the pumping rate and the thickness of the mixing zone.

The various scenarios illustrated above, indicate that the dilution, and consequently the intrusion of seawater appears to occur through different pathways. Thus depending on the abstraction rate, the thickness of the transition zone (located between freshwater and seawater) and the prevailing hydrogeological conditions such as fracturing and lithology in the vicinity of the particular abstraction point, the salinity of the invading water could be variable, making the evaluation of the real nitrate pollution very complex.

These results reveal the complexity of the aquifer system. However, dilution should be considered as a key parameter, since any planned improvement in the water quality from a salinity point of view could in certain zones be accompanied by a drastic increase in nitrate concentrations.

The chloride and nitrate content of pumping stations in the mean sea level aquifer system was further analysed. Box plots were prepared in accordance with the baseline definition, with the central tendency being identified by the median, the boxes indicating the range 16–84% and the whiskers indicating the range 2.3–97%. Outliers were not presented in the plots.

In Fig. 19.6 it is shown how the ranges in chloride vary over the years for all pumping stations grouped together. These box plots show that the chloride levels in the aquifer are in excess of the 250 mg L^{-1} chloride indicative value for drinking water outlined in the EU Drinking Water Directive (98/83/EC). The 2.3 percentile whisker does not fall

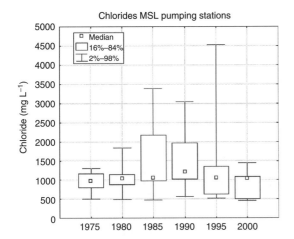

Fig. 19.6 Box plots of the Lower Coralline Limestone aquifer systems: Chloride.

below 400 mg L^{-1} in any source, or for any year.

Analogous analysis performed on the nitrate data show similarities to the results obtained in the analysis of the chloride data. Nitrate in the mean sea level aquifer systems show high levels with the 2.3 percentile whiskers exceeding 50 mg L^{-1} (as NO$_3$) nitrate for all sources except the Wied il-Ghasel pumping station. The variability in nitrate is also high, with the 97.7 percentile exceeding 125 mg L^{-1} in one case.

It is thus clear that abstraction points are not representative of this particular type of aquifer system, but only representative of the localised situation of seawater intrusion as far as chloride concentrations are concerned. The abstraction of water from such sites, in itself, modifies the quality of the water in the aquifer, and is therefore directly conflicting with the definition of baseline as far as 'excluding anthropogenic activity' is concerned.

The pumping sources, as used in this study, may be compared to control systems with feedback, where conclusions are expected to be drawn on the level of the input based on the measurement of the controlled (output) variable, as if the control mechanism did not exist (Fig. 19.6).

19.5   Geochemical controls

Further considerations on the chemical quality of groundwater in the sea level aquifer concerned an analysis of the results from a series of full chemical analysis which were performed on groundwater samples from the major pumping sources in the sea level aquifer system (WSC 2000).

These analyses indicated relatively higher fluoride values in groundwater samples from abstraction sources in the Gozo sea level aquifer, as opposed to samples from the eastern and southeastern regions of the Malta sea level aquifer, where fluoride concentration is very low. Owing to the fact that the Gozo sea level aquifers are the principal sources of potable water in this island, this fact was contributing to non-compliance with the standards set in the EU Drinking Water Directive.

It appears that the only minerals that could possibly contribute to the presence of fluoride in groundwater are carbonate mineral impurities, fluorapatite or francolite (and to a minor extent dahlite). Fluorapatite, apart from CaCO$_3$, is the most common mineral present in the phosphorite beds and the only mineral present in the Maltese rock succession that contains fluorine in substantial proportions. Table 19.5 Besides CaCO$_3$, which is the most common mineral present in Maltese limestones, other common minerals present in varying proportions are: the clay minerals: kaolinte, illite, smectite, and

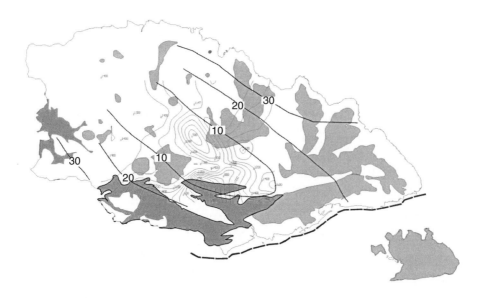

Fig. 19.7 Gozo: map showing spatial distribution of the phosphate conglomerate beds C1 and C2 and fluoride levels overlay. The dark shaded area represents the aerial distribution of the Ghajn Melel Member and the Greensand Formation extension (after Scerri 2002).

palygorskite; and glauconite. None of these minerals contains fluorine (Fig. 19.7).

The maximum concentration of francolite occurs in the phosphate conglomerate beds where analysis by XRF yielded maximum values of 19.3% P_2O_5 and 1.5% of fluoride in the outer surface of the pebbles coming from the Lower Main phosphorite conglomerate bed and 16.8% P_2O_5 and 1.4% of fluoride in the outer surface of the pebbles coming from the Upper Main phosphorite conglomerate bed.

Correspondingly, the higher levels in concentration of fluoride in the mean sea level aquifer systems lie in the area of distribution of the Upper and Lower Main Phosphorite conglomerate beds where fluoride levels measured range from 500 to 1600 µg L^{-1}; with these levels falling off to below 400 µg L^{-1} content where both conglomerate beds are absent.

Owing to the fact that the phosphorite conglomerate beds are more developed in Gozo, fluoride levels measured in the mean sea level aquifer in Gozo, are consistently high, mostly above 1000 µg L^{-1} up to a maximum of 1600 µg L^{-1}.

19.6 Conclusions

The natural baseline quality of groundwater in the Oligocene mean sea level aquifers is controlled predominantly by the lithology of the aquifer and its physical characteristics. Being underlain and surrounded by seawater, it is heavily influenced by the direct physical contact with the sea and by the processes of molecular diffusion that vary according to the hydrostatic head of the water table at a given point, whether in its natural state or as subjected to the effects of pumping. Consequently these groundwaters are highly mineralised in chloride and sodium, and other seawater related ions especially where the aquifer is exposed to dynamic conditions.

As many rock units in the Maltese lithologic column contain phosphate, albeit in smaller quantities than those found in the conglomerate beds, it was also concluded that the fluoride found in the mean sea level aquifers is a result of solution processes during slow percolation especially in the Upper and Lower Main Phosphorite conglomerate beds and hence are directly related to the vertical thickness of these interlying beds and groundwater residence times in the unsaturated zone.

It is doubtful whether the Lower Coralline Limestone of Malta can be strictly defined as a reference aquifer for purposes of baseline characterisation, given that it was impossible to correct for all anthropogenic activity. At the same time, it is true that this island aquifer constitutes a specific and relatively common type of European aquifer especially in the Mediterranean region. The study therefore indicates that intrusion of seawater related compounds is the major concern as regards water quality monetoring in this aquifer typology.

References

ATIGA Consortium (1972) Wastes disposal and water supply project in Malta. United Nations Development Programme.

Bakalowicz, M. (2002) Isotope technology for Groundwater Development. Hydrogeology of the Maltese Islands. IAEA Report MAT/8/2002.

Bakalowicz, M. and Mangion, J. (2003) The limestone aquifers of Malta. Their recharge conditions from isotope and chemical survey. IAHS Publ. no. 248, pp. 49–54.

Bosence, D.W.J., Pedley, H.M. and Rose E.P.F. (1980) *Field Guide to Mid-Tertiary Carbonate Facies of Malta*. Palentological Association, London.

Cassar, G. (2002) An analysis of chloride and nitrate in the Lower Coralline Limestone aquifer (Malta) in accordance with baseline techniques. Malta Resources Authority.

Cassar, J. (2002) Deterioration of the Globigerina Limestone of the Maltese Islands. Siegesmund, S., Weiss, T. and Vollbrecht, A. (eds). In: *Natural Stone, Weathering Phenomena Conservation Strategies and Case Studies*. Geological Society of London, Special Publication, 205, 33–49.

Chadwick, O. (1884) *Report on the Water Supply of Malta*. Government Printing Office, Malta.

Chadwick, O. (1886a) *Rainfall at Dingli and Yield of Springs*. Government Printing Office, Malta.

Chadwick, O. (1886b) *Report on the Water Supply of Malta*. Government Printing Office, Malta.

Chadwick, O. (1894) *Report on the Water Supply of Malta*. Government Printing Office, Malta.

Chadwick, O. (1896) *The Water Supply of Malta*. Government Printing office, Malta.

Chadwick, O. (1897) *Report on the Water Supply of Malta and Gozo*. Government Printing Office, Malta.

De Marsily, G. (1986) *Quantitative Hydrogeology*. Academic Press, New York.

Felix, R. (1973) Oligo-Miocene stratigraphy of Malta and Gozo. *Meded. Landbouwhogesch Wageningen* 73-20, 1–103.

Fetter, C.W. (1988) *Applied Hydrogeology*. Macmillan Publishing.

Grasso, M., Reuther, C.D. and Tortorici, L. (1992) Neotectonic Deformation in SE Sicily: The Ispica Fault, Evidence of Late Miocene-Pliocene decoupled wrenching within the

Central Mediterranean stress regime. *Journal of Geodynamics* 16, 135–46.

House, M.R., Dunham, K.C. and Wigglesworth, J.C. (1961) Geology of the Maltese islands. In: Bowen-Jones, H. Dewdney, J.C. and Fisher, W.B. (eds) *Malta: Background for Development.* University of Durham, Durham. pp. 24–33.

Hyde, H.P.T. (1955) *Geology of the Maltese Islands.* Lux Press, Malta.

Mitchell, P.K. (1958) Studies in the agrarian geography of Malta. A report to her Majesty's Colonial Economic Research Committee. University of Durham, UK, 98 pp.

Mitchell, P.K. (1963) A long term series of monthly rainfall for Malta: 1841 to 1961. *Quarterly Journal of the Royal Meteorological Society* 89 (382), 469–77.

Morris, T.O. (1952) *The Water Supply Resources of Malta.* Government of Malta, 125 pp.

Pedley, H.M., House, M.R. and Waugh, B. (1976) The geology of Malta and Gozo. *Proceedings of the Geological Association* 87, 325–41.

Pedley, H.M. (1978) A new lithostratigraphical and paleoenvironmental interpretation of the Coralline Limestone formations (Miocene) of the Maltese islands. *Overseas Geological and Mineral Resources* 54, 18.

Pedley, H.M. and Bennet, S.M. (1985) Phosphorites, hardgrounds and synedepositional subsidence: A palaeoenvironmental model from the Miocene of the Maltese Islands. *Sedimentary Geology.* 45, 1–34.

Report on the Workings of Government Departments (1977) Government Printing Office, Malta.

Scerri, S. (2002) On the correlation of fluoride levels in the Mean Sea Level Aquifer with the spatial occurrence of phosphorite conglomerate beds found in the Globigerina Limestone formation. Malta Resources Authority.

Schinas, G.C. and Chadwick, O. (1888) *Report on the Rainfall and Yield of Springs for the Season 1886–1887.* Government Printing Office, Malta.

Sutherland, J. (1867) *Sanitary Conditions of Malta and Gozo.* Government Printing Office, Malta.

Water Services Coporation (2001) Annual Report 2000/01. Water Services Corporation, Malta.

20 The Natural Inorganic Chemical Quality of Crystalline Bedrock Groundwaters of Norway

B. FRENGSTAD AND D. BANKS

One thousand, six hundred and four groundwater samples from crystalline bedrock aquifers in Norway have been analysed for pH, alkalinity, radon, major ions, and metals. A subset of 476 samples was further analysed for a range of trace elements by ICP-MS. For most major ion parameters, including pH, the median values display modest dependence on lithology, while there is a large variation in hydrochemistry within each lithological subset. Three main water types can be distinguished: Na–Cl waters, Ca–HCO_3^- waters and Na–HCO_3^- waters, of which the latter group comprises 25% of the samples. Many trace elements show a clear dependence on lithology with enrichments in aquifers of Precambrian granites relative to Precambrian anorthosites. Concentrations of fluoride, radon and uranium are above given drinking water limits for a significant part of the samples, while elevated concentrations of, for example, arsenic and nitrate seem to be rare. A comparison of two granitic aquifers in southern Britain and Norway, respectively, illustrates the importance of glacial and postglacial weathering history for the hydrochemistry.

PHREEQC modelling reveals that a common hydrochemical evolutionary pathway exists for most silicate crystalline rock aquifers, independent of lithology. The main controlling factors are the initial PCO_2 of groundwater recharge, the plagioclase composition of the host rock, the extent to which initial CO_2 has been converted to HCO_3^- by hydrolysis, whether the system is 'open' with regards to CO_2, and to which extent plagioclase hydrolysis has progressed after calcite saturation has been reached.

20.1 Geological and hydrogeological setting of Norway

Norway, with an area of 324,000 km^2, extends 1800 km from Skagerrak to the Russian border on the coast of the Barents Sea (latitudes 58°–71°N) and occupies the western margin of the Precambrian Fennoscandian shield. It straddles the NE–SW trending Caledonian orogenic belt, while the region around Oslofjord (in the southeast) has been impacted by

Permo-Carboniferous extensional rifting and volcanism. The entire country is thus underlain by 'hard rock' or 'crystalline rock' aquifers, which are igneous, metasedimentary or metamorphic in nature. These may be overlain by relatively thin Drift deposits, which can be of several types:

• In upland areas: thin deposits of moraine (often till) and peat.

• In valley bottoms and lower-lying areas: glacio-fluvial sands and gravels and alluvial sediments.

• In coastal areas: marine silts and clays, deposited prior to the postglacial emergence of the landscape from the ocean as a result of isostatic rebound.

Topography is generally steep and varied, with mountain plains of elevation around 1000 m asl (maximum elevation Galdhøpiggen 2469 m asl). Most of Norway's population (4.5 million) live in low-lying areas, usually in the vicinity of the coast, on delta surfaces where watercourses meet lakes/fjords, or in steep valleys incised into the mountain massifs.

Precipitation varies considerably, with Kirkenes, in the far northeast, experiencing 450 mm a^{-1}, with an annual average temperature of –0.2°C. Oslo, in the southeast, experiences 763 mm a^{-1} (annual average temperature 5.7°C), while Bergen, on the west coast, is drenched by 2250 mm a^{-1} (average temperature 7.6°C) at the Florida weather station (www.dnmi.no; all statistics for period 1961–1990).

Due to geologically recent glaciation (terminating c.8000–10,000 years BP), the bedrock typically presents a glacially scoured, relatively unweathered surface to the atmosphere, with native mineral assemblages largely intact.

Arable land is sparse in rural areas and farms are widely spaced. Individual drilled wells in crystalline bedrock are a popular water-supply solution in such settings. It is estimated that around 4000 such wells are drilled every year, with a total of at least 140,000 in existence, supplying around 300,000 people. Such boreholes typically yield less than 1000 L h^{-1} (median yield = 600 L h^{-1} and median depth 56 m, based on a database of 12,757 boreholes: Morland 1997; Banks et al. 2005).

In many respects, the hydrogeology and hydrogeochemistry of Norwegian hard rock aquifers is similar to neighbouring Finland and Sweden. Groundwater quality problems with Rn, F$^-$ and U are common (Lahermo et al. 1990; Aastrup et al. 1995) and, intriguingly, the Swedish hard rock well database also calculates a median yield of 600 L h^{-1} (Gustafson 2002). Some differences exist between Norway and its neighbours, however:

• Norway has generally steeper topography and hence generally lower groundwater residence times.

• Finland and Sweden are dominated to a greater extent by Precambrian Fenno-scandian shield rocks.

• Drift deposits are generally thicker and more important for water supply in Finland and Sweden.

• Specific geochemical anomalies may result in slightly different groundwater chemical distributions than in Norway. For example, a greater prevalence of arsenic in Finnish/ Swedish hard rock groundwater (Valve et al. 2005) is related to regional geochemical anomalies (Reimann et al. 2003; Backman et al. 2006).

20.2 Part I: Investigation of groundwater chemistry in Norwegian crystalline bedrock aquifers

During 1996 and 1997 the groundwater chemistry in Norwegian crystalline bedrock aquifers was investigated in a joint project between the Norwegian Radiation Protection Authority (NRPA) and the Geological Survey of Norway (NGU) and published by Banks et al. (1998a–c), Frengstad et al. (2000, 2001) and Frengstad (2002). Based on an open invitation to well owners to have their drinking water analysed, almost 2000 water samples were collected by local health authority representatives. In the field, two sample flasks were typically filled at the wellhead or from a kitchen tap via a closed pressure tank. First, a 10 mL water sample was injected (in the field) into a 20 mL glass vial, pre-filled with 10 mL Lumagel scintillation liquid. This was returned to the laboratory of NRPA after minimal delay and analysed for radon by scintillation counting.

Second, a 500 mL polyethene flask was filled with groundwater and returned to NGU's laboratory. Although filtration at a mesh size of 0.45 μm is typically recommended for analysis of groundwater samples (especially metals/cations for research purposes), in this case, a conscious decision was made not to filter the water samples. This was decided in order to simplify sampling procedures for the local health authorities and also because previous studies had indicated that most hard rock groundwaters are naturally relatively particle-free, with filtration having little impact on analysis results (Reimann et al. 1999; Frengstad 2002). On arrival of the flasks at NGU, small subsamples of water were decanted for analysis at

NGU for seven anions by ion chromatography (IC), pH, and alkalinity by titration to a bicarbonate end-point. The remaining sample was then acidified in-flask with concentrated ultrapure nitric acid to remobilise any precipitated or sorbed heavy metals and analysed for around 30 elements by inductively coupled plasma atomic emission spectrometry (ICP-AES).

Following analysis of all samples, the analytical dataset was quality controlled to remove (1) all samples which were not **bona fide** crystalline bedrock groundwaters, (2) all samples that may have been subject to water treatment, and (3) all samples with an unacceptably high particulate content. This resulted in a final dataset of 1604 analyses, henceforth referred to as 'rock_corr'.

Thereafter, the locations of all samples with valid UTM grid references were imported to a GIS environment and compared with the 1:3,000,000 scale digital bedrock map of Norway (Sigmond 1992). A predominant bedrock lithology code was assigned to each sample.

Finally, in order to ensure a more geographically and lithologically representative dataset, 476 samples were selected from 'rock_corr', as evenly distributed according to spatial location and lithological code as possible. This subset of data is referred to as 'rock_rep' (n = 476). The (acidified) flasks corresponding to these samples were sent to the Federal Institute of Geosciences and Natural Resources in Hannover (BGR) for analysis of 70 elements by inductively coupled plasma mass spectrometry (ICP-MS). Materials, methods and data quality control are documented by Banks et al. (1998a, c), Frengstad et al. (2000) and Frengstad (2002).

20.2.1 The distribution of hydrochemical parameters in crystalline bedrock

Banks et al. (2005) have argued that the distribution of groundwater within crystalline bedrock aquifers is so complex and heterogeneous, that any deterministic attempt to predict or spatially characterise hydraulic or hydrochemical parameters is doomed to failure. Rather, NGU has chosen to characterise both hydraulic parameters (Morland 1997) and hydrogeochemistry through the use of non-parametric statistics. Tables 20.1 and 20.2 document the distribution of a large variety of hydrochemical parameters in terms of median and other selected percentile values for the 'rock_corr' and 'rock_rep' datasets.

For many hydrogeochemical parameters, including most major ion parameters (such

Table 20.1 Characteristic percentile values for hydrochemical parameters determined on dataset 'rock_corr' (N = 1601-4) for crystalline bedrock groundwaters in Norway.

Parameter	25%	50%	75%	97.7%	Subset 92 50%	Subset 93 50%
pH (lab)	7.81	8.07	8.21	8.88	7.90	8.07
Alkalinity (meq L^{-1})	1.26	2.04	2.81	4.81	1.87	1.26
Bromide (mg L^{-1})	<0.1	<0.1	<0.1	0.63	<0.1	<0.1
Chloride (mg L^{-1})	5.06	9.78	18.8	124	21.8	21.6
Fluoride (mg L^{-1})	0.08	0.28	0.96	3.38	1.69	<0.05
Nitrate (mg L^{-1})	<0.05	0.46	2.52	27.4	0.23	4.65
Sulphate (mg L^{-1})	8.12	13.1	22.6	107	15.9	10.1
Radon (Bq L^{-1})	25	86	260	2634	698	<10
ICP-AES						
Aluminium (mg L^{-1})	<0.02	<0.02	0.039	0.303	0.029	<0.02
Barium (mg L^{-1})	0.004	0.013	0.035	0.17	0.012	0.003
Boron (mg L^{-1})	<0.010	0.014	0.038	0.25	0.032	0.018
Calcium (mg L^{-1})	14.2	27.2	38.8	72.2	15.6	17.3
Copper (mg L^{-1})	<0.005	0.015	0.045	0.23	0.014	0.011
Iron (mg L^{-1})	0.010	0.031	0.10	1.56	0.14	0.011
Magnesium (mg L^{-1})	1.84	3.36	5.69	16.1	2.61	3.79
Potassium (mg L^{-1})	1.33	2.36	3.87	9.12	2.59	0.91
Sodium (mg L^{-1})	6.12	12.5	34.9	175	31.1	23.1
Lithogenic sodium (mg L^{-1})	1.87	6.00	23.2	103	15.1	8.73
Silicon (mg L^{-1})	3.81	4.77	6.05	9.96	6.09	5.81
Strontium (mg L^{-1})	0.064	0.13	0.27	1.95	0.075	0.095
Zinc (mg L^{-1})	0.004	0.012	0.043	0.39	0.031	0.012

The last two columns show median concentrations for two specific lithological subsets: subset 92 = autochthonous Precambrian granite to tonalite (N = 76); subset 93 = autochthonous Precambrian charnockite to anorthosite (N = 34). Lithogenic sodium [Na*] is calculated on the basis of chloride [Cl$^-$] concentration using the formula [Na*] = [Na] − 10.6[Cl$^-$]/19.

as pH, alkalinity, calcium, magnesium, non-marine sodium) there was surprisingly little dependence of the median concentration on lithology (Frengstad and Banks 2007). Indeed, for almost all lithological subsets, the median pH was around 8.0–8.2 (Fig. 20.1). Within each lithological subset, however, there was considerable variation in water chemistry. It is considered that this reflects the evolution of water chemistry along common evolutionary pathways (e.g.

plagioclase hydrolysis) in most lithologies. The variation within lithological subsets merely reflects different stages of evolution, ultimately depending on residence times, kinetic factors, fracture geometry, and so on.

A number of parameters did display a clear lithological dependence, however. In particular, many parameters displayed elevated concentrations in Precambrian granites (lithology code 92) and low concentrations in Precambrian anorthosites (lithology code 93);

Table 20.2 Characteristic percentile values for hydrochemical parameters determined by ICP-MS on dataset 'rock_rep' (*N* = 476) for crystalline bedrock groundwaters in Norway.

Parameter	25%	50%	75%	97.7%	Subset 92 50%	Subset 93 50%
Antimony (µg L⁻¹)	0.015	0.033	0.081	0.45	0.032	0.019
Arsenic (µg L⁻¹)	0.08	0.18	0.52	5.9	0.09	0.08
Beryllium (µg L⁻¹)	<0.005	0.012	0.038	0.39	0.096	<0.005
Cadmium (µg L⁻¹)	0.009	0.017	0.033	0.53	0.028	0.010
Cerium (µg L⁻¹)	0.028	0.11	0.46	7.2	0.36	0.012
Cobalt (µg L⁻¹)	0.027	0.065	0.140	1.4	0.067	0.072
Chromium (µg L⁻¹)	0.08	0.14	0.27	1.2	0.23	0.16
Copper (µg L⁻¹)	6.8	16	47	213	26	14
Mercury (µg L⁻¹)	0.006	0.018	0.029	0.066	0.023	0.007
Iodine (µg L⁻¹)	0.29	0.60	1.2	6.6	0.85	0.63
Lanthanum (µg L⁻¹)	0.027	0.10	0.43	5.7	0.45	0.010
Lead (µg L⁻¹)	0.17	0.36	0.76	5.4	0.54	0.34
Lithium (µg L⁻¹)	1.1	2.9	4.7	18	5.0	0.44
Molybdenum (µg L⁻¹)	0.39	1.4	4.7	33	2.9	0.076
Nickel (µg L⁻¹)	0.32	0.53	0.99	12	0.52	0.67
Selenium (µg L⁻¹)	0.11	0.20	0.36	1.7	0.14	0.31
Tin (µg L⁻¹)	0.002	0.008	0.019	0.18	0.012	0.014
Thorium (µg L⁻¹)	0.002	0.006	0.023	0.40	0.025	<0.001
Titanium (µg L⁻¹)	0.29	0.59	1.6	25	0.95	0.36
Thallium (µg L⁻¹)	0.004	0.007	0.013	0.051	0.005	0.001
Tungsten (µg L⁻¹)	0.017	0.071	0.41	14	0.054	0.005
Uranium (µg L⁻¹)	0.42	2.5	11	105	8.0	0.037
Vanadium (µg L⁻¹)	0.086	0.24	0.57	2.7	0.20	0.37

The last two columns show median concentrations for two specific lithological subsets: subset 92 = autochthonous Precambrian granite to tonalite (*N* = 29); subset 93 = autochthonous Precambrian charnockite to anorthosite (*N* = 30).

for example, Be, Cd, Ce, F⁻, Hf, La, Nb, Pb, Rn, Ta, Th, Tl, U, Y, Zr and rare earth elements. Clearly, several of these (Be, F⁻, Rn, Th, Tl, U) have a potential human health impact.

Radon is an odourless, colourless, radioactive (half-life 3.9 days), rather soluble, noble gas. It occurs in excess of the NRPA guideline value (500 BqL⁻¹) in 13.9% of sampled Norwegian crystalline rock groundwaters (and >50% of all groundwaters from lithology 92: Precambrian granites) (Table 20.1 and Fig. 20.2). It has the potential to enter household domestic water systems, via the closed pressure tanks typical in Norway, and de-gas to indoor air during the use of showers, dishwashers and washing machines. The highly insulated nature of Norwegian homes means that ventilation is often poor and the gas disperses only slowly. Exposure to radon via inhalation is known to correlate with lung cancer. A risk pathway via ingestion in potable water has also been postulated. Radon is particularly prevalent in groundwater from granites and gneisses and concentrations as high as 31,900 Bq L⁻¹ have been recorded in the Precambrian Iddefjord Granite of southeast Norway (Banks et al. 2005).

Radon is ultimately derived from the radioactive decay of uranium (and then radium) in the host aquifer matrix (or fracture mineral coating). It exhibits some degree of correlation with dissolved uranium in crystalline rock groundwater (Fig. 20.2). Uranium can be very soluble and mobile under oxidising conditions. It is regarded as being more chemotoxic than radiotoxic, affecting a variety of organs in the human body. There is no current EU limit for U in drinking water although WHO (2004) suggests a guideline of 15 μgL⁻¹. Canada operates a limit of 20 μgL⁻¹ for U, and the USA 30 μgL⁻¹. Some 21% of groundwaters from 'rock_rep'

exceed the WHO guideline and around 12% exceed the US 30 μg L⁻¹ threshold. As with Rn, granite aquifers are particularly prone to excessive uranium concentrations in groundwater. Uranium concentrations range over six orders of magnitude in Norwegian crystalline rock groundwaters (Fig. 20.2), approaching the mg L⁻¹ level. Intriguingly, the frequency distribution is almost perfectly log-normal (Fig. 20.2).

Fluoride, in small amounts, is important to humans in promoting resistance to tooth decay. In excess it can, however, cause mottling or malformation of teeth in children, or even malformation of the skeleton in extreme cases (dental and skeletal fluorosis). Concentrations in excess of 1.5 mg L⁻¹ are typically deemed undesirable in most European drinking water schedules. Of the sampled Norwegian hard rock groundwaters, 16.1% exceeded 1.5 mg L⁻¹, with the most problematic lithologies being granites and gneisses. Fluoride may be derived from dissolution of fluorite fracture mineralisation, but a correlation with pH suggests it **may** be related to anion exchange on amphiboles, micas or trace apatite. Cases of dental fluorosis in children have been documented in southern and western Norway, related to consumption of fluoride-rich groundwater (Bjorvatn et al. 1992; Bårdsen et al. 1999).

Arsenic concentrations of several tens or hundreds of μg L⁻¹ have been recorded in crystalline bedrock groundwaters in certain localities in Finland (Valve et al. 2005; Backman et al. 2006) and in Sweden (Berglund et al. 2005). In Norway, the 1997–1998 survey failed to reveal a substantial arsenic problem, with only 1.5% of wells in crystalline bedrock exceeding the drinking water limit of 10 μg L⁻¹. No sample exceeded 20 μg L⁻¹ in the investigation (Fig. 20.3).

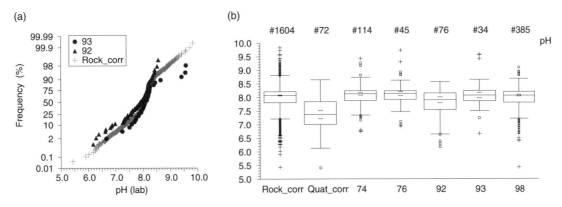

Fig. 20.1 The distribution of laboratory-determined pH in 1604 samples of Norwegian crystalline bedrock groundwater; (a) cumulative frequency curves for the entire data set 'rock_corr' (N = 1604) and for lithological subsets 92 and 93; (b) box plots showing pH distributions in 'rock_corr' and lithological subsets 74 (Cambro-Silurian meta-sediments of the Caledonian belt and Oslo rift), 76 (Cambro-Silurian greenstone, greenschist, amphibolite, and meta-andesite), 92 (autochthonous Precambrian granite to tonalite), 93 (autochthonous Precambrian charnockite to anorthosite) and 98 (Precambrian gneiss, migmatite, foliated granite, and amphibolite). These are compared with a dataset 'quat_corr' for 72 groundwaters in superficial Quaternary sedimentary aquifers (Banks et al. 1998b). After Frengstad and Banks (2007). For each plot, the 'box' represents the central 50% of the dataset, with a horizontal line at the median value (and horizontal parentheses showing the 95% uncertainty level in the median). The whiskers show the extra-quartile data, with outlying data shown as individual crosses.

The elevated As values are not clearly associated with any particular lithology, nor is there any clear correlation with iron concentrations. There is some correlation with pH, however, median arsenic concentration increases with increasing pH (Frengstad et al. 2001).

It is suspected that the chemical quality of the sampled groundwaters largely reflects natural hydrochemical processes. One might query whether there is any anthropogenic influence on the analysed quality; for example, nitrate from agricultural practice, copper and zinc from wellhead plumbing. Nitrate concentrations are generally very low in Norwegian crystalline bedrock groundwaters, with less than 0.7% of the sampled waters

exceeding 50 mgL^{-1} as NO$_3^-$ (11 mgL^{-1} as N). The median value is 0.5 mgL^{-1} as NO$_3^-$, which is lower than both Sweden and Finland (Frengstad 2002). The low nitrate concentrations are probably related to both a paucity of agriculture in Norway and the fact that fertilisers and manure are typically applied to land **after** the main spring snowmelt recharge episode, but well before the recharge event related to autumn rainfall. As regards the possible impact of wellhead 'plumbing', this cannot be excluded completely (although plastic pipework is increasingly prevalent in Norway), but the modest median concentrations of copper (15 µg L^{-1}) and zinc (12 µg L^{-1}) suggest that any influence is low.

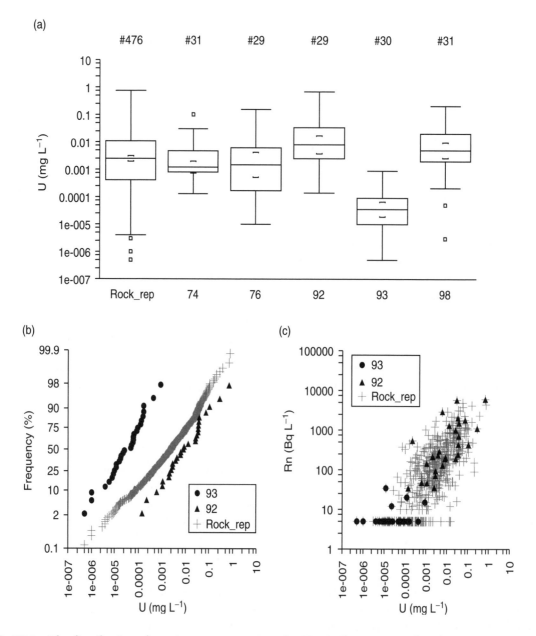

Fig. 20.2 The distribution of uranium concentrations (by ICP-MS) in 476 samples of Norwegian crystalline bedrock groundwater: (a) box plots showing U distributions for 'rock_rep' and selected lithological subsets (see Fig. 20.1 for explanation); (b) cumulative frequency curves for the entire dataset 'rock_rep' ($N = 476$) and for lithological subsets 92 and 93; (c) correlation between U and Rn in the 476 groundwaters of 'rock_rep'.

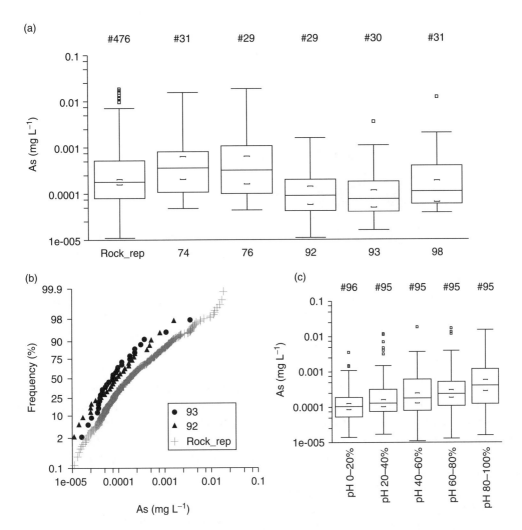

Fig. 20.3 The distribution of arsenic concentrations (by ICP-MS) in 476 samples of Norwegian crystalline bedrock groundwater: (a) box plots showing As distributions for 'rock_rep' and selected lithological subsets (see Fig. 20.1 for explanation); (b) cumulative frequency curves for the entire dataset 'rock_rep' ($N = 476$) and for lithological subsets 92 and 93; (c) box plots each representing 20% of the 'rock_rep' data-set, ranked according to increasing pH, illustrating the dependence of concentrations of arsenic on pH. The cohort pH 0–20% spans pH 6.17–7.73; pH 20–40% spans 7.74–8.01; pH 40–60% spans 8.01–8.13; pH 60–80% spans 8.14–8.22; pH 80–100% spans 8.22–9.58.

20.2.2 Groundwater types

The relative distribution of major ions (Ca, Na, Mg, SO_4^{2-}, Cl^- and HCO_3^-) in the 1604 water samples is presented as a Durov plot in Fig. 20.4, after conversion from mg L^{-1} to meq L^{-1}. The groundwater samples basically fall into three main categories (Banks et al. 1998a):

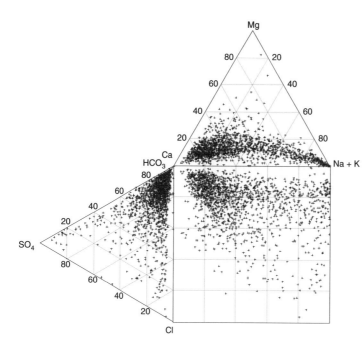

Fig. 20.4 Durov plot illustrating the major ion composition of the 1604 crystalline bedrock groundwaters comprising 'rock_corr'. Concentrations are converted to meq L^{-1} prior to plotting.

1 Na–Cl$^-$ waters (4% of the samples, median pH 8.03). These are either immature groundwaters influenced by the sea salt content of precipitation or coastal groundwaters influenced by seawater intrusion or pore waters from overlying marine clays.

2 Ca–HCO$_3^-$ waters (65% of the samples, median pH 8.03). These water samples are presumed to be dominated by weathering of calcite from the rock matrix or fracture mineralisation, or by weathering of calcic silicate rocks.

3 Na–HCO$_3^-$ waters (25% of the samples, median pH 8.26). These waters are characterised by weathering of silicate minerals with a high sodium content, or by a high degree of chemical maturity where saturation with respect to calcite has limited the accumulation of calcium (or even removed calcium, via calcite precipitation). Note the higher median pH of these waters.

The remaining 6% of the water samples had various compositions including 12 samples of Mg–HCO$_3^-$ composition and 12 samples where NO$_3^-$ was the dominant anion. A full breakdown of water types is given in Banks et al. (1998a, c).

20.2.3 The dependence of solute concentrations on borehole depth

Information on borehole depth existed for 1487 of the 1604 samples in 'rock_corr'. The sampled wells and boreholes exhibited a median depth of exactly 80 m. Only five boreholes fell within the 180–230 m interval while the shallowest well recorded was only 2 m deep. Norwegian boreholes are typically unlined in crystalline bedrock, with only shallow near-surface casing. In order to evaluate any relationship between borehole depth and borehole water chemistry, boreholes were ranked according to increasing depth and grouped into subsets of 20 m intervals. Ranked box plots were then used to exhibit the distribution of solute concentrations for the different depth subgroups (Frengstad and Banks 2003, Fig. 20.5).

There was a significant positive correlation between median groundwater pH and borehole depth. Similar correlations were also observed for Na and F⁻. It is noteworthy that pH, Na and F⁻ are generally parameters that progressively increase with increasing degree of water–rock interaction. In the case of alkalinity, an initial increase in median values with borehole depth stopped at 60 m.

Ca, Mg and Si displayed generally lower median values in the 0–19 m subset, indicating immature groundwaters with short residence time in the shallowest boreholes. Median NO_3^- concentrations decrease with increasing borehole depth, due to better protection against contamination and a probable tendency towards older groundwater and more reducing conditions in deeper boreholes. Cl^- and SO_4^{2-} concentrations appeared uncorrelated with borehole depth.

Correlations of solute concentrations with borehole depth do not, of course, **necessarily** indicate a correlation with depth of occurrence of groundwater, as the water abstracted from deep unlined boreholes will be a mixture of water from a limited number of transmissive fractures at **varying** depths. Water from a very deep borehole **may** be derived from near-surface fractures, but the water from a shallow borehole **must** be derived from shallow fractures. On average, therefore, there is likely to be a larger component of deeper groundwater in deeper boreholes. Thus a significant correlation of a solute's concentration with borehole depth

is strongly suggestive of a correlation with depth of occurrence (and hence residence time and degree of water–rock interaction) of the groundwater.

It is noted that rock fractures have a tendency to become less permeable with depth as the stress due to the rock overburden increases (Carlsson and Olsson 1978). If the amount of water required by a small house or farm is modest, borehole drilling may thus have a tendency to be terminated at shallower depths once the required water 'strike' has been achieved. On the contrary, one might expect to find deeper boreholes in poorly permeable locations where ample water 'strikes' were not achieved early in the drilling process. Indeed, Rönkä (1993) showed that the mean yield of Finnish bedrock boreholes of depth 81–100 m was less than half the mean yield of boreholes less than 40 m deep. Thus one might expect to find hydrochemically 'immature' groundwaters, with short residence times, in shallow wells fed by high-transmissivity fractures.

20.2.4 The importance of glacial and postglacial weathering history

In contrast to the Norwegian dataset, granites elsewhere may not exhibit the same characteristics. For example, the hydrochemistry of granite aquifers in southwestern Britain was radically different from that in Norway. As an example, Banks et al. (1998d) compared 10 groundwater samples derived from the Permo-Carboniferous granites of the Scilly Isles (off southwestern England) with 11 samples from the Precambrian Iddefjord granite of the Hvaler Islands in outer Oslo Fjord of Norway. Both lithologies are peraluminous S-type granites, which have been

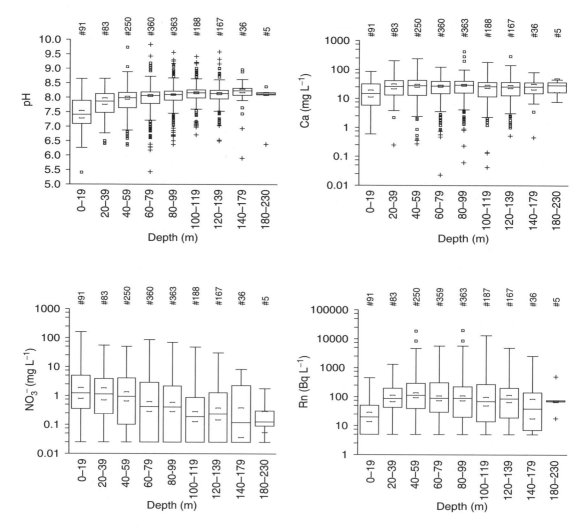

Fig. 20.5 Box plots showing the relationship of concentrations of selected hydrochemical variables (laboratory pH, calcium, nitrate [as NO_3^-] and Rn) against depth. The boxes represent cohorts of the dataset, sorted according to increasing depth.

enriched in U and Th, but the Quaternary history is diverging. The maximum extent of glaciation in the UK did not reach the Scilly Isles and the islands have thus been subject to prolonged subaerial weathering (Banks et al. 1998d). The Hvaler Islands, on the other hand, have experienced repeated glaciations and erosion during the Pleistocene and have emerged from the sea during the past 10,000 years due to isostatic rebound of the landmasses. The Scilly Isles thus present a significantly weathered geochemical profile to the atmosphere today; one that is likely to have been depleted in basic minerals such

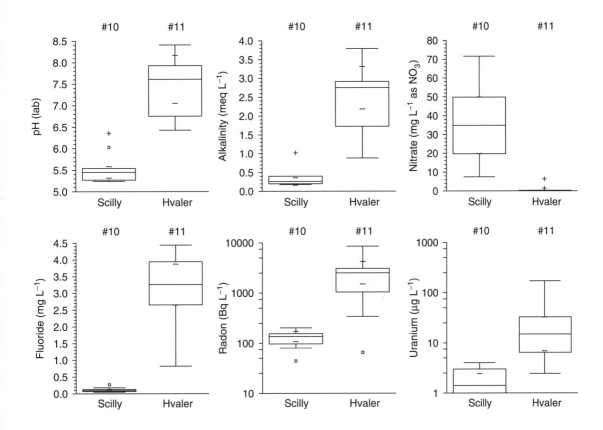

Fig. 20.6 Box plots comparing the distributions of pH, alkalinity, nitrate, fluoride, radon and uranium in groundwater from granite aquifers on the Isles of Scilly (UK) and the Hvaler Islands (Norway). Modified after Banks et al. (1998d).

as calcite (White et al. 1999, 2005) and mafic silicates and where plagioclase may already have been substantially hydrolysed along pathways of groundwater circulation. Hvaler presents a glacially scoured, geochemically 'intact' rock surface to the atmosphere and minerals such as fluorite and calcite can be observed at surface exposures of fracture planes. As a consequence, groundwaters from boreholes on the Scilly Isles exhibit low pH values and alkalinities, and low concentrations of most lithogenic solutes, including

F⁻, Rn and U, as shown in Fig. 20.6. Hvaler exhibits high-pH values and high concentrations of these solutes (in fact, the Iddefjord Granite is a member of lithology 92 in Figs. 20.1–20.3). It should be noted that typically slightly shallower borehole depths, higher hydraulic conductivities and steeper hydraulic gradients on Scilly may also provide a partial explanation for the hydrochemical differences, but these factors, too, are probably related to the recent weathering history.

One should also note the elevated concentrations of nitrate and potassium in Scilly, compared with Hvaler. Scilly enjoys a long history of floriculture, and prides itself on delivering early cut flowers to the British market. It thus has a winter agricultural activity and fertilisers are presumably applied during the main autumn–winter recharge season. Moreover, the weathered granite soils are often granular and sandy, with a strong potential for leaching of nitrate.

20.3 Part II: Groundwater chemical evolution in silicate crystalline rock aquifers

Over three decades ago, pioneering hydrogeochemists such as Garrels (1967) and Garrels and Mackenzie (1967) studied the chemistry of spring waters from crystalline rocks of the Sierra Nevada of USA, and elsewhere, and drew some surprisingly bold and simple conclusions. These conclusions (slightly modified by the findings of more recent researchers) can be summarised by the following statements:

1 If calcite is present in overburden or as a trace mineral in the crystalline rock aquifer, it will react rapidly to initially dominate groundwater chemistry, providing a $Ca-HCO_3^-$ chemical facies (White et al. 1999, 2005). However, its modal abundance is usually small and it is depleted quickly (in a geological perspective):

$$CaCO_3 + CO_2 + H_2O = Ca^{2+} + 2HCO_3^- \quad (20.1)$$

2 Plagioclase and mafic silicates are the next most rapidly reacting hydrolysable basic minerals in most crystalline rocks, reacting significantly faster than, say,

K-feldspar (White et al. 1998). As plagioclase dominates mafic silicates in many common metamorphic and igneous lithologies, groundwater chemistry is often dominated by plagioclase hydrolysis:

$$2NaCaAl_3Si_5O_{16} + 6CO_2 + 9H_2O = \\ 2Ca^{2+} + 2Na^+ + 6HCO_3^- + 4SiO_2 + \\ 3Al_2Si_2O_5(OH)_4 \downarrow \quad (20.2)$$

Sodium, calcium and bicarbonate alkalinity are released and pH is elevated;

3 In temperate climates, the typical initial alteration product of plagioclase weathering is kaolinite (as shown in equation 20.2). Garrels (1967) argued that Ca-smectite becomes more important as a secondary aluminosilicate with increasing hydrogeochemical maturity of the water;

4 In groundwater systems of low-to-moderate hydrogeochemical maturity, plagioclase hydrolysis is at the expense of CO_2. Thus, the chemical composition of a given groundwater in a crystalline rock aquifer is largely determined by

• the initial PCO_2 of groundwater recharge;
• the composition of the hydrolysed plagioclase (Ca- or Na-dominated);
• the extent to which the initial CO_2 content has been consumed and converted to HCO_3^- by hydrolysis.

In other words, the above statements predict that groundwater chemical evolution in silicate crystalline bedrock will follow a similar pathway, irrespective of lithology (in most cases). Variations in chemistry **within** lithologies will reflect differences in groundwater 'maturity' (i.e. residence time, fracture geometry, etc.). These statements thus coincide exactly with the empirical findings of the 1996–1997 Norwegian study (see above). Moreover, Banks and Frengstad (2006) quantitatively examined the Norwegian data and found it to be compatible with the conceptual model of Garrels and co-workers.

The groundwaters examined by Garrels and his colleagues were typically of a low degree of hydrogeochemical 'maturity' and had not reached a point of calcite saturation. The groundwaters of the Norwegian study contained a large subset of waters of high pH (pH 8–10), typically of sodium bicarbonate-dominated composition, and apparently depleted in calcium. These waters were not only found in granitic rocks, but also in Ca-dominated lithologies, such as the anorthosites of subset 93. In fact, a transition from the Ca–Na–HCO_3^--dominated waters (predicted by Garrels and colleagues on the basis of plagioclase hydrolysis, equation 20.2) to alkaline, sodium-dominated waters typically occurred around pH 8.0–8.2 (Fig. 20.7). Banks and Frengstad (2006) determined, via speciation modelling, that this transition also corresponded to a point where the water achieved saturation with respect to calcite (equation 20.3).

$$4NaCaAl_3Si_5O_{16} + 8CO_2 + 14H_2O = 4CaCO_3 + 4Na^+ + 4HCO_3^- + 8SiO_2 + 6Al_2Si_2O_5(OH)_4 \downarrow \quad (20.3)$$

Calcite precipitation provides a pH-buffering mechanism (equation 20.4), which probably explains the 'kink' in the cumulative frequency curve for pH (Fig. 20.1) and the apparent high frequency of pH values in the interval pH 8.0–8.3.

$$Ca^{++} + HCO_3^- = CaCO_3 \downarrow + H^+ \quad (20.4)$$

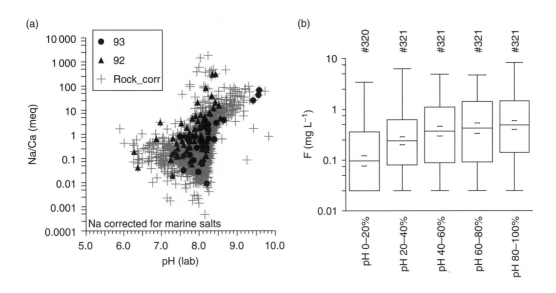

Fig. 20.7 (a) X–Y plot illustrating the dependence of the meq L^{-1} ratio of lithological sodium to calcium (Na/Ca) on pH for the entire dataset 'rock_corr' and for lithological subsets 92 and 93; (b) box plots each representing 20% of the 'rock_corr' data-set, ranked according to increasing pH, illustrating the dependence of concentrations of fluoride on pH. The cohort pH 0–20% spans pH 5.43–7.68; pH 20–40% spans 7.68–7.99; pH 40–60% spans 7.99–8.14; pH 60–80% spans 8.14–8.24; pH 80–100% spans 8.24–9.83.

Saturation and precipitation of calcite results not only in a pH-buffering effect, but also in the effective removal of Ca from the dissolved phase (equation 20.3), and thus in a progressively sodium-dominated water (if plagioclase is the predominant hydrolysable mineral).

Modelling by PHREEQC (Parkhurst 1995; Parkhurst and Appelo 1999) by Banks and Frengstad (2006) has demonstrated that plagioclase hydrolysis, coupled with: saturation and precipitation, in a system that is 'open' with respect to CO_2, will result in a circumneutral or slightly alkaline groundwater, where calcite precipitation may eventually result in a sodium-bicarbonate hydrochemistry (Fig. 20.8).

The empirical dataset for Norwegian crystalline bedrock aquifers reveals, however, that there are a number of sodium-dominated groundwaters whose pH is too high to be explained by the 'open CO_2' model illustrated in Fig. 20.8. In order to achieve the highest observed pH values, it is necessary to invoke some degree of system closure relative to CO_2. Banks and Frengstad (2006) also modelled this scenario, for differing plagioclase and initial CO_2 compositions, and found that, once available CO_2 had been effectively consumed, plagioclase hydrolysis could be envisaged as occurring through the de-protonation of water. This would drive the pH to very high values and result in the highly alkaline, Na-rich groundwaters observed in the dataset (Fig. 20.9). The simulation even predicts a buffering effect around pH 8.0–8.2 during the evolutionary pathway. In order to achieve the observed solute concentrations (Na, Ca and alkalinity) in real groundwaters, it was found necessary to invoke high initial PCO_2 values of around $10^{-1.5}$ atm. This is not atypical of continental European soil PCO_2 concentrations, but was

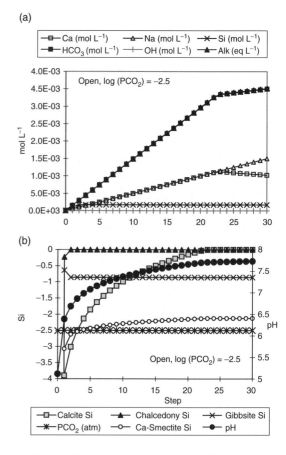

Fig. 20.8 The simulated evolution of water reacting with 1.5×10^{-3} moles of labradorite plagioclase feldspar (defined as $NaCaAl_3Si_5O_{16}$) at 7°C, in an open CO_2 system, with $PCO_2 = 10^{-2.5}$ atm, using the code PHREEQC, in 30 steps. Precipitation of kaolinite, gibbsite, calcite, chalcedony are permitted when saturated. Diagram (a) shows the concentrations of solutes, (b) the evolution of mineral saturation indices and pH. Alkalinity is calculated as the meq L^{-1} sum of bicarbonate, carbonate and hydroxide. Modified after Banks and Frengstad (2006).

initially believed to be unrealistic for poorly vegetated, postglacial Norwegian catchments. Nevertheless, literature studies and empirical field measurements (Banks and Frengstad 2006) did confirm that soil PCO_2

(a)

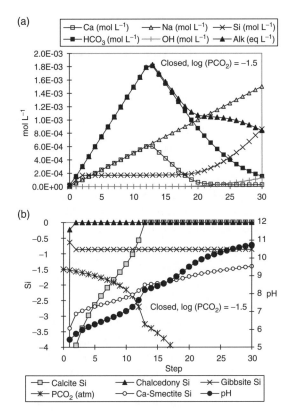

(b)

Fig. 20.9 The simulated evolution of water reacting with 1.5×10^{-3} moles of labradorite plagioclase feldspar (defined as $NaCaAl_3Si_5O_{16}$) at 7°C, in a closed CO_2 system, with an initial $PCO_2 = 10^{-1.5}$ atm, using the code PHREEQC, in 30 steps. Precipitation of kaolinite, gibbsite, calcite, chalcedony are permitted when saturated. The diagram (a) shows the concentrations of solutes, (b) the evolution of mineral saturation indices and pH. Alkalinity is calculated as the meq L^{-1} sum of bicarbonate, carbonate and hydroxide. Modified after Banks and Frengstad (2006).

values in the range 10^{-2}–$10^{-1.5}$ atm can prevail in Norwegian terrain.

The Norwegian investigations and the modelling studies of Banks and Frengstad (2006) thus allow us to extend Garrels and colleagues' conceptual model to a higher degree of hydrochemical maturity. We can now revise Statement 4 (above) to read:

4 The chemical composition of a given groundwater in a crystalline rock aquifer is largely determined by

- the initial PCO_2 of groundwater recharge;
- the composition of the hydrolysed plagioclase (Ca- or Na-dominated);
- the extent to which the system is 'open' or 'closed' with respect to CO_2;
- the extent to which the initial CO_2 content has been consumed and converted to HCO_3^- by hydrolysis;
- the extent to which plagioclase hydrolysis has progressed beyond the attainment of calcite saturation and beyond the effective consumption of available CO_2 (in a closed system).

The last two points can be used to describe the hydrochemical maturity of the groundwater. The above discussion may appear to be academic, but it has far-reaching implications for the acceptable inorganic chemical quality of groundwater in silicate crystalline rocks. Many potentially toxic inorganic species have a solubility, which is determined by the solubility of a carbonate or calcium phase, or by the pH of the water. For example, many heavy metals are soluble only in low pH waters. Other amphoteric metals and metalloids (Al, W, As) have a solubility, which increases with increasing pH in alkaline conditions (Frengstad et al. 2001). As an example, fluoride concentrations clearly seem to be related to the degree of hydrochemical maturity in crystalline bedrock groundwaters (Fig. 20.7). The increasing median fluoride concentrations with increasing pH may be related to (1) anion exchange of F⁻ for OH⁻ on apatite, sheet silicates or amphiboles, or (2) the depletion of Ca in high-pH waters, permitting the dissolution of CaF_2, without the fluorite saturation ceiling being reached.

20.4 Conclusion

Empirical 'baseline' investigations of ground-water quality in Norwegian crystalline bedrock aquifers have confirmed that a significant proportion of groundwaters breach Norwegian (and/or international) drinking water guidelines for a range of parameters, especially Rn (13.9% over 500 Bq L^{-1}), U (21% over 15 µg L^{-1}) and F^- (16.1% over 1.5 mg L^{-1}).

While these three parameters (Rn, U, F^-) are especially associated with granite and gneissic lithologies, there was found to be surprisingly little dependence of major ion chemistry on aquifer lithology. Rather than lithology being a major controlling factor for most aspects of groundwater chemistry in crystalline bedrock, the most important factors would appear to be:
• the Quaternary weathering history of the aquifer. Glacially scoured, newly emergent, mineralogically 'intact' Norwegian granite aquifers exhibit, for example, rather high pH, alkalinity, base cation, F^-, U and Rn concentrations. This contrasts strongly with their counterparts further south in Europe, which may have undergone a more prolonged subaerial weathering history,
• the initial PCO_2 of groundwater recharge and the degree of openness or closure of the system relative to CO_2,
• the extent to which plagioclase has been hydrolysed, and CO_2 converted to HCO_3^-, within the system. This will ultimately depend on residence time, kinetic factors and fracture geometry,

References

Aastrup, M., Thunholm, B., Johnson, J. et al. (1995). Groundwater chemistry in Sweden. Naturvårdsverket Förlag Report 4416, 52 pp.

Backman, B., Luoma, S., Ruskeeniemi, T. et al. (2006). *Natural Occurrence of Arsenic in the Pirkanmaa Region in Finland*. Geological Survey of Finland, Miscellaneous Publications, 82 pp.

Banks, D. and Frengstad, B. (2006) Evolution of groundwater chemical composition by plagioclase hydrolysis in Norwegian anorthosites. *Geochimica et Cosmochimica Acta* 70, 1337–55.

Banks, D., Frengstad, B., Midtgård, Aa.K. et al. (1998a) The chemistry of Norwegian groundwaters I: The distribution of radon, major and minor elements in 1604 crystalline bedrock groundwaters. *The Science of the Total Environment* 222, 71–91.

Banks, D., Midtgård, Aa.K., Frengstad, B. et al. (1998b) The chemistry of Norwegian groundwaters II: The chemistry of 72 groundwaters from Quaternary sedimentary aquifers. *The Science of the Total Environment* 222, 93–105.

Banks, D., Frengstad, B., Krog, J.R. et al. (1998c) Kjemisk kvalitet av grunnvann i fast fjell i Norge [Chemical quality of groundwater in bedrock in Norway – in Norwegian]. Norges geologiske undersøkelse (Trondheim) Report 98.058, 177 pp.

Banks, D., Reimann, C. and Skarphagen, H. (1998d) The comparative hydrochemistry of two granitic island aquifers: the Isles of Scilly, UK and the Hvaler Islands, Norway. *The Science of the Total Environment* 209, 169–83.

Banks, D., Morland, G. and Frengstad, B. (2005) Use of non-parametric statistics as a tool for the hydraulic and hydrogeochemical characterization of hard rock aquifers. *Scottish Journal of Geology* 41(1), 69–79.

Bårdsen, A., Klock, K.S. and Bjorvatn, K. (1999) Dental fluorosis among persons exposed to high- and low-fluoride drinking water in western Norway. *Community Dental and Oral Epidemiology* 27, 259–67.

Berglund, M., Ek, B.-M., Thunholm, B. et al. (2005) Nationell kartläggning av arsenikhalter i brunnsvatten samt hälsoriskbedömning. [National survey of arsenic content in well water and a health risk assessment – in Swedish.] Result Report to Naturvårdsvärket 215 0409, 15 pp.

Bjorvatn, K., Thorkildsen, A.H., Raadal, M. et al. (1992) Fluoridinnholdet i norsk drikkevann. Vann fra dype brønner kan skape helseproblemer [The fluoride content in Norwegian drinking water. Water from deep wells may cause health problems – in Norwegian]. Norsk Tannlægeforenings tidsskrift 102 m, pp. 86–89.

Carlsson, A. and Olsson, T. (1978) Variations of hydraulic conductivity in some Swedish rock types. In: Bergmann, M. (ed.) Storage in excavated rock caverns. Proceedings of the First International Symposium 'Rockstore 77', Pergamon Press, New York, pp. 301–7.

Frengstad, B. (2002) Groundwater quality of crystalline bedrock aquifers in Norway. Dr. Ing. thesis 2002:53, Department of Geology and Mineral Resources Engineering, NTNU Trondheim, Norway, 389 pp.+ appendices.

Frengstad, B. and Banks, D. (2003) Groundwater chemistry related to depth of shallow crystalline bedrock boreholes in Norway. In: Krásný, J., Hrkal, Z. and Bruthans, J. (eds) Proceedings of the International Conference on Groundwater in Fractured Rocks. Extended Abstracts. 15–19 September 2003, Prague, Czech Republic. IHP-VI, Series on groundwater No. 7, pp. 203–4.

Frengstad, B. and Banks, D. (2007). Universal controls on the evolution of groundwater chemistry in crystalline bedrock: The evidence from empirical and theoretical studies. In: Kràsny, J. and Sharp, J.M. (eds) Groundwater in Fractured Rocks. International Association of Hydrogeologists Special Publication, Taylor & Francis/Balkema, pp. 275–89.

Frengstad, B., Midtgård-Skrede, Aa.K., Banks, D. et al. (2000) The chemistry of Norwegian groundwaters III: The distribution of trace elements in 476 crystalline bedrock groundwaters, as analyed by ICP-MS techniques. The Science of the Total Environment 246, 21–40.

Frengstad, B., Banks, D. and Siewers, U. (2001) The chemistry of Norwegian groundwaters IV: The pH-dependence of element concentrations in crystalline bedrock groundwaters. The Science of the Total Environment 277, 101–17.

Garrels, R.M. (1967) Genesis of some ground waters from igneous rocks. In: Abelson, P.H. (ed.) Researches in Geochemistry. Wiley & Sons, New York, pp. 405–20.

Garrels, R.M. and Mackenzie, F.T. (1967) Origin of the chemical compositions of some springs and lakes. In: Stumm, W. (ed.) Equilibrium Concepts in Natural Water Systems. Advances in Chemistry Series 67, American Chemical Society, Washington, DC, pp. 222–42.

Gustafson, G. (2002) Strategies for groundwater prospecting in hard rocks: probabilistic approach. Norges Geologiske Undersøkelse Bulletin 439, 21–5.

Lahermo, P., Ilmasti, M., Juntunen, R. et al. (1990) The Geochemical Atlas of Finland, Part 1: The Hydrochemical Mapping of Finnish Groundwater. Geologian tutkimuskeskus, Espoo, 66 pp.

Morland, G. (1997) Petrology, lithology, bedrock structures, glaciation and sea level. Important factors for groundwater yield and composition of Norwegian bedrock boreholes? Doctoral thesis. Institut für Geowissenschaften, Montanuniversität, Leoben, 274 pp.+ appendices. NGU report 1997.122 I & II.

Parkhurst, D. (1995) User's Guide to PHREEQC– A Computer Program for Speciation, Reactive Path, Advective Transport and Inverse Geochemical Calculations. US Geological Survey, Lakewood, Colorado.

Parkhurst, D.L. and Appelo, C.A.J. (1999) User's guide to PHREEQC (Version 2) – A computer program for speciation, batch-reaction, one-dimensional transport, and inverse geochemical calculations. US Geological Survey Water Resources Investigation Report 99-4259.

Reimann, C., Siewers, U., Skarphagen, H. et al. (1999) Influence of filtration on concentrations and correlation of 62 elements analysed on crystalline bedrock groundwater samples by ICP-MS techniques. *The Science of the Total Environment* 234, 155–73.

Reimann, C., Siewers, U., Tarvainen, T. et al. (2003) Agricultural Soils in Northern Europe: A Geochemical Atlas. Geologisches Jahrbuch Sonderhefte Reihe D, Heft SD5, BGR Hannover.

Rönkä, E. (1993) Increased depth of drilled wells–advantage or drawback? In: Banks, S.B. and Banks, D. (eds) *Hydrogeology of Hard Rocks*, Mem. XXIVth Congress of International Association of Hydrogeologists (IAH), Ås, Oslo, Norway, 28 June–2 July 1993, pp. 772–9.

Sigmond, E.M.O. (1992) Berggrunnskart, Norge med havområder – Målestokk 1:3 millioner *[Bedrock map, Norway with marine areas – Scale 1:3 million]*. Norges geologiske undersøkelse.

Valve, M., Rantanen, P., Kahelin, H. et al. (2005) Point-of-use devices for arsenic removal. *The Finnish Environment* 790, 68–73.

White, A.F., Blum, A.E., Schulz, M.S. et al. (1998) Chemical weathering in a tropical watershed, Luquillo Mountains, Puerto Rico: I. Long-term versus short-term weathering fluxes. *Geochimica et Cosmochimica Acta* 62, 209–26.

White, A.F., Bullen, T.D., Vivit, D.V. et al. (1999) The role of disseminated calcite in the chemical weathering of granitoid rocks. *Geochimica et Cosmochimica Acta* 63, 1939–53.

White, A.F., Schulz, M.S., Lowenstern, J.B. et al. (2005) The ubiquitous nature of accessory calcite in granitoid rocks: Implications for weathering, solute evolution, and petrogenesis. *Geochimica et Cosmochimica Acta* 69, 1455–71.

WHO. (2004) *Guidelines for Drinking Water Quality*, 3rd edn. World Health Organisation, Geneva.

21 Natural Groundwater Quality – Summary and Significance for Water Resources Management

W. M. EDMUNDS AND P. SHAND

The main scientific conclusions of the thematic studies and reference aquifers in the book's chapters are summarised together with their significance in terms of defining natural water quality. Themes covered include inorganic and organic groundwater quality, estimation of groundwater residence times, recognition of baseline trends as well as monitoring changes in past, present and future natural quality of groundwater. The results from 14 European reference aquifers are used as illustration and to show the importance of conducting detailed studies of the baseline groundwater quality from which characterisation and monitoring may be conducted using a few simple indicators. The socio-economic relevance of access to pure natural water is considered and a practical set of guidelines for determining baseline properties is proposed.

21.1 Introduction

The chapters in this book present a cross section of thematic studies and examination of the baseline water quality in European reference aquifers. The overall objective has been to establish criteria for defining natural groundwater quality and to develop the basis for a standardised approach, which may be used in the European Water Framework Directive (WFD: Directive 2000/60/EC) and Groundwater Directive (GWD: Directive 2006/118/EC). A standardised approach, based on hydrogeological and geochemical principles has been adopted to assess the natural variations and spatial heterogeneity in groundwater quality.

Entirely natural processes may breach existing water quality limits in addition to human impacts. Baseline criteria are needed as a reference to be able to assess quantitatively and qualitatively whether or not anomalous concentrations are natural or whether anthropogenic pollution is taking place. The timescales influencing natural processes and the rates at which changes are occurring are recognised as important controls on natural water quality. The distribution of pristine natural waters in representative European aquifers has been established, together with the extent to which they are being depleted or modified by contaminated waters moving into the aquifer. As well as developing the necessary scientific tools and approaches to the characterisation of natural water quality, discussion with policy-makers and end-users has helped formulate how an improved understanding of baseline conditions can be used in water quality management and policy.

Results from 12 European countries (Estonia, Poland, Denmark, Belgium, United Kingdom, France, Switzerland, Spain, Portugal, Czech Republic, Bulgaria and Malta) are used in the investigations. Norway also has developed an approach to the study of natural water quality and was invited to contribute to this book, not least in view of the application to crystalline bedrock aquifers. The book addresses the following themes:

1 Defining criteria for establishing the baseline concentrations of a wide range of substances that occur naturally in groundwater, as well as their controls, as a basis for defining water quality status and standards in Europe;

2 Characterisation of reference aquifers across Europe that can be used to illustrate the ranges in natural groundwater quality as a basis for understanding natural groundwater quality evolution and the interfaces with modern, probably contaminated water. The baseline conditions are investigated as far as possible by cross sections along the groundwater flow gradient, in aquifers with well-understood hydrogeology, where sequential changes in the water–rock interaction (redox, dissolution–precipitation, surface reactions) as well as mixing, may be readily investigated. The interface with surface waters is also addressed;

3 Assessing long-term trends in water quality at representative localities in selected reference aquifers and to interpret these in relation to past changes in hydrochemistry (quality of water) in order to predict future changes caused by natural geochemical and anthropogenic effects;

4 Setting a scientific foundation which may be used to underpin EU water quality guideline policy, notably the GWD, with an emphasis on the protection of high-quality groundwater and contribute to its sustainable development.

Precise determination of natural concentrations is needed to distinguish the anthropogenic changes in water quality caused by both pollutants and natural geological sources induced by intensive withdrawal of water. The results of the present investigations have shown how difficult it can be to establish the baseline concentrations using single samples or from statistical treatment. A conceptual geochemical understanding of the groundwater quality in its hydrogeological context is needed to determine to what extent any water is free of pollution. This often requires some knowledge of the residence time of the groundwater.

21.2   Main scientific conclusions

The chemical properties of groundwater vary as a function of complex geological, geochemical, hydrogeological and climatic factors. This gives rise to large spatial and depth variations at a range of scales (Fig. 21.1). The baseline approach is illustrated using thematic studies dealing with inorganic and organic quality, tracer applications and investigation of trends. A series of reference aquifers, representative of major groundwater systems are then used to illustrate specific geochemical (baseline) conditions.

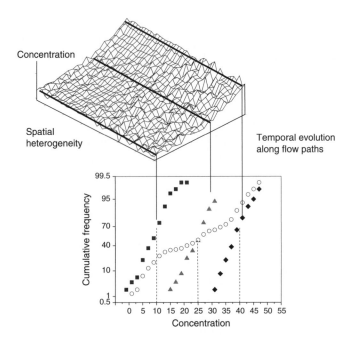

Fig. 21.1 Baseline quality is likely to vary both in space and time as a result of the variations in the landscape, climate, inputs from rainfall, and human activities and residence time, in addition to (hydro)geological and geochemical processes. Units of concentration are relative. Overall bimodal distribution reflects mixture of local lognomally distributed compositions.

21.2.1 *The significance of the reference aquifers*

The reference aquifers demonstrate scientific approaches and best practice that may be used for the characterisation of natural quality of groundwater bodies using examples from EU countries and Norway. The reference aquifers were selected as being typical of the major aquifers in each country. Results from these reference aquifers may, however, not always be directly extrapolated to other similar or adjacent aquifers in view of the intrinsic geological heterogeneity and different boundary conditions. The timescales involved for geochemical reactions vary from seconds to millions of years, therefore, the groundwater chemistry may be

variable even where the bedrock geology and mineralogy are uniform. This emphasises the need for a sound understanding of the geological setting, which to a large extent controls the baseline geochemical environments.

In many of the reference aquifers, downgradient profiles have been used to demonstrate sequential changes with distance and time. The reference aquifer studies show that within any groundwater body there are likely to be geochemical **gradients** (sequential changes in chemical properties with time and distance from recharge areas) and also geochemical **boundaries**, which may give rise to abrupt changes in water quality (solution limits such as decalcification; mineral saturation boundaries; redox boundaries;

exchangeable cation boundaries). These important characteristics need to be defined within larger management units forming the groundwater bodies (Figure 21.1). It is recommended that baseline conditions and their spatial variability be taken into account in defining the **status** of groundwater bodies.

Most of the aquifers display a range of geochemical processes working at different scales, occurring at different rates and in different environments, giving rise to large spatial and temporal variations in groundwater quality. Mineral dissolution rates vary considerably – phases such as carbonates and sulphates dissolving rapidly whilst silicate minerals dissolve much more slowly. One consequence of this is that minor reactive mineral phases may dominate groundwater chemistry. However, it is commonly found that the easily dissolved minerals have been removed from the shallow parts of the aquifer where recharge is occurring leading to the development of **reaction fronts**.

There is commonly a zonation in redox processes as groundwaters evolve within an aquifer. A redox boundary often exists close to the junction between unconfined and confined parts of the aquifer and this typically occurs where there is a change in residence time caused by the juxtaposition of different flow systems. Redox changes are found to give rise to very different baseline chemistries either side of the boundary and it has sometimes been found necessary to define these as different systems.

Ion-exchange reactions are important in aquifers where the adsorbed 'exchangeable cations' were previously at or close to equilibrium with a different type of water, but this water has now changed in terms of its baseline chemistry. This may lead to complex spatial variations, which continue to evolve over the scale of thousands of years.

Mixing with more saline water is also a dominant process in most of the reference aquifers, saline waters being present in less permeable parts of the aquifer (e.g. trapped in pore waters or clay aquitards), deeper parts of the aquifer or where seawater intrusion occurs. This is an important process, which may derogate the water in terms of drinking water or ecosystem balance. The salinity is usually natural in origin and therefore is classed as baseline.

There is a considerable range in water quality within all of the reference aquifers, with concentrations typically varying over several orders of magnitude. The ranges can be described statistically using the median (more robust than the mean) and an upper limit (97.7 percentile) imposed to remove outlying data. However, the ranges in concentration are of limited use in themselves except to highlight the order-of-magnitude variations in the aquifer or groundwater body. These must be placed in the context of spatial variability to be of use for management.

Knowledge of the spatial and depth variations is essential for the protection as well as management of groundwater resources. Stratification in water quality is common due to changes in lithological facies and residence times. The causes of stratification need to be established, for example, with the aid of hydrogeophysical logging and depth sampling to help define the conceptual model of the aquifer.

The reference aquifer studies illustrate that spatial variability varies from aquifer to aquifer, for example the presence of variable superficial deposits (such as the glacial drift in UK, Denmark and Estonia) gives rise to sometimes complex patterns of groundwater chemistry. Characterisation of such heterogeneity is a pre-requisite for establishing an effective and efficient monitoring network. It is also commonly found that for certain elements, pollution inputs are within the overall large range of baseline, therefore, quantification of pollution needs to be done at a local scale, that is, within a smaller system.

Groundwater residence time plays an important role in determining the baseline quality and, furthermore, provides information on the potential timing and impacts of pollution. The timescales of groundwater movement are typically of the order of metres per year and the influence of pollution from the past few decades can be recognised in most aquifers. Even in pristine groundwaters, the unavoidable risk from pollution still exists upgradient in the groundwater body or from waters now present in the unsaturated zone.

The natural age distribution has been modified in many of Europe's aquifers by groundwater abstraction, most evident in the unconfined parts of aquifers but may also be found in the confined aquifer, where decreases in piezometric level have occurred. Abstraction may cause mixing of different stratified layers, for example by pulling down younger, often polluted groundwater, in the cone of depression around boreholes or causing upwelling of older, deeper groundwater. In these cases it is likely that the local baseline will be modified. In some cases, this can lead to derogation of the groundwater for example up-coning of deeper, older, more saline water or seawater intrusion, or changes in redox status.

Diffusional exchange between groundwaters in an aquifer and adjacent poorly permeable layers has been discussed for several reference aquifers. This may also cause derogation of the aquifer's baseline chemistry and this process may be enhanced by abstraction.

The slow movement of groundwater in these formations also means that, if they contain pollutants, they will retain these for much longer than the groundwater residence time. Diffusional exchange may not only modify the aquifers geochemistry, but may also make it difficult to apply individual dating techniques (Chapter 5). Hence, it is concluded that the properties of aquicludes also be characterised as part of any baseline assessment.

21.2.2 Inorganic quality

There is a considerable range in the inorganic chemistry within all of the reference aquifers, with concentrations typically varying over several orders of magnitude. This is particularly the case for many trace elements reflecting both the long timescales involved and the heterogeneous sources, for example, of trace metals. The ranges have been described statistically using the median and an upper limit (97.7 percentile) to remove outlying data.

For many aquifers, baseline is unlikely to be the modern-day status since most have been affected to some degree by diffuse pollution: the data from aquifers require data manipulation (removal of outliers and anthropogenic influences) making determination of the baseline difficult. It is commonly the case that pollution inputs are masked within the overall large range of baseline, therefore, quantification of pollution needs to be done at a local scale.

Historical records provide the most important approach for determining the baseline for a number of key elements (major ions and nitrate for example). However, old records are often difficult to find, and many long-term data may have been lost as a result of organisational changes and the shift to digital records. In some countries, the data are confidential and not available for public use. Direct extrapolation using linear regression is not recommended since pollutant inputs are likely to have varied non-linearly over time.

The statistical data and plots given in this book are useful for comparison between areas, and are of strategic value. More than 80 components have been measured representing a comprehensive and strategic overview of baseline concentrations for most inorganic substances, and a reference to quantify pollution. The range of measurements also includes data for elements for which there are no existing guidelines, but which may still be harmful (e.g. Be, Se, Sb, Tl, U). The definition of these baseline ranges can be used for example to recognise contamination incidents by uncommon elements for which no guidelines exist.

Subtle differences in geological structure, facies changes and mineralogy cause the main baseline changes and care must be taken when extrapolating the results between and within aquifers. This is particularly the case for most trace elements where sources tend to be much more heterogeneous. It is recommended that data be sought from the groundwater body of interest rather than relying on the applicability of data from remote areas. There is also considerable overlap in concentration ranges for most solutes between the different aquifers regardless of lithology, due to time-dependent hydrogeochemical processes, notably mineral dissolution, redox reactions, ion-exchange reactions, mixing, adsorption–desorption and rainfall inputs. A protocol for baseline definition is given below.

The upper limits of baseline concentrations can exceed environmental-quality or drinking water standards as the latter are usually imposed on ecotoxicological or human health grounds and are divorced from observed groundwater concentration distributions. A number of solutes such as As, Ba, F and Fe often have concentrations above such standards, but the source of these is natural. Trend reversal or remediation may be either impossible or prohibitively expensive in areas where this occurs and special regulations may therefore be required. Many of these solutes, in addition, are also pollutants and a careful assessment of the baseline (natural component) needs to be undertaken to distinguish natural baseline from anthropogenic origin.

The ranges in solute concentration highlight the order-of-magnitude variations in the aquifer or groundwater body. These must be placed in the context of spatial variability to be of practical use for groundwater management. The characterisation of chemical heterogeneity is also a pre-requisite for establishing an effective and efficient monitoring network. Knowledge of the spatial and depth variations is then also essential for the protection as well as management of water resources. The baseline may be modified by disturbing groundwater stratification without the introduction of pollutants.

21.2.3 Natural organic quality

Organic carbon has been investigated in several of the reference aquifers as the most important indicator of natural organic quality, since it was beyond the scope of this book to characterise individual organic compounds. Natural organic carbon in shallow groundwaters is derived mostly from the overlying soils and consists of a mixture of macro-molecules classified as humic and fulvic substances. The remainder is a complex mixture of lower molecular weight compounds. Deeper and older waters may contain organic material, which has been mobilised from the matrix, such as lower molecular weight compounds resulting from kerogen degradation or bitumen dissolution.

Increased understanding regarding the evolution and variation of natural organic and inorganic groundwater quality is required as a basis for the recognition of human impacts on aquifers and for aquifer management. Baseline values are presented for the non-volatile organic carbon (NVOC) fraction, expressed as total organic carbon (TOC) and dissolved (0.45 μm filtered) organic carbon (DOC). Total organic carbon was measured in more than 400 groundwater samples from eight European Union countries, and DOC in filtered samples was analysed for approximately 250 groundwater samples from four of these countries. The results of the groundwater sampling show that the baseline concentrations for TOC and DOC were comparable.

TOC was present with a median concentration of 2.7 mg C L^{-1} and a range from 0.1 to 59.4 mg C L^{-1}, and DOC had a median concentration of 2.2 mg C L^{-1} with a range from 0.2 to 58.9 mg C L^{-1}, demonstrating that very high natural organic carbon values can occasionally be found locally in some pristine aquifers. A relationship between the assimilable organic carbon (AOC) utilised by bacteria and TOC was observed although the correlation is not clear. Generally, there is a linear correlation between the TOC, DOC and the chemical oxygen demand (COD) values.

TOC/DOC analysis can be an important indicator of pollution from landfills, effluent ponds and similar pollution settings with high loads of organic carbon, but in many other situations the total or dissolved organic carbon concentration may be less reliable as an indicator of contamination. Organic carbon is generally important for the evolution of different redox environments in aqueous environments, and the amount and reactivity of the organic carbon in the aquifers is therefore an important parameter when evaluating the state and trends of groundwater quality. TOC and DOC, which are standard analytes used globally for measuring the amount of organic carbon in water samples, were the focus of this study. However, the amount of TOC or DOC itself does not provide information on how reactive or reducing the environment is. Groundwater with a small but very reactive amount of young organic carbon may be much more reactive than groundwater with a much higher but old non-degradable amount of organic carbon.

Measurements of TOC and DOC on the same samples showed comparable concentrations, however, the TOC/DOC ratio varied and TOC was not always found to be significantly greater than DOC, as may have been expected. Further investigations are needed to find the reasons for this. Further research is also needed to evaluate what part of the TOC is readily available for biogeochemical processes. Increased knowledge on this issue would help to understand better the development of the different redox environments.

21.2.4 *Modelling of natural groundwater systems*

Uncontaminated aquifers frequently show large spatial variations in the baseline concentrations of various solutes. These are due to historical variations in the input concentrations, for example, due to variable climatic and environmental conditions, superimposed by water–rock interactions which are governed by distinctive mineralogical properties of the aquifer, in turn reflecting the nature of the local geology. As a result, mineral dissolution or ion-exchange fronts may be recognised which progress through the aquifer during the freshwater diagenesis reactions and the resulting solutes can be identified in sampled groundwaters.

Geochemical modelling provides a means of obtaining a quantitative understanding of the geochemical processes and their coupling to water transport, and may predict how the baseline concentrations may change over space and time. Within the present series of investigations, the code PHREEQC-2 was used for reactive transport modelling along flow paths in three main types of European aquifers: rocks containing carbonate minerals, aquifers with freshwater displacing seawater and poorly buffered siliceous aquifers.

Modelling of the East Midlands aquifer shows the effects of $CaCO_3$ dissolution on a timescale of thousands of years, de-dolomitisation as well as gypsum/anhydrite dissolution (Chapter 4). The change in carbonate dissolution mode from open to closed system evolution, as described by modelling, must have taken place somewhere between 5000 and 10,000 years ago and the resulting variation is still moving through the aquifer. De-dolomitisation correlates with in an increase in the groundwater sulphate concentration, which is coupled to increases in the magnesium and calcium concentration. De-dolomitisation is then modelled with PHREEQC by successively adding gypsum to a groundwater in equilibrium with calcite and dolomite (Fig. 4.2, Chapter 4); de-dolomitisation is found to be a main control for

the variations in water chemistry in the confined aquifer.

The example of the Valréas aquifer in France (Fig. 4.5, Chapter 4) shows an upstream fast-flow regime close to the recharge area and a downstream low-flow regime below a semi-confining Pliocene clay cover. The water chemistry along the flow path shows distinct variations in both pH and cation concentration. The model is based on freshwater replacing relict saltwater in the aquifer, the geochemical processes being ion exchange and carbonate mineral dissolution. The model trend is in good agreement with field data and suggests a quantitative understanding of the system. The timescale over which reactions have taken place is predicted by the model to be in the order of 1–2 million years!

Abstraction of groundwater in the Valréas aquifer at the present day disturbs the natural gradients. Geochemical models may be used to predict the changes in solute concentrations in time and space. The increasing water demand from local wineries has been met by installing boreholes in the slow flow part of the aquifer, accelerating the groundwater flow. The expected changes in water composition can, with some confidence be predicted from the geochemical reactive transport model. In this way predictions can be made of the future geochemical evolution of aquifers, thereby facilitating trend analysis and the set up of cost-effective monitoring programmes.

Modelling of the Aveiro aquifer in Portugal has enabled the processes of freshening (from an originally seawater-filled aquifer) to be observed and to study quantitatively the cation exchange processes. According to the model, the present-day hydrogeochemical pattern is the result of about 50 ka of geochemical evolution in the aquifer. This is in agreement with ^{14}C and chloride data for the aquifer that indicated ages of at least 35 ka, and also with the geological data. The correspondence in timescales furthermore indicates that the palaeohydrology is still dominating the aquifer geochemistry, and is not yet affected by the extensive groundwater abstraction that has taken place since the late 1960s. Thus, the reactive transport modelling of the Cretaceous Aveiro aquifer has helped to confirm the principal processes contributing to the hydrogeochemical evolution of the aquifer and has provided additional evidence about the timescale for aquifer freshening.

21.2.5 Groundwater residence times

Natural groundwater concentrations are the combined result of the composition of the water that enters the aquifer and reactions between the water and rocks in the aquifer. Variations due to natural geochemical reactions usually take place over a large range of timescales depending on the dynamics of the reactions involved. Conceptually two types of natural variations of water quality of a groundwater body can be distinguished (Fig. 21.1):

• Spatial variations of the geological environment that cause corresponding changes in the water chemistry due to water–rock interactions. The resulting concentrations of chemical and isotopic parameters are statistically distributed in a complex manner.
• Superimposed on these variations, groundwater chemistry also changes systematically in time. These changes occur along flow paths and can be very distinctive.

It is both scientifically as well as practically important to understand the chemical reactions in the aquifer and the rates at which they take place.

21.2.5.1 Range of tracers

Groundwater residence times in the investigated reference aquifers range from a few years up to many thousands of years. Therefore, a series of mainly isotope tracers (Chapter 5A, Table 5A.1) have been used to establish timescales of hydrogeochemical evolution and in detecting any modern inputs. Different processes act as 'clocks' and can be used to estimate groundwater residence times; accumulation processes (e.g. ^4He and chemical parameters), radioactive decay (e.g. ^{39}Ar, ^{14}C, ^3H, ^{85}Kr), variable inputs with known chronology (e.g. ^3H, ^{85}Kr, CFCs, SF_6, δ^2H, $\delta^{18}O$), or combinations of these processes (e.g. ^{85}Kr, ^3H). As an example, palaeowaters (waters recharged under other climatic conditions than those prevailing in the Holocene–Recent) can be identified using the isotopic composition of water molecules and the concentrations of stable noble gases in water. Furthermore, argon-39 has been used successfully to date waters of intermediate age (<1000 years), which are between the age ranges obtainable by more routinely used tracer methods.

21.2.5.2 Ranges of residence times in European aquifers

Three classes of residence times and appropriate tracers are distinguished:

Recent water (younger than 50 years). Several methods are available to date recent groundwaters in shallow and/or rapidly circulating aquifers. Such shallow systems are a particularly important part of the hydrogeological environment forming part of active recharge–discharge zones where most abstraction sites are located, and provide the main source of drinking water. Thus tracer methods that help to characterise transport in shallow aquifers are likely to play an important role in resolving baseline issues. Yet this is a difficult and sometimes contradictory task. On the one hand it is the intrinsic natural quality of shallow aquifers that is of interest, but on the other hand, such waters are often already contaminated, masking the natural properties. Dating tools provide one possibility to deal with this problem.

Ancient waters (infiltrated during pre-industrial times). The main objective in this book has been to define the quality of groundwaters that are not influenced or polluted due to human impacts. Most substances that derogate groundwater quality today were intoduced over the past 200 years since industrialisation. It can be assumed that water which infiltrated prior to this time is in pristine condition and free of pollutants. The quality of ancient groundwaters is mainly determined through time-dependent geochemical reactions between the solvent (water) and the aquifer material. The longer the groundwater residence time, the longer the duration of the contact between water and rock and the wider the range in water quality.

Palaeowaters (recharged during or prior to the last glaciation). Groundwaters that infiltrated more than 10,000 years ago are termed **palaeowaters**, originating from a time with different climatic and hydrogeological conditions. Although the quality of palaeowaters is sometimes very high, they form a minor role in European drinking water supply. However, as a consequence of increasing pollution of shallow aquifers, the pressure to exploit such high-quality reserves will increase. The identification of these waters, which have only very low renewal rates, is therefore an important task for groundwater management.

It is worth reiterating that the term **'unpolluted'** is not synonymous with **'free of anthropogenic influence'**. High groundwater exploitation may induce changes in the baseline groundwater quality due to changes in water–rock interaction. Examples include the shift of the redox boundary or the freshwater–saltwater interface in coastal areas. On the other hand, recent water components may not necessarily be contaminated. In recent groundwaters a careful separation of **natural quality** and superimposed **artificial influences** is required; in this book 'polluted' is defined by the presence of substances introduced by human activities.

21.2.5.3 Reference aquifers as examples

A subselection of reference aquifers serve as clear type models of the European aquifers for demonstrating age relationships, in order to clarify the relationships between waters of the modern era and older waters, including palaeowaters. Intercomparisons of different tracer methods performed for representative European aquifers have enabled confidence to be placed in the use of (relatively) low-cost indicators of groundwater, especially for young groundwaters potentially vulnerable to anthropogenic pollution. In addition, it may be possible to use chemical ratios or trends with, or in lieu of, isotopic tracers as qualitative residence time indicators. Environmental tracer methods combined with hydrodynamic modelling have also appeared to be very useful in the identification of modern and geologically derived chemical components and their relations to ages of water in different aquifer zones. In the study of two Palaeogene (Tertiary) sand aquifers in France (one completely confined and the other confined over most of the area), numerical hydrodynamic and transport modelling have been shown to yield ambiguous results.

The East Midlands aquifer has been the subject several intercomparison studies and the refined age structure helps to explain the observed chemistry especially the complex patterns shown by sulphate, where modern water, Holocene and late Pleistocene water all have different origins of SO_4. The combination of careful study of chemical trends along flow lines coupled with age determination and followed by modelling presents a powerful approach for understanding the baseline conditions. The reference aquifers thus serve as powerful examples for illustrating how similar hydrogeological systems in Europe and elsewhere might behave.

21.2.5.4 General conclusions regarding residence times

Dating methods exist for all timescales, and for some age ranges several methods can even be combined. The selection of the appropriate methods is determined by the age structure of the water system. The successful application of isotope methods in selected reference aquifers is described in Chapter 5. Most important for such applications is the question as to whether 'young' water components are present or not. The presence of suitable young residence time indicators, namely $^3H/^3He$, SF_6, CFCs and ^{85}Kr, answers this question. These may be used in combination with synthetic pollutants to define modern components and to resolve whether some substances (e.g. nitrate) are of recent origin.

Aquifer systems are complex and therefore difficult to describe with simple models, since it is necessary to deal with mixing of water components with different age within a complex geological framework. The age distribution of a water body is very often the result of a combination of natural mixing processes (hydrodynamic dispersion, upwelling of water from deeper parts, diffusion, etc.) and mixing induced by human impact (extraction of water from different depths in a large screen interval, changing gradients due to heavy pumping). It is not possible to determine details about the age structure of a groundwater body with only one tracer. Only the combination of several isotopic and chemical tracers at the same time provides a reliable basis for the understanding of a flow system. Furthermore, it is important to emphasise that dating is only one objective. The determined age structure has to be included as part of the general knowledge of the system, which may consist of knowing the recharge area, the flow path and flow velocities, the evolution of the chemical and isotopic components due to water–rock interaction and the reservoir size, as well as the balances of the water masses. A strong connection exists between natural baseline concentrations of chemical components and the age information obtained from isotope studies: the quality of waters older than about 100 years almost always represents baseline conditions.

21.3 Monitoring baseline quality – past, present and future

21.3.1 Identifying and interpreting baseline trends

In the preceding chapters, many examples of time series have been identified and analysed, although mainly for short timescales. An understanding of natural baseline trends is important in view of pollutant concentrations and the reversal of such trends, as required by the WFD (Directive 2000/60/EC), and it is important to establish whether anthropogenic or strictly natural processes are involved.

Baseline trends can be related to natural and induced hydrodynamic and hydrochemical changes in groundwater systems. These trends have been assessed in relation to both spatial and temporal scales and groundwater flow velocity as well as heterogeneities in groundwater bodies (including the presence of original formation and palaeowaters).

21.3.2 Recognising baseline trends in time series data

In Chapter 6, a methodology for recognition of baseline trends in time series data is proposed, based on a geostatistical approach, since a purely mathematical approach is not appropriate for understanding hydrogeochemical trends and their extrapolation. Distribution analysis, cross plots of parameters and regression analysis, nevertheless, are helpful tools. A statistically derived trend needs to be interpreted in the framework of an understanding of the hydrodynamic and geochemical controls.

The relationship between time trends of hydrogeochemical parameters and the geostatistical characteristics of groundwater sub-bodies, expressing the spatial occurrence and distribution of groundwater composition, is emphasised. A methodology has therefore been developed to evaluate time trend analysis as a function of the geostatistical characteristics of groundwater sub-bodies, by assessing the relationship between time trends and small-scale spatial variations based on the variogram.

Two main types of baseline trends can be distinguished.

Natural baseline trends. Trends in water quality may be found which are related to the intrinsic geochemical characteristics of the aquifer and which are the result, of time-dependent reactions with the host rock. Such trends may also be observed spatially along flow paths. They may therefore (1) be linked to processes causing variability at the aquifer scale. Depending on mass-transport velocities, they may occur extremely slowly (e.g. freshening of aquifers; changes across redox boundaries); and (2) be linked to small-scale spatial variability (due to small-scale heterogeneities in aquifer properties), causing fluctuations around an average level.

Baseline trends caused by exploitation. Such trends are anthropogenically induced, but solute concentrations have a natural origin, and are not due to anthropogenic input. Pumping increases mass–transport velocities, and will thus accelerate the appearance of natural trends. In addition, pumping may also change groundwater flow direction (e.g. up-coning from deeper layers) and also may induce chemical reactions in the aquifer resulting, for example, from oxidation or reduction. It is necessary to understand the origin of the observed trend in order to distinguish a baseline trend from pollution and therefore process interpretation is required.

21.3.3 Long-term evidence of changing quality

Two examples, from others dealt with in more detail in Chapter 6 summarise the types of trends that might be expected in historical changes in baseline quality.

The evolution of groundwater quality in the old well of the Carlsberg brewery (Copenhagen, Denmark) is illustrated using a long time series between 1897 and 1945 (Fig. 21.2). A sudden rise in Cl, Na, Ca and SO_4 in 1940 is due to increased salinisation by up-coning of deeper more saline water after at least 40 years of pumping. After this change in water quality, the well was not used any more. It illustrates a case of rapid natural quality deterioration induced by exploitation causing up-coning of more saline waters. This represents a common phenomenon of changes in the baseline qual-ity due to human activity where the natural stratification is disturbed. In addition to up-coning, such phenomena might also be caused by the drawdown from former lakes or oases in semi-arid areas, as well as later-ally induced seawater salinity increases in coastal areas.

Pumping may induce movement both from downgradient areas as well as from the out-crop area. Grove is a large abstraction site in the UK East Midlands Triassic Sherwood Sandstone aquifer. It is situated in the con-fined aquifer and groundwater is at least 30,000 years old. The available records from the mid-1950s to the present day (Fig. 21.3) show that nitrate is at or around detection limit (and is not shown here) and pollution is absent. Chloride decreases over the period from 15 to around 6 mg L^{-1}; sulphate con-centrations are more or less constant around 25 mg L^{-1}. The distinctive high Mg/Ca ratio of a geochemically evolved water (having undergone some reaction with dolomite and calcite) is also near-constant over the period. The trends therefore indicate that a **younger** very pure Holocene water is progressively replacing **older** but still high-quality water

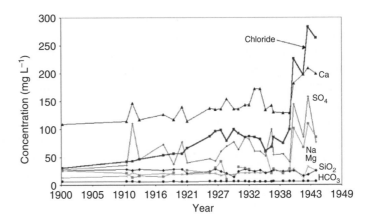

Fig 21.2 Evolution of water quality in the Carlsberg well in Denmark (1897–1945).

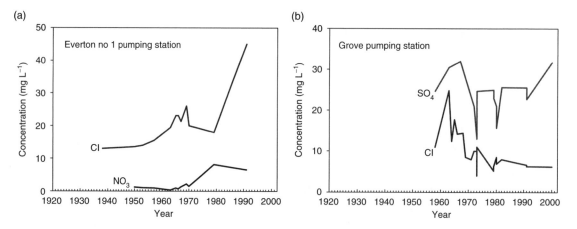

Fig. 21.3 Time series for groundwater from two boreholes from the East Midlands Triassic Aquifer: (a) Everton in the unconfined aquifer (from 1930); and (b) Grove (from 1956). These two profiles illustrate the changes that take place in the natural baseline in the pristine aquifer (e.g. decreasing Cl with time in Grove), and incoming contamination from human activity shown by increasing Cl and NO$_3$ in Everton.

with very slightly higher Cl, suggesting younger water progressively moving towards the site. Clearly there is no influence of modern (high Cl, high NO$_3$) water and this highlights how very small changes in Cl can be used to monitor baseline changes.

In the second UK example (Everton) in the same aquifer, closer to the outcrop, trends are seen since the 1930s. Initial nitrate concentrations are in the range 0–1 mg L^{-1} NO$_3$–N and represent the pre-development baseline, together with chloride, which initially lay in the range 10–15 mg L^{-1} Cl, consistent with values found in the pristine (Holocene) section of the aquifer (see Grove). The Mg/Ca ratios are lower than Grove, indicating less evolved and therefore younger waters, but this ratio remains more or less constant throughout the period. Clear and roughly parallel trends in both Cl and NO$_3$ however indicate the advance of water of the modern era. At Everton for example this

trend was marked first by an increase in Cl in the late 1950s and around two decades later by an increase in nitrate.

21.3.3.1 Improvements needed in observation and monitoring

Existing practice in many countries or regions is focused on yield, with water quality considerations often being of secondary importance. This has been the traditional case for the licensing system in European countries.

It is instructive that only few records with quality trends up to or in excess of 70 years duration are available, and in several countries it has proved difficult to find any long-term records. More archive data need to be assembled to demonstrate long-term trends and more effort is needed to conserve and interpret suitable archive data. Where such data are available, the trends with time both from natural and pollutant sources may be interpreted, the initial chemistry may be defined. The measurement of initial chemical conditions should be instigated as 'good practice' in new boreholes as a basis for assessing future change. This will probably be representative of the natural baseline quality of the water especially in confined aquifers.

Drilling penetrates stratification in age and quality, and an initial characterisation of these layers by geophysical logging and depth sampling should be adopted as a prerequisite in defining quality properties of the aquifer whenever possible (Chapter 7). It is as important to define aquitards as well as the target aquifers. This assessment of quality characterisation, although initially costly, will ultimately be cost-effective since data on water quality for first strick data are essential data for subsequent aquifer modelling and management.

Where significant baseline concentration gradients exist, changes in the hydrodynamic situation due to pumping or overexploitation of the aquifer system can lead to the displacement of groundwater bodies and to the shifting of water composition fronts. This can induce changes of groundwater quality with time in observation and production wells. Changes in the baseline quality at pumping wells can result in the loss of production sites or increase the cost of water treatment. Thus, whilst monitoring groundwater quality with time, special care needs to be taken to distinguish compositional trends (systematic increase or decrease of parameters) from fluctuations due to the natural spatial variation in the groundwater bodies.

21.3.4 Shallow groundwaters and groundwater – surface water interaction

In most areas of Europe with moderate rainfall, river base flow maintains the quality of streams and these rivers, effectively acting as an 'outcrop' of groundwater. During summer, river flow and quality may be controlled almost entirely by groundwater discharge. The purity of the groundwater and thus its baseline chemistry, derived from springs and seepages, are of importance for maintaining river quality and the biodiversity of ecosystems. Riparian zones and groundwater derived wetlands are often complex in terms not only of flux, but also in terms of geochemical processes (e.g. CO_2-cycling, N-species transformations and redox processes).

The shallow groundwater systems can also be polluted, especially during high water levels of polluted rivers and, as a consequence, high-quality groundwater normally discharging as baseflow may be rapidly degraded. Once contaminated, the shallow groundwater may represent a prolonged source of contamination to rivers (c.f. leakage of high nitrate), especially in the case of dual porosity aquifers.

Many waters recognisable as being modern and with short turnover times (for example from the presence of detectable concentrations of CFCs or tritium), may not in fact show other clear signs of pollution. Their composition may still overwhelmingly reflect a mineralisation derived from water–rock interaction and will be fully acceptable in terms of potability, providing that biological constituents do not pose a health risk. The most common indicator of incipient pollution in most groundwaters is elevated nitrate above recognisable baseline concentrations and, in order to examine this, trend data in the groundwater quality are required.

21.4 Socio-economic relevance, strategic aspects and policy

21.4.1 Public awareness of natural water quality related issues

From the dramatic growth in the mineral water and bottling water industry over the past decade, the provision of high purity water, at least for drinking, has to be viewed as a matter of high public concern, and at the same time as a lack of confidence in the water supplied by utilities. Pure and pristine groundwater is sought as the number one target for the public water supply industry. The focus of this book on natural waters attempts to raise the profile of groundwater, which in most locations has intrinsically high quality under its baseline condition. Such groundwater, being a **natural mineral water**, if handled properly often requires little treatment.

Promotion of baseline concepts is of considerable importance in raising awareness among the public that groundwater from naturally protected groundwater bodies is quite fit to drink. Thus, large sections of groundwater bodies need to be promoted and protected as sources of high-quality drinking water. Bottled waters are produced commercially, especially in regions where surface or shallow groundwaters are of poor quality, but not enough is being done in some European countries to safeguard the deeper pure resources available as public supplies.

From the survey of European opinion conducted for this study (see Chapter 7) groundwater is widely used and viewed as a reliable and continuous water supply, sometimes as the only water resource. Its generally good quality is, however, acknowledged and the protection of supplies is a paramount regulatory and managerial objective. It is recognised that there is a slow but continuous general deterioration of groundwater quality (mainly by agrochemicals), seldom are improvements reported. This is partly a consequence of improved observation and monitoring. Excess levels of natural components in groundwater (Fe, F, salinity for example) occasionally make surface water preferable for human supply, despite a greater vulnerability to pollution. Improved information about groundwater and its (baseline) quality is sought and would be welcomed by end-users. Defensive monitoring (early warning) within the catchment of the boreholes is strongly supported.

21.4.2 Groundwater management issues

New legislation such as the EU Directives are generally welcomed by those interviewed and seen as both adequate and essential to ensure stabilisation and improvements in groundwater quality. They should however relate more to real aquifer conditions and in providing practical solutions.

Groundwater quality sustainability is a key issue for all parties consulted and is recognised as a long-term issue. There is also concern that good quality groundwater (including palaeowater) is being impacted and used too often for purposes that do not need this high quality. Support exists for greater public information, more investigation, planning and regulation, with widespread support for protecting high-quality water. There is an awareness of threats to groundwater, and well-head protection zones, for example, are generally in place. Protection and improvements in groundwater quality, it is agreed, can be attained mainly by changes in land use and better farming practices.

Knowledge of hydrogeological and technical issues related to aquifer management issues is agreed to be adequate, but greater groundwater education is needed among policy-makers and producers who seem intuitively to prefer surface waters. It is noted that additional professional hydrogeologists are required to meet requirements; they are found mostly in the consultancies and some regulator communities rather than in companies. The level of professional qualification is high but no compulsory scheme currently exists for technicians to gain know-how on groundwater issues, as for instance in small drilling companies. There is an underlying requirement for better understanding by water industry personnel of how groundwater behaves and evolves, taking account of the natural processes within systems, as well as contaminant behaviour.

21.4.3   Economic issues

Aquifer pollution is considered a very important issue, but opinions differ on possible ways of financing remediation programmes, especially for cases of diffuse pollutants. Most drinking water companies are in favour of spending more on guaranteeing a good enough quality groundwater supply, provided this is recognised by the regulator and that costs are recoverable from the end-users. However, the public would expect a lot from any increase in water costs, often considered too high.

It is felt that local/regional regulators should be more proactive in catchment protection to prevent the loss of high-quality natural groundwater supplies. Quality regulation is stringent but the application often depends on the local circumstances.

It is considered that maintenance of groundwater quality should be subsidised in a mixed way from users and through general taxation, and this put into practice by water agencies. The existing differences among the diverse countries/regions might need specific in-depth studies to define a common approach on this topic.

A real, transparent and holistic view of costs and their application (who pays) is needed. There was also a request to promote a water price reflecting the true cost, with greater involvement from the EU and water companies. Substantial EU financial aid is anticipated for new Member States.

21.4.4   Administration and policy issues

From the surveys, it was agreed that current regulations already cover the main groundwater quality issues, although little evidence is available to measure how effective their implementation has been. They need to be improved, however, in areas such as monitoring and links to health problems, giving a greater weight to the specific regional conditions and technical progress. In this sense, it is widely foreseen that the new WFD and GWD will lead to improved protection and management of the groundwater environment and highlight where rules are currently lacking, especially where toxicological data are sparse and where responsibilities are not clear.

The strong influence that the main water authorities/regulators already have on the management of groundwater is sometimes perceived as helpful by public associations and as dominant by private companies, and even excessive for some irrigation activities. They play a relevant role in promoting groundwater sustainability, although this is often seen as a reactive process. Proper involvement is hindered by insufficient commitment and financial resources because problems of water quality are on the increase and staff are stretched dealing with emergency issues. At the same time, growing bureaucratic complexity tends to make decision-making a slow process – bureaucratic streamlining is essential.

It is not clear whether society is ready for regulations that assign groundwater resources to different uses according to its baseline quality and overall purity. Public awareness and debate on water quality and associated environmental issues should be improved in this regard. This factor is important in view of the rapidity with which reserves of pristine water are being reduced.

The new GWD promises to lead to many improvements, notably through formation of river basin districts, with effective surface/groundwater integration. Enforcing the existing and new regulations is seen as a priority, leading to a better assessment of groundwater resources and their quality. Greater public participation in decisions affecting water quality is also needed.

21.4.5 Monitoring requirements

Sound groundwater monitoring networks are an essential element of groundwater quality assessment programmes, yet currently many different organisations are involved in this process, at the national, regional and local levels. A multiplicity of aims and activities has led to overlapping approaches, densities and protocols for

network design and development, sometimes for the same aquifer. The current situation is confused, and this is an issue that needs to be addressed urgently and strategically at national and European levels.

Professional advice is needed within each country/region/river basin on what should be monitored. As far as understanding changes in the groundwater quality is concerned, it is vital that the current practice of monitoring, limited to threshold values, be changed. Modern analytical methods allow very low detection limits to be reached, providing early warning of changes, not just the threshold values set by drinking water standards. The health of the aquifer needs to be monitored and this requires careful monitoring of selected baseline indicators (discussed below).

The maintenance of high-quality monitoring programmes including early warning of groundwater changes requires substantial financial resources, and cost-effective application. At present, among the range of tools and sampling devices commonly used, the simplest of them clearly prevail, most important being the regular sampling for routine laboratory analysis. Remote sensors for conductivity monitoring and some other measurements may also be appropriate. However, the introduction of residence time indicators as proposed above requires a different strategic thinking and budgeting.

21.5 A practical approach to baseline determination

21.5.1 Indicators and baseline

A range of common groundwater indicators may be defined, which characterise the baseline conditions, in addition to those parameters which clearly indicate human impacts. This is in line with the approach adopted in the GWD, which rejects the definition of Standards in favour of water quality indicators. The recommended indicators are

- T, SEC;
- Eh, pH, DO, (Fe);
- Cl, SO_4, HCO_3, Ca, Mg, Na, K;
- Al, As, Cr, Cu, Fe, Ni, Zn;
- NO_3, NH_4, TOC.

Each indicator or group of indicators has a specific objective. Well-head temperature is included as a mandatory parameter since often used as a proxy for depth, whilst SEC provides initial indication of water quality; both these parameters may also be used for continuous monitoring if required. Field measurements of Eh and O_2 (DO), are used separately or together to define the redox status of the groundwater. Redox status provides a first-order subdivision of aquifer conditions controlling mobility of trace metals. Total dissolved iron (Fe), which is commonly measured, may also be used to construct the groundwater redox status. Measured in the field, pH also constitutes a master variable for all water quality studies; for geochemical interpretation, data of high precision and accuracy are required.

The primary indicators that denote change within the natural system are the major ions – Cl, SO_4, HCO_3, Ca, Mg, K, Na – which are required to assess analytical integrity as well as baseline properties and geochemical evolution. It is also stressed that Cl as a conservative element may be used as a primary indicator in natural quality assessment and the use of Br/Cl ratios generally provides a convenient way to distinguish natural from contaminant sources. Nitrogen species are of primary importance since natural concentrations may be significant (especially NO_3 in aerobic groundwater and NH_4 in reducing waters).

Other elements (As, Cr, Cu, Fe, Mn, Ni, Zn, F) should normally be included in monitoring schemes, since at high concentrations they may be the result of natural processes. Cadmium and Hg, which currently are priority target elements in monitoring schemes and are very unlikely to be present naturally in groundwaters at high concentrations, are excluded from this list, although would be included in any list for contaminant monitoring.

Aluminium (Al) is only likely to be mobile in waters below pH 5 where it may give rise to naturally high concentrations (together with other naturally available metals) and is otherwise likely to remain below 10 µg L^{-1}.

21.5.2 *A practical approach to baseline definition*

From the reference aquifer studies and thematic papers conducted and presented in this book, an outline procedure for establishing the thresholds above natural baseline groundwater quality is suggested:

1 *Initial data acquisition and processing*: Selection of sampling density, based on
- nature of aquifer system and geology;
- size of system;
- spatial heterogeneity;
- geochemical properties;

2 *Selection of indicators for baseline assessment: For each groundwater body the following minimum set of indicators (see also previous section) should be selected*:
- a core list of parameters that should be examined for all groundwater bodies: DO, pH, Eh, SEC, NO_3, DOC, K, NH_4. Note that

this group (except K, Eh and DOC) is compulsory at present for surveillance monitoring in Europe;
- major ions, so that the overall geochemical profile and conceptual understanding of the groundwater body of the groundwater may be established (HCO_3, SO_4, Na, Cl, K, Ca, Mg);
- a key set of minor and trace elements that are sensitive to change under natural conditions and which are important in relation to drinking water standards: Al, As, F, Fe.

3 *Initial assessment of data to determine baseline:*
- review data quality (determine ionic balance using the major ions);
- establish if there are historical data for any of above indicators (Cl, NO_3 usually available) showing trends;
- review data in relation to geochemical criteria (changes such as redox and chemical ratios down flow gradients);
- spatial assessment of pristine section of the aquifer and initial assessment of the extent of natural variations and likely pollutant inputs;
- discard samples that are obviously showing pollution based, for example, on high concentrations of TOC, NO_3;
- estimate baseline for parameters using historical records or studies from well-studied reference aquifers nearby, or in similar lithologies;
- establish groundwater residence time distribution; use CFCs, ^{3}H to define extent of influx of modern water as well as indirect information from chemical parameters, synthetic substances;

4 *Statistical treatment of data:*
- define the median (50%) and upper baseline (97.7 percentiles) of the above (minimum) parameter distributions as working

values using the selected samples from the groundwater body;
- assessment of the data. Concentrations at individual sampling sites should be examined to determine whether they exceed one or more of the 97.7% threshold values. If so, this should trigger further investigation, conceptualisation and, if necessary, prioritised trend assessment;

5 *Overall geochemical characterisation of the groundwater body:*
- reporting of the groundwater body, guided by reference aquifer studies with emphasis on mineralogy, flow regime, climate, vegetation influence, land use and other controls;
- review of pristine, mixed and polluted sections of the body;
- inverse modelling of the aquifer geochemical trends where appropriate;
- consideration of three-dimensional water quality distribution;

6 *Monitoring for baseline change:*
- once the natural baseline quality is defined (median and upper baseline);
- monitor for future changes in the groundwater baseline quality over time using suitable frequency of measurement;
- identify past baseline trends in the concentrations of groundwater constituents;
- provide early warning of human impacts and land management practices on groundwater quality especially through mobile constituents such as nitrate and chloride;
- provide enough hydrochemical data to establish cause–effect relationships;
- provide advice on aquifer management and/or remediation actions in order to guarantee the sustainability of groundwater resources;
- use forward modelling to predict future changes if realistic groundwater model and modelling parameters are well established.

The minimum requirements of a groundwater baseline quality monitoring network are thus: (1) detailed study of the local hydrogeology; (2) the selection and/or construction of representative sampling sites; (3) choice of the range of indicators to be monitored; (4) verification of the laboratory analytical protocols and detection limits; (5) the choice of the sampling protocol and preservation methods; (6) definition of sampling frequency and (7) the decision of which type of data analysis and interpretative approaches should be used so that early warning is provided in time for decision-making and sustainable aquifer management.

The density of monitoring wells should depend on the size of the groundwater body, the geological and hydrogeological characteristics of the aquifer, aquifer heterogeneity, proximity to potential contamination sources and aquifer vulnerability. Suggested densities are 10–100 km^2 per sampling point for regional monitoring and 100–10,000 km^2 per sampling point for national monitoring.

The sampling frequency should be a function of the nature of the groundwater body and also a compromise between the logistical and financial resources available for sample collection and laboratory analyses. Optimal frequencies have also to be adjusted according to hydrogeology (residence time, flow velocity), seasonal influences (climate, atmospheric inputs) and statistical considerations. The nature and response-time of the aquifer, and of the particular indicator, guide the choice of sampling frequency.

In porous unconsolidated sediments, the groundwater velocity is slow, in the order of a few centimetres to some tens of metres per year and, consequently, the variability of groundwater quality is also slow. In an unconfined aquifer the measurement and sampling frequency will be higher than in a confined aquifer, due to the different conditions and temporal changes in water quality.

Operational sampling and analysis frequencies could therefore vary from once every 6 years for a confined, slow-flow aquifer, to four times a year near outcrop of a fast-flowing aquifer incorporating a parameter, which is a rapidly responding indicator of groundwater quality change. For small, shallow aquifers where flow systems are in the order of months, or where suspected migration of known pollutants are of concern, more intensive monitoring may be required.

21.5.3 *Future directions*

In Europe, the GWD is now published and was approved by the European Parliament in December 2006. Taken together with the WFD, these two documents will guide water policy and management to 2015 and beyond. All countries, including new accession states, are required to formulate management policies on the basis of the Directives. It is clear that the integrated approach that has been formulated in Europe over almost a decade will be under scrutiny by other countries and is likely to be adopted as a basis for water management outside Europe. Such an approach has been conducted in England and Wales detailing baseline conditions in all the main aquifers (Shand et al. 2007).

The studies and guidelines set out in this book provide a blueprint for determining natural groundwater quality either as a definition of the status within a groundwater body, or as a target to which trend reversals may be directed. Following the EU BaSeLiNe studies on which this book is based, a further

European programme BRIDGE was launched involving a range of stakeholders, and linking the scientific and policy-making communities in all Member States. This programme has gathered scientific outputs which could be used: to set out criteria for the assessment of the chemical status of groundwater; to derive a plausible general approach, how to structure relevant criteria appropriately with the aim to set representative groundwater threshold values, scientifically sound and defined, at national river basin district or groundwater body level; to check the applicability and validity by means of case studies at European scale; to undertake additional research studies to complete the available data; and to carry out an environmental impact assessment taking into account the economic and social impacts. These studies are published synchronously in the book Groundwater Science and Policy (Quevauvallier 2007) and includes elements of the baseline studies. It is foreseen that Natural Groundwater Quality will form a sister volume, where the baseline properties of groundwaters and how they may be defined and evaluated, can be studied in detail.

References

GWD (2006) Groundwater Directive. Directive 2006/118/EC of the European Parliament.

Quevauvallier, P. (ed.) (2007) Groundwater Science and Policy. Royal Society of Chemistry.

Shand, P., Edmunds, W.M., Lawrence, A.R. et al. (2007) The Natural (Baseline) Quality of Groundwater in England and Wales. British Geological Survey. Report CR/06/258N.

WFD (2000) Directive 2000/60/CE of the European Parliament. (ECOJ 22 December 2000).

Index